ADVANCES IN EXPERIMENTAL MEDICINE AND BIOLOGY

Volume 757

For further volumes:
http://www.springer.com/series/5584

Tim Schedl
Editor

Germ Cell Development in *C. elegans*

Springer

Editor
Tim Schedl
Department of Genetics
Washington University School of Medicine
St. Louis, MO, USA

ISSN 0065-2598
ISBN 978-1-4614-4014-7 ISBN 978-1-4614-4015-4 (eBook)
DOI 10.1007/978-1-4614-4015-4
Springer New York Heidelberg Dordrecht London

Library of Congress Control Number: 2012939120

Printed on acid-free paper

Springer is part of Springer Science+Business Media (www.springer.com)

Acknowledgements

I would like to thank each of the authors for their effort in putting together the review articles. Their expertise, insight and synthesis are of great value to the community of scientists and they motivated us for further exploration and discovery. It was a pleasure working with you all. I would also like to thank the reviewers for their service and clarifications.

One does not have many opportunities to recognize those who fundamentally made it all possible. I would like to thank my father, Harold Schedl, and my brother, Paul Schedl, for sparking my interest in science and showing me that it can be a lot of fun. I would like to thank my mother, Naomi Kark Schedl, for showing me the creative process.

Contents

1 **Introduction to Germ Cell Development in *Caenorhabditis elegans*** 1
Nanette Pazdernik and Tim Schedl

2 **Germ Cell Specification** 17
Jennifer T. Wang and Geraldine Seydoux

3 **Sex Determination in the *Caenorhabditis elegans* Germline** 41
Simone Zanetti and Alessandro Puoti

4 **Stem Cell Proliferation Versus Meiotic Fate Decision in *Caenorhabditis elegans*** 71
Dave Hansen and Tim Schedl

5 **Physiological Control of Germline Development** 101
E. Jane Albert Hubbard, Dorota Z. Korta, and Diana Dalfó

6 **Meiotic Development in *Caenorhabditis elegans*** 133
Doris Y. Lui and Monica P. Colaiácovo

7 **Spermatogenesis** 171
Diana S. Chu and Diane C. Shakes

8 **Translational Control in the *Caenorhabditis elegans* Germ Line** 205
Marco Nousch and Christian R. Eckmann

9 **Germ Cell Apoptosis and DNA Damage Responses** 249
Aymeric Bailly and Anton Gartner

10 **Control of Oocyte Growth and Meiotic Maturation in *Caenorhabditis elegans*** 277
Seongseop Kim, Caroline Spike, and David Greenstein

11 **Fertilization** 321
Matthew R. Marcello, Gunasekaran Singaravelu, and Andrew Singson

12 **The Oocyte-to-Embryo Transition** 351
Scott Robertson and Rueyling Lin

13 **Epigenetic Control of Germline Development** 373
Priscilla M. Van Wynsberghe and Eleanor M. Maine

14 **Using *Caenorhabditis* to Explore the Evolution of the Germ Line** 405
Eric S. Haag and Qinwen Liu

Index 427

Contributors

Aymeric Bailly Centre de Recherche de Biochimie Macromoleculaire, Montpellier Cedex, France

Diana S. Chu Department of Biology, San Francisco State University, San Francisco, CA, USA

Monica P. Colaiácovo Department of Genetics, Harvard Medical School, Boston, MA, USA

Diana Dalfó Skirball Institute of Biomolecular Medicine, Department of Pathology, and Kimmel Center for Stem Cell Biology, New York University School of Medicine, New York, NY, USA

Christian R. Eckmann Max Planck Institute of Molecular Cell Biology and Genetics, Dresden, Germany

Anton Gartner Wellcome Trust Centre for Gene Regulation and Expression, The University of Dundee, Dundee, UK

David Greenstein Department of Genetics, Cell Biology, and Development, University of Minnesota, Minneapolis, MN, USA

Eric S. Haag Department of Biology, University of Maryland, College Park, MD, USA

Dave Hansen Department of Biological Sciences, University of Calgary, Calgary, AB, Canada

E. Jane Albert Hubbard Skirball Institute of Biomolecular Medicine, Department of Pathology, and Kimmel Center for Stem Cell Biology, New York University School of Medicine, New York, NY, USA

Seongseop Kim Department of Genetics, Cell Biology, and Development, University of Minnesota, Minneapolis, MN, USA

Dorota Z. Korta Skirball Institute of Biomolecular Medicine, Department of Pathology, and Kimmel Center for Stem Cell Biology, New York University School of Medicine, New York, NY, USA

Rueyling Lin Department of Molecular Biology, University of Texas Southwestern Medical Center, Dallas, TX, USA

Doris Y. Lui Department of Genetics, Harvard Medical School, Boston, MA, USA

Qinwen Liu Department of Biology, University of Maryland, College Park, MD, USA

Eleanor M. Maine Department of Biology, Syracuse University, Syracuse, NY, USA

Matthew R. Marcello Waksman Institute, Piscataway, NJ, USA

Department of Genetics, Rutgers University, Piscataway, NJ, USA

Department of Molecular Genetics, Microbiology and Immunology, UMDNJ-Robert Wood Johnson Medical School, Piscataway, NJ, USA

Marco Nousch Max Planck Institute of Molecular Cell Biology and Genetics, Dresden, Germany

Nanette Pazdernik Department of Genetics, Washington University School of Medicine, St. Louis, MO, USA

Alessandro Puoti Department of Biology, University of Fribourg, Fribourg, Switzerland

Scott Robertson Department of Molecular Biology, University of Texas Southwestern Medical Center, Dallas, TX, USA

Tim Schedl Department of Genetics, Washington University School of Medicine, St. Louis, MO, USA

Geraldine Seydoux Department of Molecular Biology and Genetics, Howard Hughes Medical Institute, Center for Cell Dynamics, Johns Hopkins School of Medicine, Baltimore, MD, USA

Diane C. Shakes Department of Biology, College of William and Mary, Williamsburg, VA, USA

Gunasekaran Singaravelu Waksman Institute, Piscataway, NJ, USA

Department of Genetics, Rutgers University, Piscataway, NJ, USA

Andrew Singson Waksman Institute, Piscataway, NJ, USA

Department of Genetics, Rutgers University, Piscataway, NJ, USA

Caroline Spike Department of Genetics, Cell Biology, and Development, University of Minnesota, Minneapolis, MN, USA

Priscilla M. Van Wynsberghe Department of Biology, Syracuse University, Syracuse, NY, USA

Department of Biology, Colgate University, Hamilton, NY, USA

Jennifer T. Wang Department of Molecular Biology and Genetics, Howard Hughes Medical Institute, Center for Cell Dynamics, Johns Hopkins School of Medicine, Baltimore, MD, USA

Simone Zanetti Department of Biology, University of Fribourg, Fribourg, Switzerland

Chapter 1
Introduction to Germ Cell Development in *Caenorhabditis elegans*

Nanette Pazdernik and Tim Schedl

Abstract A central feature of the continuum of life in sexually reproducing metazoans is the cycle of the germline from one generation to the next. This volume describes the cycle of the germline for *Caenorhabditis elegans* through chapters that are focused on distinct aspects or processes in germ cell development. Topics include sequential and dependent processes such as specification of germ cells as distinct from somatic cells, sex determination, stem cell proliferative fate versus meiotic development decision, recombination/progression through meiotic prophase, contemporaneous processes such as gametogenesis, meiotic development and apoptosis, and continuing the cycle into the next generation through fertilization and the oocyte-to-embryo transition. Throughout germ cell development, translational control and epigenetic mechanisms play prominent roles. These different aspects of germ cell development are seamlessly integrated under optimal conditions and are modified in the different reproductive strategies that are employed by *C. elegans* under harsh environmental conditions. In this chapter, we set the stage by providing a brief background on the *C. elegans* system and germ cell development, indicating processes in the cycle of the germline that are covered in each chapter.

Keywords *C. elegans* • Germ cell • Gametogenesis • Meiosis • Reproduction • Somatic gonad • Apoptosis • Meiotic recombination • Sex determination • Fertilization • Germline stem cell

N. Pazdernik • T. Schedl (✉)
Department of Genetics, Washington University School of Medicine,
Campus Box 8232, 4566 Scott Avenue, St. Louis, MO 63110, USA
e-mail: npazdernik@genetics.wustl.edu; ts@genetics.wustl.edu

T. Schedl (ed.), *Germ Cell Development in C. elegans*, Advances in Experimental Medicine and Biology 757, DOI 10.1007/978-1-4614-4015-4_1,

1.1 Introduction

The ability of a metazoan to create new organisms via sexual reproduction is the basis for all animal life. This volume represents the current state of knowledge in the quest for understanding the reproductive system in *C. elegans*, an organism that exists primarily as a self-fertilizing hermaphrodite. The chapters are organized from the earliest inkling of germline development in the P lineage of the embryo (Wang and Seydoux 2012, Chap. 2) to evolutionary comparisons between *C. elegans* reproduction and their closest relatives (Haag and Liu 2012, Chap. 14). Each chapter focuses upon different aspects or processes in germ cell development and differentiation, the genetic and molecular events that are known to be involved with each process, and comparisons with other organisms.

1.2 Why *C. elegans*?

In 1897, Emile Maupas, a French zoologist and botanist, described *C. elegans* as a species of nematode that lives in rich humus in which "[he] came twice across… in the surroundings of Algiers" (Maupas 1900). His sketches show remarkable detail for both forms of *C. elegans*: the hermaphrodite producing both sperm and oocytes, and the male producing sperm only. His sketches also reveal that the somatic cells are clearly divided in function. He described the hermaphrodite soma as having a female form with a vulva that is behind the middle point of the body, and the male form with a specialized tail structure with a flattened somewhat "heart-shaped bursa" containing "nine pairs of papillae." As for the germline, he described the hermaphrodite gonad as a simple S-shaped tube that is divided into "vitellogen" containing the oocytes and "germigen" containing a well-developed central rachis surrounded by a layer of germ cells (Fig. 1.1). He described the male gonad as filled with large numbers of spermatozoa with the exact shape and structure as seen in the hermaphrodites (Maupas 1900). For many subsequent years scientists largely ignored *C. elegans*. But in the 1960s, Sydney Brenner initiated the modern study of *C. elegans* as a model organism. Brenner had previously validated the hypothesis that mRNAs are read as triplet codons, and an addition or subtraction of one or two nucleotides created frameshift mutations (Brenner et al. 1965). He attributed the success of determining such basic biology to the use of bacteriophage, which he was able to grow in sufficient quantities to identify the rare mutations that led to the discoveries. He extrapolated that a simple multicellular organism would further the understanding of how cells interact with each other, and the biological basis of behavior, two key biological questions. He chose *C. elegans* for three reasons. First, like bacteriophage, the worm is small, easy to propagate in large numbers, and therefore amenable to genetic and biochemical analysis. Second, the worm is transparent, meaning the cells were visible *in situ*, and the organism would not need to be dissected into its parts before analysis. Finally, he was able to induce mutations,

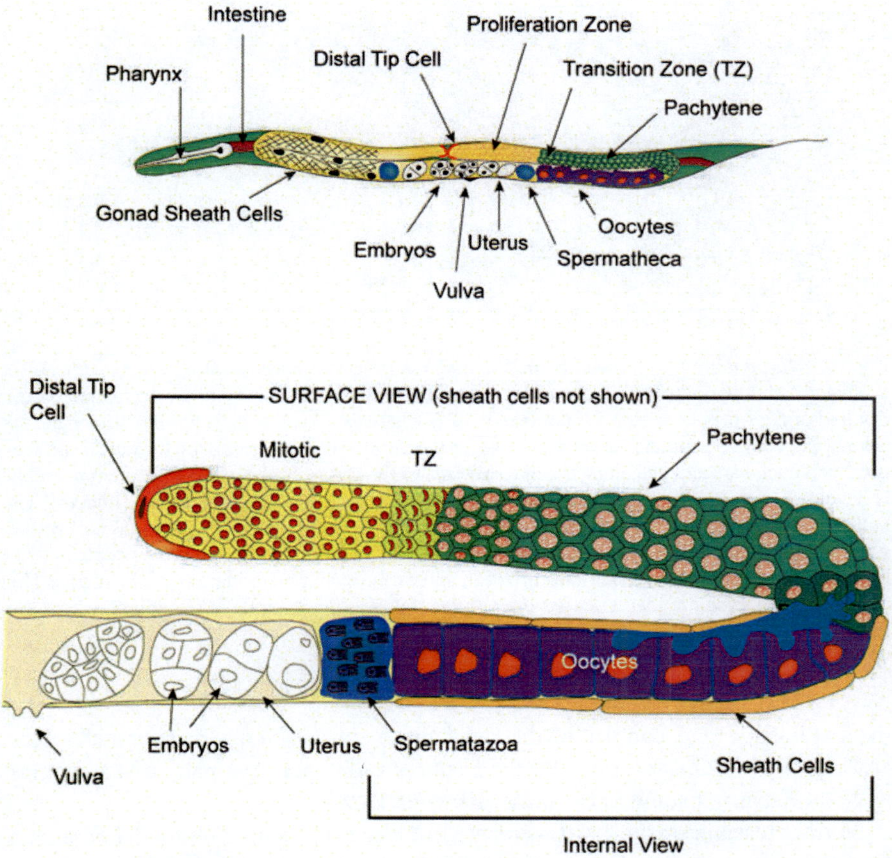

Fig. 1.1 *Schematic of adult C. elegans hermaphrodite and gonad.* (*Top*) Adult *C. elegans* hermaphrodite, highlighting the reproductive system, which contains two U-shaped gonad arms connected by a common uterus. The distal end of each gonad arm is capped by a somatic distal tip cell (DTC) that covers the distal end of the germline, containing the proliferative zone (*yellow*). A surface view of the left side U-shaped gonad arm shows the five pairs of somatic gonadal sheath cells covering the area from the transition zone to the spermatheca. On the right side, the gonad arm is shown without the sheath cells. The *green* cells represent the germ cells in meiotic prophase I, the *purple* cells represent the developing oocytes, the proximal *darker blue* area is the spermatheca, and the clear embryos are found within the uterus. (*Bottom*) A detailed view of one adult hermaphrodite gonad arm is shown. The upper part of the arm is shown as a surface view without the covering sheath cells. The transition zone is visible as a *light green color*. The lower part of the gonad is an internal view of the proximal region including the sheath cells. The oocytes closest to the loop are connected to the central rachis (*light blue*, also see Fig. 1.4), which allow cytoplasmic material to enter, while the proximal 4–5 oocytes closest to the spermatheca are fully cellularized. See text and wormatlas.org (http://www.wormatlas.org/) for details

and as hermaphrodites are the dominant form, he could create homozygous mutations by selfing. Since a small portion of the population was male, Brenner could also recombine genotypes through crosses. By using standard complementation and linkage analyses of different mutations he isolated, Brenner created

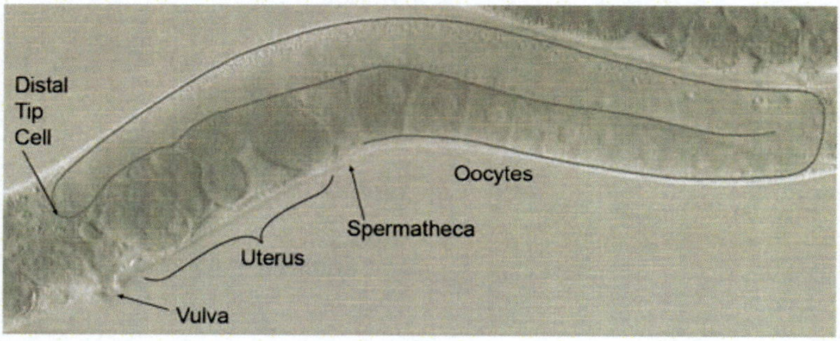

Fig. 1.2 *Morphology from Nomarski differential interference contrast (DIC) microscopy*. Live *C. elegans* adult hermaphrodite showing an interior view of the U-shaped gonad arm in the posterior half of the worm. The gonad is outlined in *gray* where the distal tip cell caps the distal end and developing oocytes populate the proximal end. In the distal half of the gonad, on the dorsal surface, germ cell nuclei are situated along the outer surface of the gonadal tube whereas the central rachis is devoid of germ cells/nuclei. Germ cells move from the very distal end to the proximal end via bulk flow. Overt oocyte development occurs from the loop region into the proximal half of the gonad on the ventral side. The spermatheca lies adjacent to the final oocyte. As an oocyte is ovulated into the spermatheca, a sperm fertilizes it, and early development initiates in the uterus. The developing embryos are then released through the vulva to the external environment, where the remainder of embryogenesis occurs. Also see Fig. 1.4

the first linkage map, confirming that the genome consists of five pairs of autosomes and a pair of X-chromosomes in the hermaphrodite and five pairs of autosomes and one X chromosome in the male (Brenner 1974).

Further research in the 1970s cemented *C. elegans* as a premier model organism for reproductive studies. Hirsh et al. (1976) used Nomarski differential interference contrast microscopy to elucidate the different structures of the hermaphrodite gonad, focusing on oogenesis (Fig. 1.2). Ward and Carrel (1979) examined spermatogenesis in both the hermaphrodite and the male, describing the cellular morphology of sperm and the developmental steps to create sperm. Kimble and Hirsh (1979) then used Nomarski microscopy-based lineage analysis to trace the development of the hermaphrodite gonad from the early primordium of four cells at hatching to the adult with two U-shaped gonad arms, each containing ~1,000 germ cells and a total of 143 somatic gonad cells (Fig. 1.3). The male gonad also begins with a four-cell primordium but follows a different pathway of development. Instead of two symmetrical arms, males have a single gonad that contains over 1,000 germ cells and 56 somatic cells (Kimble and Hirsh 1979). These studies determined that the cells in the somatic gonad follow an essentially invariant lineage as is seen in the rest of the somatic tissues, whereas germline cell fates depend upon cell position within the gonad as well as time (Kimble and Hirsh 1979; Sulston and Horvitz 1977). Since Maupas' early description of *C. elegans* and Sydney Brenner's choice to use *C. elegans* as a contemporary model, research on this organism has ballooned into a large group of scientists studying almost every aspect of biology from the view of a nematode.

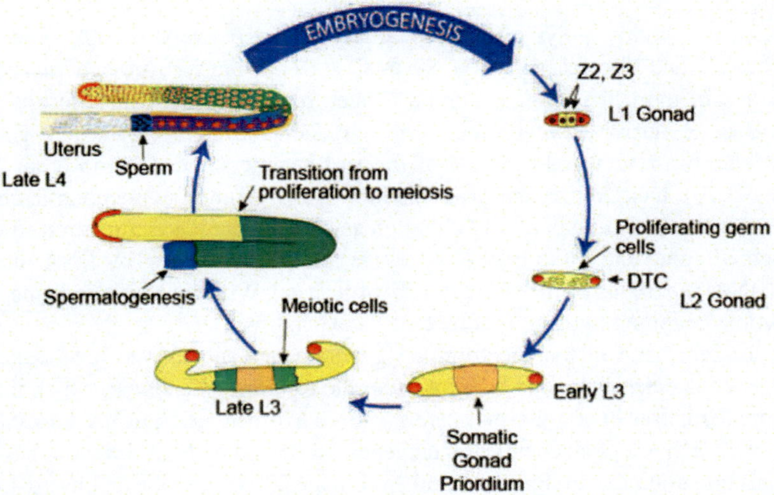

Fig. 1.3 *Cycle of the germline: hermaphrodite gonadogenesis*. Gonadogenesis begins at L1 stage with four gonad precursor cells. Z2 and Z3, the germline progenitors, are sandwiched between the somatic precursors Z1 and Z4 (*red*), which will divide and form the DTCs and the remaining somatic gonad cells. Prior to the L3 stage, all germ cells proliferate mitotically (*yellow*). Beginning in early L3, germ cells that are farthest from the DTC leave the proliferative zone, enter meiotic development (*green*), and progress through meiotic prophase and gametogenesis in an assembly-line fashion. Germ cells that switch from the proliferative fate to meiotic development at this point will become sperm by the end of the L4 stage in the proximal end of the gonad arm. Those that switch in L4 and adulthood will become oocytes. Migration of the two DTCs generates the U-shape of the gonad: starting in the L3 stage, each DTC migrates away from the centrally located gonad primordium along the ventral surface; at the L3/L4 molt, each DTC migrates centripetally to the dorsal surface; and in the L4 stage, each DTC migrates back toward the center of the animal. Germ cell proliferation and gametogenesis-mediated germ cell volume expansion fills the gonadal tube as the DTCs migrate. Gonads in the schema are for illustrative purpose and are not to scale

1.3 Experimental Approaches

Underpinning the entire body of research in *C. elegans* is the ability to identify the molecular events that control developmental and homeostatic processes via genetic and molecular analyses. Essential genetic approaches include forward genetics, reverse genetics, and creation of transgenic strains. For forward genetics or phenotype-based studies, incubating *C. elegans* with chemical mutagens produces worms with various types of reproductive phenotypes (e.g., sterility) that can be dominant, recessive, or maternal effect, which can be isolated by screening different generation of animals. Secondary screens for enhancers, suppressors or synthetic phenotypes have further expanded the collection of mutations that affect germ cell development. Molecular identification of the genes containing the phenotype-causing lesions has moved from laborious positional cloning to the use of single nucleotide polymorphisms (SNPs), as many natural isolates of *C. elegans*, for example Hawaiian isolate

CB4856, are highly polymorphic containing an SNP every ~1,000 base pairs (Jakubowski and Kornfeld 1999; Wicks et al. 2001). From forward genetic screens, mutations in more than 500 genes have been identified that affect fertility (see WormBase at http://www.wormbase.org/). In fact, the advent of whole genome sequencing has revitalized the field of forward genetic screens (Sarin et al. 2008; Hobert 2010). Discovering the exact alterations that create the phenotypic change can be done quickly and precisely although as with any gene identification strategy, orthogonal approaches like transgene rescue and phenocopying by RNA interference (RNAi) are required to confirm the identification. With expense rapidly decreasing, whole genome sequencing of mutants is expanding in use.

At the other end of the spectrum are genomic sequence-driven reverse genetic approaches to determine gene function from the resulting phenotype, which include deletion/disruption of a gene of interest and RNAi-mediated mRNA knockdown. Gene deletion by random chemical mutagenesis followed by PCR detection has been the workhorse approach (Barstead and Moerman 2006), largely facilitated by the Oklahoma-Vancouver and the NBP-Japanese consortia. More recently site-specific double strand break induction, either with the *Mos1* transposase and a large collection of mapped *Mos1* transposon insertions (Frokjaer-Jensen et al. 2008) or by designer Zn-finger or TALEN (Transcription Activator-Like Effector Nuclease) sequence specific nucleases (Wood et al. 2011), has generated gene deletions through imprecise repair by the non-homologous end-joining pathway, or gene replacements through homologous recombination via repair from a transgene template. Genetic epistasis analysis has been an important strategy for ordering genes into pathways/networks, employing null and gain-of-function mutations isolated from forward and reverse genetic approaches (e.g., Hodgkin 1986).

RNAi, another reverse genetic approach whose mechanism was uncovered in *C. elegans*, reduces the mRNA level for the gene being tested (Fire et al. 1998). Gene inactivation occurs through an evolutionary conserved pathway of defense against invading viruses, which is triggered by double-stranded RNA (dsRNA). When *C. elegans* is fed bacteria expressing dsRNA, soaked in a solution of dsRNA, or injected with dsRNA, the organism protects itself by eliminating the incoming dsRNA and any complementary endogenous mRNA (Ahringer 2006). The system self-amplifies and, therefore, is a potent method of gene suppression. Furthermore, the use of a null mutation in the gene *rrf-1*, which encodes a somatically functioning RNA-dependent RNA polymerase, allows RNAi be largely restricted to the germline (Sijen et al. 2001). When the dsRNA is complementary to an endogenous mRNA, the resulting phenotypic change provides information on the *in vivo* function of the gene. RNAi is very amenable to high throughput screens, as well as high germline phenotype content screens, and has proven particularly useful in the analysis of the role of essential genes (embryonic or larval lethal) in adult germline development. RNAi can be applied after critical embryonic or larval events have occurred and/or the *rrf-1* null mutation can be incorporated into the screening strain. If RNAi is restricted to late larval stages or young adults, the role the gene plays in adult gonad development/function can be assessed. Green et al. (2011) used a short-term RNAi treatment to examine the function of 554 essential *C. elegans* genes in the anatomy and function of the adult germline, where 116 genes were

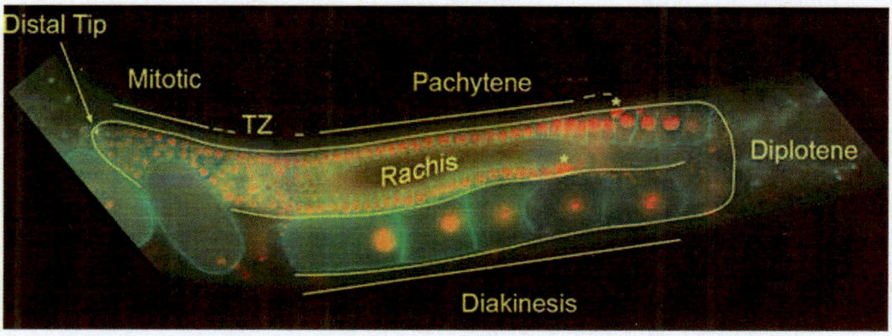

Fig. 1.4 *Fluorescence image of C. elegans adult hermaphrodite.* Live *C. elegans* hermaphrodite highlighting the U-shaped gonad arm in the posterior half of the worm. The gonad is outlined in *yellow* where the distal tip cell caps the distal end and developing oocytes populate the proximal end. The strain, OD95 (Green et al. 2011), allows visualization of plasma membranes (*green*), with a GFP tag fused to the PH-domain from PLC delta, expressed in the germline and early embryo from the pie-1 promoter, and chromatin (*red*), with an mCherry tag fused to histone H2B, also expressed from the pie-1 promoter. Interior view shows, in the distal half of the gonad, germ cells that are situated along the outer surface of the gonadal tube with openings to the central rachis that is largely devoid of germ cells/nuclei. A single file row of growing oocytes in diakinesis is shown in the proximal half of the gonad. *Asterisks* indicate highly condensed nuclei that are indicative of apoptosis

completely uncharacterized. High-content phenotypic analysis, scoring 94 phenotypic features, parsed the genes into 102 functional classes allowing the uncharacterized genes to be placed into groups with known function, such as membrane trafficking, glycosylation, fatty acid synthesis, mitochondrial function, transcription, and MAP kinase signaling, to name a few.

Another key method for genetic analysis is the expression of wild-type or altered gene products through the generation of transgenic lines. *C. elegans* germ cells have a very potent mechanism for silencing transgenes that are present as multiple tandem copies, which are generated in the course of constructing standard transgenic lines (Kelly and Fire 1998). Therefore, two methods have been employed to generate single/low copy integrated transgenic lines that are not silenced, ballistic microparticle bombardment (Merritt and Seydoux 2010) and *Mos1* transposase-mediated integration into specific *Mos1* transposable element landing sites in the genome (Frokjaer-Jensen et al. 2008). To distinguish whether gene function is required in the germline or somatic cells, genetic mosaic analysis using mitotically unstable extra-chromosomal transgene arrays can be employed (Yochem and Herman 2003). Important for genetic screens and phenotypic analysis is transgene tagging of genes and gene products to allow *in vivo* analysis of expression and localization as well as to mark cell types and various subcellular structures. Central to the *in vivo* analysis is the use of fluorescently tagged proteins like GFP, which was first employed in *C. elegans* (Chalfie et al. 1994). The combination of powerful genetic approaches with visualization of germ cell development in live worms at the level of subcellular structures, for example GFP-tagged plasma membrane and mCherry-tagged histones chromosomes (Fig. 1.4), has and will continue to rapidly advance our understanding of germ cell development.

1.4 Development and Anatomy

In the *C. elegans* adult hermaphrodite, the germline resides within two U-shaped arms of the gonad, joined at a common uterus (Fig. 1.1). The vulva lies just past the midpoint of the worm on the ventral side, and is a flat slit against the external cuticle. The tube shaped gonad contains germ cells in different stages of differentiation, sequentially developing in an assembly-line fashion from the proliferative germ cells near the somatic distal tip cell (DTC), through meiotic prophase I in the distal gonad and into the loop, and culminating with fully-formed oocytes in the proximal gonad. The hermaphrodite specifies male germ cells in the L3 stage, which differentiate into sperm in the L3/L4 stage, and then specifies female germ cells from L4 through adulthood, which differentiate into oocytes. By contrast in the male, spermatogenesis occurs continuously. The pathway that controls germ cell sexual fate is discussed by Zanetti and Puoti (2012) in Chap. 3. Hermaphrodite self-sperm, as well as sperm from a male mating, reside in the spermatheca and await ovulation of the most proximal oocyte, at which point fertilization occurs followed by initiation of embryogenesis (Figs. 1.1, 1.2 and 1.4). On the side of the spermatheca towards the vulva resides the uterus, which contains early embryos that eventually pass through the vulva and continue to develop externally. In contrast, the male has one gonad arm and only makes sperm, which passes through the seminal vesicle and vas deferens to be inserted into the vulva of the hermaphrodite via its specialized tail structure during mating. An unmated hermaphrodite is able to produce around 300 embryos, whereas a mated hermaphrodite can produce up to 1,000, demonstrating that the limiting factor for self-fertility is not oocyte production, but rather the amount of self-sperm formed by the hermaphrodite (Hodgkin and Barnes 1991). *C. elegans* propagate primarily as hermaphrodites, with two X chromosomes and a diploid set of autosomes (2X, 2A), but will produce males (genetically XO) by nondisjunction in about 0.1% of the progeny. If mating occurs, 50% of cross progeny will be male.

Wild-type *C. elegans* progress through their life cycle in about 3.5 days at 20 °C, under optimal laboratory conditions, including constant temperature and unlimited food (equivalent to "feasting at the Hilton"). Under these conditions hermaphrodites are self-fertile for about 4 days with a typical life span of about 2 weeks (Fig. 1.5). Early during embryogenesis, the germline is set aside from the somatic cells, as described by Wang and Seydoux (2012) in Chap. 2. After the hermaphrodite lays an embryo, it continues to develop from an oval mass of cells into a small larva, progressing through twofold and threefold stages of development. Following hatching from the chitin eggshell that protects the embryo, the hatchling develops through four sequential larval stages (L1–L4), molting its external cuticle with each successive stage. After the final molt emerges a sexually mature adult capable of laying new embryos, and the cycle begins again.

The reproductive tract is generated during post-embryonic development, with distinct aspects of gonadogenesis occurring during individual larval stages (Fig. 1.3). L1 larvae have two primordial germ cells (Z2 and Z3) sandwiched between two somatic gonad precursors (Z1 and Z4), surrounded by a basal lamina.

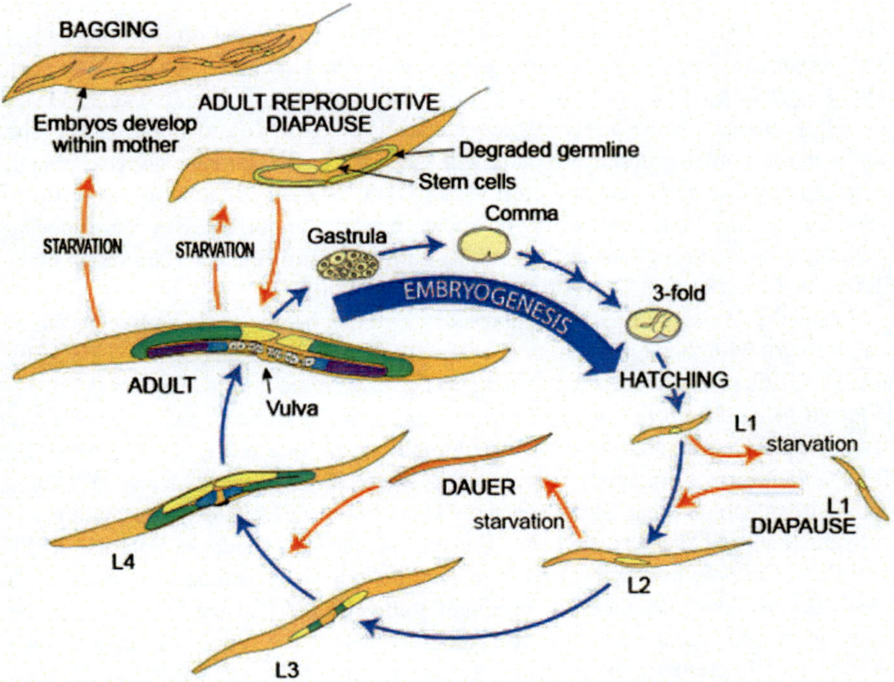

Fig. 1.5 *Life Cycle of C. elegans and Potential Diapauses. C. elegans* adults lay embryos that pass through gastrulation, comma stage, two- and threefold embryos before hatching in the L1 larval form. The larvae develop through L2, L3, and L4 molts before becoming adults. The cycle takes about 3.5 days at 20 °C under rich nutritional conditions. When the food source becomes scarce, worms at different stages of development can enter diapause where reproductive development is halted and their metabolism is slowed (*red arrows*). If L1 larvae hatch in the absence of food, they enter L1 diapause, blocking initiation of further development. When food becomes available, L1 larvae reinitiate development. If L2 worms encounter environments with reduced/no food, overcrowding and/or high temperatures, they enter an alternate developmental form called dauer, where development of the reproductive system is arrested, which is resistant to stress and desiccation but is motile and active in searching for food. If the dauer finds food, then the worm enters the developmental program in the L3 stage and proceeds to adulthood. The third diapause occurs when L4/adults find themselves without food. The adult reproductive diapause halts reproduction by degrading most of the germline, but leaving proliferative zone cells that are apparently cell cycle arrested. Upon refeeding, worms in adult reproductive diapause resume germ cell proliferation, meiotic development, and oogenesis and can become fertile. Bagging is another strategy to overcome starvation. Here the adult worm stops laying embryos, and any embryos within the mother continue to develop. At hatching, these embryos eat the mother from the inside out. The mother does not survive, but ensures that the embryos have enough food to reach larval diapauses. See Hubbard et al. (2012) (Chap. 5) for further discussions of the response of the reproductive system to alternative environments

Z2 and Z3 begin proliferating in mid-L1 to populate the gonad with germ cells. The somatic precursors also begin to proliferate and constitute 12 cells before the second molt, including two DTCs, with one capping each of the two gonad arms, and ten other cells that form the hermaphrodite somatic gonad primordium (Kimble and Hirsh 1979). The hermaphrodite DTC has two functions: a migratory/morphogenetic leader function that gives rise to the U-shape of each gonad arm, and a signaling role to promote the proliferative germ cell fate (the male DTCs only have the later function) (Hedgecock et al. 1987; Kimble and White 1981). The ten somatic gonad primordium cells will give rise to the somatic gonadal sheath cells, the spermatheca and the uterus (Fig. 1.1). The sheath cells are multifunctional, including being necessary for oocyte maturation and ovulation, promoting meiotic prophase progression, and robust germ cell proliferation (Govindan et al. 2009; McCarter et al. 1997; Killian and Hubbard 2005).

From the L4 stage the hermaphrodite germline begins to look like its adult counterpart, with the gonad arm on the dorsal side of each U-shaped tube capped by the DTC. Germ cells adjacent to the DTC proliferate mitotically (collectively referred to as the proliferative or mitotic zone), while those some distance away from the DTC enter and progress through meiotic prophase (Fig. 1.1). Pioneering laser ablation experiments by Kimble and White (1981), in which the DTC was killed, demonstrated that the DTC promotes the proliferative fate and/or inhibits the meiotic fate, and that it establishes the mitotic–meiotic prophase polarity of the germline. Cells within the proliferative zone include the germline stem cells, with the DTC providing the niche. Subsequent studies found that the DTC signals the distal germ cells via a Notch receptor, such that loss of the signal results in all proliferative cells prematurely switching to meiotic development, while hyperactivation of the Notch pathway results in a germline tumor (Austin and Kimble 1987; Berry et al. 1997) (also see Hansen and Schedl 2012, Chap. 4). In fact, the description and discovery of the *C. elegans* DTC controlled proliferative zone helped define the concept of a niche, which is described by Morrison and Spradling (2008) as a specialized local microenvironment where stem cells reside and that directly promote the maintenance of stem cells.

As germ cells move by bulk flow away from the DTC, escaping its influence, they switch to the meiotic fate, entering and progressing through meiotic prophase and forming gametes. The meiotic and gametogenic events of oogenesis and spermatogenesis are significantly dimorphic, with some of the differences conserved between species. Kim et al. (2012) describe the control of oogenesis and meiotic maturation in Chap. 10, Lui and Colaiácovo (2012) describe meiotic development during oogenesis in Chap. 6, while Chu and Shakes (2012) discuss both meiotic development and gamete formation during spermatogenesis in Chap. 7.

Spermatogenesis in both the hermaphrodite and male is rapid with meiotic prophase I, leptotene and zygotene (cells in this stage are in a region of the gonad arm called the transition zone, TZ), pachytene, diplotene, and diakinesis lasting 20–24 h (Jaramillo-Lambert et al. 2007). Homologous chromosome pairing/synapsis and initiation of meiotic recombination occurs in leptotene-zygotene, with the resolution of recombination occurring in the transition from pachytene to diplotene. In spermatogenesis all meiotic germ cells undergo two sequential meiotic divisions, without a prophase arrest, to generate four spermatids which, following spermiogenesis, form motile amoeboid sperm (Ward et al. 1981).

Meiotic prophase of oogenesis occurs over a considerably longer period, 54–60 h, presumably because of the requirement to make the very large, nutrient- and macromolecule-rich oocyte and because of controls in the production and release of the oocyte for fertilization. In oogenesis, many more germ cells enter meiotic prophase than actually become oocytes, with the excess apparently functioning as nurse cells that produce macromolecules for oogenesis, with these cells being eliminated by apoptosis during late pachytene (Fig. 1.4). In Chap. 9, Bailly and Gartner (2012) describe how germline cells undergo apoptosis as part of this developmental process, called physiological apoptosis, as well as apoptosis that occurs in response to DNA damage/unrepaired meiotic recombination.

Oogenesis has an extended pachytene stage, in which germ cells synthesize RNAs and proteins that are donated to the oocyte (Gibert et al. 1984; Schisa et al. 2001). Much of the germline is a syncytium, where the plasma membrane does not fully surround the nucleus, leaving an opening that connects each nucleus and its surrounding cytoplasm to a common cytoplasm. By convention, each nucleus and its surrounding cytoplasm and membranes is called a germ cell. While partially syncytial, trafficking of molecules is highly controlled and thus, unlike the syncytial Drosophila embryo, adjacent cells can display distinct behaviors such as being at different mitotic or meiotic cell cycle stage and display differences in molecular marker phenotypes. In pachytene, the germ cells are on the surface of the gonadal tube (Fig. 1.1) with an interior nucleus/cell free cytoplasmic region called the rachis or core (Figs. 1.2 and 1.4). As germ cells progress from pachytene to diplotene, oocyte differentiation begins in the loop region with a single file row of growing oocytes found on the external surface of the gonadal tube and the rachis found on the internal surface. RNAs and proteins made in pachytene nuclei are deposited into the rachis and delivered to the growing oocytes via cytoplasmic streaming (Nadarajan et al. 2009; Wolke et al. 2007). The proximal 4 or 5 oocytes are in diakinesis, are fully cellularized (no longer connected to the rachis) and actively uptake yolk produced by the intestine (Grant and Hirsh 1999; Hall et al. 1999; Maddox et al. 2005). The most proximal oocyte (called −1), adjacent to the spermatheca, undergoes meiotic maturation (nuclear envelope breakdown, progression to metaphase of meiosis I and rearrangement of the oocyte cortex and cytoplasm), and is ovulated into the spermatheca and fertilized (McCarter et al. 1999). Meiotic maturation/ovulation and oocyte production is regulated by the MSP signaling molecule secreted from sperm (Miller et al. 2001, 2003). In the presence of sperm (younger adult hermaphrodite or mated animal), an oocyte undergoes maturation and ovulation every ~23 min and oogenesis is continuous (Govindan et al. 2006, 2009; Lee et al. 2007; McCarter et al. 1999). In the absence of sperm (adult hermaphrodites that have exhausted their self-sperm, sex determination mutant females), oocytes arrest in diakinesis and ongoing oogenesis is inhibited. Thus, oocyte production and utilization only occurs in the presence of sperm. Fertilization triggers completion of meiosis and the initiation of zygotic development. Marecello et al. (2012) discuss fertilization in Chap. 11 and Robertson and Lin (2012), in Chap. 12, describe events in the oocyte-to-embryo transition that are essential for launching zygotic development and for the early events that distinguish the somatic and germline lineages, thus continuing the cycle of the germline into the next generation.

1.5 Physiology and the Environment

How *C. elegans* maintain reproductive fitness in their natural environment is a newer area of study. In contrast to most of experimental analysis of germline development in the laboratory, where there are unlimited nutrients, *C. elegans* in the wild live a boom or bust lifestyle, where one moment there is abundant bacteria in their environment, followed by little or no food. This lifestyle puts significant pressure on the organism to delay progeny production until the environment is optimal (Felix and Braendle 2010). In fact, three different diapauses have been identified that reduce metabolism and increase stress resistance in response to starvation conditions (Fig. 1.5). These diapauses keep the worm alive and allow it to resume normal growth, development, and/or reproduction when environmental conditions become favorable. When newly hatched L1 larvae are faced with hardship, L1 diapause blocks initiation of post-embryonic development, including blocking somatic gonad (Z1 and Z4) and germ cell (Z2 and Z3) divisions, and larvae become metabolically quiescent. This is the major form of *C. elegans* that is resistant to freezing temperatures, a feature that is exploited in the laboratory to store frozen wild-type and mutant strains. If hardship occurs during L2, an alternate developmental stage called dauer occurs. Dauer stage worms can survive with little to no food for months, where somatic gonad and germ cell development are arrested at the L2 stage. When hardship occurs in the L4/adult stage, adult reproductive diapause can maintain a small group of proliferative zone cells adjacent to the DTC (Angelo and Van Gilst 2009; Seidel and Kimble 2011). The remaining germ cells degrade and no more productive oogenesis or embryogenesis occurs until conditions improve (Fig. 1.5). Upon return of food, the proliferative zone cells resume cell division and differentiation to repopulate the gonad, supporting the view that the proliferative zone contains a true stem cell population. Starved L4/adults have another option to improve the possibility of their offspring surviving. In a process called bagging, the mother stops laying embryos. The existing embryos within the hermaphrodite hatch inside their mother, using her body as a food source. This process is terminal for the parental hermaphrodite, but gives the progeny the ability to survive through the starvation period. An understanding of what determines whether the outcome is bagging or adult reproductive diapause remains to be uncovered. However, the more mechanisms a worm can utilize to survive hardship, the more likely it will be able to propagate from one generation to the next. Hubbard et al. (2012) discuss how physiology and the availability of nutrients control germline development in Chap. 5.

1.6 Gene Expression

Transcription profiling identified significant differences in mRNA content between the germline and the soma, as well as oogenesis versus spermatogenesis: ~1,650 genes have oogenesis enriched expression, ~1,350 genes have spermatogenesis enriched expression, and ~1,250 with germline intrinsic expression (Reinke et al. 2004). Epigenetic control mechanisms, which are discussed by Van Wynsberghe and Maine (2012) in Chap. 13, play an import role in regulation of gene expression in the germline. For example, the X-chromosome is depleted of spermatogenesis expressed genes and is silenced in the germline of XO males, while the few oogenesis genes that reside on the X-chromosome are silenced except for a small burst of expression in late pachytene and diplotene of the hermaphrodite (Kelly et al. 2002). Translational regulation is used extensively in temporal/spatial control of germline gene expression in *C. elegans*, as is true in other organisms. For example, Merritt et al. (2008) created fusion genes, which each contains the same germline specific promoter, a histone H2B fused to GFP for visualization, and a unique 3′ UTR from 30 known germline expressed genes. Of the 30 constructs containing different 3′ UTRs, 24 had the same expression pattern as the endogenous gene from which the 3′ UTR was obtained. Since the 3′ UTRs is the only difference among the 24 transcripts, the results suggest that correct protein expression pattern is largely controlled via translational regulation through their 3′ UTR. In Chap. 8, Nousch and Eckmann (2012) present the current state of knowledge for translational regulation and its implications for germ cell development.

Self-fertile hermaphroditism in *C. elegans* is a recently evolved character as most sister species have a male/female mode of reproduction. Haag and Liu (2012), in Chap. 14, discuss how translational control and the rapid evolution of binding sites in mRNA targets, for RNA-binding protein regulators, likely contributed to evolutionary differences in germline sex determination in the Caenorhabditis clade.

1.7 Conclusions

The germline, from one generation to the next, is essential for the continuum of metazoan life. This volume presents a series of reviews that provide our current understanding of the various aspects of germ cell development that are necessary for this continuum in *C. elegans*. The information, insights and questions posed provide a foundation upon which future avenues of research will further our understanding of the cycle of the germline.

Acknowledgments We thank Dave Hansen for comments, and Karen Fiorino for the artwork. Nanette Pazdernik is supported by NIH T32 HD 49305. Research in the Schedl laboratory is supported by R01 GM085150.

References

Ahringer J (2006) Reverse genetics. In: Ambros V (ed) WormBook. ed. The *C. elegans* Research Community. doi:10.1895/wormbook.1.47.1

Angelo G, Van Gilst MR (2009) Starvation protects germline stem cells and extends reproductive longevity in *C. elegans*. Science 326(5955):954–958. doi:10.1126/science.1178343

Austin J, Kimble J (1987) glp-1 is required in the germ line for regulation of the decision between mitosis and meiosis in *C. elegans*. Cell 51(4):589–599. doi:0092-8674(87)90128-0

Bailly A, Gartner A (2012) Germ cell apoptosis and DNA damage responses. Advances in Experimental Medicine and Biology 757:249–276. (Chap. 9, this volume) Springer, New York

Barstead RJ, Moerman DG (2006) *C. elegans* deletion mutant screening. Methods Mol Biol 351:51–58. doi:10.1385/1-59745-151-7:51

Berry LW, Westlund B, Schedl T (1997) Germ-line tumor formation caused by activation of glp-1, a *Caenorhabditis elegans* member of the Notch family of receptors. Development 124(4):925–936

Brenner S (1974) The genetics of *Caenorhabditis elegans*. Genetics 77(1):71–94

Brenner S, Stretton AOW, Kaplan S (1965) Genetic code: the 'nonsense' triplets for chain termination and their suppression. Nature 206(4988):994–998

Chalfie M, Tu Y, Euskirchen G, Ward WW, Prasher DC (1994) Green fluorescent protein as a marker for gene expression. Science 263(5148):802–805

Chu DS, Shakes DC (2012) Spermatogenesis. Advances in Experimental Medicine and Biology 757:171–203. (Chap. 7, this volume) Springer, New York

Felix MA, Braendle C (2010) The natural history of *Caenorhabditis elegans*. Curr Biol 20(22):R965–R969. doi:10.1016/j.cub.2010.09.050

Fire A, Xu S, Montgomery MK, Kostas SA, Driver SE, Mello CC (1998) Potent and specific genetic interference by double-stranded RNA in *Caenorhabditis elegans*. Nature 391(6669): 806–811

Frokjaer-Jensen C, Davis MW, Hopkins CE, Newman BJ, Thummel JM, Olesen SP, Grunnet M, Jorgensen EM (2008) Single-copy insertion of transgenes in *Caenorhabditis elegans*. Nat Genet 40(11):1375–1383. doi:10.1038/ng.248

Gibert MA, Starck J, Beguet B (1984) Role of the gonad cytoplasmic core during oogenesis of the nematode *Caenorhabditis elegans*. Biol Cell 50(1):77–85

Govindan JA, Cheng H, Harris JE, Greenstein D (2006) Galphao/i and Galphas signaling function in parallel with the MSP/Eph receptor to control meiotic diapause in *C. elegans*. Curr Biol 16(13):1257–1268. doi:10.1016/j.cub.2006.05.020

Govindan JA, Nadarajan S, Kim S, Starich TA, Greenstein D (2009) Somatic cAMP signaling regulates MSP-dependent oocyte growth and meiotic maturation in *C. elegans*. Development 136(13):2211–2221. doi:10.1242/dev.034595

Grant B, Hirsh D (1999) Receptor-mediated endocytosis in the *Caenorhabditis elegans* oocyte. Mol Biol Cell 10(12):4311–4326

Green RA, Kao H-L, Audhya A, Arur S, Mayers JR, Fridolfsson HN, Schulman M, Schloissnig S, Niessen S, Laband K, Wang S, Starr DA, Hyman AA, Schedl T, Desai A, Piano F, Gunsalus KC, Oegema K (2011) A high-resolution *C. elegans* essential gene network based on phenotypic profiling of a complex tissue. Cell 145(3):470–482

Haag ES, Liu Q (2012) Using Caenorhabditis to explore the evolution of the germ line. Advances in Experimental Medicine and Biology 757:405–425. (Chap. 14, this volume) Springer, New York

Hall DH, Winfrey VP, Blaeuer G, Hoffman LH, Furuta T, Rose KL, Hobert O, Greenstein D (1999) Ultrastructural features of the adult hermaphrodite gonad of *Caenorhabditis elegans*: relations between the germ line and soma. Dev Biol 212(1):101–123

Hansen D, Schedl T (2012) Stem cell proliferation versus meiotic fate decision in *C. elegans*. Advances in Experimental Medicine and Biology 757:71–99. (Chap. 4, this volume) Springer, New York

Hedgecock EM, Culotti JG, Hall DH, Stern BD (1987) Genetics of cell and axon migrations in *Caenorhabditis elegans*. Development 100(3):365–382

Hirsh D, Oppenheim D, Klass M (1976) Development of the reproductive system of *Caenorhabditis elegans*. Dev Biol 49(1):200–219

Hobert O (2010) The impact of whole genome sequencing on model system genetics: get ready for the ride. Genetics 184(2):317–319. doi:10.1534/genetics.109.112938

Hodgkin J (1986) Sex determination in the nematode *C. elegans*: analysis of tra-3 suppressors and characterization of fem genes. Genetics 114:15–52

Hodgkin J, Barnes TM (1991) More is not better: brood size and population growth in a self-fertilizing nematode. Proc R Soc Lond B Biol Sci 246(1315):19–24. doi:10.1098/rspb.1991.0119

Hubbard EJ, Korta DZ, Dalfó D (2012) Physiological control of germline development. Advances in Experimental Medicine and Biology 757:101–131. (Chap. 5, this volume) Springer, New York

Jakubowski J, Kornfeld K (1999) A local, high-density, single-nucleotide polymorphism map used to clone *Caenorhabditis elegans* cdf-1. Genetics 153(2):743–752

Jaramillo-Lambert A, Ellefson M, Villeneuve AM, Engebrecht J (2007) Differential timing of S phases, X chromosome replication, and meiotic prophase in the *C. elegans* germ line. Dev Biol 308(1):206–221. doi:10.1016/j.ydbio.2007.05.019

Kelly WG, Fire A (1998) Chromatin silencing and the maintenance of a functional germline in *Caenorhabditis elegans*. Development 125(13):2451–2456

Kelly WG, Schaner CE, Dernburg AF, Lee MH, Kim SK, Villeneuve AM, Reinke V (2002) X-chromosome silencing in the germline of *C. elegans*. Development 129(2):479–492

Killian DJ, Hubbard EJ (2005) *Caenorhabditis elegans* germline patterning requires coordinated development of the somatic gonadal sheath and the germ line. Dev Biol 279(2):322–335. doi:10.1016/j.ydbio.2004.12.021

Kim S, Spike CA, Greenstein D (2012) Control of oocyte growth and meiotic maturation in *C. elegans*. Advances in Experimental Medicine and Biology 757:277–320. (Chap. 10, this volume) Springer, New York

Kimble J, Hirsh D (1979) The postembryonic cell lineages of the hermaphrodite and male gonads in *Caenorhabditis elegans*. Dev Biol 70(2):396–417

Kimble JE, White JG (1981) On the control of germ cell development in *Caenorhabditis elegans*. Dev Biol 81(2):208–219. doi:0012-1606(81)90284-0

Lee MH, Ohmachi M, Arur S, Nayak S, Francis R, Church D, Lambie E, Schedl T (2007) Multiple functions and dynamic activation of MPK-1 extracellular signal-regulated kinase signaling in *Caenorhabditis elegans* germline development. Genetics 177(4):2039–2062. doi:10.1534/genetics.107.081356

Lui DY, Colaiácovo MP (2012) Meiotic development in *C. elegans*. Advances in Experimental Medicine and Biology 757:133–170. (Chap. 6, this volume) Springer, New York

Maddox AS, Habermann B, Desai A, Oegema K (2005) Distinct roles for two *C. elegans* anillins in the gonad and early embryo. Development 132(12):2837–2848. doi:10.1242/dev.01828

Marcello MR, Singaravelu G, Singson A (2012) Fertilization. Advances in Experimental Medicine and Biology 757:321–350. (Chap. 11, this volume) Springer, New York

Maupas (1900) Modes et formes de reproduction des nematodes. Translation by Marie-Anne Felix. Archives de Zoologie Experimentale et Generale 8:463–624

McCarter J, Bartlett B, Dang T, Schedl T (1997) Soma-germ cell interactions in *Caenorhabditis elegans*: multiple events of hermaphrodite germline development require the somatic sheath and spermathecal lineages. Dev Biol 181(2):121–143. doi:10.1006/dbio.1996.8429

McCarter J, Bartlett B, Dang T, Schedl T (1999) On the control of oocyte meiotic maturation and ovulation in *Caenorhabditis elegans*. Dev Biol 205(1):111–128. doi:10.1006/dbio.1998.9109

Merritt C, Seydoux G (2010) Transgenic solutions for the germline. WormBook:1–21. doi:10.1895/wormbook.1.148.1

Merritt C, Rasoloson D, Ko D, Seydoux G (2008) 3' UTRs are the primary regulators of gene expression in the *C. elegans* germline. Curr Biol 18(19):1476–1482. doi:10.1016/j.cub.2008.08.013

Miller MA, Nguyen VQ, Lee MH, Kosinski M, Schedl T, Caprioli RM, Greenstein D (2001) A sperm cytoskeletal protein that signals oocyte meiotic maturation and ovulation. Science 291(5511):2144–2147. doi:10.1126/science.1057586291/5511/2144

Miller MA, Ruest PJ, Kosinski M, Hanks SK, Greenstein D (2003) An Eph receptor sperm-sensing control mechanism for oocyte meiotic maturation in *Caenorhabditis elegans*. Genes Dev 17(2):187–200. doi:10.1101/gad.1028303

Morrison SJ, Spradling AC (2008) Stem cells and niches: mechanisms that promote stem cell maintenance throughout life. Cell 132(4):598–611. doi:10.1016/j.cell.2008.01.038

Nadarajan S, Govindan JA, McGovern M, Hubbard EJ, Greenstein D (2009) MSP and GLP-1/Notch signaling coordinately regulate actomyosin-dependent cytoplasmic streaming and oocyte growth in *C. elegans*. Development 136(13):2223–2234. doi:10.1242/dev.034603

Nousch M, Eckmann CR (2012) Translational control in the *C. elegans* germ line. Advances in Experimental Medicine and Biology 757:205–247. (Chap. 8, this volume) Springer, New York

Reinke V, Gil IS, Ward S, Kazmer K (2004) Genome-wide germline-enriched and sex-biased expression profiles in *Caenorhabditis elegans*. Development 131(2):311–323. doi:10.1242/dev.00914

Robertson S, Lin R (2012) The oocyte-to-embryo transition. Advances in Experimental Medicine and Biology 757:351–372. (Chap. 12, this volume) Springer, New York

Sarin S, Prabhu S, O'Meara MM, Pe'er I, Hobert O (2008) *Caenorhabditis elegans* mutant allele identification by whole-genome sequencing. Nat Methods 5(10):865–867. doi:10.1038/nmeth.1249

Schisa JA, Pitt JN, Priess JR (2001) Analysis of RNA associated with P granules in germ cells of *C. elegans* adults. Development 128(8):1287–1298

Seidel HS, Kimble J (2011) The oogenic germline starvation response in *C. elegans*. PLoS One 6(12):e28074. doi:10.1371/journal.pone.0028074

Sijen T, Fleenor J, Simmer F, Thijssen KL, Parrish S et al (2001) On the role of RNA amplification in dsRNA-triggered gene silencing. Cell 107:465–476

Sulston JE, Horvitz HR (1977) Post-embryonic cell lineages of the nematode, *Caenorhabditis elegans*. Dev Biol 56(1):110–156

Van Wynsberghe PM, Maine EM (2012) Epigenetic control of germline development. Advances in Experimental Medicine and Biology 757:373–403. (Chap. 13, this volume) Springer, New York

Wang JT, Seydoux S (2012) Germ cell specification. Advances in Experimental Medicine and Biology 757:17–39. (Chap. 2, this volume) Springer, New York

Ward S, Carrel JS (1979) Fertilization and sperm competition in the nematode *Caenorhabditis elegans*. Dev Biol 73(2):304–321

Ward S, Argon Y, Nelson GA (1981) Sperm morphogenesis in wild-type and fertilization-defective mutants of *Caenorhabditis elegans*. J Cell Biol 91(1):26–44

Wicks SR, Yeh RT, Gish WR, Waterston RH, Plasterk RH (2001) Rapid gene mapping in *Caenorhabditis elegans* using a high density polymorphism map. Nat Genet 28(2):160–164. doi:10.1038/88878

Wolke U, Jezuit EA, Priess JR (2007) Actin-dependent cytoplasmic streaming in *C. elegans* oogenesis. Development 134(12):2227–2236. doi:10.1242/dev.004952

Wood AJ, Lo TW, Zeitler B, Pickle CS, Ralston EJ, Lee AH, Amora R, Miller JC, Leung E, Meng X, Zhang L, Rebar EJ, Gregory PD, Urnov FD, Meyer BJ (2011) Targeted genome editing across species using ZFNs and TALENs. Science 333(6040):307. doi:10.1126/science.1207773

Yochem J, Herman RK (2003) Investigating *C. elegans* development through mosaic analysis. Development 130(20):4761–4768. doi:10.1242/dev.00701

Zanetti S, Puoti A (2012) Sex determination in the *C. elegans* germline. Advances in Experimental Medicine and Biology 757:41–69. (Chap. 3, this volume) Springer, New York

Chapter 2
Germ Cell Specification

Jennifer T. Wang and Geraldine Seydoux

Abstract The germline of *Caenorhabditis elegans* derives from a single founder cell, the germline blastomere P_4. P_4 is the product of four asymmetric cleavages that divide the zygote into distinct somatic and germline (P) lineages. P_4 inherits a specialized cytoplasm ("germ plasm") containing maternally encoded proteins and RNAs. The germ plasm has been hypothesized to specify germ cell fate, but the mechanisms involved remain unclear. Three processes stand out: (1) inhibition of mRNA transcription to prevent activation of somatic development, (2) translational regulation of the *nanos* homolog *nos-2* and of other germ plasm mRNAs, and (3) establishment of a unique, partially repressive chromatin. Together, these processes ensure that the daughters of P_4, the primordial germ cells Z2 and Z3, gastrulate inside the embryo, associate with the somatic gonad, initiate the germline transcriptional program, and proliferate during larval development to generate ~2,000 germ cells by adulthood.

Keywords Germ plasm • Polarity • Germ granules • Cell fate • Transcriptional repression • Germline blastomeres • Primordial germ cells • P lineage • Maternal RNA

J.T. Wang • G. Seydoux (✉)
Department of Molecular Biology and Genetics, Howard Hughes Medical Institute, Center for Cell Dynamics, Johns Hopkins School of Medicine, 725 North Wolfe Street, PCTB 706, Baltimore, MD 21205, USA
e-mail: gseydoux@jhmi.edu

T. Schedl (ed.), *Germ Cell Development in C. elegans*, Advances in Experimental Medicine and Biology 757, DOI 10.1007/978-1-4614-4015-4_2,

2.1 Introduction to the Embryonic Germ Lineage (P Lineage)

2.1.1 Embryonic Origin of the Germline

P_4 arises in the 24-cell stage from a series of four asymmetric divisions starting in the zygote (P_0) (Fig. 2.1). Each division generates a larger, somatic blastomere (AB, EMS, C and D) and a smaller, germline blastomere (P_1, P_2, P_3, P_4). Laser ablation of the P_4 nucleus yields sterile worms with no germ cells (Sulston et al. 1983), confirming that P_4 is the sole founder of the germline and that no other cell can replace P_4.

In the 88-cell stage, P_4 divides once to generate two daughters: the primordial germ cells, Z2 and Z3. Soon after their birth, Z2 and Z3 gastrulate into the embryo interior (Harrell and Goldstein 2011). Z2 and Z3 do not divide further during embryogenesis, and remain close to each other and to the intestine. By the 2-fold stage, Z2 and Z3 extend protrusions towards two intestinal cells (Sulston et al. 1983). Intestinal cells have been suggested to provide sustenance to Z2 and Z3 until the gonad is formed.

In mid-embryogenesis, the somatic gonadal precursors Z1 and Z4 migrate towards Z2 and Z3 to form the gonad primordium (Sulston et al. 1983). Z2 and Z3 resume divisions only in the first (L1) larval stage after the larva begins feeding. Z2 and Z3 will eventually generate ~2,000 germ cells by adulthood (Kimble and White 1981).

2.1.2 Characteristics of the P Blastomeres

2.1.2.1 Asymmetric Divisions

P_0, P_1, P_2, P_3 all divide asymmetrically. Before each division, the spindle becomes displaced towards one side of the cell. The P granules, RNA-rich organelles specific to the germline, and several associated cytoplasmic proteins and RNAs (collectively referred to as "germ plasm"; Table 2.1) also accumulate on that same side. As a result, each division generates daughters of unequal size with the smaller daughter inheriting most of the germ plasm (Gönczy and Rose 2005; Strome 2005).

In the first two divisions, the spindle becomes displaced towards the posterior pole of the embryo, such that P_1 and P_2 are born in the posterior. The posterior pole is defined in the zygote P_0 by the position of the sperm centrosome, which orients the distribution of the PAR polarity regulators (Gönczy and Rose 2005). In the P_2 blastomere, the polarity axis is reversed by signaling from the somatic blastomere EMS, and P_3 and P_4 are born towards the anterior (Schierenberg 1987; Arata et al. 2010). As a result, P_4 is born next to the descendants of the E (intestinal) lineage. Unlike P_0–P_3, P_4 divides symmetrically into two equal size daughters (Z2 and Z3) that both inherit germ plasm.

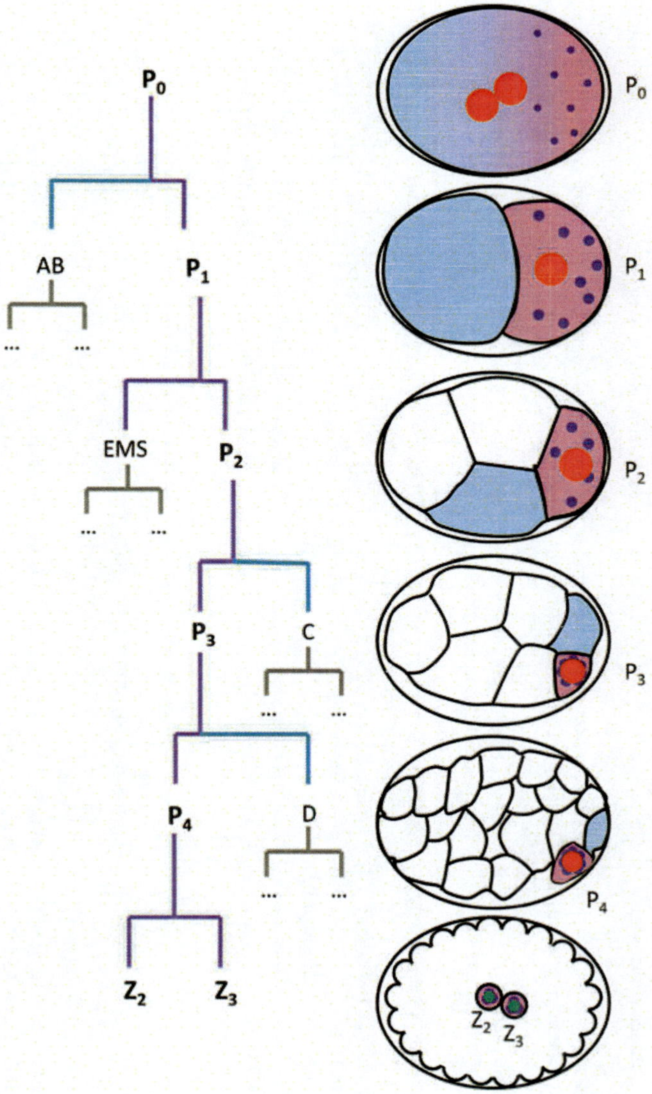

Fig. 2.1 Embryonic origin of the germline. Abbreviated embryonic lineage from the 1-cell stage to the ~88-cell stage and embryo schematics corresponding to each stage shown in the lineage tree. Germ plasm is denoted in *purple*, germ granules are *darker purple dots*. High levels of MEX-5/6 inherited by somatic blastomeres are denoted in *blue*. *Red nuclei* are not competent for mRNA transcription

2.1.2.2 Long Cell Cycle Times

P blastomeres have longer cell cycle times than their somatic sisters. For example, P_1 divides 2 min after AB, in part due to enhanced activity of a DNA replication checkpoint in P_1 (Encalada et al. 2000; Brauchle et al. 2003), and in part due to higher

Table 2.1 Proteins in germ plasm

Protein	Domains	Protein localization	Loss-of-function phenotype	Proposed function	References
PIE-1	CCCH fingers	P blastomeres Nuclear, cytoplasmic, P granules, centrosomes	Maternal effect embryonic lethality P_2 transformed to EMS fate No germ cells	Repression of RNA Polymerase II Translational control of mRNAs	Mello et al. (1992), Seydoux et al. (1996), Tenenhaus et al. (2001)
POS-1	CCCH fingers	P blastomeres Cytoplasmic, P granules	Maternal effect embryonic lethality Complex cell fate transformations No germ cells	Translational control of mRNAs	Tabara et al. (1999)
MEX-1	CCCH fingers	P blastomeres Cytoplasmic, P granules	Maternal effect embryonic lethality Complex cell fate transformations No germ cells	Unknown	Mello et al. (1992), Guedes and Priess (1997)
MEX-3	KH domain	Present in oocytes through 4-cell stage Cytoplasmic in multiple cell types P granules in P_1 through P_4	Maternal effect embryonic lethality D transformed to P_4, as a result more than two Z2/Z3 like cells are formed	Translational control of mRNAs	Draper et al. (1996)
MES-1	Receptor tyrosine kinase related	In P_2 at junction with EMS contact In P_3 at junction with E contact	Maternal effect sterility P_4 transformed to D fate No germ cells	Signaling from EMS to P_2 to reverse polarity in P_2	Strome et al. (1995), Berkowitz and Strome (2000)
Sm proteins	LSm fold	P granules	Embryonic arrest at 50- to 100-cell stage Mislocalization of PIE-1, P granules Symmetric divisions and short cell cycles	Regulation of mRNAs	Barbee et al. (2002), Barbee and Evans (2006)
PAR-1	KIN1/MARK kinase	Cortical—P blastomeres and Z2 and Z3 until morphogenesis	Maternal effect embryonic lethality No germ cells	Polarization of P blastomeres	Kemphues et al. (1988), Guo and Kemphues (1995)
MES-2/3/6	Polycomb orthologs	Nuclear in both somatic and germ cells in embryos. Persists in Z2 and Z3 in embryos and L1 larvae.	Maternal effect sterility Germ cell death at L3 and L4 stages	Repression of X chromosome transcription	Bender et al. (2004), Strome (2005)

MES-4	SET domain	Chromatin, preferentially autosomes. Present in somatic and germ cells in early embryogenesis, persists in Z2 and Z3	Maternal effect sterility Germ cell death at L3 and L4 stages	Transmission of germline transcriptional program from mother to progeny Repression of X chromosome transcription	Fong et al. (2002), Bender et al. (2006), Rechtsteiner et al. (2010)
NOS-1/2	Nanos orthologs	NOS-1: expressed in Z2 and Z3 from zygotic mRNA, cytoplasmic NOS-2: expressed in P_4 from maternal mRNA, cytoplasmic	Z2 and or Z3 outside somatic gonad Germ cell death at L3 stage	Regulation of mRNAs	Subramaniam and Seydoux (1999)
MEG-1/2	Novel	P granules from P_2 through P_4	Maternal effect sterility Germ cell death at L3 stage	Unknown	Leacock and Reinke (2008)
PGL-1/2/3	RGG box	Constitutive P granule components	Sterility with variable phenotypes	Unknown, primarily post-embryonic	Kawasaki et al. (2004)
GLH-1/2/3/4	DEAD box helicase	Constitutive P granule components	Sterility with variable phenotypes	Unknown, primarily post-embryonic	Spike et al. (2008)
Anti-germ plasm:					
MEX-5/6	CCCH fingers	High levels in somatic founder blastomeres—cytoplasmic High/low levels in Ant/Post cytoplasm of P blastomeres—cytoplasmic and enriched on P granules and centrosomes	Embryonic lethality Complex cell fate transformations No germ cells	Dispersal of P granules in dividing P blastomeres Degradation of germ plasm in somatic blastomeres	Schubert et al. (2000) Gallo et al. (2010)

levels of cell cycle regulators (PLK-1 and Cdc25.1) in AB (Rivers et al. 2008; Budirahardja and Gönczy 2008). P_4 divides about 70 min after its birth (Sulston et al. 1983). Z2 and Z3 duplicate their DNA and centrosomes, but remain arrested in G2 until after hatching (Fukuyama et al. 2006).

2.1.2.3 No mRNA Transcription

mRNA transcription begins in the 3- to 4-cell stage in somatic blastomeres, but appears to remain off in the germline blastomeres until gastrulation. In a survey of 16 mRNAs, no newly transcribed mRNAs were detected in P_0–P_4 by in situ hybridization (Seydoux et al. 1996). During the transcription cycle, the serine-rich repeats in the carboxy-terminal tail of RNA polymerase II become phosphorylated, first on Serine 5 during initiation and then on Serine 2 during elongation. These phosphoepitopes are reduced (Pser5) or completely absent (Pser2) in the germline blastomeres (Seydoux and Dunn 1997). Both phosphoepitopes appear transiently in Z2 and Z3 shortly after their birth, but return to low/background levels by the 1.5-fold stage and do not reappear until after hatching (Furuhashi et al. 2010). Z2 and Z3 also lose the active chromatin marks H3K4me2, H3K4me3, and H4K8ac (Schaner et al. 2003). Z2 and Z3 are not completely transcriptionally silent, however: zygotic expression of several germline genes have been detected in Z2 and Z3. These include P granule components (*pgl-1*, *glh-1*, and *glh-4*), the nanos ortholog *nos-1*, and meiotic genes (*htp-3*, *rec-8*) (Subramaniam and Seydoux 1999; Kawasaki et al. 2004; Takasaki et al. 2007; Spencer et al. 2011). In contrast to mRNA transcription, transcription of ribosomal RNAs has been detected in all P blastomeres with the possible exception of P_4 (Seydoux and Dunn 1997).

2.1.2.4 Maintenance of Maternal mRNAs

In situ hybridization and RNA profiling studies have uncovered two classes of maternal mRNAs in early embryos: maternal mRNAs that are maintained in all blastomeres, and maternal mRNAs that are rapidly turned over in somatic blastomeres and maintained only in germline blastomeres (Seydoux and Fire 1994; Seydoux et al. 1996; Baugh et al. 2003). Some in the latter class are also enriched in P granules. For example, the Nanos homolog *nos-2* is partitioned to both germline and somatic blastomeres during the first two divisions. Between the 4- and 8-cell stages, *nos-2* is turned over in somatic blastomeres and maintained in the P lineage, where it is enriched in P granules. By the 28-cell stage, *nos-2* RNA remains only in P_4, where it is finally translated (Subramaniam and Seydoux 1999; Tenenhaus et al. 2001).

2.2 Cellular Mechanisms of Germ Cell Specification

Two general modes of germline specification have been described in animals: induction by extracellular signals and induction by germ plasm, a specialized cytoplasm inherited from the oocyte (Seydoux and Braun 2006). In this section, we describe evidence for each of these mechanisms acting in *C. elegans*.

2.2.1 Asymmetric Segregation of the Germ Plasm

Several lines of evidence suggest that *C. elegans* embryos possess germ plasm. As described above, the germline-specific P granules and associated RNAs and RNA-binding proteins co-segregate to the same side of the P blastomere before each asymmetric cleavage (Table 2.1). P or "germ" granules have been reported in the germline of many different animals, including mammals, and are considered to be intimately associated with germ cell fate (Strome and Lehmann 1997).

Embryo manipulations support the view that at least some aspects of P cell fate are specified by factors that are asymmetrically localized in the zygote. Using a laser microbeam to create holes in the eggshell, Schierenberg (1988) extruded "partial embryos" containing cytoplasm from only the anterior or posterior of the zygote. Partial embryos containing anterior cytoplasm divided symmetrically, whereas partial embryos containing posterior cytoplasm divided asymmetrically, similar to the P blastomeres. However, mixing of posterior cytoplasm into anterior cytoplasm was not sufficient to induce asymmetric divisions. Delaying cell division eliminated the ability of posterior cytoplasm to support asymmetric divisions. Together these observations suggest that the germ plasm is required for germ cell fate but is not sufficient to induce germ cell fate when diluted with "somatic cytoplasm." In contrast, in *Drosophila*, injection of germ plasm in the anterior pole of the embryo is sufficient to create ectopic germ cells (Mahowald and Illmensee 1974).

Asymmetric distribution of the germ plasm is controlled by the PAR network of polarity regulators, which regulates anterior–posterior polarity in P_0 and most likely also in P_1, P_2, and P_3 (see below). The PAR proteins PAR-1 and PAR-2 segregate with the germ plasm, and both are maintained in the P lineage through the asymmetric divisions leading to P_4 (Guo and Kemphues 1995; Boyd et al. 1996). PAR-1 and PAR-2 become enriched at the cell periphery on the side of the germ plasm during each asymmetric division. Strong mutations in the *par* genes disrupt all polarity in the 1-cell stage and lead to embryonic lethality. Hypomorphic *par* mutations, however, lead to viable but sterile worms that lack all germ cells (Kemphues et al. 1988; Guo and Kemphues 1995; Spilker et al. 2009). These observations suggest that asymmetric segregation of the germ plasm is required to specify P_4 as the germline founder cell.

2.2.1.1 MEX-5 and MEX-6: Germ Plasm Antagonists

The PAR network regulate germ plasm asymmetry through the action of the PAR-1 kinase and its substrates MEX-5 and MEX-6, two highly related and partially redundant RNA-binding proteins that segregate opposite to the germ plasm. Phosphorylation by PAR-1 stimulates MEX-5 (and presumably MEX-6) diffusion in the posterior cytoplasm of the zygote, causing MEX-5 to become enriched in the anterior (Tenlen et al. 2008; Griffin et al. 2011). As a result, the AB blastomere inherits high levels of MEX-5/6 and low levels of PAR-1, and the P_1 blastomere inherit low levels of MEX-5/6 and high levels of PAR-1. This pattern is repeated during the divisions of P_1, P_2, and P_3 (Schubert et al. 2000; Guo and Kemphues 1995). MEX-5 and MEX-6 promote both asymmetric partitioning of the germ plasm to germ cells during cell division and asymmetric degradation of the germ plasm from the soma after cell division.

2.2.1.2 Asymmetric Partitioning of the Germ Plasm During Division

Examination of P granule dynamics in live zygotes has revealed that P granule partitioning depends both on MEX-5/6-driven granule disassembly in the anterior cytoplasm and PAR-1-driven granule assembly in the posterior cytoplasm (Cheeks et al. 2004; Brangwynne et al. 2009; Gallo et al. 2010). P granule proteins that become dispersed in the anterior cytoplasm are reincorporated into granules in the posterior cytoplasm. As a result, P_1 inherits more P granule proteins than AB (Gallo et al. 2010). After polarity reversal in P_2, P granules appear to segregate using a different mechanism involving association with the P cell nuclei (Hird et al. 1996). PAR-1 and MEX-5/6 also promote the posterior enrichment of germ plasm proteins that are only loosely associated with P granules, such as PIE-1 and POS-1 (Table 2.1), but the mechanisms involved are not known (Schubert et al. 2000). MEX-5/6 also promotes anterior enrichment of PLK-1 and CDC-25, which contribute to the fast cell cycle of the AB blastomere (Rivers et al. 2008; Budirahardja and Gönczy 2008).

2.2.1.3 Asymmetric Degradation of the Germ Plasm After Division

Asymmetric enrichment of the germ plasm during division is not absolute and low levels of germ plasm RNAs and proteins are inherited by all somatic blastomeres. These low levels are rapidly turned over, and this degradation depends on MEX-5 and MEX-6. In *mex-5;mex-6* embryos, germ plasm proteins are uniformly partitioned to all blastomeres. Heat shock-induced expression of MEX-5 in single blastomere is sufficient to degrade germ plasm proteins in that cell (Schubert et al. 2000). The potent anti-germ plasm effect of MEX-5 may explain why, in the cytoplasmic mixing experiments described above (Schierenberg 1988), anterior cytoplasm "suppresses" the potential for asymmetric division.

In somatic blastomeres, MEX-5 and MEX-6 are required for their own degradation and the degradation of other CCCH zinc finger proteins (POS-1, PIE-1, and

MEX-1). CCCH protein degradation depends on ZIF-1, a substrate recognition subunit for the CUL-2 E3 ubiquitin ligase. ZIF-1 recognizes specific CCCH fingers in MEX-5, MEX-1, POS-1, and PIE-1. A fusion between GFP and the PIE-1 first zinc finger (GFP:ZF1) is symmetrically segregated to somatic and germline blastomeres, but degraded in each somatic lineage in a ZIF-1-dependent manner (DeRenzo et al. 2003). The distribution of ZIF-1 protein is not known, but a reporter containing the *zif-1* 3′ UTR is activated in each somatic lineage, suggesting that ZIF-1 activity is restricted to somatic blastomeres by translational regulation of the *zif-1* mRNA. Recent studies indicate that *zif-1* translation is controlled combinatorially by several RNA-binding proteins that all bind directly to the *zif-1* 3′ UTR. In oocytes, *zif-1* is silenced by OMA-1 and OMA-2 (Guven-Ozkan et al. 2010; Robertson and Lin 2012, Chap. 12), two redundant RNA-binding proteins that interact with the eIF4E-binding protein and translational repressor SPN-2 (Li et al. 2009). In zygotes, OMA-1/2 are phosphorylated by the kinase MBK-2 (Nishi and Lin 2005; Shirayama et al. 2006; Stitzel et al. 2006), leading to the displacement of SPN-2 from the *zif-1* 3′ UTR and the eventual degradation of OMA-1 and OMA-2 during the first cleavage (Pellettieri et al. 2003; Nishi and Lin 2005; Shirayama et al. 2006; Guven-Ozkan et al. 2010). *zif-1* continues to be silenced, however, through the combined action of MEX-3 and SPN-4 in zygotes and POS-1 in later stages (Oldenbroek et al. 2012). This repression is lifted in somatic blastomeres by MEX-5 and MEX-6, which compete with POS-1 for binding to the *zif-1* 3′ UTR (Oldenbroek et al. 2012). Thus, MEX-5 and MEX-6 promote their own degradation and the degradation of other CCCH-binding proteins by promoting the translation of the E3 ligase subunit that targets them for ubiquitination. MEX-5 activity requires phosphorylation by the Polo kinases PLK-1 and PLK-2, which directly bind to, and segregate with, MEX-5. Phosphorylation by PLK-1 and PLK-2 is primed by MBK-2, which is active in zygotes but not oocytes. This requirement may explain why MEX-5 promotes germ plasm turnover in embryos, but not in oocytes where MEX-5 is also present (Nishi et al. 2008).

The mechanisms by which MEX-5 and MEX-6 also promote RNA degradation in somatic blastomeres are less well understood. Activation of mRNA degradation in the 4-cell stage is temporally correlated with the recruitment of LSM-1 and CCF-1 (CAF1/Pop2 subunit of the CCR4/NOT deadenylase complex) to P bodies, cytoplasmic granules that have been implicated in the decapping and deadenylation of mRNAs. In *mex-5; mex-6* (RNAi) embryos, LSM-1 is not recruited to P bodies and maternal mRNAs are stabilized. Consistent with a role for deadenylation, RNAi depletion of *let-711*/Not-1, a component of CCR4/NOT deadenylase, also interferes with LSM-1 recruitment and mRNA degradation (Gallo et al. 2008). Whether LSM-1 is required for this process, however, has not yet been examined.

2.2.1.4 Self-propagation of Germ Plasm and Anti-germ Plasm?

The properties of MEX-5 and MEX-6 suggest that in *C. elegans* the distinction between soma and germline depends both on maintenance of the germ plasm in the P lineage, and on the active degradation of germ plasm in somatic lineages (“anti-germ plasm activity"). In *par-1* mutants, MEX-5 and MEX-6 remain uniform and germ plasm RNAs and CCCH proteins are degraded in all cells by the 4-cell stage. Presumably, in wild-type embryos, PAR-1 maintains MEX-5 and MEX-6 at low enough levels in the P blastomeres to avoid degradation of the germ plasm. PAR-1 is maintained in all germline blastomeres and in Z2 and Z3, suggesting that PAR-1 is required continuously in the embryonic germ lineage to maintain the germ plasm. Intriguingly, in the zygote, MEX-5/6 activity is required for maximal enrichment of PAR-1 in the posterior (Cuenca et al. 2003). One possibility is that mutual regulation/exclusion by PAR-1 and MEX-5/6 functions in a continuous loop to ensure that germ plasm asymmetry is reestablished in each P blastomere.

2.2.2 Asymmetric Segregation of P Granules: Not Essential?

The P granules are the only components of the germ plasm that persist in all germ cells throughout the development (except in sperm, Updike and Strome 2010). P or "germ" granules have been observed in the germ plasm and/or germ cells of all animals examined (Strome and Lehmann 1997). By electron microscopy in zygotes, P granules appear as round, electron-dense structures without membranes and dispersed throughout the cytoplasm (Wolf et al. 1983). Starting in P_2, P granules associate with the cytoplasmic face of the nuclear envelope, where they will remain until gametogenesis. P granules exclude macromolecules larger than 70 kDa and greater, and have been proposed to extend the nuclear pore environment of the nuclear membrane into the cytoplasm (Updike et al. 2011).

P granules contain both constitutive components present at all stages of development and stage-specific components. Constitutive components include the RGG domain RNA-binding proteins PGL-1 and PGL-3 (Kawasaki et al. 1998, 2004) and the Vasa-related RNA helicases GLH-1,2,3 and 4 (Roussell and Bennett 1993; Kuznicki et al. 2000). PGL-1/3 are the core scaffolding components of P granules and can assemble into granules when expressed on their own in tissue culture cells (Hanazawa et al. 2011). Mutations in *pgl* and *glh* genes interfere with larval germ cell proliferation and gamete formation (Kawasaki et al. 2004; Spike et al. 2008). The most severe defects are seen when the worms are raised at high temperature or when mutations in multiple genes are combined. For example, *pgl-1* mutants are fertile at 20 °C but sterile with underproliferated germlines at 26 °C. Double loss of *pgl-1* and *pgl-3* leads to sterility even at low temperature (Kawasaki et al. 2004). In all mutant combinations, however, germ cells are still formed, suggesting that P granule proteins are required primarily for germ cell proliferation and/or differentiation, but not for germ cell fate specification (Kawasaki et al. 2004; Spike et al. 2008). The redundancy and strong maternal contribution of PGL and GLH proteins, however, has made it difficult to exclude a potential role for P granules in germ cell fate specification in embryos.

In embryos, several germ plasm proteins are enriched on P granules (e.g., PIE-1, POS-1, MEX-1, MEX-3, MEG-1, MEG-2, Sm proteins), raising the possibility that P granules organize the germ plasm. Dynamic association of PIE-1 with P granules has been suggested to drive PIE-1 partitioning into P blastomeres by slowing down PIE-1 diffusion in the cytoplasm destined for P blastomeres (Daniels et al. 2009). Mutants that mislocalize P granules to somatic blastomeres or misexpress P granule components in somatic cells, however, do not make extra germ cells, suggesting that P granules on their own are not sufficient to assemble germ plasm and/or specify germ cell fate (Strome et al. 1995; Tabara et al. 1999; Mello et al. 1992). Mutants that mislocalize P granules often fail to form primordial germ cells (i.e., *mes-1*), but because these mutants also missegregate other germ plasm components, a specific requirement for P granules could not be inferred.

Recently, a gene required specifically for the asymmetric partitioning of P granules was identified. *pptr-1* codes for a regulatory subunit of the phosphatase PP2A. In *pptr-1* mutants, P granules disassemble during each embryonic cell division. As a result, P granule components, including PGL-1/3, GLH-1/2/4 and the P granule-associated mRNAs *cey-2* and *nos-2* are partitioned equally to somatic and germline blastomeres. Surprisingly, other germ plasm components (including PAR-1, MEX-5/6 and PIE-1) still segregate asymmetrically in *pptr-1* mutants, demonstrating that P granules are in fact not essential to organize germ plasm. Consistent with normal MEX-5 and MEX-6 partitioning, *nos-2* and *cey-2* mRNAs are quickly degraded in each somatic blastomere in *pptr-1* mutants. After MEX-5 and MEX-6 turnover in the somatic lineages, PGL and GLH proteins reassemble into granules during interphase, but these granules appear in all cells and become progressively smaller with each division. By the time of the birth of Z2 and Z3, all cells have either very small or undetectable granules (Gallo et al. 2010).

The PGL granules inherited by somatic blastomeres in *pptr-1* mutants are eventually eliminated by autophagy after gastrulation (Zhang et al. 2009). During mid-embryogenesis, when zygotic transcription of P granule components begins, Z2 and Z3 assemble new P granules. At that time, Z2 and Z3 also initiate expression of the *nos-2* paralog *nos-1*, as they do in wild-type (Subramaniam and Seydoux 1999). Consistent with proper specification of Z2 and Z3, 100 % of *pptr-1* mutants are fertile when raised at 20 °C (Gallo et al. 2010). These observations demonstrate that P granule partitioning is not essential to distinguish soma from germline. If P granules harbor factors that promote germ cell fate, these factors must be quickly inactivated in somatic cells, possibly by MEX-5 and MEX-6.

When raised at 26 °C, 20 % of *pptr-1* mutants grow into sterile adults with underproliferated germlines. The *pptr-1* phenotype is reminiscent of the phenotype of *pgl* and *glh* mutants, and is exacerbated by mutations in *pgl-1*: 15 % of *pptr-1;pgl-1* double mutants are sterile at 20 °C (Gallo et al. 2010). These observations suggest that asymmetric inheritance of maternal P granules, although not essential, ensures that Z2 and Z3 have sufficient P granule material before starting to divide in the larva. Because *pptr-1* mutants *missegregate* but do not *eliminate* all maternal P granule components, the possibility remains that P granules also *contribute* to germ cell fate specification, perhaps as permissive rather than instructive cues.

2.2.3 *Cell-to-Cell Signaling: Also Required?*

Specification of the embryonic germ lineage also depends on at least one cell–cell interaction. MES-1 is a transmembrane protein that functions with SRC-1 to mediate bidirectional signaling between EMS and P_2. This signaling is required to polarize the EMS spindle and to reverse the polarity of P_2 to ensure that P_3 arises in the anterior (Strome et al. 1995; Berkowitz and Strome 2000; Bei et al. 2002). In the absence of MES-1, P_3 divides symmetrically, and P_4 adopts the somatic fate of its sister D. Both cells inherit P granules and other germ plasm components (Strome et al. 1995). The P_4 to D transformation could be due to "dilution" of the germ plasm below a certain threshold necessary to induce germ cell fate. If so, MES-1 signaling could contribute to germ cell fate indirectly by promoting P_3 polarity. Consistent with this possibility, MES-1 has been shown to be required for the proper localization of PAR-2 (Arata et al. 2010). Another possibility, however, is that signaling by MES-1 also induces other changes in P_2 and P_3 required directly to specify or maintain "germ cell fate." Because no experiment has yet shown that the germ plasm is sufficient to induce germ cell fate in *C. elegans*, the possibility that other mechanisms are involved, including induction by cell–cell interactions, cannot be excluded at this time.

2.3 Molecular Mechanisms of Germ Cell Specification

While no single molecular mechanism has been shown yet to be *sufficient* to induce germ cell fate, several have been suggested to be *required* for the proper development of P blastomeres and/or Z2 and Z3. We consider each of these in turn below.

2.3.1 *Translational Regulation of Maternal RNAs*

Several germ plasm components are RNA-binding proteins (Table 2.1). Mutations in these proteins lead to embryonic lethality and cell fate transformations affecting both somatic and germline blastomeres. POS-1 and MEX-3 regulate the translation of several mRNAs and are required to maintain germ plasm asymmetry (Tabara et al. 1999; Jadhav et al. 2008; Mello et al. 1992; Draper et al. 1996). The complex phenotypes of these mutants make it difficult to evaluate their direct contribution to germ cell fate. Because each RNA-binding protein exhibits a unique pattern of perdurance within the germ plasm, one possibility is that they function combinatorially to specify the fate of each germline blastomere and their somatic daughters.

As described above, combinatorial control involving multiple RNA-binding proteins has been demonstrated to restrict the translation of *zif-1* RNA to somatic blastomeres. Analysis of the *nos-2* mRNA supports the view that similar mechanisms cooperate to regulate the translation of mRNAs in the germ plasm. As described above, *nos-2* mRNA is maintained throughout the P lineage but translated only in P_4. Silencing of *nos-2* translation requires SPN-4, OMA-1, OMA-2, MEX-3, 5, and 6, and activation requires PIE-1 and POS-1 (Jadhav et al. 2008; Tenenhaus et al. 2001; D'agostino et al. 2006). OMA-1, OMA-2 and MEX-3 silence *nos-2* during oogenesis, whereas SPN-4 is required primarily to silence *nos-2* in embryos. POS-1 and SPN-4 compete for binding to the *nos-2* 3′ UTR; when SPN-4 levels fall below a threshold in P_4, POS-1 prevails and activates *nos-2* translation (Jadhav et al. 2008).

The role of PIE-1 in the translational activation of *nos-2* is less understood, but is distinct from PIE-1's role in transcriptional repression (described below). A *pie-1* transgene with mutations in the second zinc finger (PIE-1^{ZF2-}) rescues the transcriptional defects of a *pie-1* null mutation, but is not sufficient to activate *nos-2* translation in P_4 (see below). In embryos expressing PIE-1^{ZF2-}, Z2 and Z3 form normally, but do not gastrulate efficiently. In some embryos, Z2 and Z3 are never incorporated into the embryo proper, and are left behind when the larva crawls out of the egg shell at hatching (Tenenhaus et al. 2001).

These observations support the view that germ plasm proteins, such as PIE-1, promote the translation of mRNAs required for the proper development and/or specification of Z2 and Z3. The identity of these mRNAs is not yet known. In embryos where *nos-2* is depleted by RNAi, Z2 and Z3 gastrulate normally, and only occasionally fail to associate with the somatic gonad, suggesting that PIE-1 also regulates other mRNAs besides *nos-2*.

Analysis of MEG-1 and MEG-2 supports the view that regulation of germ plasm mRNAs is essential for the proper specification of Z2 and Z3. MEG-1 and MEG-2 are two partially redundant novel proteins that associate with P granules specifically in the P_2, P_3, and P_4 blastomeres. Loss of *meg-1* and *meg-2* leads to germ cell death in the L3 stage (Leacock and Reinke 2008). Interestingly, *meg-1* interacts genetically with *nos-2*. *nos-2(RNAi);meg-1(vr10)* animals show the most severe phenotype reported for Z2 and Z3: the cells never proliferate, lose perinuclear P granules, and die by the first larval stage in an apoptosis-independent manner (Kapelle and Reinke 2011). Since MEG-1 and NOS-2 expression overlaps only in P_4, events critical for germ cell fate specification likely occur first in this cell.

NOS-2 levels are partially reduced in *meg-1* embryos, raising the possibility that like other germ plasm components, MEG-1 regulates the expression of germ plasm RNAs. MEG-1 does not contain any recognizable RNA-binding motif, but shows complex genetic interactions with RNA-binding proteins that function during larval germline development (Leacock and Reinke 2008; Kapelle and Reinke 2011). One possibility is that RNA regulation by the MEGs and other germ plasm components initiates the network of protein–RNA regulation that drives germ cell proliferation (see Chap. 8, Nousch and Eckmann 2012).

By the mid-embryogenesis, Z2 and Z3 initiate the transcription of *nos-1*, another Nanos homolog which functions partially redundantly with *nos-2*. Embryos lacking both *nos-1* and *nos-2* do not downregulate marks of active transcription in Z2 and Z3 and all germ cells degenerate during the L3 and L4 larval stages (Subramaniam and Seydoux 1999; Furuhashi et al. 2010). Nanos family members are RNA-binding proteins that often function with the PUF family of translational regulators (Parisi and Lin 2000), so *nos-2* and *nos-1* likely function by regulating the translation of other mRNAs, but the identity of these targets is not known.

Biochemical experiments have begun to define the RNA-binding specificity of some germ plasm proteins (POS-1, MEX-3, MEX-5, Pagano et al. 2007; Farley et al. 2008; Pagano et al. 2009). These types of approaches, together with the identification of RNAs bound by germ plasm proteins in vivo, may help elucidate the complex network of protein–RNA interactions that specify the fate of Z2 and Z3.

2.3.2 *Inhibition of mRNA Transcription*

As described above, the germline blastomeres P_0–P_4 maintain many maternally inherited mRNAs, but do not transcribe any mRNAs *de novo*. RNA polymerase II is present in the P blastomeres, but kept inactive by two distinct mechanisms.

2.3.2.1 Inhibition of TAF-4 by OMA-1 and OMA-2

In addition to their role as translational regulators (see above), OMA-1 and OMA-2 also inhibit transcription in the zygote. OMA-1 and OMA-2 interact with TAF-4, a component of the TFIID transcription complex. To activate transcription, TAF-4 must bind to TAF-12 in the nucleus. OMA-1 and 2 compete with TAF-12 for binding to TAF-4, and sequester TAF-4 in the cytoplasm (Guven-Ozkan et al. 2008). OMA-1 and OMA-2 are made during oogenesis, but become competent to bind TAF-4 only in the zygote due to phosphorylation by MBK-2, a kinase activated during the oocyte-to-embryo transition (see above). Phosphorylation by MBK-2 also induces degradation of OMA-1/2 by the two-cell stage (Pellettieri et al. 2003; Stitzel et al. 2006). Regulation by MBK-2 ensures that OMA-1/2 inhibit zygotic transcription specifically in the zygote and early 2-cell stage. OMA-1/2 turnover in the 2-cell stage releases TAF-4 and activates mRNA transcription in the somatic blastomeres ABa and Abp by the three-cell stage (Guven-Ozkan et al. 2008, also see Robertson and Lin 2012, Chap. 12).

2.3.2.2 Inhibition of RNA Polymerase II Phosphorylation by PIE-1

In the germline blastomeres P_2, P_3, and P_4, transcription remains repressed through the action of PIE-1. Unlike other germ plasm components, which are primarily cytoplasmic, PIE-1 also accumulates in the nuclei of each P blastomere (Mello et al. 1996). In *pie-1* mutants, high levels of CTD phosphorylation appear prematurely in P_2, P_3, and P_4 (Seydoux and Dunn 1997). Studies in mammalian cells have shown that PIE-1 inhibits P-TEF-b, the cyclin T-Cdk9 complex that phosphorylates Serine 2 in the CTD repeats of RNA polymerase. PIE-1 binds to cyclin T and inhibits P-TEF-b kinase activity using a pseudo-substrate motif that resembles a nonphosphorylatable version of the CTD (Batchelder et al. 1999). Genetic studies have shown that this activity, although functional in the germline blastomeres, is not essential to promote germ cell fate. A *pie-1* transgene with mutations in the pseudo-substrate motif fails to repress Serine 2 phosphorylation as expected, but still inhibits Serine 5 phosphorylation and mRNA transcription. In fact, such a transgene is sufficient to rescue a *pie-1* loss-of-function mutant to viability and fertility (Ghosh and Seydoux 2008). These observations suggest that PIE-1 uses redundant mechanisms to inhibit RNA polymerase II activity and promote germ cell fate.

Why inhibit mRNA transcription in germline blastomeres? The phenotype of *pie-1* null mutants provides one clue. In *pie-1* mutants, P_2 adopts the fate of its somatic sister EMS. *pie-1* embryos die as disorganized embryos with excess intestine and pharyngeal cells (EMS fates) and no germ cells (Mello et al. 1992). This cell fate transformation depends on the transcription factor SKN-1. SKN-1 is maternally encoded and present at high levels in both P_2 and EMS (Bowerman et al. 1993). One hypothesis therefore is that repression of mRNA transcription serves to protect germline blastomeres from transcription factors like SKN-1 that would otherwise induce somatic development (Seydoux et al. 1996).

Since the original observations in *C. elegans*, inhibition of RNA polymerase II phosphorylation has been observed in the embryonic germlines of Drosophila, Xenopus, ascidians, and mice (Nakamura and Seydoux 2008; Hanyu-Nakamura et al. 2008; Shirae-Kurabayashi et al. 2011; Kumano et al. 2011; Venkatarama et al. 2010). The factors responsible have been identified in Drosophila and ascidians and, remarkably, bear no resemblance to OMA-1/2 or PIE-1 (Hanyu-Nakamura et al. 2008; Shirae-Kurabayashi et al. 2011; Kumano et al. 2011). Inhibition of RNA polymerase II appears, therefore, to be conserved characteristic of germline development that depends on multiple mechanisms that have diverged during animal evolution.

2.3.3 *Chromatin Regulation*

While the chromatin of P_0–P_3 resembles that of somatic blastomeres, the chromatin of P_4, Z2, and Z3 adopts a distinct compact configuration. PSer2 and PSer5 appear in Z2 and Z3 at birth coincident with degradation of PIE-1 at that time (Seydoux and Dunn 1997). By mid-embryogenesis, however, PSer2 and PSer5 levels are low again and Z2 and Z3 also become negative for the "active" chromatin marks H3K4me2, H3K4me3, and H4K8ac (Furuhashi et al. 2010; Schaner et al. 2003). PSer2, PSer5, and H3K4me reappear in Z2 and Z3 after hatching (Furuhashi et al. 2010). These observations suggest that Z2 and Z3 remain in a relatively transcriptionally repressed state during embryogenesis, although unlike P_0–P_4, they are capable of transcribing at least a few messages (see Sect. 2.1.2.3). Loss of H3K4me depends on *nos-1* and *nos-2* (Schaner et al. 2003). Whether the unique chromatin of Z2 and Z3 depends on their arrest in G2 is also not known (Fukuyama et al. 2006).

Genetic screens designed to identify maternal factors required for fertility identified four genes coding for chromatin regulators: MES-2, 3, 4, and 6. Mutations in these genes are maternal effect sterile (MES): homozygous mothers are fertile but give rise to sterile progeny ("grandchildless" phenotype). Z2 and Z3 cells are made in embryos derived from *mes/mes* mothers, and proliferate during the first two larval stages but die by necrosis in the L3 and L4 stages (Capowski et al. 1991; Paulsen et al. 1995). In *mes-4* mutants, Z2 and Z3 retain pSer 2 (Furuhashi et al. 2010), suggesting that these cells are already compromised during embryogenesis. *mes* germ cells are also unable to differentiate: ablation of somatic gonadal cells in the L2 stage, which causes wild-type germ cells to differentiate prematurely, only causes *mes-3* germ cells to stop proliferating (Paulsen et al. 1995).

MES-2/3/6 forms a complex related to Enhancer of Zeste that methylates Lys 27 of histone H3, a repressive mark that accumulates on the X chromosome (Xu et al. 2001; Bender et al. 2004). Consistently, the X is mostly inactive in germ cells (with the exception of oocytes; Schaner and Kelly 2006; Reinke 2006; Spencer et al. 2011). MES-4 methylates Lys 36 of histone H3, and MES-4 accumulates preferentially on autosomes (Bender et al. 2006). This specificity depends on MES-2/3/6: in *mes-2, 3,* and *6* mutants, MES-4 binds all along the X chromosome and the X is inappropriately activated in germ cells (Fong et al. 2002; Bender et al. 2006). Chromatin immunoprecipitation experiments revealed that, in embryos, MES-4 associates preferentially with genes that were active in the maternal germ line. For example, MES-4 associates with meiotic genes that are transcribed in germ cells but not in embryos, and does not associate with genes that are transcribed in embryos but not in the maternal germline (Rechtsteiner et al. 2010). H3K36 methylases typically mark genes in a transcription-dependent manner. Surprisingly, MES-4 appears unable to establish the H3K36 mark *de novo*, but is able to maintain the mark in the embryonic germ lineage even though RNA polymerase II is not active in the P blastomeres (Furuhashi et al. 2010; Rechtsteiner et al. 2010). Although further analysis is necessary to clarify the link between genes bound by *mes-4* and those that are misregulated in *mes-4* mutants, the results so far suggest that MES-4 functions as an "epigenetic memory factor" that marks genes expressed in the maternal germline for the next generation. Maternal contribution of another chromatin-associated protein, MRG-1, is also required for robust germ cell proliferation in the progeny (Takasaki et al. 2007), suggesting that inheritance of a specific chromatin state is key for germ cell development.

MES-4 is inherited maternally and segregated to all blastomeres. After the 100-cell stage, MES-4 is maintained primarily in Z2 and Z3 (Fong et al. 2002). The mechanisms that allow high levels of MES-4 to persist only in the germline are not known. Genetic evidence suggests that MES-4 is also active, at least transiently, in somatic lineages and is antagonized there by the synMuv B class of chromatin regulators. In synMuvB mutants, intestinal cells express germline genes and this ectopic expression requires MES-4 (Unhavaithaya et al. 2002; Wang et al. 2005). When grown at high temperatures, synMuv B mutants arrest as starved larvae, perhaps because germline gene expression compromises intestinal function (Petrella et al. 2011). One possibility is that maternal MES-4 initially confers competence for the germline transcriptional program to all blastomeres, including the intestinal founder cell (E blastomere).

During embryogenesis, this competence is erased by the synMuv B complex in somatic lineages, but not in the P lineage, perhaps because that lineage activates transcription later and maintains maternal MES-4 for longer.

2.3.4 *Epigenetic Licensing by Maternal RNA*

A recent report suggests that activation of the germline transcriptional program also depends on maternal inheritance of specific germline transcripts. The *fem-1* gene is required for masculinization of the germline and soma (Doniach and Hodgkin 1984). Mothers homozygous for deletions that remove the *fem-1* gene produce progeny with feminized germlines, even when these progeny inherit a wild-type copy of the *fem-1* gene from their father. This maternal effect can be rescued by injecting *fem-1* RNA in the maternal germline. Remarkably, rescue is observed even when the injected RNA lacks a start codon, spans only short sub-regions of the *fem-1* gene, or is antisense to the *fem-1* transcript, indicating that inheritance of maternal *fem-1* RNA, but not FEM-1 protein, is needed to "*license*" zygotic expression of the *fem-1* gene (Johnson and Spence 2011). One possibility is that new germline transcripts are continuously compared to maternally inherited transcripts to avoid expression of potentially toxic "intruder genes." Whether this phenomenon is specific to *fem-1* or extends to other germline genes remains to be determined.

2.4 Conclusions and Remaining Questions

While the precise molecular mechanisms that specify germ cell fate remain elusive, several themes have emerged. First key to the delineation of distinct soma and germ lineages is the PAR-1-MEX-5/6 polarity axis. MEX-5/6 promotes the disassembly and degradation of germ plasm components in somatic lineages and PAR-1 stabilizes the germ plasm in the germ lineage, in part by physically excluding MEX-5 and MEX-6. The distinction between soma and germline, therefore, involves both active turnover of the germ plasm in somatic cells and protection of the germ plasm in the P blastomeres. Second, although P granules contribute to the proliferation and viability of germ cells during post-embryonic development, P granules are unlikely to be sufficient to *specify* germ cell fate during embryogenesis. We suggest instead that germ cell fate is specified by the collective action of RNAs and RNA-binding proteins found throughout the germ plasm. In the germline blastomeres, these factors mediate two important functions: (1) inhibition of mRNA transcription which prevents somatic transcription factors from activating somatic development and (2) translation of *nos-2* and other maternal mRNAs whose products promote gastrulation of the primordial germ cells, adhesion to the intestine, and a unique partially repressive chromatin configuration. In Z2 and Z3, the chromatin regulator MES-4, perhaps with the help of "licensing RNAs" in the germ plasm, transmits the "memory" of the maternal germline transcriptional program.

The task of germ cell specification in the embryo may be viewed as a careful balancing act between the need to generate new (somatic) cell types and the need to preserve the germ cell program of the oocyte. In this context, the P_0–P_3 blastomeres may be considered an intermediate cell type, similar to the epiblast cells of the mammalian embryo, where the potential for soma and germline fates temporarily co-exist. Global silencing of transcription and of the translation of certain germline mRNAs (e.g., *nos-2*) in these cells ensures that neither program takes over. P_4 in contrast may be considered the first cell where the germ cell fate program is returned to its original state, but how this program is implemented to modify the chromatin of P_4 is not known.

We also do not yet know when P_4 and/or Z2 and Z3 first activate the germline-specific transcription program. In many studies, "germ cell fate" is evaluated using markers present in germ plasm (such as P granules), but such markers do not necessarily indicate active commitment to germ cell fate. For example, Subramaniam et al. concluded that *nos-1* and *nos-2* are not required for germ cell fate because in *nos-1;nos-2* larvae, the dying "germ cells" still expressed certain germline-specific markers, but whether these markers were maternally inherited or expressed de novo in those cells was not determined (Subramaniam and Seydoux 1999). Because maternal products can perdure in the germline into larval stages (Kawasaki et al. 1998), it will be important in future studies to use markers indicative of an "active germline program" such as germline-specific chromatin marks or zygotic transcripts (as in Schaner et al. 2003; Takasaki et al. 2007). Sequencing of RNAs isolated from Z2 and Z3 dissected from mid-stage embryos has confirmed that these cells already produce several germline-specific transcripts (Gerstein et al. 2010; Spencer et al. 2011). Analyses of the zygotic transcriptome of Z2 and Z3 may provide further insights into the molecular mechanisms that specify germ cell fate.

Acknowledgments We thank members of the Seydoux lab for helpful discussions. We gratefully acknowledge funding from NIH (T32 HD007276 to J.W. and HD037047 to G.S.) and the Howard Hughes Medical Institute.

References

Arata Y, Lee J-Y, Goldstein B, Sawa H (2010) Extracellular control of PAR protein localization during asymmetric cell division in the *C. elegans* embryo. Development 137:3337–3345. doi:10.1242/dev.054742

Barbee SA, Evans TC (2006) The Sm proteins regulate germ cell specification during early *C. elegans* embryogenesis. Dev Biol 291:132–143. doi:10.1016/j.ydbio.2005.12.011

Barbee SA, Lublin A, Evans TC (2002) A novel function for the Sm proteins in germ granule localization during *C. elegans* embryogenesis. Curr Biol 12:1502–1506

Batchelder C, Dunn MA, Choy B et al (1999) Transcriptional repression by the *Caenorhabditis elegans* germ-line protein PIE-1. Genes Dev 13:202–212

Baugh LR, Hill AA, Slonim DK et al (2003) Composition and dynamics of the *Caenorhabditis elegans* early embryonic transcriptome. Development 130:889–900. doi:10.1242/dev.00302

Bei Y, Hogan J, Berkowitz LA et al (2002) SRC-1 and Wnt signaling act together to specify endoderm and to control cleavage orientation in early *C. elegans* embryos. Dev cell 3:113–125

Bender L, Cao R, Zhang Y, Strome S (2004) The MES-2/MES-3/MES-6 complex and regulation of histone H3 methylation in *C. elegans*. Curr Biol 14:1639–1643. doi:10.1016/j

Bender LB, Suh J, Carroll CR et al (2006) MES-4: an autosome-associated histone methyltransferase that participates in silencing the X chromosomes in the *C. elegans* germ line. Development 133:3907–3917. doi:10.1242/dev.02584

Berkowitz LA, Strome S (2000) MES-1, a protein required for unequal divisions of the germline in early *C. elegans* embryos, resembles receptor tyrosine kinases and is localized to the boundary between the germline and gut cells. Development 127:4419–4431

Bowerman B, Draper BW, Mello CC, Priess JR (1993) The maternal gene skn-1 encodes a protein that is distributed unequally in early *C. elegans* embryos. Cell 74:443–452

Boyd L, Guo S, Levitan D et al (1996) PAR-2 is asymmetrically distributed and promotes association of P granules and PAR-1 with the cortex in *C. elegans* embryos. Development 122:3075–3084

Brangwynne CP, Eckmann CR, Courson DS et al (2009) Germline P granules are liquid droplets that localize by controlled dissolution/condensation. Science 324:1729–1732. doi:10.1126/science.1172046

Brauchle M, Baumer K, Gönczy P (2003) Differential activation of the DNA replication checkpoint contributes to asynchrony of cell division in *C. elegans* embryos. Curr Biol 13:819–827. doi:10.1016/S

Budirahardja Y, Gönczy P (2008) PLK-1 asymmetry contributes to asynchronous cell division of *C. elegans* embryos. Development 135:1303–1313. doi:10.1242/dev.019075

Capowski EE, Martin P, Garvin C, Strome S (1991) Identification of grandchildless loci whose products are required for normal germ-line development in the nematode *Caenorhabditis elegans*. Genetics 129:1061–1072

Cheeks RJ, Canman JC, Gabriel WN et al (2004) *C. elegans* PAR proteins function by mobilizing and stabilizing asymmetrically localized protein complexes. Curr Biol 14:851–862. doi:10.1016/j

Cuenca AA, Schetter A, Aceto D et al (2003) Polarization of the *C. elegans* zygote proceeds via distinct establishment and maintenance phases. Development 130:1255–1265. doi:10.1242/dev.00284

D'Agostino I, Merritt C, Chen P-L et al (2006) Translational repression restricts expression of the *C. elegans* Nanos homolog NOS-2 to the embryonic germline. Dev Biol 292:244–252. doi:10.1016/j.ydbio.2005.11.046

Daniels BR, Perkins EM, Dobrowsky TM et al (2009) Asymmetric enrichment of PIE-1 in the *Caenorhabditis elegans* zygote mediated by binary counter diffusion. J Cell Biol 184:473–479. doi:10.1083/jcb.200809077

DeRenzo C, Reese KJ, Seydoux G (2003) Exclusion of germ plasm proteins from somatic lineages by cullin-dependent degradation. Nature 424:685–689. doi:10.1038/nature01887

Doniach T, Hodgkin J (1984) A sex-determining gene, fem-1, required for both male and hermaphrodite development in *Caenorhabditis elegans*. Dev Biol 106:223–235

Draper BW, Mello CC, Bowerman B et al (1996) MEX-3 is a KH domain protein that regulates blastomere identity in early *C. elegans* embryos. Cell 87:205–216

Encalada SE, Martin PR, Phillips JB et al (2000) DNA replication defects delay cell division and disrupt cell polarity in early *Caenorhabditis elegans* embryos. Dev Biol 228:225–238. doi:10.1006/dbio.2000.9965

Farley BM, Pagano JM, Ryder SP (2008) RNA target specificity of the embryonic cell fate determinant POS-1. RNA 14:2685–2697. doi:10.1261/rna.1256708

Fong Y, Bender L, Wang W, Strome S (2002) Regulation of the different chromatin states of autosomes and X chromosomes in the germ line of *C. elegans*. Science 296:2235–2238. doi:10.1126/science.1070790

Fukuyama M, Rougvie AE, Rothman JH (2006) *C. elegans* DAF-18/PTEN mediates nutrient-dependent arrest of cell cycle and growth in the germline. Curr Biol 16:773–779. doi:10.1016/j.cub.2006.02.073

Furuhashi H, Takasaki T, Rechtsteiner A et al (2010) Trans-generational epigenetic regulation of *C. elegans* primordial germ cells. Epigenetics chromatin 3:15. doi:10.1186/1756-8935-3-15

Gallo CM, Munro E, Rasoloson D et al (2008) Processing bodies and germ granules are distinct RNA granules that interact in *C. elegans* embryos. Dev Biol 323:76–87. doi:10.1016/j.ydbio.2008.07.008

Gallo CM, Wang JT, Motegi F, Seydoux G (2010) Cytoplasmic partitioning of P granule components is not required to specify the germline in *C. elegans*. Science 330:1685–1689. doi:10.1126/science.1193697

Gerstein MB, Lu ZJ, Van Nostrand EL et al (2010) Integrative analysis of the *Caenorhabditis elegans* genome by the modENCODE project. Science 330:1775–1787. doi:10.1126/science.1196914

Ghosh D, Seydoux G (2008) Inhibition of transcription by the *Caenorhabditis elegans* germline protein PIE-1: genetic evidence for distinct mechanisms targeting initiation and elongation. Genetics 178:235–243. doi:10.1534/genetics.107.083212

Gönczy P, Rose LS (2005) Asymmetric cell division and axis formation in the embryo. Wormbook, ed. The C elegans Research Community, Wormbook 1–20. doi: 10.1895/wormbook.1.30.1

Griffin EE, Odde DJ, Seydoux G (2011) Regulation of the MEX-5 gradient by a spatially segregated kinase/phosphatase cycle. Cell 146:955–968. doi:10.1016/j.cell.2011.08.012

Guedes S, Priess JR (1997) The *C. elegans* MEX-1 protein is present in germline blastomeres and is a P granule component. Development 124:731–739

Guo S, Kemphues K (1995) par-1, a gene required for establishing polarity in *C. elegans* embryos, encodes a putative Ser/Thr kinase that is asymmetrically distributed. Cell 81:611

Guven-Ozkan T, Nishi Y, Robertson SM, Lin R (2008) Global transcriptional repression in *C. elegans* germline precursors by regulated sequestration of TAF-4. Cell 135:149–160. doi:10.1016/j.cell.2008.07.040

Guven-Ozkan T, Robertson SM, Nishi Y, Lin R (2010) zif-1 translational repression defines a second, mutually exclusive OMA function in germline transcriptional repression. Development 137:3373–3382. doi:10.1242/dev.055327

Hanazawa M, Yonetani M, Sugimoto A (2011) PGL proteins self associate and bind RNPs to mediate germ granule assembly in C. elegans. J Cell Biol 192:929–937

Hanyu-Nakamura K, Sonobe-Nojima H, Tanigawa A et al (2008) Drosophila Pgc protein inhibits P-TEFb recruitment to chromatin in primordial germ cells. Nature 451:730–733. doi:10.1038/nature06498

Harrell JR, Goldstein B (2011) Internalization of multiple cells during *C. elegans* gastrulation depends on common cytoskeletal mechanisms but different cell polarity and cell fate regulators. Dev Biol 350:1–12. doi:10.1016/j.ydbio.2010.09.012

Hird SN, Paulsen JE, Strome S (1996) Segregation of germ granules in living *Caenorhabditis elegans* embryos: cell-type-specific mechanisms for cytoplasmic localisation. Development 122:1303–1312

Jadhav S, Rana M, Subramaniam K (2008) Multiple maternal proteins coordinate to restrict the translation of *C. elegans* nanos-2 to primordial germ cells. Development 135:1803–1812. doi:10.1242/dev.013656

Johnson CL, Spence AM (2011) Epigenetic licensing of germline gene expression by maternal RNA in *C. elegans*. Science 333:1311–1314. doi:10.1126/science.1208178

Kapelle WS, Reinke V (2011) *C. elegans* meg-1 and meg-2 differentially interact with nanos family members to either promote or inhibit germ cell proliferation and survival. Genesis 49:380–391. doi:10.1002/dvg.20726

Kawasaki I, Shim YH, Kirchner J et al (1998) PGL-1, a predicted RNA-binding component of germ granules, is essential for fertility in *C. elegans*. Cell 94:635–645

Kawasaki I, Amiri A, Fan Y et al (2004) The PGL family proteins associate with germ granules and function redundantly in *Caenorhabditis elegans* germline development. Genetics 167:645–661. doi:10.1534/genetics.103.023093

Kemphues KJ, Priess JR, Morton DG, Cheng NS (1988) Identification of genes required for cytoplasmic localization in early *C. elegans* embryos. Cell 52:311–320

Kimble JE, White JG (1981) On the control of germ cell development in *Caenorhabditis elegans*. Dev Biol 81:208–219

Kumano G, Takatori N, Negishi T et al (2011) A maternal factor unique to ascidians silences the germline via binding to P-TEFb and RNAP II regulation. Curr Biol 21:1308–1313. doi:10.1016/j.cub.2011.06.050

Kuznicki KA, Smith PA, Leung-Chiu WM et al (2000) Combinatorial RNA interference indicates GLH-4 can compensate for GLH-1; these two P granule components are critical for fertility in *C. elegans*. Development 127:2907–2916

Leacock SW, Reinke V (2008) MEG-1 and MEG-2 are embryo-specific P-granule components required for germline development in *Caenorhabditis elegans*. Genetics 178:295–306. doi:10.1534/genetics.107.080218

Li W, DeBella LR, Guven-Ozkan T, Lin R, Rose LS (2009) An eIF4E-binding protein regulates katanin protein levels in C. elegans embryos. J Cell Biol 187:33–42

Mahowald A, Illmensee K (1974) Transplantation of posterior polar plasm in drosophila. Induction of germ cells at the anterior pole of the egg. PNAS 71:1016–1020

Mello CC, Draper BW, Weintraub H, Priess JF (1992) The pie-1 and mex-1 genes and maternal control of blastomere in early *C. elegans* embryos. Cell 70:163–176

Mello CC, Schubert C, Draper B et al (1996) The PIE-1 protein and germline specification in *C. elegans* embryos. Nature 382:710–712

Nakamura A, Seydoux G (2008) Less is more: specification of the germline by transcriptional repression. Development 135:3817–3827

Nishi Y, Lin R (2005) DYRK2 and GSK-3 phosphorylate and promote the timely degradation of OMA-1, a key regulator of the oocyte-to-embryo transition in *C. elegans*. Dev Biol 288: 139–149. doi:10.1016/j.ydbio.2005.09.053

Nishi Y, Rogers E, Robertson SM, Lin R (2008) Polo kinases regulate *C. elegans* embryonic polarity via binding to DYRK2-primed MEX-5 and MEX-6. Development 135:687–697. doi:10.1242/dev.013425

Nousch M, Eckmann CR (2012) Translational control in the *C. elegans* germ line. Advances in Experimental Medicine and Biology 757:205–247. (Chap. 8, this volume) Springer, New York

Oldenbroek M, Robertson SM, Guven-Ozkan T et al (2012) Multiple RNA-binding proteins function combinatorially to control the soma-restricted expression pattern of the E3 ligase subunit ZIF-1. Dev Biol 363:388–398. doi:10.1016/j.ydbio.2012.01.002

Pagano JM, Farley BM, McCoig LM, Ryder SP (2007) Molecular basis of RNA recognition by the embryonic polarity determinant MEX-5. J Biol Chem 282:8883–8894. doi:10.1074/jbc.M700079200

Pagano JM, Farley BM, Essien KI, Ryder SP (2009) RNA recognition by the embryonic cell fate determinant and germline totipotency factor MEX-3. Proc Natl Acad Sci USA 106: 20252–20257. doi:10.1073/pnas.0907916106

Parisi M, Lin H (2000) Translational repression: a duet of Nanos and Pumilio. Curr Biol 10:R81–R83

Paulsen JE, Capowski EE, Strome S (1995) Phenotypic and molecular analysis of mes-3, a maternal-effect gene required for proliferation and viability of the germ line in *C. elegans*. Genetics 141:1383–1398

Pellettieri J, Reinke V, Kim SK, Seydoux G (2003) Coordinate activation of maternal protein degradation during the egg-to-embryo transition in *C. elegans*. Dev Cell 5:451–462

Petrella LN, Wang W, Spike CA et al (2011) synMuv B proteins antagonize germline fate in the intestine and ensure *C. elegans* survival. Development 138:1069–1079. doi:10.1242/dev.059501

Rechtsteiner A, Ercan S, Takasaki T et al (2010) The histone H3K36 methyltransferase MES-4 acts epigenetically to transmit the memory of germline gene expression to progeny. PLoS Genetics. doi:10.1371/journal.pgen.1001091

Reinke V (2006) Germline genomics. WormBook, ed. The C elegans Research Community, Wormbook 1–10. doi: 10.1895/wormbook.1.74.1

Rivers DM, Moreno S, Abraham M, Ahringer J (2008) PAR proteins direct asymmetry of the cell cycle regulators Polo-like kinase and Cdc25. J Cell Biol 180:877–885. doi:10.1083/jcb.200710018

Robertson S, Lin R (2012) The oocyte-to-embryo transition. Advances in Experimental Medicine and Biology 757:351–372. (Chap. 12, this volume) Springer, New York

Roussell DL, Bennett KL (1993) glh-1, a germ-line putative RNA helicase from *Caenorhabditis*, has four zinc fingers. Proc Natl Acad Sci USA 90:9300–9304

Schaner CE, Kelly WG (2006) Germline chromatin. Wormbook, ed. The C elegans Research Community, Wormbook 1–14. doi: 10.1895/wormbook.1.73.1

Schaner CE, Deshpande G, Schedl PD, Kelly WG (2003) A conserved chromatin architecture marks and maintains the restricted germ cell lineage in worms and flies. Dev Cell 5:747–757

Schierenberg E (1987) Reversal of cellular polarity and early cell-cell interaction in the embryo of *Caenorhabditis elegans*. Dev Biol 122:452–463. doi:10.1016/0012-1606(87)90309-5

Schierenberg E (1988) Localization and segregation of lineage-specific cleavage potential in embryos of *Caenorhabditis elegans*. Roux's Arch Dev Biol 197:282–293

Schubert CM, Lin R, de Vries CJ et al (2000) MEX-5 and MEX-6 function to establish soma/germline asymmetry in early *C. elegans* embryos. Mol Cell 5:671–682

Seydoux G, Braun RE (2006) Pathway to totipotency: lessons from germ cells. Cell 127:891–904. doi:10.1016/j.cell.2006.11.016

Seydoux G, Dunn MA (1997) Transcriptionally repressed germ cells lack a subpopulation of phosphorylated RNA polymerase II in early embryos of *Caenorhabditis elegans* and *Drosophila melanogaster*. Development 124:2191–2201

Seydoux G, Fire A (1994) Soma-germline asymmetry in the distributions of embryonic RNAs in *Caenorhabditis elegans*. Development 120:2823–2834

Seydoux G, Mello CC, Pettitt J et al (1996) Repression of gene expression in the embryonic germ lineage of *C. elegans*. Nature 382:713–716

Shirae-Kurabayashi M, Matsuda K, Nakamura A (2011) Ci-Pem-1 localizes to the nucleus and represses somatic gene transcription in the germline of Ciona intestinalis embryos. Development 138:2871–2881. doi:10.1242/dev.058131

Shirayama M, Soto MC, Ishidate T et al (2006) The conserved kinases CDK-1, GSK-3, KIN-19, and MBK-2 promote OMA-1 destruction to regulate the oocyte-to-embryo transition in *C. elegans*. Curr Biol 16:47–55. doi:10.1016/j.cub.2005.11.070

Spencer WC, Zeller G, Watson JD et al (2011) A spatial and temporal map of *C. elegans* gene expression. Genome Res 21:325–341. doi:10.1101/gr.114595.110

Spike C, Meyer N, Racen E et al (2008) Genetic analysis of the *Caenorhabditis elegans* GLH family of P-granule proteins. Genetics 178:1973–1987. doi:10.1534/genetics.107.083469

Spilker AC, Rabilotta A, Zbinden C et al (2009) MAP kinase signaling antagonizes PAR-1 function during polarization of the early *Caenorhabditis elegans* embryo. Genetics 183:965–977. doi:10.1534/genetics.109.106716

Stitzel ML, Pellettieri J, Seydoux G (2006) The *C. elegans* DYRK kinase MBK-2 marks oocyte proteins for degradation in response to meiotic maturation. Curr Biol 16:56–62. doi:10.1016/j.cub.2005.11.063

Strome S (2005) Specification of the germ line. Wormbook, ed. The *C elegans* Research Community, Wormbook 1–10. doi: 10.1895/wormbook.1.9.1

Strome S, Lehmann R (1997) Germ versus soma decisions: lessons from flies and worms. Science 316:392–393. doi:10.1126/science.1140846

Strome S, Martin P, Schierenberg E, Paulsen J (1995) Transformation of the germ line into muscle in mes-1 mutant embryos of *C. elegans*. Development 121:2961–2972

Subramaniam K, Seydoux G (1999) nos-1 and nos-2, two genes related to Drosophila nanos, regulate primordial germ cell development and survival in *Caenorhabditis elegans*. Development 126:4861–4871

Sulston JE, Schierenberg E, White JG, Thomson JN (1983) The embryonic cell lineage of the nematode *Caenorhabditis elegans*. Dev Biol 100:64–119

Tabara H, Hill RJ, Mello CC et al (1999) pos-1 encodes a cytoplasmic zinc-finger protein essential for germline specification in *C. elegans*. Development 126:1–11

Takasaki T, Liu Z, Habara Y et al (2007) MRG-1, an autosome-associated protein, silences X-linked genes and protects germline immortality in *Caenorhabditis elegans*. Development 134:757–767. doi:10.1242/dev.02771

Tenenhaus C, Subramaniam K, Dunn MA, Seydoux G (2001) PIE-1 is a bifunctional protein that regulates maternal and zygotic gene expression in the embryonic germ line of *Caenorhabditis elegans*. Genes Dev 15:1031–1040. doi:10.1101/gad.876201

Tenlen JR, Molk JN, London N et al (2008) MEX-5 asymmetry in one-cell *C. elegans* embryos requires PAR-4- and PAR-1-dependent phosphorylation. Development 135:3665–3675. doi:10.1242/dev.027060

Unhavaithaya Y, Shin TH, Miliaras N et al (2002) MEP-1 and a homolog of the NURD complex component Mi-2 act together to maintain germline-soma distinctions in *C. elegans*. Cell 111:991–1002

Updike D, Strome S (2010) P granule assembly and function in *Caenorhabditis elegans* germ cells. J Androl 31:53–60. doi:10.2164/jandrol.109.008292

Updike DL, Hachey SJ, Kreher J, Strome S (2011) P granules extend the nuclear pore complex environment in the *C. elegans* germ line. J Cell Biol 192:939–948. doi:10.1083/jcb.201010104

Venkatarama T, Lai F, Luo X et al (2010) Repression of zygotic gene expression in the Xenopus germline. Development 137:651–660. doi:10.1242/dev.038554

Wang D, Kennedy S, Conte D et al (2005) Somatic misexpression of germline P granules and enhanced RNA interference in retinoblastoma pathway mutants. Nature 436:593–597. doi:10.1038/nature04010

Wolf N, Priess J, Hirsh D (1983) Segregation of germline granules in early embryos of *Caenorhabditis elegans*: an electron microscopic analysis. J Embryol Exp Morphol 73:297–306

Xu L, Fong Y, Strome S (2001) The *Caenorhabditis elegans* maternal-effect sterile proteins, MES-2, MES-3, and MES-6, are associated in a complex in embryos. Proc Natl Acad Sci USA 98:5061

Zhang Y, Yan L, Zhou Z et al (2009) SEPA-1 mediates the specific recognition and degradation of P granule components by autophagy in *C. elegans*. Cell 136:308–321. doi:10.1016/j.cell.2008.12.022

Chapter 3
Sex Determination in the *Caenorhabditis elegans* Germline

Simone Zanetti and Alessandro Puoti

Abstract How is sex determined? In the animal kingdom, there are diverse sets of mechanisms for determining organismal sex, with the predominant ones being chromosomally based, either a dominant-acting sex chromosome or the ratio of the number of X chromosome to autosomes, which lead to oocyte-producing females and sperm-producing males. The resulting germline sexual phenotype is often the logical consequence of somatic sex determination. In this respect however, the *Caenorhabditis elegans* hermaphrodite is different from mammals and *Drosophila*. In fact in the *C. elegans* hermaphrodite germline, male gametes are transiently produced in a female body during larval development. To override chromosomal signals, sex determination of germ cells strongly depends on post-transcriptional regulation. A pivotal role for male gamete production (spermatogenesis) is played by the *fem-3* mRNA, which is controlled through FBF and other RNA-binding proteins or splicing factors. Thanks to its powerful genetics, transparent body, small size, and the ability to make sperm and oocytes within one individual, *C. elegans* represents an excellent system to investigate cellular differentiation and post-transcriptional control.

Keywords Germ cells • Sex determination • RNA processing • *C. elegans* • Gametogenesis

S. Zanetti • A. Puoti (✉)
Department of Biology, University of Fribourg,
Chemin du Musée 10, 1700 Fribourg, Switzerland
e-mail: simone.zanetti@unifr.ch; alessandro.puoti@unifr.ch

T. Schedl (ed.), *Germ Cell Development in C. elegans*, Advances in Experimental Medicine and Biology 757, DOI 10.1007/978-1-4614-4015-4_3,

3.1 Introduction

Reproduction is one of the life's essential features. While vegetative growth or fission generates clonal progeny, sexual reproduction enhances genetic variation and therefore offers an additional possibility for adaptation. Sexual reproduction uses gametes that are generally of two types: sperm and oocytes. These specialized cells are produced through meiosis, allowing two different haploid genomes to combine at fertilization. This chapter describes the processes of sex determination in the germline.

3.1.1 *Sex Determination Among Species*

Although sex determination leads to a clear-cut decision, male or female, the mechanisms that are behind this process vary widely. These are directed through chromosomal signals as well as environmental cues such as temperature, hormones, and nutrition. In this chapter we focus on the mechanisms in *Caenorhabditis elegans*. Of course, much is also known in other model organisms, such as the fruit fly *Drosophila melanogaster* and vertebrates. These will be shortly introduced to point at possible links with *C. elegans*. In this chapter we have not included *Danio rerio* (zebrafish), a well-studied vertebrate model organism. In fact, in contrast to other fish species with male (XY) or female (ZW) heterogamety, zebrafish has no sex chromosomes. Instead, its sex determination is influenced by environmental cues and the presence of the gonad, which is crucial for female sex (Slanchev et al. 2005).

3.1.1.1 *Mammals*

As in many cases, important findings have come from investigation of disorders in humans. The role of the Y chromosome in mammalian sex determination is a classical example. More than 50 years ago, Ford and colleagues identified an abnormal karyotype in patients with Turner's syndrome (Ford et al. 1959). Individuals affected by this syndrome lack parts of, or the entire Y heterosome. Moreover, all patients are phenotypically female, indicating that the Y chromosome is not required for female development. Nevertheless, such genotypic X0 females are sterile due to gonadal dysfunction (Ford et al. 1959). Additional evidence for Y chromosome function came from patients with Klinefelter syndrome (XXY karyotype; Jacobs and Strong 1959). In spite of an additional X chromosome, Klinefelter individuals develop as males. Both disorders led to the identification of the Y chromosome as a major determinant for male sex determination in mammals.

Thirty years later, the discovery of the *Sry* (*S*ex determining *r*egion on *Y*) gene further illuminated the dominantly acting function of the Y chromosome in sex determination (Sinclair et al. 1990; Gubbay et al. 1990). The role of the *Sry* gene as a major effector for sex determination was confirmed with transgenic XX mice which developed into phenotypic males when bearing a copy of this gene (Koopman et al. 1991). SRY is a transcription factor of the SOX family (SRY-related HMG box). The HMG box is crucial for its action because mutations in this domain lead to XY females (reviewed in Harley and Goodfellow 1994).

The general process of sex determination in mammals features two main events. Primary sex determination is strictly chromosomal and establishes the sexual identity of the gonads. At a later step, the sexual phenotype is shaped through hormones secreted by the sexually determined gonad. This second event is referred to as secondary sex determination. Mammalian gonads first arise as bipotential organs which later develop into either ovaries or testes. In response to *Sry*, the bipotential gonad is switched to the male fate. The female fate instead does not require *Sry* and is therefore considered as a default state (Ford et al. 1959; Sinclair et al. 1990; Jacobs and Strong 1959). In males, *Sry* activation results in upregulation of *Sox9* (SRY box containing gene 9) and repression of *Wnt4,* which would otherwise activate the female sex-determining pathway in the genital ridge (Kent et al. 1996; Tamashiro et al. 2008; Vaiman and Pailhoux 2000). Therefore, different genes are expressed thus leading to development of testis instead of ovary (for a review see Kashimada and Koopman 2010). In mammals, secondary sex determination is triggered by sex-specific hormones secreted by the testis or the ovary. It leads to somatic male or female phenotypes (reviewed in Ostrer 2000).

In mammals, germ cells are indistinguishable prior to colonization of the genital ridges (McLaren 1984; Donovan et al. 1986). Germ cell sex is determined through the sexual phenotype of the somatic environment, rather than their own heterochromosomal content (for review, see Ewen and Koopman 2010).

3.1.1.2 Drosophila melanogaster

Almost 100 years ago, Calvin Bridges observed that genotypic XXY flies developed as females and X0 flies developed as somatic sterile males (Bridges 1916). He concluded that the Y chromosome was required for male fertility in *Drosophila*, but that it was not necessary for sex determination. Thus, contrary to mammals, the *Drosophila* Y chromosome does not function in sex determination. Rather, the ratio between the number of X chromosomes and sets of autosomes (X:A) proved to be the main determinant for *Drosophila* sex (Bridges 1916). It took 60 years to identify the *sex-lethal* gene as the "reader" of the X:A ratio (Cline 1978). The expression of *Sxl* depends on at least four X-linked genes (*scute*, *sisA*, *runt*, and *unpaired*) which encode transcription factors that activate the early promoter of *Sxl* (*SxlPe*). The current working model postulates that activation of *SxlPe* is initiated when a threshold concentration of X-linked signal elements (XSE) is reached (reviewed in Salz and Erickson 2010). Upon activation of *SxlPe*, the late promoter of *Sxl (SxlPm)* is activated in all somatic cells of both sexes (Gonzalez et al. 2008). The activation of this second promoter ensures the maintenance of *Sxl* action in females (Gonzalez et al. 2008). Furthermore, action of the Sxl protein is restricted to females through alternative splicing of its pre-mRNA: an autoregulatory feedback mechanism leads to skipping of the third exon, which contains a stop codon (Cline 1984; Bell et al. 1991).

Sxl protein binds near the 3′ splice site and modulates splicing efficiency of specific introns in other pre-mRNAs that code for downstream effectors. For example, Sxl causes exon skipping in the female *transformer* (*tra*) mRNA, which results in female-specific splicing of *doublesex* (Hoshijima et al. 1991). The *tra* genes regulate the alternative splicing of the switch gene *doublesex* (*dsx*), which produces either the male or female isoform of the Doublesex protein (Baker and Wolfner 1988). *dsx* is responsible for somatic sexual development in most cells, including the somatic gonad (Coschigano and Wensink 1993; Ryner et al. 1996).

In *Drosophila,* how somatic tissues influence the sexual fate of germ cells is not well understood. The Jak/Stat pathway appears to be involved, perhaps being largely responsible for establishing the male fate in male gonads. Signals emanating from the female soma remain to be identified (Wawersik et al. 2005). While the somatic gonad participates in determining germ cell sexual fate, germ cells also determine their fate autonomously. In this respect, *Sxl* and other genes such as *Ovo* and *Otu* (*ovarian tumor*) contribute directly to female identity in XX germ cells (reviewed in Casper and Van Doren 2009).

3.1.1.3 C. elegans

The primary signals for chromosomal sex determination are similar in *C. elegans* and in *Drosophila*. The balance between the number of X chromosomes and the number of sets of autosomes also determines the sexual phenotype in worms (Nigon 1952). Albeit essential for male fertility, the Y chromosome in *Drosophila* does not control sex determination (Bridges 1916). In *C. elegans* there is no Y chromosome. The "heterogametic" sex is therefore denoted as X0, as opposed to the XX "homogametic" sex. In *C. elegans*, low X dosage (X0) leads to activation of the male-specific switch gene *xol-1* (*X0 l*ethal) (Miller et al. 1988). *xol-1* is on top of the sex determination pathway and promotes male development. In contrast, *C. elegans* XX embryos undergo female development as a result of *xol-1* inactivation through the X-linked determinants *sex-1* and *fox-1* (Carmi et al. 1998; Hodgkin et al. 1994) and possibly others. It should be noted at this point that *C. elegans* females are truly protandric hermaphrodites because they transiently produce sperm (Miller et al. 1988).

The absence of *xol-1* in females allows *sdc-2* expression (*s*ex and *d*osage *c*ompensation) (Nusbaum and Meyer 1989). *sdc-2* functions together with *sdc-1* and *sdc-3* in hermaphrodites (DeLong et al. 1993; Villeneuve and Meyer 1987; Trent et al. 1991). In addition to their role in sex determination, *sdc* genes also function in dosage compensation (Lieb et al. 2000). The remaining part of the sex determination pathway consists of a cascade of negative regulatory interactions that alternately activate or repress male- or female-specific genes. More precisely, the SDC proteins repress *her-1* to prevent male development (Trent et al. 1991). In the absence of HER-1, TRA-2 blocks the action of the FEM proteins and, as a result, the active TRA-1 transcription factor determines the female fate. In contrast, X0 individuals have active HER-1, which represses TRA-2. Further downstream, the FEM proteins inactivate TRA-1 and therefore control male development by activating *fog-1* and *fog-3* (reviewed in Zarkower 2006). Activation of *fog-1* and *fog-3*

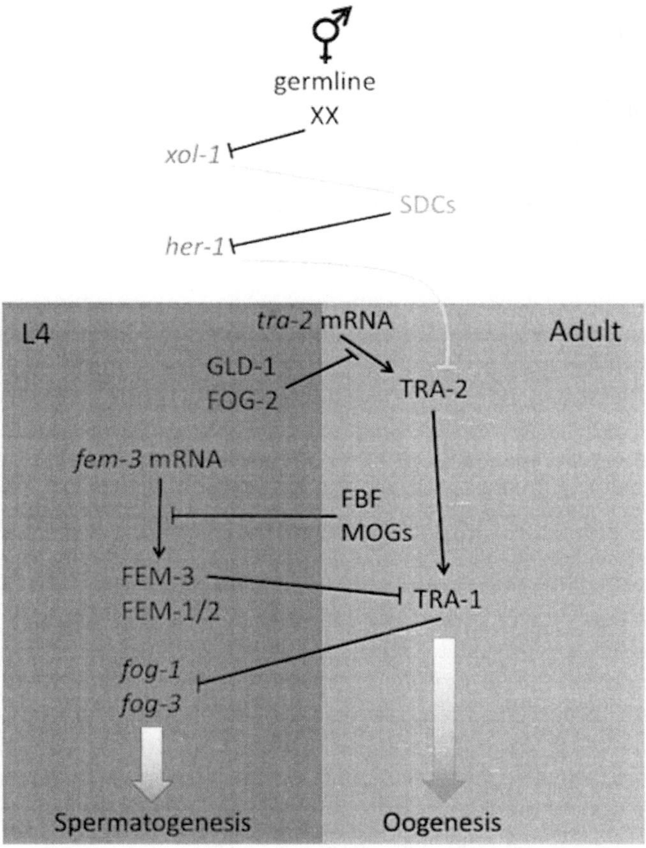

Fig. 3.1 Principal genes controlling sex determination in the *C. elegans* XX hermaphrodite germline. In response to a double dose of X signal elements, the *xol-1* gene is inactivated, thus allowing *sdc* to repress *her-1* and to promote dosage compensation. Post-transcriptional regulation for transient spermatogenesis and the sperm-oocyte switch characterize the *tra-2* and *fem-3* mRNAs. *tra-2* is repressed by GLD-1 and FOG-2, which are required for spermatogenesis. *fem-3* is repressed by the FBF and MOG proteins to allow the switch to oogenesis. While TRA-2 activates TRA-1, the FEM proteins repress its action. TRA-1 drives oogenesis and inhibits spermatogenesis by repressing *fog-1* and *fog-3*. Only the main interactions are shown

takes place either indirectly, through *tra-1*, or directly through the *fem* genes. In fact *fem-3;tra-1* double mutants produce oocytes and not sperm (Hodgkin 1986).

In *C. elegans*, the ratio of sex chromosomes to autosomes influences sex determination of the germ cells, as X0 animals exclusively make sperm and adult XX individuals continuously produce oocytes. Interestingly, the developmental balance is transiently shifted to the opposite sex during the “female” development, so that XX animals become self-fertile hermaphrodites. The determination of germ cell sex in the *C. elegans* hermaphrodite thus represents an extremely interesting system for the study of cell fate specification (Fig. 3.1).

3.1.2 Sexual Dimorphism in *C. elegans*

Sexual dimorphism in *C. elegans*, as in many other animals, is generated through the action of genes acting in the sex determination pathway. Germ cells in *C. elegans* X0 males are not responsible for their own sexual fate but depend on HER-1, which is mainly produced in the intestine and acts as a secreted ligand to shift the balance towards the male fate (Kuwabara 1996; Perry et al. 1993). HER-1 is only produced in X0 males. In the germline, HER-1 is responsible for the indirect activation of the *fem* genes, *fog-1*, and *fog-3,* which are needed for spermatogenesis (Hodgkin 1986; Ellis and Kimble 1995; Doniach and Hodgkin 1984; Barton and Kimble 1990). *tra-1* is the major effector of somatic sex determination in the hermaphrodite. It also promotes sexual dimorphism of the male germline. In fact, loss of *tra-1* results in transient spermatogenesis and continuous oogenesis in adult X0 individuals (Schedl et al. 1989; Hodgkin 1987). This indicates that *tra-1* is required for maintenance of spermatogenesis in males. Furthermore, partial absence of *tra-1* activity in males results in expression of *dmd-3* and *mab-3,* which contribute to the morphogenesis of the male copulatory organ (Mason et al. 2008). Recent studies have shown that in adults, male and hermaphrodite germlines not only differ in their content of mature gametes, but also in terms of the maintenance of germline stem cells (Morgan et al. 2010). Actually, the mitotic region in males is longer and thinner than in hermaphrodites. Moreover, the extent of the cell-cycle length appears to be shorter in males than in hermaphrodites (Morgan et al. 2010). This aspect of sexual dimorphism is caused by the influence of the somatic gonad on the distal germline and suggests that the mitotic region might play a role in early events of germline sex determination and commitment for meiosis (Morgan et al. 2010).

3.1.3 Cell Fate Decisions in the *C. elegans* Germline

Albeit it is interesting to study how the male fate is maintained throughout the development of X0 individuals, it is much more intriguing to understand how spermatogenesis is achieved in a somatic female environment of a hermaphrodite. Germ cells in *C. elegans* hermaphrodites develop either into male or female gametes, depending on the worm's developmental stage. The first three larval stages of *C. elegans* males and hermaphrodites are characterized by mitotic proliferation. During the L3 stage, germ cells that are located at the proximal end of the gonad begin to differentiate and thus enter the meiotic cell cycle. These cells continue differentiating during the fourth larval stage and are destined to undergo spermatogenesis in both males and hermaphrodites. In contrast, germ cells that enter meiosis in the L4 stage and beyond are destined to female development and differentiate as oocytes in hermaphrodites (reviewed in Schedl 1997 and Kimble and Crittenden

2007). Intriguingly, spermatozoans and oocytes are completely different in size and function, but develop from a common pool of undifferentiated cells.

In summary, the development of gametes in *C. elegans* hermaphrodites is the result of three fundamental decisions. First, during embryogenesis, cells are specified as germ cells rather than somatic blastomeres. Second, cells proliferate, not only during larval stages L2 and L3, but also throughout later developmental stages, when they keep dividing only in the distal portion of the germline. Third, when germ cells differentiate by entering the meiotic cell cycle, they can develop either as spermatids or as oocytes (Fig. 3.2).

3.2 The Players in *C. elegans* Sex Determination

3.2.1 *xol-1 Is a Master Switch Gene Controlling Sex Determination in C. elegans*

As mentioned above, the sex of *C. elegans* embryos is determined by the ratio between the number of X chromosomes and sets of autosomes (Nigon 1952). Studies by Madl and Herman determined that an X:A ratio of 0.67 leads to males, while a ratio higher than 0.75 results in female development (Madl and Herman 1979). The signals from the X chromosome correspond to at least *fox-1* and *sex-1* (Carmi et al. 1998; Hodgkin et al. 1994). Other signals originate from the autosomes (the *sea* genes (Powell et al. 2005; reviewed in Wolff and Zarkower 2008). An X:A ratio equal or below 0.75 results in the activation of *xol-1* and therefore leads to the male fate. Conversely, an XX genotype leads to inactivation of *xol-1* (Miller et al. 1988). Therefore, *xol-1* functions as a master switch gene that controls sex determination, analogous to *Sxl* in *Drosophila*. In addition to its role in sex determination, *xol-1* inhibits dosage compensation in males by repressing the *sdc* genes (Miller et al. 1988; Rhind et al. 1995; Nusbaum and Meyer 1989). *xol-1* is regulated both transcriptionally by the SEA proteins and SEX-1, and post-transcriptionally by the RNA-binding protein FOX-1 (Skipper et al. 1999; Carmi et al. 1998; Powell et al. 2005). Although the exact mechanism of *xol-1* regulation is not fully understood, two lines of evidence indicate that *xol-1* is the unique target of chromosomal signals. In fact, *xol-1* null mutations only affect male development, while hermaphrodites develop normally (Miller et al. 1988). Furthermore, XX animals that ectopically express *xol-1* develop as males, indicating that *xol-1* acts as an early switch factor for sex determination (Rhind et al. 1995). A recent study proposes that *xol-1* is also transcriptionally regulated by the feminizing protein TRA-1, which sits at the end of the sex determination cascade. Therefore, at least in somatic tissues, regulation of *xol-1* may include a feedback loop (Hargitai et al. 2009).

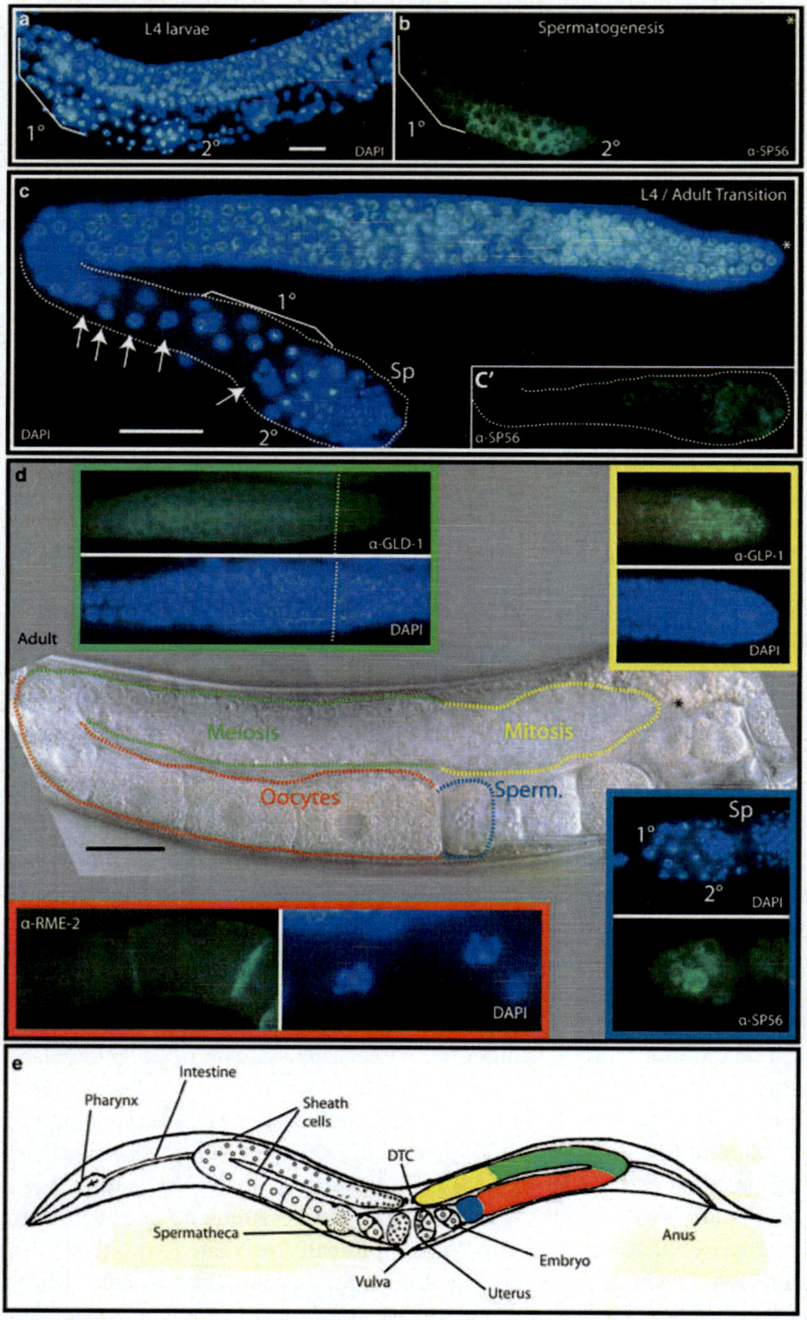

Fig. 3.2 Morphology and protein expression throughout hermaphrodite germline development. **(a, b)** germline arm of an L4 larva. Chromosomal staining with DAPI shows primary (1°) and secondary (2°) spermatocytes. No spermatids are visible at this step. Immunolocalization of sperm-specific polypeptides in the cytoplasm using anti-SP56 antibodies (Ward et al. 1986). **(c)** Staining of a dissected gonad at the transition from L4 to adult. Spermatids and spermatozoans are shown (Sp).

3.2.2 Dosage Compensation and Somatic Sex Determination

Although sex determination in *C. elegans* is initiated by the dose of X chromosomes, normal development requires identical amounts of X-chromosomal gene products in both sexes. Dosage compensation in *C. elegans* is controlled by three *sdc* genes that act downstream of *xol-1* (Rhind et al. 1995). Among these, *sdc-2* plays a predominant role in dosage compensation, as well as promoting hermaphrodite sex determination (Chu et al. 2002; Dawes et al. 1999). SDC-2 belongs to a protein complex containing SDC-1, SDC-3, DPY-21, DPY-26, DPY-27, DPY-28, and DPY-30 (reviewed in Meyer 2005). This dosage compensation complex functions as repressive machinery that reduces by half the transcription on both X chromosomes in hermaphrodites (Meyer and Casson 1986). The mechanism that controls X0 specific *her-1* transcription and therefore sex determination requires the same complex, but lacks DPY-21. It also differs in other aspects. First SDC-2 recognizes elements on the X chromosome, while *her-1* is located on an autosome and is recognized by SDC-3. Second, in hermaphrodites, the transcription rate of *her-1* is reduced by 20-fold and not by 2-fold, as for X-linked genes (Yonker and Meyer 2003). By repressing *her-1*, *sdc* genes affect the activity of all genes that are further downstream in the sex determination pathway. They therefore play a dual role in dosage compensation and sex determination.

3.2.3 Germline Sex Determination Genes

HER-1 acts as a secreted ligand for male sex determination by inhibiting the function of the transmembrane receptor TRA-2A (Perry et al. 1993; Hamaoka et al. 2004; Kuwabara 1996). The current model postulates that TRA-2A represses the FEM proteins (FEM-1, -2, -3) in XX animals, thereby leaving transcription factor TRA-1 active. TRA-1 is the terminal activator for female development in somatic tissues (for review, see Zarkower 2006). In X0 individuals, HER-1 binds and represses TRA-2A. As a consequence, the FEM proteins are active and repress TRA-1 to achieve male development (Hodgkin 1986). To do so, the FEM proteins, together with CUL-2, direct proteasome-mediated degradation of TRA-1 (Starostina et al. 2007).

Fig. 3.2 (continued) The first oocytes are visible (*white arrows*). The distal end of the germline is indicated by an *asterisk*. (**d**) DIC image of an adult germline with the localization of markers for mitosis (GLP-1), meiosis (GLD-1), spermatogenesis (SP56), and oogenesis (RME-2). The *inset* (*bottom, right*) with spermatocytes and spermatids comes from a younger worm in which oocytes have not yet pushed all male germ cells into the spermatheca. The dotted line in the inset with GLD-1 expression indicates the transition zone between mitosis (*right*) and meiosis (*left*). (**e**) Drawing of an adult hermaphrodite. The germline has the same color code as in panel (**d**): *yellow* = mitosis; *green* = different stages of the meiotic sequence; *red* = mature and maturing oocytes; *blue* = sperm. The somatic gonad is made up of the distal tip cell (DTC), the sheath cells, spermatheca and uterus. Nuclear morphology is shown by DAPI staining. Bar: 10 μm

In the germline of hermaphrodites, which produce both spermatids and oocytes, sex determination is more complex because additional factors are required for this process. Nevertheless, *C. elegans* hermaphrodites offer the possibility to study regulation of spermatogenesis and oogenesis within the same individual. Moreover, both gametes are produced from a single pool of germline stem cells (for a review, see Schedl 1997). This additional regulation requires the control of *tra-2* and *fem-3* (for a review, see Puoti et al. 2001; Ellis 2010). As a first step, *tra-2* is repressed for the onset of spermatogenesis in the L3 larva (Doniach 1986). After a transient period of spermatogenesis, the onset of oogenesis coincides with the repression of spermatogenesis. *fem-3* regulation plays a pivotal role in the sperm-oocyte switch (Kimble et al. 1984; Ahringer and Kimble 1991; Barton et al. 1987). While the sex determination pathway culminates with the activation or repression of *tra-1* in somatic tissues, *fog-1* and *fog-3* are also required in the germline for spermatogenesis in both males and hermaphrodites (reviewed in Ellis and Schedl 2007). Moreover, *fbf* and the *mog* genes are required for oogenesis by repressing *fem-3*, but do not influence somatic sex determination (Gallegos et al. 1998; Graham and Kimble 1993; Graham et al. 1993; Zhang et al. 1997). The sexual fate of germ cells in the hermaphrodite therefore depends on a balance between *tra-2* and *fem-3* activity. In fact *tra-2(gf)* mutations result in feminized germlines, whereas *fem-3(gf)* alleles lead to masculinization. Furthermore, gain-of-function mutations in *tra-2* and *fem-3* suppress each other (Barton et al. 1987).

3.3 Regulation of Germline Sex Determination in *C. elegans*

3.3.1 Transcriptional Control of her-1

After *xol-1,* which is controlled by the X:A ratio, *her-1* is the next gene in line for male sex determination. Ectopic expression of *her-1* in XX animals results in partial masculinization (Trent et al. 1991). How does XOL-1 determine the activity of *her-1* in males? While the activation of *her-1* in X0 animals remains mysterious, there is evidence that *her-1* must be repressed in hermaphrodites. When comparing X0 and XX animals, *her-1* mRNA levels are clearly reduced in the latter, indicating that *her-1* is transcriptionally regulated (Trent et al. 1991). Furthermore, *her-1* gain-of-function alleles correspond to mutations in the *her-1* promoter region and lead to increased transcription (Trent et al. 1991; Perry et al. 1994). Finally, elevated levels of *her-1* RNA are also observed in *sdc-1* or *sdc-2* mutants, suggesting that the latter are involved in transcriptional repression of *her-1* in hermaphrodites (Trent et al. 1991). Together with SDC-1 and SDC-2, SDC-3 preferentially recognizes the *her-1* promoter region (Yonker and Meyer 2003). After recognition by SDC-3, the whole dosage compensation machinery, except DPY-21, is recruited by SDC-2 to the regulatory region of *her-1*. This results in *her-1* silencing (Yonker and Meyer 2003; Chu et al. 2002). In the male germline, *her-1* expression is constantly maintained to ensure continuous spermatogenesis (Hodgkin 1980).

3.3.2 Role of the Transcription Factor TRA-1 in the Germline

tra-1 loss-of-function (lf) XX mutant animals are somatically transformed into males (Hodgkin and Brenner 1977). *tra-1* is thus required for the female fate. It regulates, among other genes, two homologs of the *Drosophila melanogaster doublesex* gene: *mab-3* and *dmd-3* (Raymond et al. 1998; Mason et al. 2008). In the germline the situation is less clear since *tra-1 (lf)* XX or X0 gonads transiently produce sperm and then switch to oogenesis (Schedl et al. 1989). Therefore, XX *tra-1* null mutants have normal germlines making both sperm and oocytes. In contrast, *tra-1* X0 mutants produce ectopic oocytes, indicating that *tra-1* is required for continuous production of sperm in males (Hodgkin 1987). The *tra-1* gene encodes two mRNAs that are translated into two transcription factors containing either two or five zinc-finger motifs (Zarkower and Hodgkin 1992). The larger transcript is present at constant levels throughout development, while the shorter peaks during the second larval stage (Zarkower and Hodgkin 1992). Nevertheless, the levels of both *tra-1* mRNAs are similar in males and hermaphrodites, indicating that *tra-1* is regulated post-transcriptionally (Zarkower and Hodgkin 1992). All *tra-1* activity is attributed to the larger transcript that encodes the TRA-1A protein, while the shorter TRA-1B isoform has no obvious function (Zarkower and Hodgkin 1992). TRA-1A, referred to as TRA-1 in this report, belongs to the GLI family of transcriptional repressors, but it also includes potential activators of transcription (Koebernick and Pieler 2002). TRA-1 is activated by TRA-2 and TRA-3, and repressed through CUL-2 and three FEM proteins, which target TRA-1 for proteasome-mediated degradation (Starostina et al. 2007; Schvarzstein and Spence 2006; Sokol and Kuwabara 2000). TRA-1 activation by TRA-2 is achieved through the intracellular part of TRA-2, which binds to TRA-1 (Lum et al. 2000; Wang and Kimble 2001). In addition, TRA-3 can cleave the intracellular part of TRA-2 to generate an active peptide that binds and represses FEM-3, thus preventing degradation of TRA-1 (Starostina et al. 2007; Sokol and Kuwabara 2000). Nuclear TRA-1 represses the transcription of *fog-3*, a terminal regulator for spermatogenesis (Chen and Ellis 2000). Additional results show that TRA-1 is also involved in positive regulation of *fog-3*, possibly explaining the requirement of TRA-1 for continuous spermatogenesis in males (Chen and Ellis 2000). These results place TRA-1 as a regulator for both somatic and germline sex determination in males and hermaphrodites (Chen and Ellis 2000).

TRA-1 is also regulated at the level of its protein, which is cleaved differently in males and in hermaphrodites (Schvarzstein and Spence 2006). First, the TRA-1 levels are higher in hermaphrodites as compared to males. Second, a shorter product, TRA-1^{100} accumulates in adult hermaphrodites, while in masculinized animals, which produce only sperm, the larger isoform is predominant (Schvarzstein and Spence 2006). Taken together, full-length TRA-1 promotes spermatogenesis in adult males, and the shorter TRA-1^{100} is required for oogenesis (Schvarzstein and Spence 2006, reviewed in Ellis 2008). Finally, in feminized *fem-1* and *fem-3* mutants both isoforms of TRA-1 accumulate, indicating that the FEM complex controls TRA-1 degradation (Schvarzstein and Spence 2006).

3.3.3 Post-transcriptional Regulation: 3′UTR-Mediated Control in the Germline

Many germline genes encode RNA-binding proteins, and most of them are required for germline development (reviewed in Lee and Schedl 2006; Nousch and Eckmann 2012, Chap. 8). In addition, the 3′untranslated regions (UTR) are primary regulatory elements for the expression of many germline mRNAs (Merritt et al. 2008). In fact, based on GFP reporter genes flanked by specific 3′UTRs, most germline genes are faithfully expressed, therefore suggesting that post-transcriptional mechanisms play a major role in the germline. This does not, however, apply to genes that are specifically expressed in sperm (Merritt et al. 2008). One of the best characterized translational regulators is the STAR/KH domain protein GLD-1, which has three distinct functions in the germline: initiation of meiosis, progression through the meiotic prophase in oocytes, and spermatogenesis in hermaphrodites (Kadyk and Kimble 1998; Jones et al. 1996; Francis et al. 1995a,b). The GLD-1 protein physically interacts with several mRNAs, including the *tra-2* mRNA (Jan et al. 1999; Jungkamp et al. 2011; Lee and Schedl 2001, 2004).

3.3.4 Transitory Spermatogenesis and Translational Control of tra-2 by GLD-1 and FOG-2

tra-2 promotes female fates in both germline and somatic tissues (Hodgkin and Brenner 1977). The discovery of *tra-2* gain-of-function *(gf)* mutant alleles led to the identification of *cis*-acting negative regulatory elements in its 3′UTR. These elements (TGE, Tra and GLI element; also known as DRE, direct repeat element) correspond to two 28-nucleotide tandem repeats (Goodwin et al. 1993). Dominant mutations in the TGEs transform XX animals, but not X0 males, into females producing only oocytes (Doniach 1986). *tra-2(gf)* mutations do not affect steady-state levels of *tra-2* transcripts suggesting that the TGEs neither control transcription nor stability of their mRNA. Rather, *tra-2(gf)* mRNAs are preferentially associated with polyribosomes indicating that gain-of-function mutated *tra-2* mRNAs are more actively translated than the wild-type transcripts (Goodwin et al. 1993). To allow transient spermatogenesis during the fourth larval stage, the TGEs are bound and repressed by FOG-2 and a dimer of GLD-1 (Jan et al. 1999; Clifford et al. 2000; Ryder et al. 2004). Repression of *tra-2* and transient spermatogenesis is abrogated in both *tra-2(gf)* and *fog-2(lf)* mutants, indicating that the binding of GLD-1/FOG-2 to the TGEs is abolished (Schedl and Kimble 1988; Jan et al. 1999; Clifford et al. 2000). Since HER-1 is not transcribed in hermaphrodites, *tra-2* expression must be post-transcriptionally repressed in hermaphrodites for spermatogenesis. In males, *her-1* is active and represses *tra-2*, allowing continuous spermatogenesis (Kuwabara 1996). Therefore post-transcriptional regulation of *tra-2* is critical for transient spermatogenesis in hermaphrodites.

3.3.5 The Sperm-Oocyte Switch and Repression of fem-3 by FBF, MOGs, DAZ-1

The role of *fem-3* in the germline of XX animals is particularly intriguing. On the one hand, in the absence of *fem-3*, XX animals develop as fertile females (Hodgkin 1986). On the other hand, *fem-3* gain-of-function alleles lead to masculinized germlines that are genotypically XX (Barton et al. 1987; Hodgkin 1986). Again, *fem-3* is regulated post-transcriptionally through its 3′UTR. More specifically, a five-nucleotide element (Point Mutation Element, PME) acts *in cis* to achieve repression of *fem-3* and therefore the switch from spermatogenesis to oogenesis (Ahringer and Kimble 1991). The PME is bound by the conserved RNA-binding proteins FBF-1 and FBF-2 (Zhang et al. 1997). *fem-3(gf)* mutations abolish FBF-binding and could therefore abrogate repression of *fem-3* (Zhang et al. 1997; Zanetti et al. 2012). The FBF proteins belong to the Puf family and include members in all eukaryotes. The most prominent example is *Drosophila* Pumilio, which prevents the translation of the *hunchback* mRNA (Murata and Wharton 1995; Wharton et al. 1998). Additional factors that are involved in regulation of *fem-3* are the translational activator DAZ-1 and the MOG proteins (Otori et al. 2006; Graham et al. 1993). DAZ-1 (*Deleted in Azoospermia*) binds and increases FBF translation and is therefore required for oogenesis.

Six *mog* genes (masculinization of the germline) are also strictly implicated in the post-transcriptional regulation of the *fem-3* mRNA via its 3′UTR. *mog* loss-of-function mutations lead to masculinized germlines, but do not affect the XX soma (Graham and Kimble 1993; Graham et al. 1993). X0 *mog* mutant males are largely normal in that they produce sperm, but have occasional somatic defects. Two observations suggest that the *mog* genes are possible regulators of *fem-3*. *fem-3* is epistatic to *mog* because *mog(lf); fem-3(lf)* double mutants produce oocytes and *fem-3(gf)* and *mog(lf)* have masculinized germlines (Graham and Kimble 1993). Therefore, *mog* genes are not essential for oogenesis, but rather for the switch from spermatogenesis to oogenesis. Additional evidence comes from somatic expression of a reporter transgene flanked by the *fem-3* 3′UTR, which is derepressed in the absence of *mog* (Gallegos et al. 1998).

3.4 *fem-3*: A Paradigm to Study RNA Regulation

3.4.1 The RNA-Binding Protein FBF

FBF-1 and FBF-2 specifically bind to the wild-type *fem-3* 3′UTR (Zhang et al. 1997). Both proteins are highly similar and contain a conserved RNA-binding motif composed of eight tandem repeats that are found throughout the Puf family members (Zhang et al. 1997; Zamore et al. 1997). *Drosophila* Pumilio plays an essential role in establishing an anterior–posterior gradient of *hunchback* mRNA throughout the syncytium of the embryo (Murata and Wharton 1995). It binds to the Nanos

Responsive Element (NRE) in the *hunchback* 3′UTR and represses its translation via deadenylation in the presence of the Nanos protein, which is restricted to the posterior of the embryo (Wreden et al. 1997; Wharton et al. 1998; Sonoda and Wharton 1999). Another deadenylation-independent mechanism has been proposed but also leads to translational repression of the *hunchback* mRNA (Chagnovich and Lehmann 2001; Wreden et al. 1997). To date, FBF has been found to be a broad-spectrum gene regulator potentially targeting 7 % of all protein coding genes in *C. elegans* (Kershner and Kimble 2010). In addition to its role in RNA binding, FBF interacts with poly(A) polymerase GLD-2 and with deadenlyase CCF-1/Pop2p, therefore suggesting two different roles of FBF in mRNA regulation: activation through poly(A) tail extension and repression through deadenylation (Suh et al. 2009). At least in vitro, the action of FBF on the *gld-1* RNA is predicted to occur through deadenylation and translational silencing (Suh et al. 2009; see also Nousch and Eckmann 2012, Chap. 8). FBF binds to the *fem-3* 3′UTR to achieve the switch from spermatogenesis to oogenesis in hermaphrodites (Zhang et al. 1997). This switch requires, in addition, the binding of NOS-3, a worm homolog of *Drosophila* Nanos (Kraemer et al. 1999). Therefore, parts of the mechanism that governs embryonic polarity in *Drosophila* and germ cell sex determination in *C. elegans* have been conserved throughout evolution (Kraemer et al. 1999; Zhang et al. 1997). Even if the mechanism of FBF action is not fully understood, it is hypothesized that FBF reduces the expression of *fem-3* via repression of translation, as shown for Pumilio and Nanos on *hunchback* mRNA (Crittenden et al. 2002; Hansen et al. 2004b; Wharton et al. 1998; Wreden et al. 1997).

To date, two additional mechanisms of *fem-3* regulation have been proposed. FEM-1 and FEM-3 might be subjected to degradation by the proteasome when targeted by the F-box protein SEL-10 (Jager et al. 2004). Furthermore, *sel-10* functions upstream of *fem-3* and loss-of-function mutations for *sel-10* cause weak masculinization of hermaphrodites. Therefore SEL-10 might be required for female fates via degradation of FEM-3, at least in somatic tissues (Jager et al. 2004). Another interesting possibility is that *fem-3* might be regulated through the stabilization of its mRNA (Zanin et al. 2010; Zanetti et al. 2012). In fact, *fem-3* mRNA levels are increased in the absence of *larp-1*, suggesting that *larp-1* controls the switch from spermatogenesis to oogenesis by decreasing *fem-3* mRNA levels (Zanin et al. 2010). *larp-1* codes for a protein with an ancient La RNA-binding motif, and functions in oogenesis by lowering Ras-MAPK signaling (Nykamp et al. 2008). Masculinization of the germline is more pronounced in *larp-1(lf);fem-3(q22)* double mutants than in *fem-3(q22)* animals (Zanin et al. 2010). In addition, in masculinized *fem-3(gf)*/+ heterozygotes *fem-3(gf)* mRNA is more abundant than its wild-type counterpart indicating that the former is stabilized (Zanetti et al. 2012).

3.4.2 Do Splicing Factors Regulate *fem-3*?

As mentioned above, the MOG proteins are implicated in the post-transcriptional regulation of *fem-3* (Graham and Kimble 1993; Graham et al. 1993; Gallegos et al. 1998). Many MOG proteins are homologs of well-known splicing factors:

MOG-1, -4, and -5 are, for example, the worm homologs of yeast PRP16, PRP2, and PRP22, respectively (Puoti and Kimble 1999; Puoti et al. 2001). This finding suggests that MOG might regulate *fem-3* through the processing of *fem-3* itself, or via other transcripts, which in turn are required for *fem-3* repression (Kasturi et al. 2010; Belfiore et al. 2002; Puoti and Kimble 1999). General splicing, however, and in particular the splicing of *fem-3,* is normal in most *mog* null mutants (Kasturi et al. 2010). The first direct evidence for *mog* action in splicing came with *mog-2*, which encodes the worm homolog of the well-conserved U2A′ protein (Zanetti et al. 2011). The U2A′ protein is a component of the spliceosomal U2snRNP complex, which includes the protein U2B″ and the U2snRNA (Mattaj et al. 1986; Scherly et al. 1990; Sillekens et al. 1989). *C. elegans mog-2* mutants are not defective in general splicing, but are less efficient in processing cryptic splice sites (Zanetti et al. 2011). Nonetheless, *fem-3* mRNA splicing is unaffected in *mog-2* animals. It is therefore possible that the Mog phenotype is caused by splicing detects in mRNAs that code for repressors of *fem-3*. Although the splicing targets of *mog* remain to be found, it is tempting to compare MOG-2 with *Drosophila* SNF (*Sans-fille*), the unique fly homolog of both U1A and U2B″ (Nagengast and Salz 2001). SNF functions as an accessory factor in somatic and germ cell sex determination (Oliver et al. 1988; Steinmann-Zwicky 1988).

Finally, an RNAi-based survey identified homologs of splicing factors that control germline proliferation and sex determination in *C. elegans* (Kerins et al. 2010). For example, *teg-4*, *ddx-23* and *prp-17* mutants are defective in germline proliferation and sex determination, indicating that reduced splicing efficiency may affect germ cell development, with few consequences on somatic development (Kerins et al. 2010; Zanetti et al. 2011; Konishi et al. 2008; Mantina et al. 2009).

3.5 Making Appropriate Amounts of Sperm: Fine-Tuned Control of the Sperm-Oocyte Switch

In order to produce the appropriate number of sperm per gonadal arm, *fem-3* has to be exclusively active at a given time, and in the correct set of cells. GLD-3 is another RNA-binding protein regulator of spermatogenesis (Zhang et al. 1997; Eckmann et al. 2002). *gld-3* encodes a homolog of *Bicaudal-C* family of RNA-Binding proteins with two KH domains (Eckmann et al. 2002). A closer look at the phenotype of *gld-3* mutants indicates that GLD-3 is required for continuous spermatogenesis and repression of oogenesis. In hermaphrodites the number of sperms produced is strongly reduced (Eckmann et al. 2002). The possibility of GLD-3 regulating *fem-3* in this process has been explored. GLD-3 binds to FBF and functions upstream of FBF (Eckmann et al. 2002). Importantly, this interaction antagonizes the binding of FBF to the *fem-3* mRNA. Therefore, it has been proposed that a competition between GLD-3 and FBF promotes spermatogenesis by release of the *fem-3* mRNA from FBF (Eckmann et al. 2002). However, additional studies have shown that GLD-3 does not only bind to FBF-1 and FBF-2. In fact GLD-3 has also been found to associate with

GLD-2 to form an active cytoplasmic poly(A) polymerase (Wang et al. 2002). This finding leads to an alternative hypothesis for GLD-3: it could antagonize the repressing activity of FBF by promoting polyadenylation of target mRNAs, perhaps of *fem-3*, and thus favor spermatogenesis (Suh et al. 2009; Wang et al. 2002).

In addition to binding to GLD-3, GLD-2 associates with RNP-8, another RNA-binding protein, and forms a different cytoplasmic poly(A) polymerase that functions in oogenesis (Kim et al. 2009). *rnp-8* mutants produce more sperm than normal because the switch from spermatogenesis to oogenesis is delayed (Kim et al. 2009). Depending on its partner, GLD-2 either promotes spermatogenesis with GLD-3 or favors oogenesis when bound to RNP-8 (Kim et al. 2009; Suh et al. 2009). Moreover, GLD-2 also functions in sex determination by binding to either FBF-1 or FBF-2 (Suh et al. 2009). In fact, *gld-2; fbf-1* mutants are masculinized, while *gld-2; fbf-2* animals are feminized in their germlines (Kim et al. 2009). Putative target mRNAs of the GLD-2/RNP-8 poly(A) polymerase were obtained by co-immunoprecipitation (Kim et al. 2010). These targets include many maternal mRNAs that function in oogenesis of which the most prominent are *tra-2*, *gld-1*, *gld-3*, *gld-2*, and *rnp-8*. Remarkably, masculinizing mRNAs such as *fem-3, gld-3,* and *fog-1* were also identified as possible GLD-2/RNP-8 targets, suggesting that polyadenylation is not the only driving force for transient spermatogenesis in hermaphrodites (Kim et al. 2010).

In addition to RNP-8 and GLD-3, which compete for GLD-2 binding (Kim et al. 2009), GLS-1 (germline sterile) antagonizes FBF for GLD-3 binding (Rybarska et al. 2009). At least in vitro, the novel protein GLS-1 recruits GLD-3 from the FBF/GLD-3 duplex (Rybarska et al. 2009). Moreover, *gls-1* functions in oogenesis and is epistatic to *gld-3* and other genes that are required for spermatogenesis (Rybarska et al. 2009). Taken together, GLS-1 is on the top of a pathway that regulates GLD-3, which in turn counters FBF-directed repression of *fem-3* mRNA for the control of appropriate amounts of sperm in hermaphrodites (Rybarska et al. 2009).

3.5.1 FOG-1 and FOG-3 Are Terminal Regulators for Spermatogenesis

fog-1 function is restricted to the germline as it causes transformation of all germ cells into oocytes in both XX and X0 animals, without affecting the somatic phenotype (Barton and Kimble 1990). Additional phenotypic characterization revealed a dose-dependent function of *fog-1* and genetic interaction with *fbf* (Thompson et al. 2005; Barton and Kimble 1990; Ellis and Kimble 1995). In fact, *fog-1* is epistatic to *fbf* and the FBF proteins repress *fog-1* mRNA by binding to regulatory elements in its 3′UTR to promote oogenesis (Thompson et al. 2005). Therefore, *fog-1* is a target of FBF. High doses of FOG-1 promote spermatogenesis, whereas low FOG-1 levels are required for germ cell proliferation (Thompson et al. 2005). In fact *fog-1;fbf-1fbf-2* triple mutants are feminized but only make very few oocyte-like cells. *fbf-1;fbf-2* double mutants are masculinized and also largely underproliferative (Crittenden et al. 2002). However, with one dose of *fog-1, fog-1/+;fbf-1 fbf-2* germlines are masculinized but show much more proliferation than *fbf-1 fbf-2* or *fog-1; fbf-1 fbf-2* germlines (Thompson et al. 2005). FOG-1 is a member of the CPEB family of RNA-binding proteins and might therefore control translation by regulating the poly(A) tails of target mRNAs (Luitjens et al. 2000; Mendez and Richter 2001). FOG-1 is found in germ cells that are committed to spermatogenesis, but not in spermatocytes and spermatids, indicating that high doses of FOG-1 are required for the onset on spermatogenesis (Lamont and Kimble 2007). A model for *fog-1* action suggests that low doses of FOG-1, which are necessary for germ cell proliferation, are maintained in the distal region of the germline by the combined action of *fbf* and *glp-1/Notch* signaling. Moving proximally, the level of FOG-1 increases in germ cells destined to spermatogenesis (Thompson et al. 2005). The different germ cell decisions thus appear closely related and the amount of sperm produced is under the control of FOG-1 (Lamont and Kimble 2007). In addition to being post-transcriptionally regulated by FBF, *fog-1* is also transcriptionally repressed by TRA-1 (Chen and Ellis 2000). Nevertheless, FOG-1 targets that function in spermatogenesis have not been identified to date.

Additional *fog* genes also direct spermatogenesis. *fog-3* codes for a putative deadenylase and functions in parallel with *fog-1* at the end of the sex determination cascade (Chen and Ellis 2000; Ellis and Kimble 1995). FOG-2 is located more upstream as it cooperates with GLD-1 for *tra-2* mRNA-binding and repression for the onset of spermatogenesis in hermaphrodites (Table 3.1) (Jan et al. 1999; Schedl and Kimble 1988; Clifford et al. 2000).

3.6 Sex Determination and Germline Proliferation: Same Players, Similar Mechanisms, but Different Outcomes

The previous section pointed out an additional role of *fog-1* in germ cell proliferation (Thompson et al. 2005). This applies to many other genes involved in germline sex determination (also see Hansen and Schedl 2012, Chap. 4). Proliferation of germ cells and maintenance of germline stem cells are primarily controlled by *glp-1/Notch* signaling in *C. elegans* (Austin and Kimble 1987; Morgan et al. 2010). *glp-1(0)* germlines produce only 8–16 spermatids, indicating that proliferating germ cells enter meiosis instead of expanding through mitotic divisions (Austin and Kimble 1987). Conversely, *glp-1* gain-of-function germlines are tumorous: they produce neither spermatids nor oocytes but contain thousands of mitotic nuclei (Berry et al. 1997). Tumorous germlines are also found in *gld-1(lf) gld-2(lf)* double and *glp-1(0); gld-1(lf) gld-2(lf)* triple mutants indicating that *glp-1* functions upstream and represses both *gld-1* and *gld-2* (Kadyk and Kimble 1998). *gld-1* alleles cause a variety of mutant phenotypes, such as masculinization, feminization, and germline tumors (Francis et al. 1995b). Its intriguing genetics therefore indicate multiple functions. For example, the tumorous phenotype of *gld-1* mutants is strongly enhanced in the absence of *gld-2* (Kadyk and Kimble 1998). Moreover, *gld-1* is also synthetically required for meiosis with *gld-3*, another sex determination gene (Kadyk and Kimble

Table 3.1 Comprehensive list of genes involved in germline sex determination. Genes are listed alphabetically. The relative molecular identity and role in sex determination are based on the reference allele(s) and principal publications. Some of the genes listed are not described in this review

Gene	Alleles	Protein function/identity	Role in sex determination	References
atx-2	*tm3397*	Translational regulator	Acts downstream of FOG-2; promotes female fate	Maine et al. (2004), Ciosk et al. (2004)
cpb-1	*tm2821*	Cytoplasmic polyadenylation element binding (CPEB) protein	Spermatogenesis	Luitjens et al. (2000)
cpb-2	*ok1772*	CPEB protein	Spermatogenesis	Luitjens et al. (2000)
cpb-3	*tm1746*	CPEB protein	Spermatogenesis	Luitjens et al. (2000)
cul-2	*t1664*	Ubiquitin ligase	Targets TRA-1 for degradation	Starostina et al. (2007)
ddx-23	*tj21*	Splicing factor PRP28	Sperm-oocyte switch	Konishi et al. (2008)
fbf-1	*ok91*	PUF RNA-binding protein, homolog of Pumilio	*fem-3* repression for oogenesis	Zhang et al. (1997), Kim et al. (2009)
fbf-2	*q704*	PUF RNA-binding protein, homolog of Pumilio	Redundant with FBF-1; spermatogenesis	Zhang et al. (1997), Kim et al. (2009)
fem-1	*e1965*	Ankyrin repeat-containing protein	Spermatogenesis	Doniach and Hodgkin (1984)
fem-2	*e2105*	Protein phosphatase	Spermatogenesis	Chin-Sang and Spence (1996)
fem-3	*e1996lf, q20gf*	Novel Protein	Spermatogenesis	Hodgkin (1986), Barton et al. (1987)
fog-1	*e2121*	CPEB protein	Spermatogenesis	Schedl and Kimble (1988)
fog-2	*q71*	N-terminal F-box and C-terminal FOG-2 homology domain	Spermatogenesis	Schedl and Kimble (1988)
fog-3	*tm4376*	N-terminal similar to vertebrate Tob	Spermatogenesis	Ellis and Kimble (1994)
daz-1	*tj3*	RNA recognition motif	*fbf* translational activator and oogenesis	Karashima et al. (2000), Otori et al. (2006)
gld-1	*q485*	KH RNA-binding protein	Maintenance of meiotic prophase in oocytes	Ellis and Kimble (1994), Francis et al. (1995a,b)
gld-2	*q497*	Cytoplasmic poly(A) polymerase subunit	Promotes spermatogenesis and/or oogenesis	Wang et al. (2002), Kim et al. (2009)
gld-3	*q730*	KH RNA-binding protein; PAP subunit with GLD-2	Promotes spermatogenesis by antagonizing FBF	Eckmann et al. (2002)
gls-1	*ef8*	Cytoplasmic RNA regulatory protein	Represses GLD-3, promotes oogenesis	Rybarska et al. (2009)

her-1	*e1518, n695*	Secreted cystein-rich protein	Determines male fate in the germline and in the soma	Hodgkin (1980)
larp-1	*q783*	La-related RNA-binding protein	Promotes oogenesis by repressing *fem-3*	Zanin et al. (2010)
lin-45	*sy96, dx19*	RAF protein homolog	Spermatogenesis	Hsu et al. (2002), Lee et al. (2007)
mag-1	*tm645*	Exon–exon junction complex, homolog of Mago nashi	Oogenesis	Kawano et al. (2004)
mek-2	*n2516, q425*	MAP kinase kinase	Spermatogenesis	Hsu et al. (2002), Lee et al. (2007)
mep-1	*q660*	Zinc-finger DNA binding protein	Sperm/oocyte switch; soma-germline distinction	Belfiore et al. (2002), Unhavaithaya et al. (2002)
mog-1	*q223*	Splicing factor PRP16	Sperm/oocyte switch; oogenesis	Graham and Kimble (1993), Puoti and Kimble (1999)
mog-2	*q75*	Spliceosomal protein U2A′	Sperm/oocyte switch; oogenesis	Zanetti et al. (2011)
mog-3	*q74*	Splicing factor sharing homologies with Cwf25 and Cwc25	Sperm/oocyte switch; oogenesis	Kasturi et al. (2010)
mog-4	*q233*	Splicing factor PRP2	Sperm/oocyte switch; oogenesis	Puoti and Kimble (2000)
mog-5	*q449*	Splicing factor PRP22	Sperm/oocyte switch; oogenesis	Puoti and Kimble (2000)
mog-6	*q465*	Nuclear cyclophilin-like protein	Sperm/oocyte switch; oogenesis	Belfiore et al. (2004)
mpk-1	*ga117*	Mitogen-activated protein kinase	Spermatogenesis	Hsu et al. (2002), Lee et al. (2007)
nos-3	*oz231*	Homolog of *Drosophila* NANOS	*fem-3* repression; oogenesis	Kraemer et al. (1999)
prp-17	*oz273*	Splicing factor PRP17/CDC40	Oogenesis	Kerins et al. (2010)
puf-8	*q725*	RNA-binding protein related to Pumilio	Switch to oogenesis	Bachorik and Kimble (2005)
rnp-8	*tm2435*	RRM RNA-binding protein; PAP subunit with GLD-2	Determines the oocyte fate	Kim et al. (2009)
rnp-4	*ok622*	Exon–exon junction complex, hortholog of Y14	Oogenesis	Kawano et al. (2004)
rpn-10	*tm1349*	Proteosomal ubiquitin receptor	Prevents spermatogenesis	Shimada et al. (2006)
sel-10	*ar41*	F-Box protein	Targets FEM proteins for degradation; oogenesis	Jager et al. (2004)

(continued)

Table 3.1 (continued)

Gene	Alleles	Protein function/identity	Role in sex determination	References
sdc-1	*n485*	Zinc-finger protein	Hermaphrodite and female development	Villeneuve and Meyer (1987)
sdc-2	*y55*	Novel protein	Hermaphrodite and female development	Nusbaum and Meyer (1989)
sdc-3	*y132*	Zinc-finger protein	Hermaphrodite and female development	DeLong et al. (1993)
teg-4	*oz210*	Splicing factor SAP130	Oogenesis	Mantina et al. (2009)
tra-1	*e1099, e1575gf*	GLI transcription factor protein	Female fates; continuous spermatogenesis in males	Hodgkin and Brenner (1977), Hodgkin (1987)
tra-2	*e1095lf, e2046gf*	Transmembrane receptor	Oogenesis	Hodgkin and Brenner (1977), Doniach (1986)
tra-3	*e1107*	Calpain regulatory protease	Cleavage of TRA-2; oogenesis	Hodgkin and Brenner (1977), Sokol and Kuwabara (2000)
xol-1	*y9*	GHMP kinase	Male development; inhibits dosage compensation	Miller et al. (1988)

1998; Hansen et al. 2004a,b; Eckmann et al. 2004). Although not as strong as in *glp-1(gf)* or *gld-1 gld-2* mutants, germline tumors have also been observed as proximally proliferating nuclei in *gld-3 nos-3* double mutants (Hansen et al. 2004b). Similarly, *mog; gld-3* double mutants are also overproliferative, indicating an additional role of *mog* in the decision between mitosis and meiosis (Belfiore et al. 2004; Kasturi et al. 2010; Zanetti et al. 2011). As mentioned above, both *fbf* genes are required for robust germline proliferation, and genetically interact with *fog-1* in this process. Therefore, germline proliferation and sex determination are tightly linked to each other. Both processes follow each other and share many common players.

As in sex determination, many regulatory events controlling mitosis or meiosis rely upon post-transcriptional mechanisms (Marin and Evans 2003; Crittenden et al. 2003; Ogura et al. 2003; Puoti et al. 2001; Clifford et al. 2000). Finally, since meiosis is different in oocytes and sperms, the sex of germ cells must be determined before meiosis has started. Even more dramatically, hermaphrodite and male germlines are sexually dimorphic in their distal region indicating that germ cells are sexually determined at a time when they still divide mitotically (Morgan et al. 2010). Nevertheless, their sexual fate remains labile since it can be redirected to oogenesis in *fog, her-1,* or *fem* mutants or to spermatogenesis in *mog(0) or fem-3(gf)* mutants (Barton et al. 1987; Chen and Ellis 2000; Graham and Kimble 1993).

3.7 The Role of Maternal RNAs, New Regulatory Mechanisms

In *C. elegans*, some of the genes that control sexual fate maternally supply mRNA and/or protein through the oocyte for zygotic sex determination (Ahringer et al. 1992; Doniach 1986; Doniach and Hodgkin 1984; Graham et al. 1993; Rosenquist and Kimble 1988). In some cases, maternal mRNAs are translated upon fertilization and permit protein synthesis before zygotic transcription sets in. A new role for maternal RNA has now been proposed for *fem-1*, which is required for spermatogenesis and male development. Heterozygous *fem-1(0)/+* progeny can be obtained by mating a wild-type male with a feminized *fem-1* null mutant. Because *fem-1* null alleles are recessive, such progenies are expected to be phenotypically wild type (Doniach and Hodgkin 1984). Andrew Spence's group has found that progeny descended from females that produce no *fem-1* RNA, and therefore are totally lacking maternal *fem-1* RNA, are unable to use the paternal wild-type copy of *fem-1*. As a consequence, their germlines are feminized even if one wild-type copy of *fem-1* is available (Johnson and Spence 2011). Intriguingly, the open reading frame of the *fem-1* RNA does not need to be intact, indicating that the maternal effect is caused by the *fem-1* RNA and not its protein product. Moreover, the effect is heritable thus suggesting that epigenetic control licenses the expression of zygotic *fem-1* (Johnson and Spence 2011). This mechanism uses sense RNA and is independent of *rde-1*, indicating that it is distinct from RNA interference (Tabara et al. 1999; Fire et al. 1998). Similarly to what has been proposed for RNA interference and epigenetic control, we notice again that the germline makes great efforts to avoid expressing foreign gene products (Kelly and Fire 1998).

3.8 Summary Paragraph

Sex determination has been studied in numerous species. Among these, the nematode *C. elegans* has a prominent role because it led to numerous genes that function in various aspects of gamete formation. Here we describe most of these genes and put them into their regulatory and functional contexts.

Acknowledgments We thank Tim Schedl and Nivedita Awasthi for comments on the manuscript. This work was partially supported by the Novartis Foundation for Medical and Biological Research, grant 10B34.

References

Ahringer J, Kimble J (1991) Control of the sperm-oocyte switch in *Caenorhabditis elegans* hermaphrodites by the *fem-3* 3' untranslated region. Nature 49:346–348

Ahringer J, Rosenquist TA, Lawson DN, Kimble J (1992) The *Caenorhabditis elegans* sex determining gene *fem-3* is regulated post-transcriptionally. EMBO J 11:2303–2310

Austin J, Kimble J (1987) *glp-1* is required in the germ line for regulation of the decision between mitosis and meiosis in *C. elegans*. Cell 51:589–599

Bachorik JL, Kimble J (2005) Redundant control of the *Caenorhabditis elegans* sperm/oocyte switch by PUF-8 and FBF-1, two distinct PUF RNA-binding proteins. Proc Natl Acad Sci USA 102:10893–10897

Baker BS, Wolfner MF (1988) A molecular analysis of *doublesex*, a bifunctional gene that controls both male and female sexual differentiation in *Drosophila melanogaster*. Genes Dev 2:477–489

Barton MK, Kimble J (1990) *fog-1*, a regulatory gene required for specification of spermatogenesis in the germ line of *Caenorhabditis elegans*. Genetics 125:29–39

Barton MK, Schedl TB, Kimble J (1987) Gain-of-function mutations of *fem-3*, a sex-determination gene in *Caenorhabditis elegans*. Genetics 115:107–119

Belfiore M, Mathies LD, Pugnale P, Moulder G, Barstead R, Kimble J, Puoti A (2002) The MEP-1 zinc-finger protein acts with MOG DEAH box proteins to control gene expression via the *fem-3* 3 ' untranslated region in *Caenorhabditis elegans*. RNA 8:725–739

Belfiore M, Pugnale P, Saudan Z, Puoti A (2004) Roles of the *C. elegans* cyclophilin-like protein MOG-6 in MEP-1 binding and germline fates. Development 131:2935–2945

Bell LR, Horabin JI, Schedl P, Cline TW (1991) Positive Autoregulation of Sex-Lethal by Alternative Splicing Maintains the Female Determined State in *Drosophila*. Cell 65:229–239

Berry LW, Westlund B, Schedl T (1997) Germ-line tumor formation caused by activation of *glp-1*, a *Caenorhabditis elegans* member of the Notch family of receptors. Development 124(4):925–936

Bridges CB (1916) Non-disjunction as proof of the chromosome theory of heredity (concluded). Genetics 1:107–163

Carmi I, Kopczynski JB, Meyer BJ (1998) The nuclear hormone receptor SEX-1 is an X-chromosome signal that determines nematode sex. Nature 396:168–173

Casper AL, Van Doren M (2009) The establishment of sexual identity in the *Drosophila* germline. Development 136:3821–3830

Chagnovich D, Lehmann R (2001) Poly(A)-independent regulation of maternal hunchback translation in the *Drosophila* embryo. Proc Natl Acad Sci USA 98:11359–11364

Chen P, Ellis RE (2000) TRA-1A regulates transcription of *fog-3*, which controls germ cell fate in *C. elegans*. Development 127:3119–3129

Chin-Sang ID, Spence AM (1996) *Caenorhabditis elegans* sex-determining protein FEM-2 is a protein phosphatase that promotes male development and interacts directly with FEM-3. Genes Dev 10:2314–2325

Chu DS, Dawes HE, Lieb JD, Chan RC, Kuo AF, Meyer BJ (2002) A molecular link between gene-specific and chromosome-wide transcriptional repression. Genes Dev 16:796–805

Ciosk R, DePalma M, Priess JR (2004) ATX-2, the *C. elegans* ortholog of ataxin 2, functions in translational regulation in the germline. Development 131:4831–4841

Clifford R, Lee MH, Nayak S, Ohmachi M, Giorgini F, Schedl T (2000) FOG-2, a novel F-box containing protein, associates with the GLD-1 RNA binding protein and directs male sex determination in the *C. elegans* hermaphrodite germline. Development 127:5265–5276

Cline TW (1978) Two closely linked mutations in *Drosophila* melanogaster that are lethal to opposite sexes and interact with daughterless. Genetics 90:683–698

Cline TW (1984) Autoregulatory functioning of a *Drosophila* gene-product that establishes and maintains the sexually determined state. Genetics 107:231–277

Coschigano KT, Wensink PC (1993) Sex-specific transcriptional regulation by the male and female doublesex proteins of *Drosophila*. Genes Dev 7:42–54

Crittenden SL, Bernstein DS, Bachorik JL, Thompson BE, Gallegos M, Petcherski AG, Moulder G, Barstead R, Wickens M, Kimble J (2002) A conserved RNA-binding protein controls germline stem cells in *Caenorhabditis elegans*. Nature 417:660–663

Crittenden SL, Eckmann CR, Wang L, Bernstein DS, Wickens M, Kimble J (2003) Regulation of the mitosis/meiosis decision in the *Caenorhabditis elegans* germline. Philos Trans R Soc Lond B Biol Sci 358:1359–1362

Dawes HE, Berlin DS, Lapidus DM, Nusbaum C, Davis TL, Meyer BJ (1999) Dosage compensation proteins targeted to X chromosomes by a determinant of hermaphrodite fate. Science 284:1800–1804

DeLong L, Plenefisch JD, Klein RD, Meyer BJ (1993) Feedback control of sex determination by dosage compensation revealed through *Caenorhabditis elegans sdc-3* mutations. Genetics 133:875–896

Doniach T (1986) Activity of the sex-determining gene *tra-2* is modulated to allow spermatogenesis in the *C. elegans* hermaphrodite. Genetics 114:53–76

Doniach T, Hodgkin J (1984) A sex-determining gene, *fem-1*, required for both male and hermaphrodite development in *Caenorhabditis elegans*. Dev Biol 106:223–235

Donovan PJ, Stott D, Cairns LA, Heasman J, Wylie CC (1986) Migratory and postmigratory mouse primordial germ cells behave differently in culture. Cell 44:831–838

Eckmann CR, Kraemer B, Wickens M, Kimble J (2002) GLD-3, a bicaudal-C homolog that inhibits FBF to control germline sex determination in *C. elegans*. Dev Cell 3:697–710

Eckmann CR, Crittenden SL, Suh N, Kimble J (2004) GLD-3 and control of the mitosis/meiosis decision in the germline of *Caenorhabditis elegans*. Genetics 168:147–160

Ellis RE (2008) Sex determination in the *Caenorhabditis elegans* germ line. Curr Top Dev Biol 83:41–64

Ellis R (2010) The sperm/oocyte decision, a *C. elegans* perspective. In: Verlhac M-H, Villeneuve A (eds) Oogenesis: the universal process. Wiley, New York

Ellis RE, Kimble J (1994) Control of germ cell differentiation in *Caenorhabditis elegans*. Ciba Found Symp 182:179–188

Ellis RE, Kimble J (1995) The *fog-3* gene and regulation of cell fate in the germ line of *Caenorhabditis elegans*. Genetics 139:561–577

Ellis R, Schedl T (2007) Sex determination in the germ line. WormBook: 1–13

Ewen KA, Koopman P (2010) Mouse germ cell development: from specification to sex determination. Mol Cell Endocrinol 323:76–93

Fire A, Xu S, Montgomery MK, Kostas SA, Driver SE, Mello CC (1998) Potent and specific genetic interference by double-stranded RNA in *Caenorhabditis elegans*. Nature 391:806–811

Ford CE, Jones KW, Polani PE, De Almeida JC, Briggs JH (1959) A sex-chromosome anomaly in a case of gonadal dysgenesis (Turner's syndrome). Lancet 1:711–713

Francis R, Barton MK, Kimble J, Schedl T (1995a) *gld-1*, a tumor suppressor gene required for oocyte development in *Caenorhabditis elegans*. Genetics 139:579–606

Francis R, Maine E, Schedl T (1995b) Analysis of the multiple roles of *gld-1* in germline development: interactions with the sex determination cascade and the *glp-1* signaling pathway. Genetics 139:607–630

Gallegos M, Ahringer J, Crittenden S, Kimble J (1998) Repression by the 3' UTR of *fem-3*, a sex-determining gene, relies on a ubiquitous mog-dependent control in *Caenorhabditis elegans*. EMBO J 17:6337–6347

Gonzalez AN, Lu H, Erickson JW (2008) A shared enhancer controls a temporal switch between promoters during *Drosophila* primary sex determination. Proc Natl Acad Sci USA 105:18436–18441

Goodwin EB, Okkema PG, Evans TC, Kimble J (1993) Translational regulation of *tra-2* by its 3' untranslated region controls sexual identity in *C. elegans*. Cell 75:329–339

Graham PL, Kimble J (1993) The Mog-1 gene is required for the switch from spermatogenesis to oogenesis in *Caenorhabditis Elegans*. Genetics 133:919–931

Graham PL, Schedl T, Kimble J (1993) More Mog genes that influence the switch from spermatogenesis to oogenesis in the hermaphrodite germ-line of *Caenorhabditis elegans*. Dev Genet 14:471–484

Gubbay J, Collignon J, Koopman P, Capel B, Economou A, Munsterberg A, Vivian N, Goodfellow P, Lovell-Badge R (1990) A gene mapping to the sex-determining region of the mouse Y chromosome is a member of a novel family of embryonically expressed genes. Nature 346:245–250

Hamaoka BY, Dann CE 3rd, Geisbrecht BV, Leahy DJ (2004) Crystal structure of *Caenorhabditis elegans* HER-1 and characterization of the interaction between HER-1 and TRA-2A. Proc Natl Acad Sci USA 101:11673–11678

Hansen D, Schedl T (2012) Stem cell proliferation versus meiotic fate decision in *C. elegans*. Advances in Experimental Medicine and Biology 757:71–99. (Chap. 4, this volume) Springer, New York

Hansen D, Hubbard EJ, Schedl T (2004a) Multi-pathway control of the proliferation versus meiotic development decision in the *Caenorhabditis elegans* germline. Dev Biol 268:342–357

Hansen D, Wilson-Berry L, Dang T, Schedl T (2004b) Control of the proliferation versus meiotic development decision in the *C. elegans* germline through regulation of GLD-1 protein accumulation. Development 131:93–104

Hargitai B, Kutnyanszky V, Blauwkamp TA, Stetak A, Csankovszki G, Takacs-Vellai K, Vellai T (2009) *xol-1*, the master sex-switch gene in *C. elegans*, is a transcriptional target of the terminal sex-determining factor TRA-1. Development 136:3881–3887

Harley VR, Goodfellow PN (1994) The biochemical role of *Sry* in sex determination. Mol Reprod Dev 39:184–193

Hodgkin J (1980) More sex-determination mutants of *Caenorhabditis elegans*. Genetics 96:649–664

Hodgkin J (1986) Sex determination in the nematode *C. elegans*: analysis of *tra-3* suppressors and characterization of *fem* genes. Genetics 114:15–52

Hodgkin J (1987) A genetic analysis of the sex-determining gene, *tra-1*, in the nematode *Caenorhabditis elegans*. Genes Dev 1:731–745

Hodgkin JA, Brenner S (1977) Mutations causing transformation of sexual phenotype in the nematode *Caenorhabditis elegans*. Genetics 86:275–287

Hodgkin J, Zellan JD, Albertson DG (1994) Identification of a candidate primary sex determination locus, *fox-1*, on the X chromosome of *Caenorhabditis elegans*. Development 120:3681–3689

Hoshijima K, Inoue K, Higuchi I, Sakamoto H, Shimura Y (1991) Control of *doublesex* alternative splicing by *transformer* and *transformer-2* in *Drosophila*. Science 252:833–836

Hsu V, Zobel CL, Lambie EJ, Schedl T, Kornfeld K (2002) *Caenorhabditis elegans lin-45* raf is essential for larval viability, fertility and the induction of vulval cell fates. Genetics 160:481–492

Jacobs PA, Strong JA (1959) A case of human intersexuality having a possible XXY sex-determining mechanism. Nature 183:302–303

Jager S, Schwartz HT, Horvitz HR, Conradt B (2004) The *Caenorhabditis elegans* F-box protein SEL-10 promotes female development and may target FEM-1 and FEM-3 for degradation by the proteasome. Proc Natl Acad Sci USA 101:12549–12554

Jan E, Motzny CK, Graves LE, Goodwin EB (1999) The STAR protein, GLD-1, is a translational regulator of sexual identity in *Caenorhabditis elegans*. EMBO J 18:258–269

Johnson CL, Spence AM (2011) Epigenetic licensing of germline gene expression by maternal RNA in *C. elegans*. Science 333:1311–1314

Jones AR, Francis R, Schedl T (1996) GLD-1, a cytoplasmic protein essential for oocyte differentiation, shows stage- and sex-specific expression during *Caenorhabditis elegans* germline development. Dev Biol 180:165–183

Jungkamp AC, Stoeckius M, Mecenas D, Grun D, Mastrobuoni G, Kempa S, Rajewsky N (2011) In vivo and transcriptome-wide identification of RNA binding protein target sites. Mol Cell 44:828–840

Kadyk LC, Kimble J (1998) Genetic regulation of entry into meiosis in *Caenorhabditis elegans*. Development 125:1803–1813

Karashima T, Sugimoto A, Yamamoto M (2000) *Caenorhabditis elegans* homologue of the human azoospermia factor DAZ is required for oogenesis but not for spermatogenesis. Development 127:1069–1079

Kashimada K, Koopman P (2010) *Sry*: the master switch in mammalian sex determination. Development 137:3921–3930

Kasturi P, Zanetti S, Passannante M, Saudan Z, Muller F, Puoti A (2010) The *C. elegans* sex determination protein MOG-3 functions in meiosis and binds to the CSL co-repressor CIR-1. Dev Biol 344:593–602

Kawano T, Kataoka N, Dreyfuss G, Sakamoto H (2004) Ce-Y14 and MAG-1, components of the exon-exon junction complex, are required for embryogenesis and germline sexual switching in *Caenorhabditis elegans*. Mech Dev 121:27–35

Kelly WG, Fire A (1998) Chromatin silencing and the maintenance of a functional germline in *Caenorhabditis elegans*. Development 125:2451–2456

Kent J, Wheatley SC, Andrews JE, Sinclair AH, Koopman P (1996) A male-specific role for SOX9 in vertebrate sex determination. Development 122:2813–2822

Kerins JA, Hanazawa M, Dorsett M, Schedl T (2010) PRP-17 and the pre-mRNA splicing pathway are preferentially required for the proliferation versus meiotic development decision and germline sex determination in *Caenorhabditis elegans*. Dev Dyn 239:1555–1572

Kershner AM, Kimble J (2010) Genome-wide analysis of mRNA targets for *Caenorhabditis elegans* FBF, a conserved stem cell regulator. Proc Natl Acad Sci USA 107:3936–3941

Kim KW, Nykamp K, Suh N, Bachorik JL, Wang L, Kimble J (2009) Antagonism between GLD-2 binding partners controls gamete sex. Dev Cell 16:723–733

Kim KW, Wilson TL, Kimble J (2010) GLD-2/RNP-8 cytoplasmic poly(A) polymerase is a broad-spectrum regulator of the oogenesis program. Proc Natl Acad Sci USA 107:17445–17450

Kimble J, Crittenden SL (2007) Controls of germline stem cells, entry into meiosis, and the sperm/oocyte decision in *Caenorhabditis elegans*. Annu Rev Cell Dev Biol 23:405–433

Kimble J, Edgar L, Hirsh D (1984) Specification of male development in *Caenorhabditis elegans*: the *fem* genes. Dev Biol 105:234–239

Koebernick K, Pieler T (2002) Gli-type zinc finger proteins as bipotential transducers of Hedgehog signaling. Differentiation 70:69–76

Konishi T, Uodome N, Sugimoto A (2008) The *Caenorhabditis elegans* DDX-23, a homolog of yeast splicing factor PRP28, is required for the sperm-oocyte switch and differentiation of various cell types. Dev Dyn 237:2367–2377

Koopman P, Gubbay J, Vivian N, Goodfellow P, Lovellbadge R (1991) Male development of chromosomally female mice transgenic for *Sry*. Nature 351:117–121

Kraemer B, Crittenden S, Gallegos M, Moulder G, Barstead R, Kimble J, Wickens M (1999) NANOS-3 and FBF proteins physically interact to control the sperm-oocyte switch in *Caenorhabditis elegans*. Curr Biol 9:1009–1018

Kuwabara PE (1996) A novel regulatory mutation in the *C. elegans* sex determination gene *tra-2* defines a candidate ligand/receptor interaction site. Development 122:2089–2098

Lamont LB, Kimble J (2007) Developmental expression of FOG-1/CPEB protein and its control in the *Caenorhabditis elegans* hermaphrodite germ line. Dev Dyn 236:871–879

Lee MH, Schedl T (2001) Identification of in vivo mRNA targets of GLD-1, a maxi-KH motif containing protein required for *C. elegans* germ cell development. Genes Dev 15:2408–2420

Lee MH, Schedl T (2004) Translation repression by GLD-1 protects its mRNA targets from nonsense-mediated mRNA decay in *C. elegans*. Genes Dev 18:1047–1059

Lee MH, Schedl T (2006) RNA-binding proteins. WormBook: 1–13

Lee MH, Ohmachi M, Arur S, Nayak S, Francis R, Church D, Lambie E, Schedl T (2007) Multiple functions and dynamic activation of MPK-1 extracellular signal-regulated kinase signaling in Caenorhabditis elegans germline development. Genetics 177:2039–2062

Lieb JD, de Solorzano CO, Rodriguez EG, Jones A, Angelo M, Lockett S, Meyer BJ (2000) The *Caenorhabditis elegans* dosage compensation machinery is recruited to X chromosome DNA attached to an autosome. Genetics 156:1603–1621

Luitjens C, Gallegos M, Kraemer B, Kimble J, Wickens M (2000) CPEB proteins control two key steps in spermatogenesis in *C. elegans*. Genes Dev 14:2596–2609

Lum DH, Kuwabara PE, Zarkower D, Spence AM (2000) Direct protein-protein interaction between the intracellular domain of TRA-2 and the transcription factor TRA-1A modulates feminizing activity in *C. elegans*. Genes Dev 14:3153–3165

Madl JE, Herman RK (1979) Polyploids and sex determination in *Caenorhabditis elegans*. Genetics 93:393–402

Maine EM, Hansen D, Springer D, Vought VE (2004) *Caenorhabditis elegans atx-2* promotes germline proliferation and the oocyte fate. Genetics 168:817–830

Mantina P, MacDonald L, Kulaga A, Zhao L, Hansen D (2009) A mutation in *teg-4,* which encodes a protein homologous to the SAP130 pre-mRNA splicing factor, disrupts the balance between proliferation and differentiation in the *C. elegans* germ line. Mech Dev 126:417–429

Marin VA, Evans TC (2003) Translational repression of a *C. elegans* Notch mRNA by the STAR/KH domain protein GLD-1. Development 130:2623–2632

Mason DA, Rabinowitz JS, Portman DS (2008) *dmd-3,* a *doublesex*-related gene regulated by *tra-1*, governs sex-specific morphogenesis in *C. elegans*. Development 135:2373–2382

Mattaj IW, Habets WJ, van Venrooij WJ (1986) Monospecific antibodies reveal details of U2 snRNP structure and interaction between U1 and U2 snRNPs. EMBO J 5:997–1002

McLaren A (1984) Meiosis and differentiation of mouse germ cells. Symp Soc Exp Biol 38:7–23

Mendez R, Richter JD (2001) Translational control by CPEB: a means to the end. Nat Rev Mol Cell Biol 2:521–529

Merritt C, Rasoloson D, Ko D, Seydoux G (2008) 3'UTRs are the primary regulators of gene expression in the *C. elegans* germline. Curr Biol 18:1476–1482

Meyer BJ (2005) X-Chromosome dosage compensation. WormBook: 1–14

Meyer BJ, Casson LP (1986) *Caenorhabditis elegans* compensates for the difference in X chromosome dosage between the sexes by regulating transcript levels. Cell 47:871–881

Miller LM, Plenefisch JD, Casson LP, Meyer BJ (1988) *xol-1*: a gene that controls the male modes of both sex determination and X chromosome dosage compensation in *C. elegans*. Cell 55:167–183

Morgan DE, Crittenden SL, Kimble J (2010) The *C. elegans* adult male germline: stem cells and sexual dimorphism. Dev Biol 346:204–214

Murata Y, Wharton RP (1995) Binding of pumilio to maternal *hunchback* mRNA is required for posterior patterning in *Drosophila* embryos. Cell 80:747–756

Nagengast AA, Salz HK (2001) The *Drosophila* U2 snRNP protein U2A' has an essential function that is SNF/U2B" independent. Nucleic Acids Res 29:3841–3847

Nigon V (1952) Experimental modifications of the sex ratio in a pseudogamous nematode. C R Hebd Seances Acad Sci 234:2568–2570

Nousch M, Eckmann CR (2012) Translational control in the *C. elegans* germ line. Advances in Experimental Medicine and Biology 757:205–247. (Chap. 8, this volume) Springer, New York

Nusbaum C, Meyer BJ (1989) The *Caenorhabditis elegans* gene *sdc-2* controls sex determination and dosage compensation in XX animals. Genetics 122:579–593

Nykamp K, Lee MH, Kimble J (2008) *C. elegans* La-related protein, LARP-1, localizes to germline P bodies and attenuates Ras-MAPK signaling during oogenesis. RNA 14:1378–1389

Ogura K, Kishimoto N, Mitani S, Gengyo-Ando K, Kohara Y (2003) Translational control of maternal *glp-1* mRNA by POS-1 and its interacting protein SPN-4 in *Caenorhabditis elegans*. Development 130:2495–2503

Oliver B, Perrimon N, Mahowald AP (1988) Genetic-evidence that the *Sans-Fille* locus is involved in *Drosophila* sex determination. Genetics 120:159–171

Ostrer H (2000) Sexual differentiation. Semin Reprod Med 18:41–49

Otori M, Karashima T, Yamamoto M (2006) The *Caenorhabditis elegans* homologue of deleted in azoospermia is involved in the sperm/oocyte switch. Mol Biol Cell 17:3147–3155

Perry MD, Li W, Trent C, Robertson B, Fire A, Hageman JM, Wood WB (1993) Molecular characterization of the *her-1* gene suggests a direct role in cell signaling during *Caenorhabditis elegans* sex determination. Genes Dev 7:216–228

Perry MD, Trent C, Robertson B, Chamblin C, Wood WB (1994) Sequenced alleles of the *Caenorhabditis elegans* sex-determining gene *her-1* include a novel class of conditional promoter mutations. Genetics 138:317–327

Powell JR, Jow MM, Meyer BJ (2005) The T-box transcription factor SEA-1 is an autosomal element of the X:A signal that determines *C. elegans* sex. Dev Cell 9:339–349

Puoti A, Kimble J (1999) The *Caenorhabditis elegans* sex determination gene *mog-1* encodes a member of the DEAH-box protein family. Mol Cell Biol 19:2189–2197

Puoti A, Kimble J (2000) The hermaphrodite sperm/oocyte switch requires the *Caenorhabditis elegans* homologs of PRP2 and PRP22. Proc Natl Acad Sci USA 97:3276–327681

Puoti A, Pugnale P, Belfiore M, Schlappi AC, Saudan Z (2001) RNA and sex determination in *Caenorhabditis elegans*. Post-transcriptional regulation of the sex-determining *tra-2* and *fem-3* mRNAs in the *Caenorhabditis elegans hermaphrodite*. EMBO Rep 2:899–904

Raymond CS, Shamu CE, Shen MM, Seifert KJ, Hirsch B, Hodgkin J, Zarkower D (1998) Evidence for evolutionary conservation of sex-determining genes. Nature 391:691–695

Rhind NR, Miller LM, Kopczynski JB, Meyer BJ (1995) *xol-1* acts as an early switch in the *C. elegans* male/hermaphrodite decision. Cell 80:71–82

Rosenquist TA, Kimble J (1988) Molecular cloning and transcript analysis of *fem-3*, a sex-determination gene in *Caenorhabditis elegans*. Genes Dev 2:606–616

Rybarska A, Harterink M, Jedamzik B, Kupinski AP, Schmid M, Eckmann CR (2009) GLS-1, a novel P granule component, modulates a network of conserved RNA regulators to influence germ cell fate decisions. PLoS Genet 5:e1000494

Ryder SP, Frater LA, Abramovitz DL, Goodwin EB, Williamson JR (2004) RNA target specificity of the STAR/GSG domain post-transcriptional regulatory protein GLD-1. Nat Struct Mol Biol 11:20–28

Ryner LC, Goodwin SF, Castrillon DH, Anand A, Villella A, Baker BS, Hall JC, Taylor BJ, Wasserman SA (1996) Control of male sexual behavior and sexual orientation in *Drosophila* by the *fruitless* gene. Cell 87:1079–1089

Salz HK, Erickson JW (2010) Sex determination in *Drosophila*: the view from the top. Fly 4:60–70

Schedl T (1997) Developmental genetics of the germ line. In: Riddle DL, Blumenthal T, Meyer BJ, Priess JR (eds) *C. elegans* II. Cold Spring Harbor Laboratory Press, New York

Schedl T, Kimble J (1988) *fog-2*, a germ-line-specific sex determination gene required for hermaphrodite spermatogenesis in *Caenorhabditis elegans*. Genetics 119:43–61

Schedl T, Graham PL, Barton MK, Kimble J (1989) Analysis of the role of *tra-1* in germline sex determination in the nematode *Caenorhabditis elegans*. Genetics 123:755–769

Scherly D, Boelens W, Dathan NA, van Venrooij WJ, Mattaj IW (1990) Major determinants of the specificity of interaction between small nuclear ribonucleoproteins U1A and U2B" and their cognate RNAs. Nature 345:502–506

Schvarzstein M, Spence AM (2006) The *C. elegans* sex-determining GLI protein TRA-1A is regulated by sex-specific proteolysis. Dev Cell 11:733–740

Shimada M, Kanematsu K, Tanaka K, Yokosawa H, Kawahara H (2006) Proteasomal ubiquitin receptor RPN-10 controls sex determination in *Caenorhabditis elegans*. Mol Biol Cell 17:5356–5371

Sillekens PT, Beijer RP, Habets WJ, van Verooij WJ (1989) Molecular cloning of the cDNA for the human U2 snRNA-specific A' protein. Nucleic Acids Res 17:1893–1906

Sinclair AH, Berta P, Palmer MS, Hawkins JR, Griffiths BL, Smith MJ, Foster JW, Frischauf AM, Lovell-Badge R, Goodfellow PN (1990) A gene from the human sex-determining region encodes a protein with homology to a conserved DNA-binding motif. Nature 346:240–244

Skipper M, Milne CA, Hodgkin J (1999) Genetic and molecular analysis of *fox-1*, a numerator element involved in *Caenorhabditis elegans* primary sex determination. Genetics 151:617–631

Slanchev K, Stebler J, de la Cueva-Mendez G, Raz E (2005) Development without germ cells: the role of the germ line in *zebrafish* sex differentiation. Proc Natl Acad Sci USA 102:4074–4079

Sokol SB, Kuwabara PE (2000) Proteolysis in Caenorhabditis elegans sex determination: cleavage of TRA-2A by TRA-3. Genes Dev 14:901–906

Sonoda J, Wharton RP (1999) Recruitment of Nanos to *hunchback* mRNA by Pumilio. Genes Dev 13:2704–2712

Starostina NG, Lim JM, Schvarzstein M, Wells L, Spence AM, Kipreos ET (2007) A CUL-2 ubiquitin ligase containing three FEM proteins degrades TRA-1 to regulate *C. elegans* sex determination. Dev Cell 13:127–139

Steinmann-Zwicky M (1988) Sex determination in *Drosophila* - the X-chromosomal gene *Liz* is required for *Sxl* activity. EMBO J 7:3889–3898

Suh N, Crittenden SL, Goldstrohm A, Hook B, Thompson B, Wickens M, Kimble J (2009) FBF and its dual control of *gld-1* expression in the *Caenorhabditis elegans* germline. Genetics 181:1249–1260

Tabara H, Sarkissian M, Kelly WG, Fleenor J, Grishok A, Timmons L, Fire A, Mello CC (1999) The *rde-1* gene, RNA interference, and transposon silencing in *C. elegans*. Cell 99:123–132

Tamashiro DA, Alarcon VB, Marikawa Y (2008) Ectopic expression of mouse *Sry* interferes with Wnt/beta-catenin signaling in mouse embryonal carcinoma cell lines. Biochim Biophys Acta 1780:1395–1402

Thompson BE, Bernstein DS, Bachorik JL, Petcherski AG, Wickens M, Kimble J (2005) Dose-dependent control of proliferation and sperm specification by FOG-1/CPEB. Development 132:3471–3481

Trent C, Purnell B, Gavinski S, Hageman J, Chamblin C, Wood WB (1991) Sex-specific transcriptional regulation of the *C. elegans* sex-determining gene *her-1*. Mech Dev 34:43–55

Unhavaithaya Y, Shin TH, Miliaras N, Lee J, Oyama T, Mello CC (2002) MEP-1 and a homolog of the NURD complex component Mi-2 act together to maintain germline-soma distinctions in *C. elegans*. Cell 111:991–1002

Vaiman D, Pailhoux E (2000) Mammalian sex reversal and intersexuality: deciphering the sex-determination cascade. Trends Genet 16:488–494

Villeneuve AM, Meyer BJ (1987) *sdc-1*: a link between sex determination and dosage compensation in *C. elegans*. Cell 48:25–37

Wang S, Kimble J (2001) The TRA-1 transcription factor binds TRA-2 to regulate sexual fates in *Caenorhabditis elegans*. EMBO J 20:1363–1372

Wang L, Eckmann CR, Kadyk LC, Wickens M, Kimble J (2002) A regulatory cytoplasmic poly(A) polymerase in *Caenorhabditis elegans*. Nature 419:312–316

Ward S, Roberts TM, Strome S, Pavalko FM, Hogan E (1986) Monoclonal antibodies that recognize a polypeptide antigenic determinant shared by multiple *Caenorhabditis elegans* sperm-specific proteins. J Cell Biol 102:1778–1786

Wawersik M, Milutinovich A, Casper AL, Matunis E, Williams B, Van Doren M (2005) Somatic control of germline sexual development is mediated by the JAK/STAT pathway. Nature 436:563–567

Wharton RP, Sonoda J, Lee T, Patterson M, Murata Y (1998) The Pumilio RNA-binding domain is also a translational regulator. Mol Cell 1:863–872

Wolff JR, Zarkower D (2008) Somatic sexual differentiation in *Caenorhabditis elegans*. Curr Top Dev Biol 83:1–39

Wreden C, Verrotti AC, Schisa JA, Lieberfarb ME, Strickland S (1997) Nanos and pumilio establish embryonic polarity in *Drosophila* by promoting posterior deadenylation of *hunchback* mRNA. Development 124:3015–3023

Yonker SA, Meyer BJ (2003) Recruitment of *C. elegans* dosage compensation proteins for gene-specific versus chromosome-wide repression. Development 130:6519–6532

Zamore PD, Williamson JR, Lehmann R (1997) The Pumilio protein binds RNA through a conserved domain that defines a new class of RNA-binding proteins. RNA 3:1421–1433

Zanetti S, Meola M, Bochud A, Puoti A (2011) Role of the *C. elegans* U2 snRNP protein MOG-2 in sex determination, meiosis, and splice site selection. Dev Biol 354:232–241

Zanetti S, Grinschgl S, Meola M, Belfiore M, Rey S, Bianchi P, Puoti A (2012) The sperm-oocyte switch in the *C. elegans* hermaphrodite is controlled through abundance of the *fem-3* mRNA. RNA 18:1385–1394

Zanin E, Pacquelet A, Scheckel C, Ciosk R, Gotta M (2010) LARP-1 promotes oogenesis by repressing *fem-3* in the *C. elegans* germline. J Cell Sci 123:2717–2724

Zarkower D (2006) Somatic sex determination. WormBook: 1–12

Zarkower D, Hodgkin J (1992) Molecular analysis of the *C. elegans* sex-determining gene *tra-1*: a gene encoding two zinc finger proteins. Cell 70:237–249

Zhang B, Gallegos M, Puoti A, Durkin E, Fields S, Kimble J, Wickens MP (1997) A conserved RNA-binding protein that regulates sexual fates in the *C. elegans* hermaphrodite germ line. Nature 390:477–484

Chapter 4
Stem Cell Proliferation Versus Meiotic Fate Decision in *Caenorhabditis elegans*

Dave Hansen and Tim Schedl

Abstract The *C. elegans* germ line has emerged as an important model for understanding how a stem cell population is maintained throughout the life of the animal while still producing the gametes necessary for propagation of the species. The stem cell population in the adult hermaphrodite is relatively large, with stem cells giving rise to daughters that appear intrinsically equivalent; however, some of the daughters retain the proliferative fate while others enter meiotic prophase. While machinery exists for cells to progress through the mitotic cell cycle and machinery exists for cells to progress through meiotic prophase, central to understanding germ line development is identifying the genes and regulatory processes that determine whether the mitotic cell cycle or meiotic prophase machinery will be utilized; in other words, the genes that regulate the switch of germ cells from the proliferative stem cell fate to the meiotic development fate. Whether a germ cell self-renews or enters meiotic prophase is largely determined by its proximity to the distal tip cell (DTC), which is the somatic niche cell that caps the distal end of the gonad. Germ cells close to the DTC have high levels of GLP-1 Notch signaling, which promotes the proliferative fate, while cells further from the DTC have high activity levels of the GLD-1 and GLD-2 redundant RNA regulatory pathways, as well as a third uncharacterized pathway, each of which direct cells to enter meiotic prophase. Other factors and pathways modulate this core genetic pathway, or work in parallel to it, presumably to ensure that a tight balance is maintained between proliferation and meiotic entry.

D. Hansen (✉)
Department of Biological Sciences, University of Calgary,
2500 University Drive, Calgary, AB, Canada
e-mail: dhansen@ucalgary.ca

T. Schedl
Department of Genetics, Washington University School of Medicine,
Campus Box 8232, 4566 Scott Avenue, St. Louis, MO, USA
e-mail: ts@genetics.wustl.edu

T. Schedl (ed.), *Germ Cell Development in C. elegans*, Advances in Experimental Medicine and Biology 757, DOI 10.1007/978-1-4614-4015-4_4,

Keywords *C. elegans* • Germ line • Stem cells • Proliferative fate • Meiotic fate • GLP-1 Notch • GLD-1 pathway • GLD-2 pathway

4.1 Introduction

Studies focused on understanding the behavior and control of stem cells have garnered much interest over the past few decades due to the fundamental roles that stem cells play in development and tissue homeostasis, as well as for their therapeutic potential. Certain stem cell systems have arisen as important models in pursuing fundamental questions relating to stem cell function, such as the *Drosophila* germ line (Losick et al. 2011), the mouse hematopoietic stem cell system (Bianco 2011), and as discussed here, the *Caenorhabditis elegans* hermaphrodite germ line. Studying these models has, for example, helped in understanding the molecular factors and physiological attributes that regulate whether a given daughter of a dividing stem cell will remain a stem cell (self-renew), or enter a path to differentiation. While differences have emerged between stem cell systems, such as the utilization of different signaling pathways, or the number of stem cells maintained, similarities between these systems have also become apparent. Perhaps one of the most unifying themes of these stem cell systems is their reliance on a stem cell niche to regulate stem cell behavior.

The *C. elegans* germ line stem cell niche was first identified ~30 years ago when ablation of the *d*istal *t*ip *c*ell (DTC), a somatic cell that caps the distal end of the gonad, resulted in all proliferative cells switching to the meiotic fate (Kimble and White 1981). Much of the work since has involved identifying the molecular factors that emanate from the niche, the factors in the germ cell that perceive and execute this signal, the factors that promote meiotic entry, and determining how these factors work together to control the balance between the proliferative stem cell fate and meiotic differentiated fate [for prior reviews, see (Seydoux and Schedl 2001; Hansen and Schedl 2006; Hubbard 2007; Crittenden et al. 2003; Kimble and Crittenden 2007)]. The factors and mechanisms that regulate the balance between the proliferative and the meiotic fates can be separated from those that are necessary for mitotic cell cycle progression or meiotic prophase progression. In other words, there is a difference between executing the switch that determines the cell fate (which is the primary focus of this review) and fulfilling the normal functions of cells adopting a given fate, although some genes may function separately in the two. The factors involved in regulating the balance between the proliferative and meiotic fates have primarily been identified through the use of powerful genetic screens, generating mutations that contribute either to a loss of the stem cell population, due to premature meiotic entry, or to over-proliferation of the stem cell population, often resulting in tumor formation, due to a failure of meiotic entry. A significant theme that has emerged is the existence of substantial redundancy within the genetic network controlling this balance. This redundancy has made it difficult to identify factors and tease apart their regulatory relationships. However, it has also revealed a system that that can be finely tuned. This redundancy may be common to other stem cell systems, providing a barrier to identify and characterize all involved factors, especially in systems that do not have the genetic tools that are available in the *C. elegans* model.

While genetic analysis has revealed significant insight into mechanisms regulating the proliferation vs. meiotic entry decision in *C. elegans*, cell biological analysis of the stem cell population has proved more difficult. This difficulty is due, in part, to the syncytial nature of the germ line; each nucleus and surrounding cytoplasm is not fully enclosed by membranes (although collectively these are commonly referred to as a cell). Therefore, cells within the gonad are interconnected, making transplantation and repopulation experiments virtually impossible, as well as complicating lineage tracing approaches. However, recent work using S-phase labeling approaches to examine cell kinetics has helped in understanding cell behaviors within the proliferative region. These studies have reinforced the idea that, unlike stem cell systems that contain a very small number of asymmetrically dividing stem cells, the *C. elegans* germ line contains a relatively large population of stem cells that give rise to equivalent daughters, but which may eventually assume different fates depending on their position relative to the niche signal. Here we discuss our current understanding of the molecular mechanisms that control the balance between the proliferative fate and meiotic entry in the adult *C. elegans* germ line. Although mechanisms that control this decision appear very similar between the male and hermaphrodite, in this review we will primarily focus on the adult hermaphrodite germ line. Additionally, we focus on how this balance is maintained under optimal laboratory growth conditions ("feasting") that provides a baseline upon which environmental conditions modify this balance for optimal reproduction, which is discussed in Chap. 5, (Hubbard et al. 2012).

4.2 Anatomy/Cytology of the Gonad

As the *C. elegans* hermaphrodite larvae hatches, the gonad primordium consists of the Z2 and Z3 germ cells, the Z1 and Z4 somatic cells, and a surrounding basement membrane. Midway through the first larval stage the germ cells begin to proliferate, with all cells remaining proliferative until midway through the third larval stage, when the first cells to enter meiosis can be detected. By this time two gonad arms have begun to develop, each capped by the somatic DTC (descendants of Z1 and Z4), with the meiotic cells found at the proximal end of each gonad arm, thereby establishing a polarity that will be maintained throughout the remaining reproductive life of the animal. The timing of the onset of germ cell differentiation is tightly controlled relative to somatic gonad development in order for this polarity to be properly established and maintained (McGovern et al. 2009). By the time the worm is nearing the end of larval development, the first differentiated cells have developed as sperm. Approximately 150 sperm are made in each gonad arm, with all subsequent germ cells differentiating as oocytes.

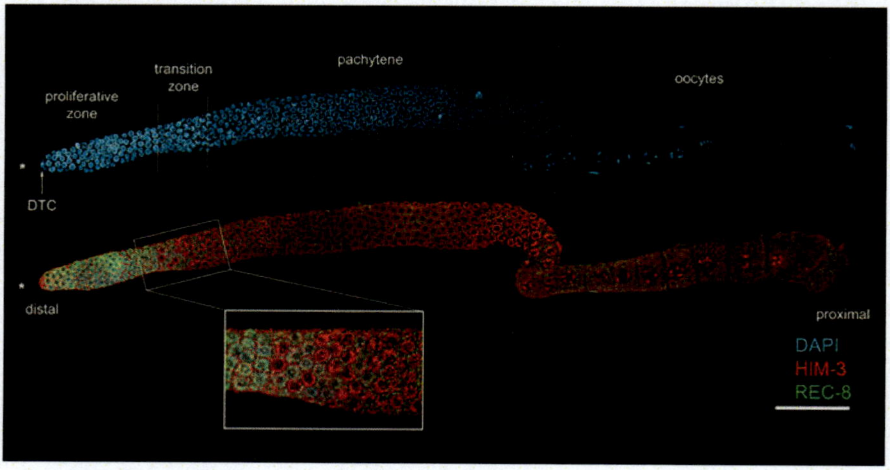

Fig. 4.1 Dissected adult hermaphrodite gonad. A dissected hermaphrodite adult gonad stained with DAPI (*blue*) to visualize nuclear morphology, anti-REC-8 antibodies (*green*) to mark mitotic cells, and anti-HIM-3 antibodies to mark meiotic cells. The distal end is to the left (*asterisk*). Germ cells in the distal end of the gonad are proliferative, but enter meiotic prophase as they move out of the proliferative zone and into the transition zone, where they display pairing and synapsis indicative of leptotene/zygotene (Chap. 6, Lui and Colaiácovo 2012). The somatic *d*istal *t*ip cell (DTC) is at the very distal end of the gonad arm (*arrow*). Cells at similar distances from the distal end of the gonad do not necessarily enter meiosis at precisely the same location. Rather, some cells show evidence of having entered into meiotic prophase while some neighboring cells do not (*inset*)

The adult gonad is an assembly line, with germ cells progressing from an undifferentiated stem cell fate in the distal end to a fully differentiated gamete at the proximal end. This linear progression of gamete formation simplifies the analysis of the progression of germ cells through the various stages of gamete formation. For convenience, the hermaphrodite gonad has been divided into different regions, or zones, in which cells are at specific stages of germ cell development (Fig. 4.1). Cells in the distal end of the gonad arm are in the proliferative, or mitotic, zone, which extends ~20 germ cell diameters from the distal tip (Crittenden et al. 1994; Hansen et al. 2004a). As cells move proximally, out of the proliferative zone, they enter the transition zone, where cells first show evidence of entering into meiotic prophase. Transition zone nuclei have a distinct crescent moon-shaped DNA morphology as the chromosomes pair in leptotene/zygotene of meiotic prophase (Dernburg et al. 1998; Francis et al. 1995a; MacQueen and Villeneuve 2001). Not all cells enter meiotic prophase at precisely the same location in the gonad arm. For example, transition zone nuclei can be found at the same distance from the distal end as cells that are undergoing progression through the mitotic cell cycle, as judged by M-phase figures (Fig. 4.1) (Hansen et al. 2004a; Crittenden et al. 2006; Fox et al. 2011). The transition zone has been defined as the region containing the most distal transition zone nucleus to the most proximal transition zone nucleus (Crittenden et al. 1994; Hansen et al. 2004a). Cells that move proximally from the transition zone enter into the pachytene region, followed by gamete formation.

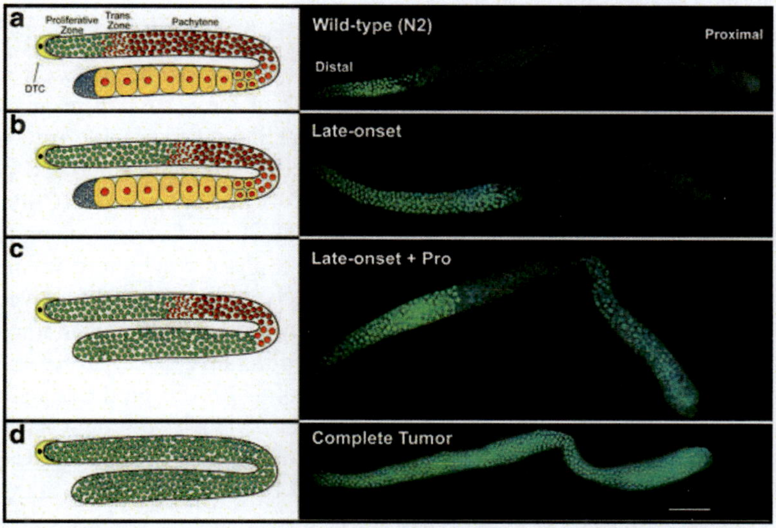

Fig. 4.2 Different classes of germline tumors. Diagrams (*left*) and pictures (*right*) of hermaphrodite gonad arms illustrating different classes of germline tumors. The distal tip cell (DTC) caps the distal end of the gonad. In wild-type animals (**a**) proliferative cells (*green*) are restricted to an ~20 cell diameter region in length from the distal end of the gonad, then enter into meiosis (*red*) as they move proximally, progress through meiotic prophase, and eventually differentiate into sperm (*blue*) or oocytes (*orange*). In gonads with a late-onset tumor (**b, c**), the region containing proliferative cells is much longer than in wild type. A Pro (**c**) tumor consists of proliferating cells in the proximal end of the gonad, while a complete tumor (**d**) contains proliferative cells throughout the entire gonad arm. Dissected gonad arms are stained with anti-REC-8 antibodies (*green*) to mark proliferative cells. Genotypes used (**a**) Wild-type (N2) (**b**) *glp-1(gf)/glp-1(0)* [Actual genotype *unc-32(e189) glp-1(oz112)/unc-36(e251) glp-1(q175)]* (**c**) *glp-1(gf) teg-1(0)* [Actual genotype *unc-32(e189) glp-1(ar202) teg-1(oz230)*] (**d**) *glp-1(gf)/glp-1(gf)/ glp-1(+)* [Actual genotype *dpy-19(e1259) unc-32(e189) glp-1(oz112)/dpy-19(e1259) unc-32(e189) glp-1(oz112); qDp3[dpy-19(+) glp-1(+)]*]. Scale bar = 20 μ

Some markers have aided in the analysis of cells in these zones. For example, antibodies against the REC-8 protein, when used with mild fixation conditions, stain the nucleoplasm of proliferative cells (Fig. 4.1) (Pasierbek et al. 2001; Hansen et al. 2004a). Likewise, antibodies specific to HIM-3 highlight the meiotic chromosome axis, marking cells that have entered into meiotic prophase (Zetka et al. 1999). The use of these and other markers has emphasized the uneven border between the proliferative and transition zones in wild-type animals (Fig. 4.1), as well as allowed assessment of germ cell fate in gonads with various phenotypic defects (Fig. 4.2) or whose DNA morphology is abnormal, such as occurs in cell cycle arrest.

4.3 The Adult Proliferative/Mitotic Zone

The adult proliferative zone is a steady state system of ~230 cells (200–250 range), with M-phase nuclei found throughout the proliferative zone, although at lower frequency more proximally (Crittenden et al. 1994; Hansen et al. 2004a;

Francis et al. 1995b; Maciejowski et al. 2006; Jaramillo-Lambert et al. 2007). The cell-cycle kinetics of cells within the proliferative zone has not been closely examined in real-time in living animals due to the lack of the necessary reagents and techniques; however, experiments using various labeling schemes with BrdU, EdU, and fluorescent-nucleotides, each of which is incorporated into DNA during S-phase, have provided insight into the cycling behavior of this cell population (Crittenden et al. 2006; Jaramillo-Lambert et al. 2007; Fox et al. 2011). Currently there are no markers to distinguish between cells in mitotic S-phase from those in meiotic S-phase; however, kinetic analysis of the ~230 cells within the proliferative zone suggests that ~130–160 of these cells are actively cycling, with the remaining ~70–90 cells, residing in the proximal part of the proliferative zone, are in meiotic S-phase (Fox et al. 2011). Overall, the population of ~230 cells produces an output of ~20 cells that enter meiosis each hour (Fox et al. 2011). Two groups have analyzed the total cell cycle length, using similar methods; however, different measures of cell cycle length were obtained, 6.5–8 h or 16–24 h (Crittenden et al. 2006; Fox et al. 2011). It is unclear as to the cause of this difference in estimates. Cell cycle kinetics analysis indicates that G1-phase is either very short, or absent, in cycling adult proliferative zone cells. G2 may thus be a major phase for cell cycle regulation of *C. elegans* germ cells. Consistent with this idea, germ cells in L1 diapause and dauer worms, following starvation, are arrested in G2 (Narbonne and Roy 2006; Fukuyama et al. 2006).

In the *Drosophila* ovary and testis, the stem cell daughter displaced from the niche and committed to differentiation undergoes four stereotypic cystoblast/gonialblast transit-amplifying divisions prior to their entry into meiosis (Losick et al. 2011). Of the ~130–160 mitotically cycling cells within the *C. elegans* adult hermaphrodite proliferative zone, it is not known whether there are two populations of proliferating cells (stem cells and transit-amplifying cells) or a single developmentally equivalent population of cells [see (Seydoux and Schedl 2001)]. There are no obvious cytological features, such as asymmetric cell divisions or stereotypic division patterns, as well as companion somatic cells, which might distinguish stem cells from possible transit-amplifying cells (Crittenden et al. 2006). In some stem cell systems (e.g., mammalian hair follicle), the stem cells are slow cycling relative to their transit-amplifying counterparts (Fuchs 2009); however, there are no significant differences in cell cycle length among proliferative zone cells (Hansen et al. 2004a; Maciejowski et al. 2006; Jaramillo-Lambert et al. 2007; Crittenden et al. 2006; Fox et al. 2011). There are no markers that distinguish between specific cells in the proliferative zone, although GLD-1 protein (see below) is very low/absent in germ cells in the distal part of the proliferative zone and then rises to a high level as the cells enter the transition zone (Jones et al. 1996). Recently, suggestive evidence that there are two pools of cells with distinct behaviors has been reported from analysis of the proliferative zone following blocking germ cell movement through cell cycle arrest (Cinquin et al. 2010). However, it is unclear to what extent the phenomena observed following cell cycle arrest can be extended to the behavior of proliferative zone cells in a reproductively active germ line.

The lack of markers that differentiate cells within the proliferative zone has also made it difficult to determine whether putative transit-amplifying cells or meiotic S-phase cells are capable of returning to a stem cell fate if they come back in contact with the niche cell (DTC), as has been observed in the Drosophila germ line (Kai and Spradling 2004; Brawley and Matunis 2004). Through processes that extend between or around germ cells, the DTC is only in close contact with the most distal three or four rows of germ cells, which together comprise approximately 30 germ cells (Fitzgerald and Greenwald 1995; Crittenden et al. 2006; Hall et al. 1999), although the influence of the DTC appears to extends farther than the region of germ cells it immediately surrounds. The close contact between the germ cells and the DTC processes is likely important for cells to self-renew; however, the factors that mediate this interaction are yet to be identified. The interaction between the DTC and germ cells is likely quite dynamic, which could explain why junctional contacts have not been detected between the DTC and germ cells (Hall et al. 1999). Given the large amount of bulk flow from distal to proximal, with 20 new cells generated per hour, any return of cells to a more distal position is probably a low frequency event.

Where in the distal germ line does the decision to enter meiotic prophase occur? Cells in meiotic S-phase are largely REC-8 positive, HIM-3 negative (Jaramillo-Lambert et al. 2007; Fox et al. 2011), indicating that the bulk of HIM-3 loading onto meiotic chromosome axes occurs after meiotic S-phase. In budding yeast and mouse, the decision to enter meiotic prophase can be viewed as a switch from a mitotic S-phase to a meiotic S-phase (Baltus et al. 2006; Honigberg and Purnapatre 2003). The decision point is likely the same for the *C. elegans*. Thus, cytological markers such as HIM-3 or crescent-shaped nuclear DNA morphology assess fate based on when cells enter leptotene, which is temporally displaced from when the switch was initiated at entry into meiotic S-phase. Therefore, a major challenge for the field is to generate molecular markers that subdivide proliferative zone cells into those that are mitotically cycling, in meiotic S-phase, stem cells and transit-amplifying cells (if they exist).

The adult hermaphrodite proliferative zone containing 130–160 mitotically cycling cells represents a large stem cell system, relative to those found in other organisms, even if some of these cells are transit-amplifying cells. This large number of stem cells is likely a consequence of the requirement to produce many meiotic cells, most of which function as nurse cells for the forming oocytes prior to their undergoing physiological germ cell apoptosis in late pachytene (Chap. 9, Bailly and Gartner 2012; Chap. 10, Kim et al. 2012). Analysis of the gametogenesis output during oogenesis under optimal growth conditions leads to estimates that 90 % or more of pachytene germ cells undergo physiological apoptosis (McCarter et al. 1999; Jaramillo-Lambert et al. 2007; Fox et al. 2011). Thus, the large, rapidly dividing, proliferative zone may be necessary to generate the ~20 cells entering meiotic prophase per hour, 90 % of which will undergo apoptosis after providing cytoplasmic constituents to the forming oocytes.

4.4 Phenotypic Consequence of a Disruption in the Balance

The primary method used to study the mechanism by which the balance is maintained between specification of the proliferative fate and meiotic entry is to disrupt this balance. Physical manipulation of the somatic gonad, or disruption of normal gene function, can cause the balance to be shifted to either excessive proliferation or premature meiotic entry. In the most extreme cases of excess proliferation, a germ line tumor is formed that only consists of proliferative cells, with no cells entering meiotic prophase (Berry et al. 1997; Hansen et al. 2004a). Conversely, animals with premature meiotic entry, occurring very early in larval development, have a complete depletion of proliferative cells and only ~16 sperm being formed per gonad arm (Kimble and White 1981; Austin and Kimble 1987). Study of the physical manipulations and genetic mutations that result in these extreme phenotypes, as well as more moderate phenotypes, has been essential in teasing apart the molecular mechanisms that control the tight balance between proliferation and meiotic entry.

4.4.1 Premature Meiotic Entry Results in a Glp Phenotype

Significant insight into the proliferative fate vs. meiotic entry decision was obtained through ablating the DTC with a laser microbeam (Kimble and White 1981). Ablating the DTC in a given gonad arm resulted in all germ cells prematurely entering meiotic prophase and concomitant loss of the proliferative cell population. Whether ablation occurred in early larval development or later in adulthood, all the proliferative cells entered into meiotic prophase resulting in meiotic/differentiated cells extending to the distal end of the gonad. This phenotype is referred to as "*G*erm *l*ine *p*roliferation abnormal," or Glp. That ablation of the DTC resulted in Glp animals suggests that the DTC is necessary to promote the proliferative fate and/or inhibit the meiotic fate. Additionally, mis-positioning of the DTC to a more proximal position in the gonad arm resulted in the establishment of a proliferative population of cells adjacent to the new location of the DTC (Kimble and White 1981), suggesting that the DTC is also sufficient to promote the proliferative fate.

Various genetic mutations have been isolated that cause a Glp phenotype, as discussed below. Another class of mutations, as well as environmental treatments, results in an adult proliferative zone that contains fewer cells than wild type. In some cases this is due to a disruption in the proliferation vs. meiotic entry cell fate decision (e.g., a *glp-1* temperature sensitive (ts) allele at the permissive temperature). In other cases it is not due to a disruption in this decision and additional experiments are necessary to tease apart the process that is disrupted. For example, RNAi knockdown of numerous cell cycle genes results in a smaller proliferative zone, but this appears to be due directly to a disruption in the mitotic cell cycle (Fox et al. 2011). From the L3 stage to adulthood, the proliferative zone expands from ~20 cells to more than 200 (Killian and Hubbard 2005). The larval proliferative zone expansion is regulated by food status, controlled in part through DAF-2 insulin-like signaling and TGF-beta signaling (Chap. 5, Hubbard et al. 2012). Reduced food or disruption of either pathway leads to an adult containing fewer proliferative zone cells. Mutations that directly or indirectly affect pathways involved in environmental signals controlling larval expansion could thus produce adults with a smaller proliferative zone.

4.4.2 Over-proliferation Results in a Germ Line Tumor

As mentioned, a germ line tumor is formed when the balance between the proliferative fate and the meiotic fate is shifted toward excess specification of the proliferative fate. However, the severity of the tumor can differ depending on a number of factors, including the degree to which a shift in the balance occurs, or when during development the shift is most pronounced (Fig. 4.2). The most severe tumor, or a complete tumor, contains only proliferative cells with no evidence of any cells entering into meiotic prophase (Berry et al. 1997; Hansen et al. 2004a). Fully tumorous germ lines are thought to be due to a complete failure of cells to switch from the proliferative fate to meiotic prophase. Classification of a tumor as complete generally requires monitoring the germ line cells during the development of the animal to ensure that some cells do not enter meiosis early during development. Incomplete tumors can be sufficiently robust that even though some cells enter meiosis early during development, proliferative cells quickly fill the entire gonad, making it difficult to see interspersed meiotic or differentiated cells in the adult.

Animals that have excess germ line proliferation, but also contain some cells that enter meiotic prophase, often have what is referred to as a late-onset tumor. Although the severity of a late-onset tumor can vary, in most cases germ cells in these animals initially enter meiosis normally during larval and young adult development, but then a disruption occurs such that proliferative self-renewal outpaces entry into meiosis, resulting in a progressively increasing distal proliferative zone (Berry et al. 1997). The size of the distal proliferative zone is often measured as the number of cell diameters from the distal end of the gonad until the first (most distal) meiotic cell; however, counting the total number of proliferative cells within the proliferative zone is thought to be more accurate.

Mutant animals can also have tumors in the *pro*ximal end of the gonad, referred to as Pro tumors (Fig. 4.2). Pro tumors are usually due to a disruption of the proliferation vs. meiotic entry decision in larval development. Some mutant animals can have both late-onset and Pro tumors, resulting in a region of over-proliferation in the distal end of the gonad, followed by cells in meiotic prophase or possibly differentiated gametes (usually sperm), and finally another region of over-proliferation in the proximal end of the gonad, next to the spermatheca. Pro tumors are often very robust, quickly filling the entire proximal end of the gonad with proliferative cells. While there appear to be diverse molecular and physiological origins for proximal tumor formation, most, if not all, seem to result in inappropriate contact between undifferentiated proximal germ cells and gonadal sheath cells (Killian and Hubbard

2004; Pepper et al. 2003; Seydoux et al. 1990; Voutev et al. 2006). During normal larval development, there is a period when all germ cells are proliferative, prior to when some proximal cells enter meiotic prophase. The timing of when differentiation first occurs, relative to the timing of somatic gonad formation, is crucial for distal to proximal polarity to be established. If there is a delay in when proximal cells first begin to differentiate, the somatic sheath cells develop to a point in which they produce a signal (GLP-1 Notch ligands) that promotes the proliferative fate (McGovern et al. 2009). These sheath cells function as a "latent niche," in that normally these sheath cells would not come in contact with proliferating cells, and therefore would not promote the proliferative fate (McGovern et al. 2009). However, since there is delayed differentiation in these mutants, proliferative cells do contact the sheath cells and receive a signal to continue to proliferate, resulting in the formation of a Pro tumor.

The germ line tumors discussed thus far are due to a disruption in the balance between specification of the proliferative fate and meiotic entry; therefore, studying the causative mutations has provided tremendous insight into the molecular mechanisms involved in controlling this balance (see below). However, it should be emphasized that germ line tumors can be formed that are unrelated to the proliferative fate vs. meiotic entry decision. For example, animals that are mutant for *gld-1* alone, or *puf-8* alone, can form de-differentiation tumors (Francis et al. 1995a; Subramaniam and Seydoux 2003). In these mutant animals, a proliferative population of cells exists in the distal end of the gonad arm that is very similar in size to wild-type animals, and cells appear to enter into meiotic prophase normally. However, after entry into meiotic prophase the cells fail to properly progress through meiosis (oogenesis for *gld-1* and spermatogenesis for *puf-8*), and return to a mitotic cell cycle. This results in a proliferative population of cells in the proximal end of the gonad. These proximal mitotic cells receive a signal from the "latent niche," described above, that contributes to the robustness of this proximal tumor, but is not required for its formation (Francis et al. 1995b). As the animal ages, the tumor continues to grow such that it is no longer restricted to the proximal end. Therefore, if the gonads are analyzed when the animals are quite old, it is possible to misinterpret the cause of the tumor as being due to a defect in the proliferative fate vs. meiotic entry decision.

4.5 The Core Genetic Pathway

4.5.1 GLP-1 Notch Signaling Promotes the Proliferative Fate

The DTC ablation experiments, which resulted in a Glp phenotype, identified the DTC as the major regulator of the proliferative fate in the germ line (Kimble and White 1981). Genetic screens were performed to determine the molecular signal that promotes the proliferative fate. From these screens, components of the conserved

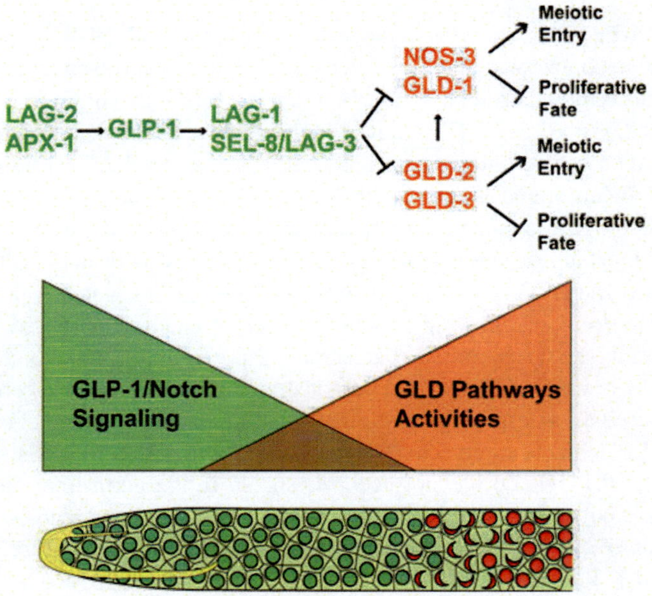

Fig. 4.3 Genetic control of the proliferation vs. differentiation decision. Core genetic pathway controlling the proliferation vs. differentiation decision (*top*) with factors that promote the proliferative fate and/or inhibit meiotic entry shown in green, while factors that promote meiotic entry and/or inhibit the proliferative fate shown in *red*. In this model, the LAG-2 and APX-1 ligands, which are expressed on the distal tip cell (DTC), interact with the GLP-1 Notch receptor, which is expressed on the germ cells. This causes the intracellular portion of GLP-1 to translocate to the nucleus and interact with the LAG-1 CSL transcription factor and SEL-8 Mastermind co-activator. These are all conserved components of the Notch signaling pathway. The GLP-1 pathway inhibits the activities of the redundant GLD-1 and GLD-2 genetic pathways, whose components all appear to regulate mRNA stability and/or translatability. An additional redundant pathway that is inhibited by GLP-1 signaling has been uncovered genetically (not shown), although genes that reside in this pathway have not been identified. The basic model of how signaling from the DTC regulates the proliferation vs. differentiation decision is shown (*bottom*). While germ cells are in the distal-most end of the gonad arm, close to the DTC (*yellow*), GLP-1 Notch signaling levels are high (*green*), keeping the activities of the GLD-1 and GLD-2 pathways low (*red*), resulting in cells remaining in the proliferative fate. As cells move proximally, GLP-1 Notch signaling levels decrease, causing the activities of the GLD-1 and GLD-2 pathways to increase, resulting in cells entering into meiotic prophase

Notch signaling pathway were identified as being the molecular mechanism through which the DTC exerts its influence (Fig. 4.3). Disruption in the activities of the DSL (*D*elta/*S*errate/*L*AG-2) LAG-2 and APX-1 ligands (Henderson et al. 1994; Tax et al. 1994; Nadarajan et al. 2009), GLP-1 Notch receptor (Austin and Kimble 1989; Yochem and Greenwald 1989), or CSL (*C*BF1/*S*uppressor of Hairless/*L*AG-1) LAG-1 transcription factor (Christensen et al. 1996), all conserved components of the Notch signaling pathway, result in Glp animals (Lambie and Kimble 1991; Austin and Kimble 1987; Yochem and Greenwald 1989). The basic model of how GLP-1 Notch signaling promotes the proliferative fate is as follows. When LAG-2 and APX-1, which are expressed in the DTC (Henderson et al. 1994; Nadarajan et al. 2009), come in contact with the GLP-1 receptor, which is expressed on the surface of the germ cells (Crittenden et al. 1994), the *i*ntra*c*ellular *d*omain of GLP-1, referred to as GLP-1(ICD) or GLP-1(INTRA), is thought to move to the nucleus and complex with the LAG-1 transcription factor and the SEL-8 (a.k.a. LAG-3) transcriptional co-activator (Mumm and Kopan 2000; Petcherski and Kimble 2000; Doyle et al. 2000). This complex is then thought to activate transcription of genes necessary for the proliferative fate. As germ cells migrate proximally, away from the DTC, they no longer contact the DTC and bound ligand; therefore, GLP-1 Notch signaling levels are thought to decrease, allowing for cells to enter into meiotic prophase.

However, it should be noted that certain aspects of GLP-1 Notch signaling described above have not fully been demonstrated in the *C. elegans* germ line, but rather are based on a description of Notch signaling in other systems. For example, although GLP-1 has been detected on the surface of distal germ cells (Crittenden et al. 1994), GLP-1(ICD) has not been visualized in the germ line; therefore, its temporal distribution is currently unknown. Furthermore, a direct read-out of GLP-1 Notch signaling in the germ line has not yet been established. Therefore, the spatial pattern of GLP-1 Notch signaling is not precisely known.

4.5.2 Redundant Genetic Pathways Function Downstream of GLP-1 Notch Signaling

The major mechanism by which GLP-1 Notch signaling maintains cells in the proliferative fate is through inhibiting the activities of RNA regulatory pathways in the distal end of the gonad that serve to promote meiotic entry and/or inhibit the proliferative fate (Fig. 4.3). The two main pathways are the GLD-1 and GLD-2 pathways, named for the first genes identified in each pathway (Kadyk and Kimble 1998; Francis et al. 1995a,b). The GLD-1 and GLD-2 pathways function redundantly, such that if the activity of only one of the two pathways is reduced or eliminated, the balance between proliferation and meiotic entry is similar to wild type; however, if the activities of both pathways are reduced or eliminated, overproliferation occurs resulting in a germ line tumor (Kadyk and Kimble 1998; Hansen et al. 2004b; Eckmann et al. 2004). Therefore, either pathway alone is sufficient to cause germ cells to enter meiosis. *gld-1* encodes a translational inhibitor that is homologous to mouse Quaking (Jones and Schedl 1995; Lee and Schedl 2010), and has been shown to bind the 3'UTRs of mRNAs to inhibit their translation (Jan et al. 1999; Lee and Schedl 2001; Marin and Evans 2003; Lee and Schedl 2004; Wright et al. 2011; Jungkamp et al. 2011). Also within the *gld-1* pathway is the *nos-3* gene (Hansen et al. 2004b), which encodes a protein related to the Drosophila translational regulator Nanos (Kraemer et al. 1999). Within the *gld-2* pathway are *gld-2* and *gld-3* (Kadyk and Kimble 1998; Eckmann et al. 2004); *gld-2* encodes the catalytic portion of a poly(A) polymerase (Kadyk and Kimble 1998), while *gld-3* encodes a BicC homologue (Eckmann et al. 2002). GLD-2 and GLD-3 bind to one another and GLD-3 enhances GLD-2 activity (Eckmann et al. 2004).

The overall model of how the GLD-1 and GLD-2 pathways regulate the proliferation vs. meiotic entry decision is that their activities are thought to be low or absent in the distal end of the gonad where GLP-1 Notch signaling levels are high, but as cells move more proximally, away from the influence of the DTC, GLP-1 Notch signaling levels decrease, causing GLD-1 and GLD-2 pathway activity to increase, which in turn allows cells to enter into meiotic prophase (Fig. 4.3). Perhaps the regulation of this reciprocal level of activity between GLP-1 Notch signaling and the GLD-1 and GLD-2 pathways is best described for GLD-1. GLD-1 protein accumulation levels are low or absent in the distal end of the gonad, but rise gradually in the proliferative zone until reaching maximum levels in the distal portion of the transition zone, where cells show the first signs of having entered into meiotic prophase (Jones et al. 1996). If GLP-1 Notch signaling levels are eliminated, GLD-1 levels increase more than tenfold in the distal portion of the gonad (Hansen et al. 2004b), suggesting that GLP-1 Notch signaling suppresses GLD-1 accumulation in the distal end. Furthermore, increasing GLD-1 levels in the distal end causes cells to enter meiosis prematurely (Hansen et al. 2004b), suggesting that the low level of GLD-1 in the distal end allows cells to remain proliferative. This also suggests that the rise in GLD-1 levels in wild-type animals promotes the transit of cells from the proliferative fate to meiotic prophase. While GLD-2 and GLD-3 also show nonuniform accumulation patterns in the gonad (Wang et al. 2002; Eckmann et al. 2002, 2004), the variations in their spatial distribution patterns is significantly more subtle than the GLD-1 accumulation pattern, and the importance of their spatial distributions in regulating the proliferative fate vs. meiotic entry decision has not been established.

GLD-1 and NOS-3 are RNA-binding translational inhibitors, while GLD-2 is a poly(A) polymerase catalytic subunit that appears to be targeted to certain mRNAs by GLD-3 (Chap. 8, Nousch and Eckmann 2012). Therefore, genes downstream of these pathways are likely to be regulated at the level of translational control and/or mRNA stability. However, the identification of genes that function downstream of these pathways is still in its early stages. Also, there is cross-talk between the two pathways with GLD-2/GLD-3 promoting the activity of *gld-1* (Hansen et al. 2004b; Suh et al. 2006). However, GLD-2/GLD-3 must have targets in addition to *gld-1*, otherwise a *gld-2* or *gld-3* single mutant would have the same phenotype as a *gld-2 gld-1* or *gld-1; gld-3* double mutant. Immunoprecipitation of GLD-1 has identified many mRNAs that are bound by GLD-1 (Lee and Schedl 2001; Wright et al. 2011; Jungkamp et al. 2011); however, only a limited number of these are likely to be involved in the proliferative fate vs. meiotic entry decision. GLD-1 has many functions in germ line development, including sex-determination and meiotic prophase progression of female germ cells (Francis et al. 1995a,b). Therefore, any given mRNA may have a role specific to only one of GLD-1's functions in germ line development. Given our current understanding of the molecular functions of GLD-1 and GLD-2/GLD-3, with GLD-1 being an RNA-binding translational inhibitor and GLD-2/GLD-3 functioning as a poly(A) polymerase that likely stabilizes mRNAs, GLD-1 and GLD-2/GLD-3 likely have RNA targets with opposite functions: GLD-1 likely represses the translation of mRNAs that promote the proliferative fate, while GLD-2 likely promotes the translation of mRNAs that promote the meiotic fate.

4.5.3 A Third Genetic Pathway Functions Redundantly with the *gld-1* and *gld-2* Pathways

Although the GLD-1 and GLD-2 pathways are the major known players functioning downstream of GLP-1 Notch signaling to regulate the balance between proliferation and meiotic entry, GLP-1 Notch signaling inhibits the activity of an additional unknown factor(s) that inhibits the proliferative fate and/or promotes meiotic entry. We suggest that there is a third redundant pathway that functions in parallel to the GLD-1 and GLD-2 pathways (Hansen et al. 2004a; Fox et al. 2011). Perhaps the strongest evidence for this third pathway comes from analysis with the CYE-1/CDK-2 cell cycle regulator, which will be discussed in more detail below. In brief, CYE-1/CDK-2 promotes the proliferative fate in the *C. elegans* germ line (Fox et al. 2011; Jeong et al. 2011). In animals that have mutations in both the GLD-1 and GLD-2 pathways, and also have reduced *cye-1/cdk-2* activity, significantly more meiotic entry is observed than in just the GLD-1 and GLD-2 pathway double mutant (Fox et al. 2011). However, meiotic cells are not observed in the most distal end of the gonad in these triple mutants, where GLP-1 Notch signaling is thought to be at its highest level. When GLP-1 Notch signaling is also removed in these triple mutants, meiotic cells become apparent in the distal end of the gonad. Therefore, even when the activities of the GLD-1 and GLD-2 pathways are eliminated, GLP-1 Notch signaling still promotes the proliferative fate in the distal end of the gonad by inhibiting the activity of this third pathway. However, the relative strength of this third pathway may be lower than the other two pathways because its activity alone is not sufficient to cause cells to enter meiotic prophase; *gld-3 nos-3* double mutants have completely tumorous germ lines, with no evidence of cells entering meiotic prophase, even though the third pathway should still be active (Fox et al. 2011). No components of this third pathway have yet been identified.

4.5.4 GLP-1 Notch Signaling Functions, in Part, Through FBF

Since Notch signaling is thought to *activate* the transcription of downstream genes, but GLP-1 Notch signaling *inhibits* the activities of the GLD-1 and GLD-2 pathways, this inhibition must be indirect. Inhibition of the GLD pathways is accomplished, in part, through the activities of the *fbf-1* and *fbf-2* genes. FBF-1 and FBF-2, collectively referred to as FBF, are homologous to the *Drosophila* Pumilio translation regulator (Zhang et al. 1997) and have largely redundant functions relative to the proliferation vs. meiotic entry decision. *fbf-2* is directly regulated by GLP-1 Notch signaling; it contains LAG-1 binding sites in its promoter region that can be bound by LAG-1 in vitro, and FBF-2 protein accumulates in the distal region of the gonad arm, including where GLP-1 Notch signaling levels are thought to be the highest (Crittenden et al. 2002). Interestingly, even though the functions of *fbf-1* and *fbf-2* are largely redundant, only *fbf-2* appears to be directly regulated by LAG-1. When the activities of both *fbf-1* and *fbf-2* are eliminated, germ cells enter into meiosis prematurely (Crittenden et al. 2002). However, the premature meiotic entry, or Glp, phenotype is not as severe as when GLP-1 Notch signaling is eliminated; the proliferative population of cells is not fully depleted until adulthood, resulting in many more differentiated gametes in *fbf-1 fbf-2* double null mutants than in *glp-1* null mutants, and depletion does not occur at all at higher temperatures (Crittenden et al. 2002). This weaker Glp phenotype suggests that other factors function redundantly with FBF, or at different points in development, to promote the proliferative fate. One of these redundant factors is FOG-1 (Thompson et al. 2005), which is an RNA-binding protein of the CPEB family (Jin et al. 2001; Luitjens et al. 2000). While the Glp phenotype of *fbf-1 fbf-2* double mutants is less severe than when GLP-1 Notch signaling is eliminated, *fog-1; fbf-1 fbf-2* triple mutants display the more severe Glp phenotype (Thompson et al. 2005). This strong Glp phenotype reveals a redundant function for FBF in promoting the proliferative fate during early larval development, when germ cells are undergoing male development. However, FOG-1 is absent in late larvae and adults when germ cells are undergoing female development, so there must be additional genes that are redundant with *fbf-1 fbf-2*. While FBF functions redundantly with FOG-1, it also appears to promote FOG-1 activity by maintaining FOG-1 levels at low level: if FOG-1 levels increase above a threshold, FOG-1 no longer promotes the proliferative fate (Thompson et al. 2005).

The *fbf-1 fbf-2* double mutant Glp phenotype is consistent with these genes functioning redundantly to promote the proliferative fate; however, the single mutant phenotype of *fbf-2* is not consistent with the predicted phenotype of a direct LAG-1 target. As mentioned, GLP-1 Notch signaling is thought to transcriptionally activate target genes that promote the proliferative fate. Therefore, the predicted loss-of-function phenotype of LAG-1 target genes would be a decrease in proliferation. However, *fbf-2* single mutants have a larger proliferative zone as compared to wild-type animals, implying that *fbf-2* may function to inhibit the proliferative fate (Lamont et al. 2004). This inconsistent phenotype may have to do with inhibitory feedback between FBF-1 and FBF-2, as they bind to each other's 3'UTRs (Lamont et al. 2004). Therefore, in an *fbf-2* mutant, FBF-1 levels may increase above normal levels, thereby causing an increase in the proliferative fate. Alternatively, FBF-2 may function to both promote the proliferative fate and inhibit the proliferative fate. Indeed, *gld-1; fbf-1 fbf-2* triple null mutants have a tumorous germ line (Crittenden et al. 2002), suggesting that *fbf* may function in the GLD-2 pathway to inhibit the proliferative fate (Hansen and Schedl 2006). Additionally, data suggests that FBF also *promotes* GLD-1 accumulation (Suh et al. 2009). Recently, >1,000 mRNA targets of FBF were identified, although most still need to be verified (Kershner and Kimble 2010). Interestingly, identification of certain targets has revealed that FBF not only has targets involved in regulating the switch between the proliferative and meiotic fates (see below), but also has targets directly involved in the execution of meiotic development. For example, FBF binds to and regulates the translation of *him-3*, *htp-1*, *htp-2*, *syp-2,* and *syp-3* mRNAs, each of which encodes structural components of the synaptonemal complex that forms between paired meiotic chromosomes (Merritt and Seydoux 2010).

4.5.5 FBF Inhibits *gld-1* mRNA Translation

FBF appears to promote the proliferative fate, at least in part, by inhibiting the activities of the GLD-1 and GLD-2 pathways, thereby providing the link between GLP-1 Notch signaling and the downstream GLD-1 and GLD-2 redundant pathways. FBF binds to the 3'UTR of *gld-1* mRNA (Suh et al. 2009), and reducing GLD-1 dose partially suppresses the FBF Glp phenotype (Crittenden et al. 2002). Furthermore, GLD-1 levels increase in the distal end in *fbf-1* single mutants. Together, these data suggest that FBF binds to *gld-1* mRNA in the distal end of the gonad, preventing the accumulation of GLD-1 protein. Likewise, FBF appears to inhibit the activity of the GLD-2 pathway by binding to the 3'UTR of *gld-3* and preventing the accumulation of GLD-3 protein in at the distal end of the gonad in larvae (Eckmann et al. 2004), although it is unclear if this regulation occurs in the adult. Given that both the GLD-1 and GLD-2 pathways are each sufficient for cells to enter meiotic prophase, complete activation of either of these pathways should result in a strong Glp phenotype. Therefore, since the *fbf-1 fbf-2* double mutant Glp phenotype only occurs in adults, FBF cannot be the sole inhibitor of these pathways. Furthermore, in certain genetic backgrounds that include a lack of FBF activity (e.g., *fog-3; fbf-1 fbf-2 nos-3* animals), germ cells proliferate, enter meiosis, and control GLD-1 levels similar to what is observed in wild-type animals (Hansen et al. 2004b), providing additional support for the existence of players, in addition to FBF, in inhibiting the GLD-1 and GLD-2 pathways.

4.6 Connection Between the Mitotic Cell Cycle and the Proliferation vs. Meiotic Entry Decision

Two studies have implicated the cell cycle regulators cyclin E (CYE-1) and Cdk2 (CDK-2) in regulating the proliferative fate vs. meiotic entry decision (Fox et al. 2011; Jeong et al. 2011). RNAi against *cye-1* or *cdk-2* enhances the premature meiotic entry defect of a weak *glp-1* loss-of-function allele (Fox et al. 2011). RNAi against other cell cycle regulators leads to cell cycle arrest and a smaller proliferative zone size but does not enhance *glp-1*, suggesting that it is not simply a disruption of the cell cycle that causes premature meiotic entry, but rather that *cye-1/cdk-2* are specifically involved in regulating this process (Fox et al. 2011). GLD-1 is a likely phosphorylation target of CYE-1/CDK-2, and reduction of *cye-1* activity results in an increase in GLD-1 accumulation in the distal end of the gonad (Jeong et al. 2011), suggesting that *cye-1/cdk-2* may function through *gld-1* to regulate the proliferative fate vs. meiotic entry decision. However, reduction of *cye-1/cdk-2* causes premature meiotic entry even in the absence of *gld-1* and *gld-2* activity (Fox et al. 2011). Therefore, *cye-1/cdk-2* cannot solely function through *gld-1* in regulating the proliferation vs. meiotic entry decision; *cye-1/cdk-2* must have an additional target downstream, or parallel to, the GLD-1 and GLD-2 pathways. One proposed model is that CYE-1/CDK-2 may limit the amount of time in certain parts of the cycle when cells may be receptive to differentiation factors. Alternatively, CYE-1/CDK-2 may have phosphorylation targets, like GLD-1, that are involved in regulating the proliferative fate vs. meiotic entry decision, but which do not have a direct role in controlling the cell cycle.

CYE-1 protein levels and activity (as judged by pCDC-6 staining) are high throughout the proliferative zone and fall abruptly as cells enter the transition zone (Brodigan et al. 2003; Fox et al. 2011). The loss of CYE-1 appears to be through redundant mechanisms of GLD-1-mediated translational repression (Biedermann et al. 2009) and SCF-mediated proteosomal degradation (Fox et al. 2011). However, ectopic CYE-1 in transition zone cells due to removal of both inhibitory processes did not result in continued proliferation, suggesting that yet an additional inhibitory mechanism is in play. The additional mechanism may involve CKI-2, a Cip/Kip protein that has been shown to inhibit Cyclin E/Cdk2 in other systems (Kalchhauser et al. 2011). CKI-2 levels are low or absent in the proliferative zone, due to repression by FBF, but increase as cells enter meiotic prophase, suggesting a model in which CKI-2 inhibits CYE-1/CDK-2 activity to allow cells to enter into meiotic prophase, although inhibition of CYE-1/CDK-2 by CKI-2 has not yet been shown to occur in the germ line. Supporting this model, loss of *cki-2* activity suppresses the premature meiotic entry phenotype of *fbf-2 fbf-1* double mutants (Kalchhauser et al. 2011). However, *cki-2* single mutants do not have a proliferative fate vs. meiotic entry defect (Buck et al. 2009; Kalchhauser et al. 2011), consistent with the presence of multiple redundant mechanisms. A key remaining puzzle is that the fall in CYE-1 level and activity, as well as rise in CKI-2 level, occurs when cells enter meiotic prophase at the transition zone, after completion of meiotic S-phase, and presumably, after implementation of the cell fate switch. Perhaps alterations in yet another factor within the proliferative zone, which controls CYE-1/CDK-2 substrate specificity, for example, triggers the switch to meiotic S-phase in the proximal region of the proliferative zone.

4.7 Splicing Cascade Functions in the Proliferative Fate vs. Meiotic Entry Decision

A number of splicing factors have been implicated in regulating the proliferative fate vs. meiotic entry decision (Belfiore et al. 2004; Kerins et al. 2010; Mantina et al. 2009; Zanetti et al. 2011; Puoti and Kimble 1999, 2000; Wang et al. 2012). Mutants in genes encoding these factors all show similar interactions with the genetic pathway regulating this decision. A complete reduction of the activity of many splicing factors is lethal; however, a partial reduction of splicing factor activity, or more complete reduction of the activities of nonessential splicing factors, enhances the over-proliferation phenotype of *glp-1(gf)* mutants. Additionally, splicing factor mutants show a synthetic over-proliferation phenotype with mutants of the GLD-2 pathway genes, *gld-2* and *gld-3*, suggesting that they likely function, at least in part, in the GLD-1 pathway. Consistent with them functioning in the GLD-1 pathway, downstream of GLP-1 Notch signaling, the synthetic over-proliferation phenotype with *gld-3* null is epistatic to a *glp-1* null mutant.

An RNAi screen of 114 splicing factors identified 31 that enhance the over-proliferation phenotype of *glp-1(gf)*, in addition to some that cause a synthetic tumor in *gld-3(0)* animals (Kerins et al. 2010). These splicing factors are found throughout the splicing cascade, suggesting that there is not a single aspect of splicing that has been co-opted for a function in the proliferative fate vs. meiotic entry decision. Indeed, it is likely that many more splicing factors are involved in this decision but were not identified in the RNAi screen because there is likely a "sweet spot" of gene reduction in which over-proliferation can be observed. Splicing factor activity would need to be reduced enough to cause a phenotype; however, if it is reduced too much, many cell functions may be disrupted and mask a proliferative fate vs. meiotic entry phenotype. Therefore, it is likely that many more splicing factors than those identified are involved in regulating the proliferative fate vs. meiotic entry balance.

Certainly, the primary question regarding the involvement of splicing factors in regulating this decision pertains to the precise mechanism by which they exert their influence. Perhaps the most straightforward model is that the reduction in splicing factor activity results in reduced splicing efficiency, and that certain genes functioning in the proliferative fate vs. meiotic entry decision are particularly susceptible to lower levels of splicing efficiency. Lowering of splicing efficiency could result in mis-splicing, or inappropriate alternative splicing. However, attempts to identify mis-splicing in these mutants have thus far been unsuccessful. These attempts have primarily involved a candidate gene approach; therefore, it is possible that the right candidate(s) has not yet been tested. It is also possible that the level of mis-splicing of a target needed to cause the over-proliferation phenotype may be quite low, and that the techniques used are not sensitive enough to detect this low level of mis-spliced product.

An alternative model for the involvement of splicing factors in regulating the proliferative fate vs. meiotic entry decision is that lowering of splicing factor activity could affect aspects of mRNA function other than splicing, such as the ability of the mRNA to be translated or translationally regulated in the cytoplasm. It is known that splicing factors are involved in formation of ribonucleo-protein complexes on the mRNA, such as the Exon Junction Complex (EJC). If splicing factor activity is reduced, this may affect the ability of such protein complexes to form and the mRNA to be properly transported, translated, or translationally regulated.

4.8 Additional Factors and/or Pathways Involved

The basic model for how the proliferation vs. meiotic entry decision is regulated consists of high GLP-1 Notch signaling in the distal end promoting the proliferative fate, and high levels of RNA regulatory genes more proximally allowing for meiotic entry (Fig. 4.3). However, many other factors and pathways have been identified that interact with this basic genetic network. For many of these factors, it is only in highly sensitized genetic backgrounds that a role in the proliferative fate vs. meiotic entry decision is revealed.

For example, from a mutant screen that utilized a *gld-2* null sensitized background a partial loss-of-function allele of a proteasome subunit was isolated (MacDonald et al. 2008). Genetic analysis revealed that the proteasome functions in at least two places in the proliferative fate vs. meiotic entry genetic pathway to inhibit the proliferative fate. First, it inhibits GLP-1 Notch signaling, presumably to degrade GLP-1(ICD). Proteasomal degradation of the intracellular portion of Notch has been identified as an important means of decreasing Notch signaling levels in many systems (Lai 2002; Andersson et al. 2011). Second, it functions in the GLD-1 pathway. As mentioned above, GLD-1 protein levels increase as cells enter meiotic prophase, and it functions as a translational inhibitor. Therefore, GLD-1 likely acts to repress the translation of the genes that are necessary for the proliferative fate. However, since cells are moving along the gonad arm, proliferative fate promoting proteins, which were made while the cell was in the proliferation zone, may persist in the cell as it moves proximally to the transition zone, even though GLD-1 prevents new translation of these proliferative proteins. Therefore, the current model is that the proteasome degrades these proliferative proteins that persist as the cell moves proximally, allowing the cell to enter meiotic prophase. Indeed, this model holds true for at least one GLD-1 target, *cye-1*, as described above.

Many of the factors that have been implicated in regulating the proliferation vs. meiotic entry decision also have other roles in germ line development and function. For example, HIM-17 was first identified due to its role in the execution of double strand breaks necessary for meiotic recombination (Reddy and Villeneuve 2004). While the single mutant appears to have a normal balance between proliferation and meiotic entry, *him-17* mutants enhance the tumorous phenotype of a *glp-1* gain-of-function allele (Bessler et al. 2007). The position of HIM-17 in the cell fate decision is uncertain, but appears not to function in the GLD-1 or GDL-2 pathways as a synthetic tumorous phenotype was not observed in double mutants with canonical null alleles of *gld-1* or *gld-2* pathway genes (A. Mohammad and TS, unpublished).

METT-10, the ortholog of vertebrate METT10D methyltransferase, also interacts with the core genetic pathway to regulate the proliferative fate vs. meiotic entry decision. A function for *mett-10* in this decision was first identified through the novel *mett-10(oz36)* allele, whose poison protein product results in a partially penetrant late-onset germ line tumor (Dorsett et al. 2009). This tumor is dependent on GLP-1 Notch signaling, suggesting that METT-10 normally inhibits the proliferative fate by negatively regulating GLP-1 Notch signaling. Loss-of-function alleles of *mett-10* do not cause a germ line tumor to form; however, they do enhance over-proliferation in animals carrying weak *glp-1* gain-of-function alleles (Dorsett et al. 2009). Therefore, it is only through a poison protein product, or in a sensitized genetic background, that a role for *mett-10* in the proliferative fate vs. meiotic entry decision is apparent, further emphasizing the redundancy that exists in the genetic pathway regulating this decision.

puf-8 was recently identified as being an inhibitor of the proliferative fate, as loss of PUF-8 activity strongly enhanced the tumorous phenotype of weak *glp-1* gain-of-function alleles (H. Racher and DH, unpublished). PUF-8 belongs to the same family of Pumilio homologues as FBF-1 and FBF-2 (Crittenden et al. 2002), which function to promote the proliferative fate and/or promote meiotic entry; therefore, it is intriguing that Pumilio homologs appear to function in both directions in the balance between the proliferative fate and meiotic entry. As mentioned earlier, in a proportion of *puf-8* single mutants male germ cells fail to properly progress through meiotic prophase, dedifferentiate and enter the mitotic cell cycle, forming a proximal tumor (Subramaniam and Seydoux 2003). This role of *puf-8* is distinct from its role in regulating the balance between the proliferative fate and meiotic entry. Therefore, *puf-8* inhibits the proliferative fate in at least two points in germ line development.

While the factors discussed thus far in this section all enhance over-proliferation when their activity is reduced, suggesting that they normally function to inhibit the proliferative fate and/or promote meiotic entry, other factors have been identified that normally function in the opposite direction, to promote the proliferative fate and/or inhibit meiotic entry. For example, five *ego* genes were identified in a genetic screen for mutations that enhance the Glp phenotype of a *glp-1* partial loss-of-function allele (Qiao et al. 1995). EGO-1 is an RNA-dependent RNA polymerase and functions in parallel to GLP-1 Notch signaling (Vought et al. 2005; Smardon et al. 2000). EGO-1 also has roles in other aspects of germ line development, including proper P-granule and nuclear pore complex assembly (Vought et al. 2005). Additionally, *ego-1* mutants disrupt the effectiveness of RNA interference and proper heterochromatin assembly (Maine et al. 2005; Smardon et al. 2000). It is currently unclear whether each of these functions for EGO-1 is independent, or whether EGO-1 has one (or few) function that influences the other processes.

Like the *ego* genes, *atx-2* also promotes the proliferative fate. Dominant mutations in the human homologue of *atx-2* cause the neurodegenerative disorder spinocerebellar ataxia type 2 (SCA2) (Sanpei et al. 1996); however, its molecular function is not well understood. A reduction in *atx-2* function enhances the Glp phenotype of a weak *glp-1* loss-of-function allele (Maine et al. 2004); however, ATX-2 does not appear to function as a positive regulator of GLP-1 Notch signaling (Ciosk et al. 2004; Maine et al. 2004). Rather, it may function parallel to GLP-1 Notch signaling, or it could be involved in the translational regulation of mRNA targets of GLD-1 (Ciosk et al. 2004).

These are just some of the factors that influence the activity of, or work in parallel to, the core genetic pathway that regulates the proliferation vs. differentiation decision. While the activities of some of these factors may be spatially regulated and directly involved in regulating the transition from the proliferative fate to meiotic entry, others may ensure that the field of cells is competent to respond the regulatory signals provided by the core genetic pathway.

4.9 Teratoma, Totipotency, and Reprogramming

Germ cells are totipotent; from germ cells, all cells and tissues of offspring are formed. In order for totipotency to be maintained in the germ line, the activities of differentiation genes must be suppressed. However, after fertilization, the activities of many of these differentiation genes must increase in order to drive embryogenesis. Therefore, many genes necessary for embryogenesis are transcribed in the gonad, but are masked by RNA-binding translational repressors until the activities of these genes are needed for embryogenesis. GLD-1, in addition to its role in regulating the proliferative fate vs. meiotic entry decision, is also involved in masking maternal mRNAs of a number of differentiation genes. Loss of *gld-1* activity results in female germ cells failing to properly progress through meiotic prophase, and subsequently reentering the mitotic cell cycle (Francis et al. 1995a). These mitotic cells then proliferate, forming a proximal tumor. Using somatic cell markers, it was found that a number of the mitotic cells within a *gld-1* tumor were not undifferentiated totipotent cells, but rather were differentiating as somatic cells, analogous to those in a human teratoma (Biedermann et al. 2009; Ciosk et al. 2006). The extent of teratoma formation is increased in *gld-1* mutants if the activity of *mex-3*, another translational regulating RNA binding protein, is also reduced (Ciosk et al. 2006). It is thought that MEX-3 and GLD-1 normally bind to and repress target mRNAs in order to repress differentiation genes, thereby preventing teratoma formation. At least one of these mRNA targets is *pal-1*, which encodes a homeobox transcription factor that promotes muscle development (Ciosk et al. 2006). Both MEX-3 and GLD-1 inhibit *pal-1* translation, and PAL-1 ectopic expression in *mex-3 gld-1* double mutants contributes to teratoma formation (Mootz et al. 2004; Ciosk et al. 2006).

The cells that become the teratoma in *mex-3 gld-1* animals successfully entered into meiotic prophase, but then failed to progress through meiosis and reentered the mitotic cell cycle. However, cells within the proliferative zone also must maintain their totipotency. Recently, it was found that overexpression of the transcription factors CHE-1, UNC-3, or UNC-30 resulted in the reprogramming of mitotic germ cells to glutamatergic, cholinergic, or GABAergic neurons, respectively (Tursun et al. 2011). However, this conversion only occurred if the activity of LIN-53, a histone chaperone, was also removed. Therefore, it is thought that one mechanism to maintain totipotency in germ cells is through chromatin factors that render the DNA inaccessible to differentiation promoting transcription factors.

4.10 Compare and Contrast with the *Drosophila* Germ Line

In order for dividing stem cells to be maintained in any system, a balance must be maintained between the number of cells that self-renew, and the number of cells that enter a differentiation pathway. As we have described, in *C. elegans* this balance is

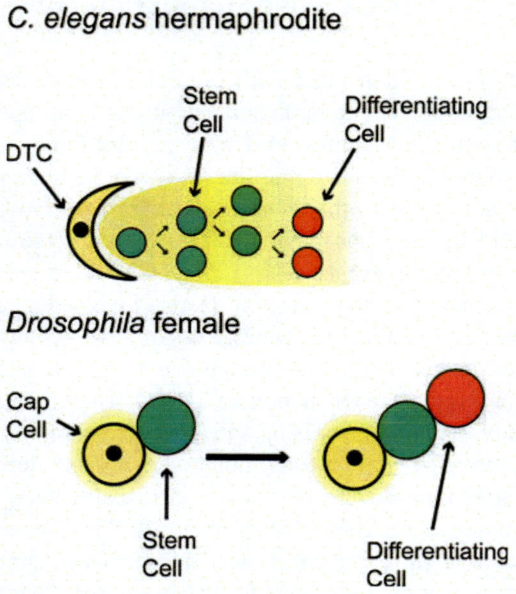

Fig. 4.4 Comparison of *Drosophila* ovarian and *C. elegans* hermaphrodite gonad stem cell systems. In the *C. elegans* germ line, the distal tip cell (DTC) functions as the niche cell, and its zone of influence (*yellow gradient*) extends beyond its physical location. Within this system, stem cells (*green*) are thought to divide symmetrically, giving rise to intrinsically similar daughter cells. It is not until cells reach a certain distance from the DTC that they begin to differentiate by entering meiotic prophase (*red*). In *Drosophila*, the zone of influence (*yellow gradient*) of the niche Cap cell extends only to the cells that it directly contacts. The dividing stem cell (*green*) gives rise to intrinsically equivalent daughter cells; however, the daughter cell (*red*) that does not contact the Cap cell is outside of the zone of influence; therefore, it adopts the differentiating fate. The two daughter cells receiving different extrinsic signals results in an asymmetric cell division

achieved on a population basis, with the proximity to the DTC niche being the primary determinant of the proliferative or meiotic fates (Fig. 4.4). While this system may appear quite different than many other systems that rely on asymmetric division to specify fate, there are common themes between these systems. The *Drosophila* female germ line is a prime example of a stem cell system that relies on asymmetric cell division (Fig. 4.4). In this system, the niche consists of a Cap cell, working with the terminal filament and escort cell. The GSC directly contacts the Cap cell, and is held in place via adherins junctions. The Cap cell utilizes the Jak/Stat pathway to signal the GSC, preventing the transcription of the *bag-of-marbles* differentiation gene. When the GSC divides, one daughter (usually the proximal) remains attached to the Cap cell, while the other daughter is positioned on the opposite side of its sister from the Cap cell. This difference in contact to the Cap cell results in an asymmetric outcome; with the sister contacting the Cap cell remaining a GSC, while the other cell enters a differentiation pathway. However, it should be emphasized, and as has been recently discussed (Losick et al. 2011), the asymmetry between sister cells does not appear to be due to intrinsic differences between the cells, but rather due to differences, based on position, in extrinsic signals. Space constraints around the Cap cell usually limit only one of the sister cells to remain attached to the Cap cell (Fig. 4.4). The *Drosophila* male germ line stem cells also utilize asymmetric division to maintain the balance between the proliferative fate and differentiation. A regulated spindle orientation ensures that one daughter cell of the dividing stem cell remains attached to the hub cells, which make up the niche, while the other daughter cell is one cell diameter removed from the hub cells. However, as with the female, the proliferating vs. differentiating fate of the cells appears to rely on differences in external signals rather than on intrinsic differences.

Comparing these *Drosophila* stem cell systems to the *C. elegans* system reveals some differences. For example, in the *C. elegans* proliferation zone the total number of stem cells appears to be larger, the stem cells do not appear to require direct contact with the niche cell to self-renew, and dividing stem cells do not result in asymmetric outcomes (Fig. 4.4). However, these differences should not distract from the overarching themes that are common to these, and perhaps most, stem cell systems. For example, that the proximity to the niche is the primary determinant of whether daughters of a dividing stem cell self-renew or differentiate. In *Drosophila*, the niche zone of influence is only one cell diameter (Fig. 4.4), with many factors, including the extracellular matrix, ensuring that its influence does not extend beyond this boundary. In *C. elegans*, the DTC does not appear to require direct contact with cells to specify their fate, although the presence of extended DTC membrane and processes is a distinguishing feature. Additionally, factors have been identified, such as the proteasome, which appear to limit the range of signaling. As a second example of similarity between these systems, both stem cell daughters are capable of self-renewal, and that it is differences in external signals that determine fate. In *Drosophila*, that both daughters are intrinsically capable of self-renewal is best exemplified by cells that are in the early stages of differentiation regaining their stem cell properties if they come in contact with the niche cells (Brawley and Matunis 2004; Sheng et al. 2009). In *C. elegans*, any given stem cell may give rise to two self-renewing cells, or two differentiating cells, depending on their proximity to the niche (Fig. 4.4).

Therefore, although differences exist between the asymmetric and symmetric stem cell systems exemplified by the *Drosophila* and *C. elegans* germ lines, respectively, both utilize the proximity of equally competent daughter cells to the niche to determine cell fate. While these systems differ in the number of stem cells maintained, the signaling pathways utilized, and the zone of influence of the niche, each system is able to maintain a balance between the proliferative fate and differentiation.

4.11 Conclusions

The *C. elegans* germ line has emerged as an important model in understanding how a balance is achieved between the proliferative fate and differentiation. The transparency of the animal has allowed the system to be visualized in living animals, which has brought significant power to genetic screens. These screens have aided in providing a molecular understanding of how the DTC signals to the proliferative zone cells and how these cells interpret this signal, as well as identified many of the factors involved in differentiation once the cell is out of the range of the signal. Characterization of the factors involved has revealed a significant level of genetic redundancy in the system. Perhaps this level of redundancy is needed to fine-tune the signal emanating from the niche in a relatively large stem cell system, where the signal is likely received over a range of cells, rather than just received by cells in direct contact with the niche cell. Additionally, as conditions encountered by *C. elegans* in the environment can drastically alter reproductive strategies (Chap. 5, Hubbard et al. 2012), the redundant mechanisms provide a diverse set of nodes by which environmental signaling can modify the proliferation vs. meiotic cell fate decision and mitotic cell cycle progression. Certainly, major advances in our understanding of this system will come once we are able to specifically identify true stem cells, putative transit-amplifying cells, and cells in meiotic S-phase. Distinguishing between these cell types will aid in determining what cells are truly committed to differentiation, and help to refine our understanding of the molecular mechanisms involved in this commitment.

Acknowledgments Research in DH's lab is funded by grants from the Natural Sciences and Engineering Research Council of Canada, and the Canadian Institutes of Health Research. TS is funded by R01 GM085150 from the National Institutes of Health. We thank Lindsay Leahul for assistance with the figures.

References

Andersson ER, Sandberg R, Lendahl U (2011) Notch signaling: simplicity in design, versatility in function. Development 138(17):3593–3612. doi:10.1242/dev.063610

Austin J, Kimble J (1987) glp-1 is required in the germ line for regulation of the decision between mitosis and meiosis in *C. elegans*. Cell 51(4):589–599

Austin J, Kimble J (1989) Transcript analysis of glp-1 and lin-12, homologous genes required for cell interactions during development of *C. elegans*. Cell 58(3):565–571

Bailly A, Gartner A (2012) Germ cell apoptosis and DNA damage responses. Advances in Experimental Medicine and Biology 757:249–276. (Chap. 9, this volume) Springer, New York

Baltus AE, Menke DB, Hu YC, Goodheart ML, Carpenter AE, de Rooij DG, Page DC (2006) In germ cells of mouse embryonic ovaries, the decision to enter meiosis precedes premeiotic DNA replication. Nat Genet 38(12):1430–1434. doi:10.1038/ng1919

Belfiore M, Pugnale P, Saudan Z, Puoti A (2004) Roles of the *C. elegans* cyclophilin-like protein MOG-6 in MEP-1 binding and germline fates. Development 131(12):2935–2945

Berry LW, Westlund B, Schedl T (1997) Germ-line tumor formation caused by activation of glp-1, a *Caenorhabditis elegans* member of the Notch family of receptors. Development 124:925

Bessler JB, Reddy KC, Hayashi M, Hodgkin J, Villeneuve AM (2007) A role for *Caenorhabditis elegans* chromatin-associated protein HIM-17 in the proliferation vs. meiotic entry decision. Genetics 175(4):2029–2037. doi:10.1534/genetics.107.070987

Bianco P (2011) Minireview: the stem cell next door: skeletal and hematopoietic stem cell "niches" in bone. Endocrinology 152(8):2957–2962. doi:10.1210/en.2011-0217

Biedermann B, Wright J, Senften M, Kalchhauser I, Sarathy G, Lee MH, Ciosk R (2009) Translational repression of cyclin E prevents precocious mitosis and embryonic gene activation during *C. elegans* meiosis. Dev Cell 17(3):355–364. doi:10.1016/j.devcel.2009.08.003

Brawley C, Matunis E (2004) Regeneration of male germline stem cells by spermatogonial dedifferentiation in vivo. Science 304(5675):1331–1334

Brodigan TM, Liu J, Park M, Kipreos ET, Krause M (2003) Cyclin E expression during development in *Caenorhabditis elegans*. Dev Biol 254(1):102–115

Buck SH, Chiu D, Saito RM (2009) The cyclin-dependent kinase inhibitors, cki-1 and cki-2, act in overlapping but distinct pathways to control cell cycle quiescence during *C. elegans* development. Cell Cycle 8(16):2613–2620

Christensen S, Kodoyianni V, Bosenberg M, Friedman L, Kimble J (1996) lag-1, a gene required for lin-12 and glp-1 signaling in *Caenorhabditis elegans*, is homologous to human CBF1 and Drosophila Su(H). Development 122(5):1373–1383

Cinquin O, Crittenden SL, Morgan DE, Kimble J (2010) Progression from a stem cell-like state to early differentiation in the *C. elegans* germ line. Proc Natl Acad Sci USA 107(5):2048–2053. doi:10.1073/pnas.0912704107

Ciosk R, DePalma M, Priess JR (2004) ATX-2, the *C. elegans* ortholog of ataxin 2, functions in translational regulation in the germline. Development 131(19):4831–4841

Ciosk R, DePalma M, Priess JR (2006) Translational regulators maintain totipotency in the *Caenorhabditis elegans* germline. Science 311(5762):851–853

Crittenden SL, Troemel ER, Evans TC, Kimble J (1994) GLP-1 is localized to the mitotic region of the *C. elegans* germ line. Development 120:2901

Crittenden SL, Bernstein DS, Bachorik JL, Thompson BE, Gallegos M (2002) A conserved RNA-binding protein controls germline stem cells in *Caenorhabditis elegans*. Nature 417:660

Crittenden SL, Eckmann CR, Wang L, Bernstein DS, Wickens M, Kimble J (2003) Regulation of the mitosis/meiosis decision in the *Caenorhabditis elegans* germline. Philos Trans R Soc Lond B Biol Sci 358(1436):1359–1362

Crittenden SL, Leonhard KA, Byrd DT, Kimble J (2006) Cellular analyses of the mitotic region in the *Caenorhabditis elegans* adult germ line. Mol Biol Cell 17(7):3051–3061

Dernburg AF, McDonald K, Moulder G, Barstead R, Dresser M, Villeneuve AM (1998) Meiotic recombination in *C. elegans* initiates by a conserved mechanism and is dispensable for homologous chromosome synapsis. Cell 94(3):387–398

Dorsett M, Westlund B, Schedl T (2009) METT-10, a putative methyltransferase, inhibits germ cell proliferative fate in *Caenorhabditis elegans*. Genetics 183(1):233–247. doi:10.1534/genetics.109.105270

Doyle TG, Wen C, Greenwald I (2000) SEL-8, a nuclear protein required for LIN-12 and GLP-1 signaling in *Caenorhabditis elegans*. Proc Natl Acad Sci USA 97(14):7877–7881

Eckmann CR, Kraemer B, Wickens M, Kimble J (2002) GLD-3, a bicaudal-C homolog that inhibits FBF to control germline sex determination in *C. elegans*. Dev Cell 3(5):697–710

Eckmann C, Crittenden SL, Suh N, Kimble J (2004) GLD-3 and control of the mitosis/meiosis decision in the germline of *Caenorhabditis elegans*. Genetics 168:147

Fitzgerald K, Greenwald I (1995) Interchangeability of *Caenorhabditis elegans* DSL proteins and intrinsic signalling activity of their extracellular domains in vivo. Development 121(12):4275–4282

Fox PM, Vought VE, Hanazawa M, Lee MH, Maine E, Schedl T (2011) Cyclin E and CDK-2 regulate proliferative cell fate and cell cycle progression in the *C. elegans* germline. Development 138:2223–2234. doi:10.1242/dev.059535

Francis R, Barton MK, Kimble J, Schedl T (1995a) gld-1, a tumor suppressor gene required for oocyte development in *Caenorhabditis elegans*. Genetics 139(2):579–606

Francis R, Maine E, Schedl T (1995b) Analysis of the multiple roles of gld-1 in germline development: interactions with the sex determination cascade and the glp-1 signaling pathway. Genetics 139(2):607–630

Fuchs E (2009) The tortoise and the hair: slow-cycling cells in the stem cell race. Cell 137(5):811–819. doi:10.1016/j.cell.2009.05.002

Fukuyama M, Rougvie AE, Rothman JH (2006) *C. elegans* DAF-18/PTEN mediates nutrient-dependent arrest of cell cycle and growth in the germline. Curr Biol 16(8):773–779. doi:10.1016/j.cub.2006.02.073

Hall DH, Winfrey VP, Blaeuer G, Hoffman LH, Furuta T, Rose KL, Hobert O, Greenstein D (1999) Ultrastructural features of the adult hermaphrodite gonad of *Caenorhabditis elegans*: relations between the germ line and soma. Dev Biol 212(1):101–123

Hansen D, Schedl T (2006) The regulatory network controlling the proliferation-meiotic entry decision in the *Caenorhabditis elegans* germ line. Curr Top Dev Biol 76:185–215

Hansen D, Hubbard EJ, Schedl T (2004a) Multi-pathway control of the proliferation versus meiotic development decision in the *Caenorhabditis elegans* germline. Dev Biol 268(2): 342–357

Hansen D, Wilson-Berry L, Dang T, Schedl T (2004b) Control of the proliferation versus meiotic development decision in the *C. elegans* germline through regulation of GLD-1 protein accumulation. Development 131(1):93–104. doi:10.1242/dev.00916

Henderson ST, Gao D, Lambie EJ, Kimble J (1994) lag-2 may encode a signaling ligand for the GLP-1 and LIN-12 receptors of *C. elegans*. Development 120(10):2913–2924

Honigberg SM, Purnapatre K (2003) Signal pathway integration in the switch from the mitotic cell cycle to meiosis in yeast. J Cell Sci 116(Pt 11):2137–2147. doi:10.1242/jcs.00460

Hubbard EJA (2007) *Caenorhabditis elegans* germ line: a model for stem cell biology. Dev Dyn 236(12):3343–3357

Hubbard EJ, Korta DZ, Dalfó D (2012) Physiological control of germline development. Advances in Experimental Medicine and Biology 757:101–131. (Chap. 5, this volume) Springer, New York

Jan E, Motzny CK, Graves LE, Goodwin EB (1999) The STAR protein, GLD-1, is a translational regulator of sexual identity in *Caenorhabditis elegans*. EMBO J 18:258

Jaramillo-Lambert A, Ellefson M, Villeneuve AM, Engebrecht J (2007) Differential timing of S phases, X chromosome replication, and meiotic prophase in the *C. elegans* germ line. Dev Biol 308(1):206–221. doi:10.1016/j.ydbio.2007.05.019

Jeong J, Verheyden JM, Kimble J (2011) Cyclin E and Cdk2 control GLD-1, the mitosis/meiosis decision, and germline stem cells in *Caenorhabditis elegans*. PLoS Genet 7(3):e1001348. doi:10.1371/journal.pgen.1001348

Jin SW, Kimble J, Ellis RE (2001) Regulation of cell fate in *Caenorhabditis elegans* by a novel cytoplasmic polyadenylation element binding protein. Dev Biol 229(2):537–553

Jones AR, Schedl T (1995) Mutations in gld-1, a female germ cell-specific tumor suppressor gene in *Caenorhabditis elegans*, affect a conserved domain also found in Src-associated protein Sam68. Genes Dev 9(12):1491–1504

Jones AR, Francis R, Schedl T (1996) GLD-1, a cytoplasmic protein essential for oocyte differentiation, shows stage- and sex-specific expression during *Caenorhabditis elegans* germline development. Dev Biol 180:165

Jungkamp AC, Stoeckius M, Mecenas D, Grun D, Mastrobuoni G, Kempa S, Rajewsky N (2011) In vivo and transcriptome-wide identification of RNA binding protein target sites. Mol Cell 44(5):828–840. doi:10.1016/j.molcel.2011.11.009

Kadyk LC, Kimble J (1998) Genetic regulation of entry into meiosis in *Caenorhabditis elegans*. Development 125(10):1803–1813

Kai T, Spradling A (2004) Differentiating germ cells can revert into functional stem cells in Drosophila melanogaster ovaries. Nature 428(6982):564–569

Kalchhauser I, Farley BM, Pauli S, Ryder SP, Ciosk R (2011) FBF represses the Cip/Kip cell-cycle inhibitor CKI-2 to promote self-renewal of germline stem cells in *C. elegans*. EMBO J. doi:10.1038/emboj.2011.263

Kerins JA, Hanazawa M, Dorsett M, Schedl T (2010) PRP-17 and the pre-mRNA splicing pathway are preferentially required for the proliferation versus meiotic development decision and germline sex determination in *Caenorhabditis elegans*. Dev Dyn 239(5):1555–1572. doi:10.1002/dvdy.22274

Kershner AM, Kimble J (2010) Genome-wide analysis of mRNA targets for *Caenorhabditis elegans* FBF, a conserved stem cell regulator. Proc Natl Acad Sci USA. doi:1000495107 [pii] 10.1073/pnas.1000495107

Killian DJ, Hubbard EJ (2004) *C. elegans* pro-1 activity is required for soma/germline interactions that influence proliferation and differentiation in the germ line. Development 131(6):1267–1278

Killian DJ, Hubbard EJ (2005) *Caenorhabditis elegans* germline patterning requires coordinated development of the somatic gonadal sheath and the germ line. Dev Biol 279(2):322–335

Kim S, Spike CA, Greenstein D (2012) Control of oocyte growth and meiotic maturation in *C. elegans*. Advances in Experimental Medicine and Biology 757:277–320. (Chap. 10, this volume). Springer, New York

Kimble J, Crittenden SL (2007) Controls of germline stem cells, entry into meiosis, and the sperm/oocyte decision in *Caenorhabditis elegans*. Annu Rev Cell Dev Biol 23:405–433

Kimble JE, White JG (1981) On the control of germ cell development in *Caenorhabditis elegans*. Dev Biol 81(2):208–219

Kraemer B, Crittenden S, Gallegos M, Moulder G, Barstead R, Kimble J, Wickens M (1999) NANOS-3 and FBF proteins physically interact to control the sperm-oocyte switch in *Caenorhabditis elegans*. Curr Biol 9(18):1009–1018

Lai EC (2002) Protein degradation: four E3s for the notch pathway. Curr Biol 12(2):R74–R78

Lambie EJ, Kimble J (1991) Two homologous regulatory genes, lin-12 and glp-1, have overlapping functions. Development 112(1):231–240

Lamont LB, Crittenden SL, Bernstein D, Wickens M, Kimble J (2004) FBF-1 and FBF-2 regulate the size of the mitotic region in the *C. elegans* germline. Dev Cell 7(5):697–707

Lee MH, Schedl T (2001) Identification of in vivo mRNA targets of GLD-1, a maxi-KH motif containing protein required for *C. elegans* germ cell development. Genes Dev 15(18):2408–2420

Lee MH, Schedl T (2004) Translation repression by GLD-1 protects its mRNA targets from nonsense-mediated mRNA decay in *C. elegans*. Genes Dev 18(9):1047–1059

Lee MH, Schedl T (2010) *C. elegans* star proteins, GLD-1 and ASD-2, regulate specific RNA targets to control development. Advances in Experimental Medicine and Biology 693:106–122

Losick VP, Morris LX, Fox DT, Spradling A (2011) Drosophila stem cell niches: a decade of discovery suggests a unified view of stem cell regulation. Dev Cell 21(1):159–171. doi:10.1016/j.devcel.2011.06.018

Lui DY, Colaiácovo MP (2012) Meiotic development in *C. elegans*. Advances in Experimental Medicine and Biology 757:133–170. (Chap. 6, this volume) Springer, New York

Luitjens C, Gallegos M, Kraemer B, Kimble J, Wickens M (2000) CPEB proteins control two key steps in spermatogenesis in *C. elegans*. Genes Dev 14(20):2596–2609

MacDonald LD, Knox A, Hansen D (2008) Proteasomal regulation of the proliferation vs. meiotic entry decision in the *C. elegans* germ line. Genetics 180(2):905–920

Maciejowski J, Ugel N, Mishra B, Isopi M, Hubbard EJ (2006) Quantitative analysis of germline mitosis in adult *C. elegans*. Dev Biol 292(1):142–151

MacQueen AJ, Villeneuve AM (2001) Nuclear reorganization and homologous chromosome pairing during meiotic prophase require *C. elegans* chk-2. Genes Dev 15(13):1674–1687

Maine EM, Hansen D, Springer D, Vought VE (2004) *Caenorhabditis elegans* atx-2 promotes germline proliferation and the oocyte fate. Genetics 168(2):817–830

Maine EM, Hauth J, Ratliff T, Vought VE, She X, Kelly WG (2005) EGO-1, a putative RNA-dependent RNA polymerase, is required for heterochromatin assembly on unpaired dna during *C. elegans* meiosis. Curr Biol 15(21):1972–1978

Mantina P, Macdonald L, Kulaga A, Zhao L, Hansen D (2009) A mutation in teg-4, which encodes a protein homologous to the SAP130 pre-mRNA splicing factor, disrupts the balance between proliferation and differentiation in the *C. elegans* germ line. Mech Dev 126(5–6):417–429. doi:S0925-4773(09)00007-0 [pii] 10.1016/j.mod.2009.01.006

Marin VA, Evans TC (2003) Translational repression of a *C. elegans* Notch mRNA by the STAR/KH domain protein GLD-1. Development 130(12):2623–2632

McCarter J, Bartlett B, Dang T, Schedl T (1999) On the control of oocyte meiotic maturation and ovulation in *Caenorhabditis elegans*. Dev Biol 205(1):111–128

McGovern M, Voutev R, Maciejowski J, Corsi AK, Hubbard EJ (2009) A "latent niche" mechanism for tumor initiation. Proc Natl Acad Sci USA 106(28):11617–11622. doi:10.1073/pnas.0903768106

Merritt C, Seydoux G (2010) The Puf RNA-binding proteins FBF-1 and FBF-2 inhibit the expression of synaptonemal complex proteins in germline stem cells. Development 137(11):1787–1798. doi:10.1242/dev.050799

Mootz D, Ho DM, Hunter CP (2004) The STAR/Maxi-KH domain protein GLD-1 mediates a developmental switch in the translational control of *C. elegans* PAL-1. Development 131(14):3263–3272. doi:10.1242/dev.01196

Mumm JS, Kopan R (2000) Notch signaling: from the outside in. Dev Biol 228(2):151–165

Nadarajan S, Govindan JA, McGovern M, Hubbard EJ, Greenstein D (2009) MSP and GLP-1/Notch signaling coordinately regulate actomyosin-dependent cytoplasmic streaming and oocyte growth in *C. elegans*. Development 136(13):2223–2234. doi:10.1242/dev.034603

Narbonne P, Roy R (2006) Inhibition of germline proliferation during *C. elegans* dauer development requires PTEN, LKB1 and AMPK signalling. Development 133(4):611–619. doi:10.1242/dev.02232

Nousch M, Eckmann CR (2012) Translational control in the *C. elegans* germ line. Advances in Experimental Medicine and Biology 757:205–247. (Chap. 8, this volume) Springer, New York

Pasierbek P, Jantsch M, Melcher M, Schleiffer A, Schweizer D, Loidl J (2001) A *Caenorhabditis elegans* cohesion protein with functions in meiotic chromosome pairing and disjunction. Genes Dev 15(11):1349–1360

Pepper AS, Lo TW, Killian DJ, Hall DH, Hubbard EJ (2003) The establishment of *Caenorhabditis elegans* germline pattern is controlled by overlapping proximal and distal somatic gonad signals. Dev Biol 259(2):336–350

Petcherski AG, Kimble J (2000) LAG-3 is a putative transcriptional activator in the *C. elegans* Notch pathway. Nature 405(6784):364–368

Puoti A, Kimble J (1999) The *Caenorhabditis elegans* sex determination gene mog-1 encodes a member of the DEAH-Box protein family. Mol Cell Biol 19(3):2189–2197

Puoti A, Kimble J (2000) The hermaphrodite sperm/oocyte switch requires the *Caenorhabditis elegans* homologs of PRP2 and PRP22. Proc Natl Acad Sci USA 97(7):3276–3281

Qiao L, Lissemore JL, Shu P, Smardon A, Gelber MB, Maine EM (1995) Enhancers of glp-1, a gene required for cell-signaling in *Caenorhabditis elegans*, define a set of genes required for germline development. Genetics 141(2):551–569

Reddy KC, Villeneuve AM (2004) *C. elegans* HIM-17 links chromatin modification and competence for initiation of meiotic recombination. Cell 118(4):439–452. doi:10.1016/j.cell.2004.07.026

Sanpei K, Takano H, Igarashi S, Sato T, Oyake M, Sasaki H, Wakisaka A, Tashiro K, Ishida Y, Ikeuchi T, Koide R, Saito M, Sato A, Tanaka T, Hanyu S, Takiyama Y, Nishizawa M, Shimizu N, Nomura Y, Segawa M, Iwabuchi K, Eguchi I, Tanaka H, Takahashi H, Tsuji S (1996) Identification of the spinocerebellar ataxia type 2 gene using a direct identification of repeat expansion and cloning technique, DIRECT. Nat Genet 14(3):277–284. doi:10.1038/ng1196-277

Seydoux G, Schedl T (2001) The germline in *C. elegans*: origins, proliferation, and silencing. Int Rev Cytol 203:139–185

Seydoux G, Schedl T, Greenwald I (1990) Cell-cell interactions prevent a potential inductive interaction between soma and germline in *C. elegans*. Cell 61(6):939–951

Sheng XR, Brawley CM, Matunis EL (2009) Dedifferentiating spermatogonia outcompete somatic stem cells for niche occupancy in the Drosophila testis. Cell Stem Cell 5(2):191–203. doi:10.1016/j.stem.2009.05.024

Smardon A, Spoerke JM, Stacey SC, Klein ME, Mackin N, Maine EM (2000) EGO-1 is related to RNA-directed RNA polymerase and functions in germ-line development and RNA interference in *C. elegans*. Curr Biol 10(4):169–178

Subramaniam K, Seydoux G (2003) Dedifferentiation of primary spermatocytes into germ cell tumors in *C. elegans* lacking the pumilio-like protein PUF-8. Curr Biol 13(2):134–139

Suh N, Jedamzik B, Eckmann CR, Wickens M, Kimble J (2006) The GLD-2 poly(A) polymerase activates gld-1 mRNA in the *Caenorhabditis elegans* germ line. Proc Natl Acad Sci USA 103(41):15108–15112

Suh N, Crittenden SL, Goldstrohm A, Hook B, Thompson B, Wickens M, Kimble J (2009) FBF and its dual control of gld-1 expression in the *Caenorhabditis elegans* germline. Genetics 181(4):1249–1260. doi:10.1534/genetics.108.099440

Tax FE, Yeargers JJ, Thomas JH (1994) Sequence of *C. elegans* lag-2 reveals a cell-signalling domain shared with Delta and Serrate of Drosophila. Nature 368(6467):150–154

Thompson BE, Bernstein DS, Bachorik JL, Petcherski AG, Wickens M, Kimble J (2005) Dose-dependent control of proliferation and sperm specification by FOG-1/CPEB. Development 132(15):3471–3481

Tursun B, Patel T, Kratsios P, Hobert O (2011) Direct conversion of *C. elegans* germ cells into specific neuron types. Science 331(6015):304–308. doi:10.1126/science.1199082

Vought VE, Ohmachi M, Lee MH, Maine EM (2005) EGO-1, a putative RNA-directed RNA polymerase, promotes germline proliferation in parallel with GLP-1/notch signaling and regulates the spatial organization of nuclear pore complexes and germline P granules in *Caenorhabditis elegans*. Genetics 170(3):1121–1132. doi:10.1534/genetics.105.042135

Voutev R, Killian DJ, Ahn JH, Hubbard EJ (2006) Alterations in ribosome biogenesis cause specific defects in *C. elegans* hermaphrodite gonadogenesis. Dev Biol 298(1):45–58

Wang L, Eckmann CR, Kadyk LC, Wickens M, Kimble J (2002) A regulatory cytoplasmic poly(A) polymerase in *Caenorhabditis elegans*. Nature 419(6904):312–316

Wang C, Wilson-Berry L, Schedl T, Hansen D (2012) TEG-1 CD2BP2 regulates stem cell proliferation and sex determination in the *C. elegans* germ line and physically interacts with the UAF-1 U2AF65 splicing factor. Dev Dynamics 241(3):505–521. doi:10.1002/dvdy.23735

Wright JE, Gaidatzis D, Senften M, Farley BM, Westhof E, Ryder SP, Ciosk R (2011) A quantitative RNA code for mRNA target selection by the germline fate determinant GLD-1. EMBO J 30(3):533–545. doi:10.1038/emboj.2010.334

Yochem J, Greenwald I (1989) glp-1 and lin-12, genes implicated in distinct cell-cell interactions in *C. elegans*, encode similar transmembrane proteins. Cell 58(3):553–563

Zanetti S, Meola M, Bochud A, Puoti A (2011) Role of the *C. elegans* U2 snRNP protein MOG-2 in sex determination, meiosis, and splice site selection. Dev Biol 354(2):232–241. doi:10.1016/j.ydbio.2011.04.001

Zetka MC, Kawasaki I, Strome S, Muller F (1999) Synapsis and chiasma formation in *Caenorhabditis elegans* require HIM-3, a meiotic chromosome core component that functions in chromosome segregation. Genes Dev 13(17):2258–2270

Zhang B, Gallegos M, Puoti A, Durkin E, Fields S, Kimble J, Wickens MP (1997) A conserved RNA-binding protein that regulates sexual fates in the *C. elegans* hermaphrodite germ line. Nature 390(6659):477–484

Chapter 5
Physiological Control of Germline Development

E. Jane Albert Hubbard, Dorota Z. Korta, and Diana Dalfó

Abstract The intersection between developmental programs and environmental conditions that alter physiology is a growing area of research interest. The *C. elegans* germ line is emerging as a particularly sensitive and powerful model for these studies. The germ line is subject to environmentally regulated diapause points that allow worms to withstand harsh conditions both prior to and after reproduction commences. It also responds to more subtle changes in physiological conditions. Recent studies demonstrate that different aspects of germ line development are sensitive to environmental and physiological changes and that conserved signaling pathways such as the AMPK, Insulin/IGF, TGFβ, and TOR-S6K, and nuclear hormone receptor pathways mediate this sensitivity. Some of these pathways genetically interact with but appear distinct from previously characterized mechanisms of germline cell fate control such as Notch signaling. Here, we review several aspects of hermaphrodite germline development in the context of "feasting," "food-limited," and "fasting" conditions. We also consider connections between lifespan, metabolism and the germ line, and we comment on special considerations for examining germline development under altered environmental and physiological conditions. Finally, we summarize the major outstanding questions in the field.

Keywords Stem cell • *C. elegans* • Nutrient • Diapause • Lifespan • TOR • TGFβ • Insulin • Notch • AMPK

E.J.A. Hubbard (✉) • D.Z. Korta • D. Dalfó
Skirball Institute of Biomolecular Medicine, Department of Pathology,
and Kimmel Center for Stem Cell Biology, New York University School of Medicine,
540 First Avenue, New York, NY 10016, USA
e-mail: jane.hubbard@med.nyu.edu

T. Schedl (ed.), *Germ Cell Development in C. elegans*, Advances in Experimental Medicine and Biology 757, DOI 10.1007/978-1-4614-4015-4_5,

5.1 Introduction

Natural selection favors species that evolve mechanisms to reproduce within the limits of available resources. It is not surprising, therefore, that reproduction, including the development and function of the germ line, might be intimately linked to aspects of physiology that are influenced by environmental conditions. Successful species might also optimize reproduction when environmental resources are abundant and conserve reproductive energy expenditure when resources are scarce.

The nematode *C. elegans* likely represents an extreme case with respect to a boom-and-bust economy of resources and the evolution of developmental and reproductive plasticity to cope with these ecological challenges. One of the most striking examples of this plasticity is the multiple stages at which worms can undergo a reversible developmental arrest or diapause, allowing for dispersal while delaying or suspending reproduction. Prior to a commitment to reproduction, worms can enter a diapause at least two times during development. If conditions are extremely poor, they arrest in the first larval stage (L1) just after hatching (L1 diapause) (Johnson et al. 1984). If they pass this point, after an assessment of poor conditions in the late L1, they can enter an alternate L2 stage (L2d) that in the face of continued deprivation and high population density leads to dauer formation. Dauer (German for "enduring") larvae are characterized by changes in the cuticle, pharynx, mouth, metabolism, and behavior that facilitate dispersal while delaying reproduction and aging (see Fielenbach and Antebi 2008; Bargmann 2006; Hu 2007, for reviews). Even after the commitment to and commencement of gametogenesis, starved animals can survive in a reversible "adult reproductive diapause" (ARD) state in which their germ line is severely diminished but remains poised to regenerate once conditions improve (Angelo and Van Gilst 2009; Seidel and Kimble 2011). Finally, as adults, reduced food abundance has several effects on reproduction. Not only has *C. elegans* evolved developmental and reproductive strategies to deal with boom-or-bust resource allocation, but recent findings suggest that animals also respond to intermediate conditions, tailoring the rate and progress of germline development to environmental and metabolic cues.

From the early days, the *C. elegans* field has considered the effects of unfavorable conditions (such as starvation, crowding, elevated temperature) versus favorable conditions on dauer formation (Cassada and Russell 1975; Klass and Hirsh 1976) and on lifespan (Klass 1977). However, the effects of physiological changes on reproduction in general and on different stages of germline development in particular have gained increasing attention only recently. Here we focus on the impact of environmental resources on the development of the germ line during larval stages and on the sustained "development" that is required for continuous gametogenesis in adulthood. For the purpose of comparing and contrasting the effects of different resource scenarios on the germ line, we refer to "feasting" boom conditions (abundant food, low population density), "food-limited" intermediate conditions of food availability, and "fasting" bust/ starvation conditions.

C. elegans is a very powerful system to study the intersection between genetically encoded and environmentally regulated effects on germline development. Historically, these influences on germline development have been studied in relative isolation. Initially *C. elegans* reproduction, nutrition and ageing were examined simultaneously (Klass 1977). However, the study of germline development as a field developed separately from fields focused on sensory perception, feeding behavior, metabolism, and aging. Recently, these areas have converged. Because so much of *C. elegans* biology is conserved across evolution, this convergence offers a remarkable opportunity to investigate the mechanisms of physiological impact on the germ line in a whole-organism context.

Several themes are emerging that will likely engage the research community for many years to come. First, the germ line is highly sensitive: "feasting" worms quickly produce full self-progeny broods, while "food-limited" and "fasting" worms modulate reproduction to different degrees. This sensitivity likely reflects *C. elegans* evolutionary history and the fact that reproduction is not all-or-nothing. As a result, physiology can impact several points of germline development and function in different ways. Subtle changes in food abundance or quality, for example, provoke a different response from all-out starvation. Second, the timing of physiological changes alters the reproductive response. In particular, mechanisms that enable worms to withstand poor conditions prior to the commitment to reproduction differ from those that are engaged after the commitment to reproduction. Third, cellular mechanisms vary depending on the specific changes and on the responding tissues. Finally, contrary to the view that the modulation of development and reproduction by environmental factors such as food availability is just a matter of sufficient material for cell division, growth and reproduction, there appear to be specific signaling pathways through which reproduction can be influenced by environmental conditions that alter physiology. These highly conserved signaling pathways, some of which have been studied in other phenotypic contexts in *C. elegans*, are important for similar responses in other organisms as well.

Here, we focus on aspects of hermaphrodite germline development that are altered by changes in food or feeding (feasting, food-limited and fasting) from hatching through larval life and adulthood. We also consider connections between lifespan and germline development. Finally, we comment on special considerations for examining germline development and on the major outstanding questions in the field. We refer the reader to Kim et al. (2012) (Chap. 10) for the influence of sperm on the germ line. Due to the broad scope of this subject we do not cover additional environmental influences such as physiological stress or pathogenesis, but we refer the reader to recent reviews (see, for example, Gartner et al. 2008; Darby 2005; Braeckman et al. 2009).

5.2 Feasting

For *C. elegans*, the true "feast" is likely an apple in an orchard in late autumn littered with dense patches of rotting fruit and a high bacterial load (Félix and Braendle 2010; Kiontke and Sudhaus 2006). The key feature of the feast experience is that good food is continuously present in excess given the population density of the worms. These conditions allow worms to avoid diapause and continue rapidly through four larval stages to the reproductive adult. Laboratory conditions under which three worms are typically transferred to a fresh bacterial lawn before the previous lawn is exhausted also represent feast-like conditions, even if laboratory strains of OP50 bacteria (at concentrations of ~10^9 or greater colony forming units per ml) are not necessarily the most preferred food source (Shtonda and Avery 2006; Kiontke and Sudhaus 2006). Thus, much of our current knowledge of germline development comes from analyses performed under what will be described here as "feasting" conditions.

5.2.1 Larval Germline Development While Feasting

Under feast conditions, worms hatch from their eggshell as L1 larvae and immediately feed. At hatching, the two somatic gonad precursor cells Z1 and Z4 flank the two germline precursors Z2 and Z3. These four cells are encased in a basement membrane and remain quiescent until midway through the first larval stage (L1) when, provided the worm feeds, they resume mitotic cell division cycles and continue development. Under these conditions, both germ cells and somatic gonad cells divide in the L1 with the latter cells arresting after the generation of ten cells, until the L3 stage when they resume divisions (Hirsh et al. 1976; Kimble and Hirsh 1979).

During the late L2 stage, the somatic gonad cells reorganize, forming a central primordium flanked by two gonad "arms," each of which is each capped by a single distal tip cell (DTC). The DTCs guide gonad migration and express ligands that activate the germline-expressed GLP-1/Notch receptor (Kimble and Hirsh 1979; Kimble and White 1981; Austin and Kimble 1987, see also Hansen and Schedl 2012, Chap. 4). Notch pathway activity maintains the distal germ cells in a proliferative (versus differentiated) fate. Complete removal of GLP-1/Notch signaling at any stage causes all germ cells to differentiate (Austin and Kimble 1987), while reducing GLP-1/Notch signaling changes the relative balance of proliferative and differentiated germ cells without reducing the cell cycle rate of the remaining proliferative cells (Michaelson et al. 2010). Hyperactive Notch signaling prevents specification of the differentiated fate, causes continuous specification of the proliferative fate, and can elevate cell cycle (Berry et al. 1997; Maciejowski et al. 2006). Additional signals promote robust larval proliferation to expand the larval germline progenitor pool (Korta and Hubbard 2010, and references therein).

Several critical events in hermaphrodite germline development take place in the L3 stage. First, two separate mechanisms move the anterior and posterior distal tip cells toward the head and tail of the animal, respectively. These mechanisms are (1) a centrifugal DTC migration program (dependent on *gon-1* and *mig-24/hlh-12*;

Tamai and Nishiwaki 2007; Blelloch et al. 1999) and (2) robust germline proliferation (McGovern et al. 2009). This latter mechanism is impaired in sub-optimal environmental conditions, as will be discussed below. In feasting conditions with proper DTC migration and robust larval germline proliferation, the proximal germ cells in the mid-L3 stage take on a nuclear morphology and expression of markers characteristic of early meiotic prophase when they reach a distance of 13 cell diameters from the distal end (Hansen et al. 2004). Once this initial differentiation event occurs, the proliferative zone of the germ line is demarcated as the area distal to the mitosis/meiosis border. Proximal to this border, all germ cells are differentiated (in prophase of meiosis I and undergoing gametogenesis). Importantly, while the worms are in feasting conditions, the proliferative zone continues to expand rapidly during the L3 and L4 stages, reaching a total of ~200 cells per gonad arm and stretching over ~20 cell diameters by the late L4 stage (Hansen et al. 2004; Killian and Hubbard 2005).

If the germline stem/progenitor pool does not rapidly expand in the L3 and L4 stages, the number of cells in the adult germline proliferative zone is reduced, affecting fecundity. Recent studies indicate that many factors—genetic, anatomical, and physiological—can influence this vital expansion of the larval proliferative zone. For example, mutations in *pro-1* (Killian and Hubbard 2004), ablation of the distal-most pair of sheath cells (Killian and Hubbard 2005), or reduced food abundance (Korta et al. 2012) all interfere with the accumulation of larval proliferative germ cells. This physiologically sensitive progenitor cell accumulation thus serves as a good model for understanding developmental and physiological control of cell proliferation. Using somewhat counterintuitive genetic screening assays, genes were identified that are required for expansion of the larval proliferative germ cell population under well-fed conditions (Michaelson et al. 2010; Hubbard 2011, Dalfó and Hubbard, unpublished). These screens took advantage of an inappropriate cell-cell interaction that occurs when larval germ cells maintain an undifferentiated state but do not proliferate properly (see Hubbard 2011, for further explanation). These and related approaches identified genes required for robust larval germline proliferation—that is, key genes that mediate developmental or physiological responses—while avoiding genes required for more general aspects of cell proliferation. Among these, the effects of the insulin/IGF pathway on this accumulation will be discussed here in the context of "feasting," since the sensitivity of this pathway in the germ line to dietary changes not been documented directly. Two other pathways, TGFβ and TOR, will be discussed below under "food-limited" since they have been associated with responses to food limitation.

5.2.2 The Insulin/IGF-Like Signaling Pathway Promotes the Larval Germline Cell Division Cycle

The many roles of the sole insulin/IGF-like receptor (IIR) DAF-2 in *C. elegans* have been reviewed elsewhere (Kenyon 2010; Kleemann and Murphy 2009; Landis and Murphy 2010; Fielenbach and Antebi 2008; Taguchi and White 2008; Hubbard 2011).

Briefly, the *daf-2* gene was initially discovered and named for its role in the dauer decision. Certain reduction-of-function *daf-2* mutations cause worms to enter the dauer pathway constitutively, regardless of the environmental conditions, suggesting that the activity of this pathway signals an environment sufficient to support reproductive development. Insulin/IGF-like signaling (IIS) modulates metabolism, response to toxins, hypoxia, immunity, post-reproductive lifespan, and reproductive timing. IIS acts through a highly conserved PI3K-dependent pathway and, in most contexts, negatively regulates the activity of the FOXO transcription factor DAF-16. Two exceptions are the control of germline cell cycle arrest in both the L1 and in dauer diapause that depend on *daf-18*/PTEN but not *daf-16*/FOXO (see Sect. 5.4, below).

The accumulation of proliferative germ cells during the L3 and L4 stages under "feast" conditions is sensitive to *daf-2*/IIR activity in a manner dependent on *daf-18*/PTEN and *daf-16*/FOXO (Michaelson et al. 2010). Of the 40 putative insulin-like ligand genes in the *C. elegans* genome, two were identified for their marked role in larval germline expansion: *ins-3* and *ins-33*. Additional ligands are likely also involved, since the *ins-3* and *ins-33* double mutant phenotype is not as strong as that of the *daf-2* mutant. Nevertheless, the activities of these ligands display several interesting characteristics. First, although the germline proliferation defect resulting from reducing *ins-3* or *ins-33* activity can be completely suppressed by loss of *daf-16*/FOXO, reduction of these ligand-encoding genes does not cause nuclear localization of intestinal DAF-16::GFP, nor does it activate target gene expression in the intestine. In other words, a reduction of *daf-2*, *ins-3*, or *ins-33* exerts similar phenotypic effects on the germ line, but different effects on the intestine. Somehow, the animal distinguishes between activities of these ligands on different target tissues. Second, reporters for *ins-3* and *ins-33* are expressed in head neurons, and head and uterine cells, respectively, and they are required in the soma for their role in larval germline cell cycle progression. Third, the reduction of activity of each ligand gene alone yields the same phenotype as reducing both together. This intriguing result is hard to reconcile with a simple ligand–receptor interaction model, and future studies will be required to determine precisely how these ligands are working. One hypothesis is that a physical interaction occurs between them; an alternative hypothesis is that a ligand–receptor relay exists, possibly including an alternate receptor.

One model to explain the observation that *daf-2* mutants fail to properly expand the larval germ line is that they retain dauer-like characteristics that prevent germline proliferation (see below "food-limited"). However, the data do not support this hypothesis and are consistent with a general model in which the DAF-2/IIR signaling cascade is reused after the window of opportunity for the dauer decision has passed (Michaelson et al. 2010). This conclusion is based on several observations. First, in the role of DAF-2 signaling in expanding the larval germ line, *daf-16* activity is required primarily within the germ line and expression of *daf-16* in neurons (a predominant site of action for the dauer role) has no effect. Second, temperature-shift and timed RNAi experiments demonstrated a post-dauer-decision period of sensitivity. Third, reducing *ins-3* and *ins-33* activity does not induce dauer formation, even at high temperatures, nor does it extend the time of reproduction (the time over which progeny are produced in adulthood) (Michaelson et al. 2010), as is observed upon reduction in *daf-2* activity (Dillin et al. 2002).

While this influence of IIS on larval germline progenitor cell accumulation has yet to be directly linked to environmental changes, the expression of *ins-3* in neurons suggests a possible physiological connection similar to the well-documented role for insulin signaling in *Drosophila* to promote germline cell cycle in response to diet (Drummond-Barbosa and Spradling 2001; LaFever and Drummond-Barbosa 2005; Hsu et al. 2008). This is an area for future investigation. In addition, it will be of interest to identify the targets for DAF-16 activity in the germ line that are important for promoting robust larval germline proliferation and to determine how they influence the cell cycle.

5.2.3 *The Adult Germ Line After a Life of Feasting*

In feasting conditions, before the adult molt, spermatogenesis ends and subsequent gametes that form are oocytes. In adult hermaphrodites, oocytes mature and are ovulated approximately every 23 min per gonad arm in an assembly-line fashion (McCarter et al. 1999, see also Kim et al. 2012, Chap. 10). In the spermatheca, they are fertilized either by stored self-sperm or by sperm introduced from a male by copulation. Embryos develop in the uterus to the ~30-cell stage when eggs are laid through the vulva out onto the substrate (Fig. 5.1; also see Pazdernik and Schedl 2012, Chap. 1).

This pace of reproduction in feasting conditions requires tremendous resources to generate the estimated 200-fold volumetric increase from germ cell to oocyte at a pace that produces 3–6 embryos per hour (Hirsh et al. 1976; McCarter et al. 1999). Germ cells retain an opening to a cytoplasmic core prior to spermatogensis or to late stages of oocyte development (Hirsh et al. 1976). Therefore many cells contribute to material that flows through this core and is loaded into oocytes (Wolke et al. 2007). Germ cell nucleoli are relatively large and germ cytoplasm is densely packed with ribosomes, suggesting high translational capacity (Hirsh et al. 1976). In addition, raw material for oocyte growth likely comes from the contents of female germ cells that undergo programmed cell death during the pachytene of prophase of meiosis I. Estimates of as few as 2 (Gumienny et al. 1999) and as many as 30 (Jaramillo-Lambert et al. 2007) germ cells may die per oocyte generated. Interestingly, the rate of programmed cell death appears to include a physiological component since stressors can elevate it (Gartner et al. 2008). In addition, energy must be directed to the synthesis of lipid-rich yolk and other lipids that are transferred from the intestine to oocytes (Kimble and Sharrock 1983; Grant and Hirsh 1999).

5.3 Food-Limited Conditions

There are myriad ways that food limitation can be imposed on worms either in the wild or in the laboratory. Laboratory methods vary from generally reducing food abundance, altering the overall quality of food, or restricting food intake, to changing specific diet components or reducing uptake of specific nutrients, altering key

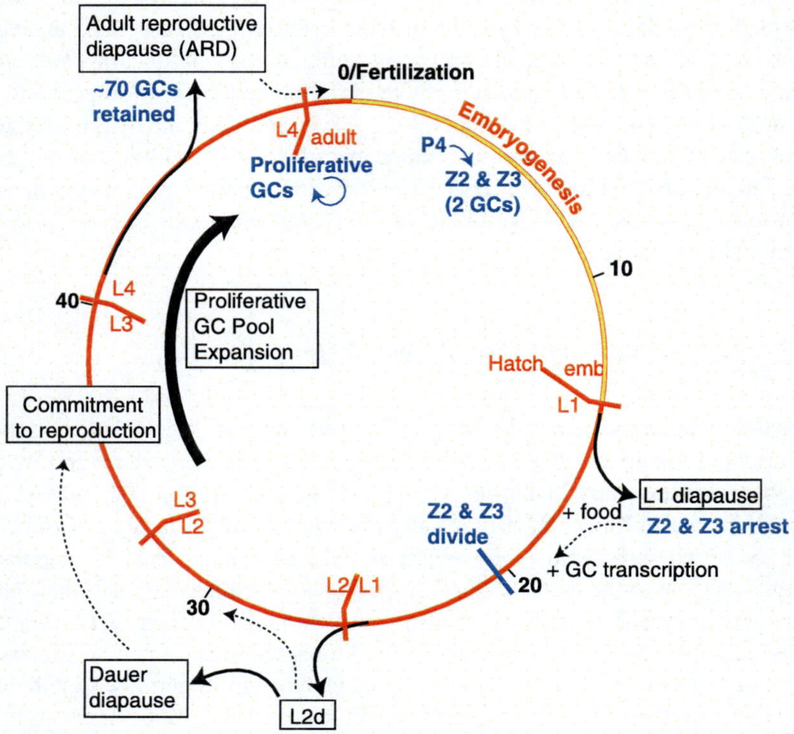

Fig. 5.1 The *C. elegans* lifecycle, diapause, and germline development. Figure adapted from (Wood et al. 1980) and modified from (Korta and Hubbard 2010). Time (hours at 25°), hatching and molts are indicated. Physiologically sensitive diapause and larval germline expansion are indicated. *Circular arrow* indicates renewal potential of adult germ cells

metabolic enzyme activities, or genetic changes within specific cell types such that they can no longer respond to nutrients or nutrient signaling (see below). In addition to well-characterized environmental sensory triggers for dauer diapause that couple physiology to a developmental program, many different adult laboratory restrictions and feeding regimes have been described in the context of studies on lifespan extension due to "dietary restriction" or "caloric restriction" (see below). Generally, these restrictions are defined as reducing caloric intake without compromising nutrition. For the purposes of this discussion, the more general "food-limited" term is used to encompass a broad range of conditions that reduce food availability (including regimes that extend lifespan) and that lie as intermediate between starvation and feasting.

For details of intermediary metabolism in the worm, see (Braeckman et al. 2009). One of the interesting results of metabolic studies is that unlike changes that occur during dauer, restricting food does not appear to cause a general slowing of metabolism but rather changes flux through different metabolic pathways (Houthoofd et al. 2002b; Houthoofd and Vanfleteren 2006, and references therein). Future studies are required to link these changes to specific germline responses.

5.3.1 Larval Germline Development Under Conditions of Food Limitation

Worms possess robust and complex sensory and metabolic systems to monitor their environment. That these systems may be deployed in circumstances other than the few known diapause stages is a relatively new and interesting general area of investigation (see Edison 2009; Macosko et al. 2009). Germline development appears to be particularly sensitive to environmental changes. The potential for gamete production extends far into the lifetime of the animal (Mendenhall et al. 2011; Angelo and Van Gilst 2009), and, though energetically expensive, it is vital for reproductive success. Therefore, it “makes sense,” anthropomorphically speaking, for the germ line to be both sensitive and responsive to environmental changes to balance reproduction with available resources. Altering the sensory environment and/or metabolic status of the animal can influence the germ line at several points during larval germline development, prior to commitment to gametogenesis.

5.3.1.1 Dauer, a Special Case

If worms hatch into suboptimal (but not total starvation) conditions, development of the first larval stage begins. Late in the L1 stage, however, reassessment of environmental conditions determines whether animals proceed immediately into a reproductive mode or enter the nonfeeding dauer diapause stage. Here, we consider dauer as a special case since compared with other diapause states, it involves radical changes, both morphological and physiological, that only occur in response to prolonged poor conditions. While a detailed discussion of dauer regulation is beyond the scope of this review (the reader is directed to Fielenbach and Antebi 2008; Bargmann 2006; Hu 2007), a few salient features are mentioned here that are relevant to germline development in dauer and to the reuse of dauer-decision pathways in post-dauer-decision stages of germline development.

The combined assessment of cues reporting ambient worm population density, food, and temperature (Golden and Riddle 1984; Bargmann and Horvitz 1991) determines whether an animal will enter dauer or will proceed to the L3 and L4 stages toward becoming reproductive adults. The dauer larva is a remarkable adaptation: during dauer entry, animals secrete a desiccation-resistant cuticle and store fat to sustain them as nonfeeding dauer larvae which exhibit motile behaviors to facilitate dispersal. Dauer larvae can thereby suspend development and extend lifespan for months. When conditions improve, they undergo a special molt and reenter a reproductive developmental pathway.

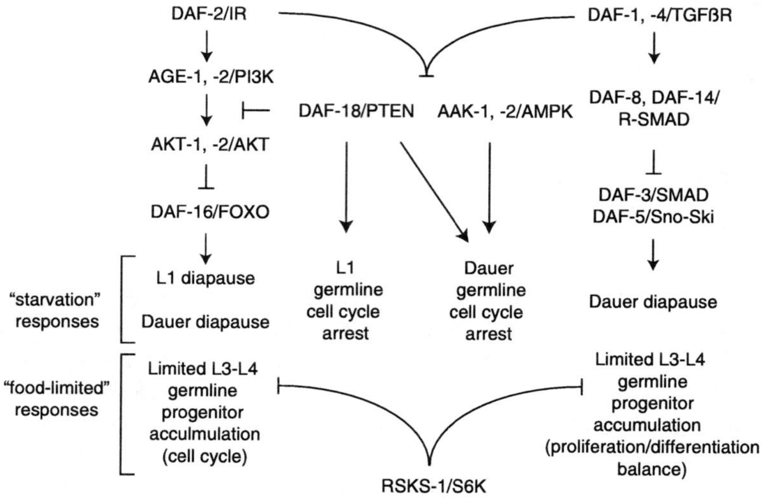

Fig. 5.2 Signaling for developmental and germline response to poor conditions by PI3K, TGFβ, AMPK, and S6K pathways. See text for details and references

"Dauer pheromone," a main cue for population density, was identified ~30 years ago as a crude preparation that could induce dauer entry and inhibit dauer exit (Golden and Riddle 1982). Worms are now known to secrete at least ten different ascarosides (derivatives of the dideoxy sugar ascarylose) (Edison 2009) many of which are active in nondauer contexts such as mating or social behavior (Jeong et al. 2005; Srinivasan et al. 2008; Butcher et al. 2007, 2008, 2009; Pungaliya et al. 2009; Macosko et al. 2009). The precise identities and concentrations of ascarosides in the immediate environment likely convey a tremendous amount of information since ascarosides are often most potent in mixtures (Srinivasan et al. 2008; Butcher et al. 2007, 2008). Moreover, in addition to different baseline potencies observed in different assays, distinct combinations can be more potent in one assay over another. Ascarosides may also provide information about the composition and condition of the population since the ascarosides secreted by worms differ with stage and culture conditions (Kaplan et al. 2011). Interestingly, some ascarosides (ascr#1/C7 and ascr#3/C9) peak in L3-L4 stages (Kaplan et al. 2011), a time when the germline proliferative zone is expanding. Because ascaroside abundance is altered when worms are partially starved, "food-limited" conditions likely produce both sensory and metabolic responses.

The decision to enter and exit dauer involves no fewer than four highly conserved signaling pathways, including GPCR/cGMP signaling thought to act upstream of parallel Insulin/IGF and TGFβ pathways (Fig. 5.2), both of which converge on the activity of a nuclear hormone receptor (related to LXR and Vitamin D receptor) DAF-12 (see Fielenbach and Antebi 2008; Hu 2007).

The relationship between the dauer decision, specific neurons and sensory-responsive signaling is well understood for the paired sensory ASI neurons and TGFβ. Two largely nonoverlapping TGFβ pathways in *C. elegans* correspond to TGFβ-like and BMP-like signaling and affect dauer and body size, respectively. The DAF-7/TGFβ pathway regulates dauer formation (Savage-Dunn 2005). ASI-specific expression of a *daf-7*/TGFβ reporter is elevated under nondauer or dauer exit conditions (Schackwitz et al. 1996; Ren et al. 1996) in response to DAF-11/PKG signaling (Murakami et al. 2001). Recently, two ASI-specific, redundantly acting, GPCRs were identified as specific receptors for one of three potent dauer-inducing ascarosides (McGrath et al. 2011).

Importantly, during the protracted dauer entry process, metabolism shifts to storage in preparation for starvation which occurs during the nonfeeding dauer stage. In contrast to more sudden starvation responses, in anticipation of initiating the dauer pathway, the pre-dauer second larval stage is extended (L2d), allowing the animal to prepare for future nutrient deprivation by slowing development and metabolism, while storing energy (see Riddle and Albert 1997, and references therein). Dauer-stage animals display metabolic (Wadsworth and Riddle 1989; Burnell et al. 2005) and gene-expression changes (Jones et al. 2001; Holt and Riddle 2003; Wang and Kim 2003) that favor utilization of stored fats, suggesting that fat reservoirs function as a nutrient source for the dauer stage. These alterations (see Braeckman et al. 2009, for additional primary references) are primarily a shift away from the direct use of biosynthetic precursors such as amino acids and nucleic acids obtained from dietary sources, to the generation of these precursors from fat (via fatty acid β oxidation, gluconeogenesis, and the glyoxylate cycle). Energy (ATP) generation from acetyl CoA, via the citric acid cycle and oxidative phosphorylation is also altered as acetyl CoA production is shifted away from glycolysis with more reliance on fatty acid β oxidation. Moreover, the proper rationing and utilization of stored material is important for dauer survival. Dauer survival is greatly reduced in the absence of AMPK signaling, a defect that has been linked to abnormally rapid use of stored lipid reserves in the hypodermis (Narbonne and Roy 2009).

5.3.1.2 The Germ Line During Dauer

Dauer formation leads to a progressive establishment of cell cycle arrest, including the germ line. Dauer precedes substantial growth of the germline stem cell pool, and dauer larvae ultimately arrest germline proliferation, likely in the G2 stage (Narbonne and Roy 2006). Importantly, since the meiotic fate is not yet specified prior to or during dauer, elevated Notch activity, which causes a failure in specification of the differentiated germ cell fate, would not be expected to drive hyperproliferation during dauer. Interestingly, like the inhibition of germline proliferation in the L1 diapause (see below) the inhibition of germline proliferation in dauer can be genetically separated from the general somatic cell cycle inhibition. To identify the genetic mechanism by which the germ line limits its proliferative capacity during dauer, Narbonne and Roy (2006) screened for mutations that would permit germline proliferation during *daf-2*-mutation-induced dauer conditions. Interestingly, they found that germline proliferation is kept in check by the activity of the AMPK orthologs *aak-1* and *aak-2* (Narbonne and Roy 2006). AMPK activity is exquisitely sensitive to levels and ratios of ATP, ADP, and AMP, being most active under elevated AMP conditions that signal energy deficit (Hardie 2011). Therefore, from a purely energetic standpoint, one interpretation of these results is that if AMPK levels are low, regardless of the actual cellular levels of AMP and ATP, the germ cells interpret the energy climate as "good" and will proceed to inappropriately proliferate the germ line. These results might also suggest that the ATP/AMP ratio (as opposed to other metabolic signals) may be paramount for the control of germline proliferation in response to dauer-promoting signals. Alternatively, the phenotype may be related to the role of AMPK activity in lipid rationing (Narbonne and Roy 2009). For the dauer-induced germline proliferation arrest, AMPK and its upstream regulator PAR-4/LKB1 coordinate proliferation with somatic development during dauer, in response to input from both the TGFβ and insulin-like pathways (Fig. 5.2).

In terms of the metabolic and germline responses to the environment, dauer differs from "food-limited" conditions defined by a chronic low-food condition or a qualitative change in food intake (see below). The differences likely reflect the special preparation for dauer and its timing prior to reproductive commitment. How these different sensory, physiological, and metabolic conditions and the specific dynamics of their onset affect germline development remains to be determined.

5.3.2 *Sensory and Nutritional Control of Post-Dauer-Decision Germline Expansion*

Once animals bypass the developmental window during which they might enter dauer, they are committed to "reproductive" development. Under replete conditions, the timing of this commitment is accompanied by rapid accumulation of proliferative germ cells in the L3 and L4 stages that eventually constitute the adult germline stem cell pool. This accumulation requires the *daf-2*/Insulin-IGF-like (IIS) pathway, as described above. Although IIS control of larval germline cell cycle has not yet been linked directly to nutritional or metabolic cues as in other organisms, two additional highly conserved pathways have been found recently to link physiology and larval germline expansion in response to sensory and metabolic cues: TGFβ and TOR-S6K.

5.3.2.1 Post-Dauer-Decision Expansion of the Germline Progenitor Pool: A Sensory Link via TGFβ

The same TGFβ pathway ligand and receptors that are required for the dauer decision (*daf-7/TGFβ*, *daf-1*/TGFβRI and *daf-4/TGFβRII*) were identified in large-scale RNAi screens for genes that affect larval germline development (Dalfó et al. 2012;

Dalfó and Hubbard, unpublished). Depletion of these gene activities by RNAi or mutation, or depletion of the relevant R-Smads, causes a reduction of both the number of cells in the adult progenitor pool and brood size by up to ~50 %, similar to what is observed with IIS pathway mutants (Table 5.1). Since IIS acts in parallel with the TGFβ pathway for the dauer decision (Fig. 5.2), converging on DAF-12, a plausible hypothesis was that these pathways might constitute a parallel cassette that acts similarly for dauer and for reproductive germline proliferation control. However, this hypothesis proved incorrect: while IIS is important germline-autonomously to promote robust larval cell cycle progression, the TGFβ receptor pathway is required germline-nonautonomously in the DTC and has no effect on cell cycle (Dalfó et al. 2012). As in the dauer pathway (Vowels and Thomas 1992), the DAF-3/co-Smad and DAF-5/Sno-Ski are downstream negatively regulated targets of TGFβ signaling in the control of germline progenitor expansion, and they, too, are required in the DTC for this role. Moreover, although germline proliferation is greatly reduced during dauer, TGFβ pathway activity for expansion of the larval proliferative zone does not require the activity of the nuclear hormone receptor DAF-12, as it does for dauer. That is, the *daf-7* mutant dauer-constitutive phenotype (25°) is suppressed by loss of *daf-12*, while the larval germline defect persists. These results, and others, support the hypothesis that the TGFβ roles in L1/L2 regulation of dauer entry and in L3/L4 germline progenitor accumulation are separate (Dalfó et al. 2012).

The effects of reduced TGFβ signaling share some similarities with reduced *glp-1*/Notch activity: TGFβ mutants display neither inappropriate cell death in the distal germ line nor any deviation from wild-type cell cycle parameters (L4 mitotic- and S-phase indexes) (Dalfó et al. 2012; Michaelson et al. 2010). In addition, reducing TGFβ signaling causes a further severe reduction in proliferative germ cells in the background of reduced *(rf) glp-1/Notch* activity. Depletion of *daf-3/CoSmad* or *daf-5/Sno-Ski* suppresses this enhancement back to *glp-1(rf)* levels. Thus the TGFβ pathway acts differently from the IIS pathway and more similarly to GLP-1/Notch in that it promotes the proliferative fate and/or interferes with differentiation, without influencing cell cycle (Table 5.1). Given this Notch-like role and the requirement for TGFβ receptor (TGFβR) signaling in the DTC, TGFβ signaling might be postulated to affect the production or activity of the Notch ligands LAG-2 and APX-1 produced in the DTC (Henderson et al. 1994; Nadarajan et al. 2009). However, no change in the levels of ligand reporters or transcripts were observed, and reduced TGFβ signaling still lowered proliferative germ cell numbers in the absence of GLP-1 (in a triple null mutant *gld-2 gld-1; glp-1* strain that produces proliferative germ cells despite the absence of *glp-1*). These results suggest that the TGFβ pathway acts in parallel with Notch signaling to inhibit differentiation (Dalfó et al. 2012).

DAF-7/TGFβ itself is expressed by the paired ASI ciliated chemosensory neurons (Ren et al. 1996; Schackwitz et al. 1996). Surprisingly, these neurons and the environmental cues they mediate also influence the expansion of the L3-L4 larval germline progenitor pool (as measured by early adult total germline progenitors) in a *daf-3* and *daf-5*-dependent manner (Dalfó et al. 2012). Using conditions or timing that circumvented dauer formation, genetic ablation of cilia, physical ablation of

Table 5.1 Genetic and nutritional control of the accumulation of the larval germline progenitor cells

Pathway or condition disrupted	Adult prolif number[a]	Cellular mechanism(s) of defect	Additional defect on low food?[b]	Tissue-autonomy	Enhances Glp-1 phenotype?	Refs
none (wild type)	~200 n/a	–	major[c]	–	–	Michaelson et al. (2010), Korta et al. (2012), Dalfó et al. (2012)
GLP-1 (Notch)	~100 0	↓Prolif/↑diff	major[c]	germ line	–	Michaelson et al. (2010), Korta et al. (2012)
DAF-2 (Insulin/IGF)	~130 ~70	↓Cell cycle	ND	IIR pathway germ line	no	Michaelson et al. (2010)
DAF-1 (TGFβ)	~125 ~90	↓Prolif/↑diff	none[d]	TGFβR pathway DTC	yes[e]	Dalfó et al. (2012)
RSKS-1 (TOR-S6K)	~100 n/a	↓Cell cycle ↓Prolif/↑diff	minor[c]	germ line	yes[f]	Korta et al. (2012)
PRO-1 (rRNA processing)	~70 n/a	ND	ND	distal sheath	yes	Killian and Hubbard (2004), Voutev et al. (2006)
Food (bacterial) Concentration	~50[c] n/a	↓Cell cycle	–	–	yes[c]	Korta et al. (2012)

[a]Top number: approximate number of proliferative zone nuclei in early adult hermaphrodite germ lines in the following mutant or food conditions, as assayed at 20°: *glp-1(e2141)*, *daf-2(e1370)*, *daf-1(m40)*, *rsks-1(sv31)*, *pro-1(na48)*, OP50 bacteria on solid media plated at 1×10^8 colony forming units. These same conditions were used, where relevant, to assess enhancement of *glp-1(rf)* as indicated in the 6th column

Bottom number: indicates the number of proliferative zone nuclei under more stringent conditions: *glp-1(null)*, *daf-2(e1370)* after shift to 25°, and *daf-7(e1372); daf-12(m20)* at 25°. "*n/a*" indicates "not applicable"; "*ND*" indicates "not determined"

[b]Reduced bacteria concentration on solid media

[c]Bacteria concentration 1×10^8 on solid media

[d]Bacteria concentration 5×10^8 on solid media

[e]Early adult phenotype is a further-reduced number of proliferative zone cells, but not an "all-meiotic" total loss of proliferative zone as seen in the most severe loss of *glp-1*

[f]Early adult enhancement phenotype includes elevated penetrance of "all-meiotic" phenotype

ASI neurons, application of crude dauer pheromone, or reduced food (5×10^8 colony forming units/ml) were found to reduce proliferative germ cell numbers in a *daf-5*-dependent manner. These results suggest that, similar to the cues that alter *daf-7/TGFβ* reporter expression during the dauer decision, sensing of low pheromone levels and high food concentration during L3-L4 stages is required to expand the progenitor cell pool, and that this information is passed from cilia in ASI to the DTC germline stem cell niche via TGFβ (Dalfó et al. 2012).

5.3.2.2 Post-Dauer-Decision Expansion of the Germline Progenitor Pool: A Nutrition Link via TOR and S6K

Restricting food either by reducing the bacterial concentration (to 1×10^8 colony forming units/ml) or by protein deprivation (by loss of *pept-1*) severely inhibits the accumulation of germline progenitors in larval stages (Korta et al. 2012). Importantly, this effect involves cell cycle control and occurs independently of continued signaling by *glp-1*/Notch. For example, reducing bacterial concentration causes a ~65–75 % decrease in the number of young–adult proliferative germ cells in wild type (from ~200 to 50 cells) despite the presence of GLP-1/Notch signaling. Even when *glp-1* activity is reduced (e.g., in *glp-1(rf)* mutants, ~100 proliferative cells are retained), the proliferative germ cell pool is further reduced by ~65 % to ~ 35 cells under the same food restriction (Korta et al. 2012).

The conserved TOR (Target of Rapamycin) pathway links nutrition to cell growth and aspects of cell division in organisms from yeast to mammals. In mammals, the TOR/RAPTOR complex (TORC1) responds to PI3K-mediated growth factor signaling as well as nutrient and energy load, and much of this response is mediated by the conserved substrate S6K (p70 ribosomal S6 kinase) (for a general review, see Wullschleger et al. 2006). Recent work indicates that TOR and S6K are important in *C. elegans* for accumulation of larval germline progenitors and for the control of this accumulation by diet (Korta et al. 2012). Interfering with TOR pathway activity causes numerous defects (Long et al. 2002). Most strikingly, mutations in the *C. elegans* TORC1 complex components TOR or RAPTOR (*let-363* and *daf-15*, respectively) cause an L3-stage larval arrest phenotype that is similar to but distinct from dauer (Long et al. 2002; Jia et al. 2004). TOR and S6K also influence the accumulation of larval germ line progenitors, and the somatic effects of TOR and S6K can be separated from germline-autonomous roles of this pathway (Korta et al. 2012). Among the major findings are that *rsks-1/S6K*, the sole S6K-encoding gene in *C. elegans*, acts germline-autonomously to influence the expansion of the larval germline progenitor pool by both promoting cell cycle progression and inhibiting differentiation. Similar to reduced *daf-2*-mediated IIS, loss of *rsks-1* slows the germline cell cycle in larvae but not adults (Korta et al. 2012). Also similar to IIS (Pinkston et al. 2006), reducing S6K markedly reduces the size (cell number) of germline tumors (Korta et al. 2012). However, unlike the effects of reducing IIS, loss of S6K does not specifically slow the G2 phase of the cell cycle and its germline phenotype does not depend on the activity of *daf-16/*

FOXO. Also, in contrast to reduced IIS, loss of *rsks-1* potently enhances the Glp-1 meiotic entry phenotype (total loss of proliferative germ cells) of *glp-1(rf)* mutants and suppresses *glp-1(gf)* phenotypes. In addition, germline reduction of TOR or RAPTOR causes a dramatic reduction in the number of germline progenitors, a phenotype that is partially mediated by S6K and eIF4E. Finally, the larval germline role of *rsks-1/S6K* appears distinct from its role in longevity since the germline phenotype is not suppressed by the same genetic alterations that suppress the longevity phenotype. These results suggest that regulation of larval germline progenitors and lifespan by S6K occurs via distinct S6K targets (Korta et al. 2012).

Interestingly, S6K is also required for the effects of certain aspects of diet on the growth of the larval germline progenitor pool. The same food-limiting regimes (reduced bacterial concentration, pharyngeal pumping or amino acid uptake) that cause a severe decrease in the number of wild-type or *glp-1* germline progenitors amassed by the early adult cause only a mild further decrease in the absence of *rsks-1*. These results suggest that the S6K (and likely TOR) is a key mediator of the effect of nutrition on establishment of the germline progenitor pool (Korta et al. 2012). It will be of interest to determine the targets of S6K in this role and the precise molecular pathways that influence germline proliferation and differentiation in response to dietary restriction.

5.3.3 Reproduction Under Conditions of Food Limitation

It remains to be determined whether dietary restriction affects additional aspects of adult germline development. Because reducing food intake or altering food composition of adult worms (and other organisms) has a marked effect on longevity (Fontana et al. 2010), multiple methods of dietary restriction (DR) have been developed in *C. elegans* for analysis of the effects of caloric restriction on lifespan (compared in Greer and Brunet 2009). Not surprisingly, the vast majority of manipulations that restrict food intake also reduce fecundity of unmated adult hermaphrodites. Fecundity or brood size is the culmination of multiple aspects of reproductive biology including earlier cumulative effects on germline development including the proliferation and maintenance of germline stem cells, meiosis, cell death, sex determination, gametogenesis, and fertilization. In addition, studies of brood size can be confounded by defects in egg-laying. Therefore, detailed analyses of the effects of altering diet on specific aspects of germline development will advance the field.

Specific DR protocols that extend lifespan and negatively influence brood size include (1) genetic methods such as mutations in *eat-2* that reduce pharyngeal pumping rate (Avery 1993; Lakowski and Hekimi 1998) or *pept-1* that reduces amino acid absorption (Meissner et al. 2004), (2) diluted bacteria in liquid culture (Klass 1977; Houthoofd 2003; Panowski et al. 2007), (3) chemically defined liquid media (Houthoofd et al. 2002a; Szewczyk et al. 2006), (4) serial dilution of bacteria on solid media (Greer et al. 2007), (5) total absence of bacteria on plates (Kaeberlein et al. 2006; Lee et al. 2006) (that is, fasting, where brood size is also limited by retention of embryos and subsequent matricide), and (6) metformin treatment (Onken and Driscoll 2010), which may mimic DR. Two methods of DR that influence lifespan but for which brood size has not been reported are intermittent fasting (Honjoh et al. 2009) and treatment with the glucose analog 2DG (Schulz et al. 2007). Interestingly, two treatments that extend lifespan but do not reduce brood size are reduced peptone in agar plates (Hosono et al. 1989) and treatment with resveratrol (Wood et al. 2004). Therefore, these two aspects of life history are not always linked.

What is, perhaps, surprising from the analyses of dietary restriction in the context of lifespan extension is that while pathways that mediate these effects overlap and interact, specific methods to restrict caloric intake can be associated primarily with the activity of specific pathways (Greer and Brunet 2009). That is, in many cases, the activity of a single effector can dictate whether the dietary restriction regimen extends lifespan. For example loss of *pha-4* suppresses lifespan extension caused by DR imposed by reduced pumping in the *eat-2* mutant background (Panowski et al. 2007), but not as imposed by reducing food on solid media (Greer and Brunet 2009), while *aak-2* affects the latter (Greer et al. 2007) and not the former (Curtis et al. 2006). This is very exciting since it suggests that a limited number of specific pathways may also link dietary restriction to germline development and that they mediate different aspects of the restriction or the response. A reduction in reproductive capacity could be simply a secondary effect of a simple lack of critical building blocks necessary for basic cellular processes such as germ cell proliferation or oocyte growth. As with lifespan, to argue a pathway-specific response, as opposed to a simple lack of "stuff," the genetic "suppressibility" of nutritional manipulations will be important to establish. In cases where a suppressor is not yet identified, the effect of a pathway in the presence or absence of pathway components might suggest a dependency relationship (e.g., Korta et al. 2012).

5.4 Fasting

Severe food deprivation has many consequences. Depending on the stage at which animals experience starvation, the response of the germ line and, indeed, the animal's overall response differs. While most of these consequences have been investigated at diapause points, it bears mention that worms are exquisitely sensitive to and respond rapidly to such changes throughout larval life (Liu et al. 1997). Substantial changes in the transcription levels of key metabolic genes have been observed within 1 h of food removal in the L4 stage. Also, both differential stage-dependent and shared transcriptional responses occur (Van Gilst et al. 2005). Like mammals, both larval and adult worms quickly mobilize fat upon starvation (McKay et al. 2003). One highly conserved and critical response to fasting that facilitates this mobilization is downregulation of the sterol regulatory element binding protein (SREBP) transcription factor by a sirtuin, SIR-2.1 (Walker et al. 2010).

5.4.1 Larval and Adult Starvation

Total starvation of hatchlings causes L1 arrest, whereas near-starvation of hatchlings (e.g., axenic culture conditions—see Johnson et al. 1984) permits delayed development. Both L1 arrest and delay are reversible upon refeeding. L1 larvae that hatch in the absence of food arrest both somatic and germline cell cycle progression and development in a *daf-18/PTEN*-dependent manner (Baugh and Sternberg 2006; Fukuyama et al. 2006). Somatic cell arrest occurs in the G1 and is dependent on the FOXO transcription factor DAF-16 that is antagonized by PI3K signaling through the IIS and DAF-2. Gene expression profiling during L1 arrest and recovery indicates that RNAPol II is poised on the promoters of development and growth genes during L1 starvation-induced arrest, allowing for rapid recovery once conditions improve (Baugh et al. 2009). Interestingly, germline cell cycle arrest has somewhat different control, reminiscent of the arrest at dauer: it occurs in the G2 and is dependent on *daf-18/*PTEN, but not *daf-16/*FOXO (Fukuyama et al. 2006).

From the sensory standpoint, L1 starvation involves neuronal signaling in response to a subset of amino acids. Neuronally required metabotropic G protein-coupled glutamate receptors respond to an amino acid signal, particularly leucine. These responses then modulate the activity of specific neurons (AIY an AIB) to inhibit and promote the starvation response, respectively (Kang and Avery 2009). The same receptors were found to modulate fat accumulation in the worm (Greer et al. 2008). This type of global starvation-signaling system is likely important in ensuring an appropriate whole-organism response to changing environmental conditions.

How different tissues such as the L1 germ line integrate and respond to specific starvation cues—both sensory and metabolic—to control arrest and recovery is an area for future study. Very recently, the activities of intestinal microRNAs were implicated in survival during L1 arrest. In addition, *miR-71* was identified as important for recovery from L1 arrest, affecting only a subset of tissues (including the intestine) and mediating both DAF-16-dependent and -independent processes (Zhang et al. 2011).

Total food deprivation of L4-stage hermaphrodites results in different outcomes: L4 or adult arrest, adult matricide ("bagging"), or a distinct survival-enhancing "adult reproductive diapause" (ARD) (Angelo and Van Gilst 2009). The relative proportion of these outcomes varies depending on the exact time of starvation (Angelo and Van Gilst 2009; Seidel and Kimble 2011). The matricide phenotype among comparative older animals (those already competent to produce embryos when starvation occurs) is not unexpected: worms removed from food will immediately reduce the rate of egg-laying causing progeny to develop internally where they hatch and devour their mother if they cannot escape to the outside (Trent et al. 1983). This control of egg-laying is mediated by neuropeptides and cGMP signaling and TGFβ pathway components (see Schafer 2005, for a review). The ARD outcome, however, was unanticipated. Strikingly, ARD is characterized by germline atrophy that occurs over the first 10 days of starvation, ultimately leveling off with ~35 germ cells per gonad arm. Remarkably, even after 30 days of starvation, ARD animals survive and can reestablish the germ line with its full distal-proximal developmental pattern over the subsequent 72 h once returned to food. These starved and recovered animals are capable of producing progeny either from surviving self-sperm or when mated (Angelo and Van Gilst 2009). The germline atrophy is partially dependent on the programmed cell death pathway (Angelo and Van Gilst 2009), and appears to be associated with oogenesis (Seidel and Kimble 2011). Further, when the matricide phenotype is prevented by interfering with embryonic development, all animals are capable of ARD (Seidel and Kimble 2011).

Interestingly, oocytes are continuously produced in starved ARD animals, albeit at a much-reduced rate of one per 8 h (in contrast to ~3 per hour per gonad arm in fed adults) (Seidel and Kimble 2011). This ability to make one oocyte at a time likely reflects an adaptation to preferentially direct dwindling energetic resources to the continued production of progeny, regardless of how few (see below, Sect. 5.5 for further discussion). Other aspects of growth conditions (amount or quality of food, degree of crowding) may or may not influence ARD (Seidel and Kimble 2011), but the extent to which these possible influences can be attributed to differences in worm or bacteria strains or differences in growth conditions remains to be determined. Finally, the proportion of worms that exhibit ARD after starvation is partially dependent on the activity of a nuclear hormone receptor, NHR-49, an HNF-4a ortholog that, in mammals, has been implicated in promoting fatty acid oxidation and gluconeogenesis in response to food deprivation (Angelo and Van Gilst 2009). It will be important to establish the mechanism by which the distal-most cells are protected from degradation, the proliferation status of the distal-most cells under ARD, and the cellular and molecular mechanisms underlying germ cell renewal.

The unexpected germline atrophy and capacity for regrowth that are characteristic of ARD are very exciting to the field. ARD demonstrates a remarkable plasticity of germ line in response to the environment and the renewal of the germ line in *C. elegans* from a greatly diminished pool of distal germ cells. The results of Angelo and Van Gilst (2009) demonstrated that ARD is indeed a reproductive diapause state (MacRae 2010) in that ARD reversibly alters gametogenesis and reproductive capacity in response to environmental adversity. In this case, it also leads to an extension of reproductive competence and lifespan (Angelo and Van Gilst 2009). Additional insight into mechanisms underlying ARD and the environmentally regulated plasticity of the germ line that it reveals is indeed an attractive area for further investigation.

5.5 Fat Metabolism, Reproduction, and Aging

5.5.1 Two Examples Connecting Fat, Reproduction, and/or Aging

Lipid biology is essential to many aspects of cellular function and both cell-cell communication and global organismal signaling, including the germ line. Two examples here illustrate the utility of *C. elegans* to understand these conserved metabolic processes and their impact at the cellular and organismal levels.

One example is the effect of dietary supplementation of specific polyunsaturated fatty acids (PUFAs) on the germ line (Brock et al. 2006). A particularly interesting class is the C20 PUFAs since they are precursors to eicosanoids and other bioactive moieties, and they play a prominent role in membrane biology. Dietary supplementation and subsequent uptake of dihommogamma-linolenic acid (DGLA), but not other related fatty acids, causes dramatic dose-dependent sterility. Supplementation in the L2/L3 stages causes larval germline degradation and elevated levels of cell death in the adult germ line. Both genetic and biochemical results argue for the specificity and sensitivity of the effect and support the hypothesis that exogenous DGLA may interfere with signaling critical for germline development. Alternatively a byproduct of DGLA metabolism could be cytotoxic. The precise nature of the germline loss and the relationship between this effect and endogenous fatty acid function will be interesting to determine.

Another example involves one of the earliest known lifespan-altering genes, *clk-1*, which encodes an enzyme required for the ubiquinone biosynthesis (Lakowski and Hekimi 1996). Loss of *clk-1* is highly pleiotropic and causes a delay in germline development relative to somatic development (Shibata et al. 2003). A mutation in *dsc-4* suppresses the germline delay, but not the rate of overall postembryonic development. DSC-4 encodes a protein related to the large subunit of the microsomal triglyceride transfer protein that is required for the secretion of LDLs in mammals. In worms, it is expressed in the intestine, the site of yolk production, and *dsc-4* appears to act in the same pathway as a subset of the vitellogenins. Consistent with the notion that cholesterol reduction would reduce LDL levels in worms, as in mammals, and that LDL production is an important aspect of the *clk-1* germline phenotype, depletion of cholesterol also suppresses the germline delay in *clk-1* mutants with less effect on other phenotypes. In addition, consistent with the hypothesis that the reduction of reactive oxygen species (ROS) production is also involved, elevating ROS by mutation of a superoxide dismutase, *sod-1*, also suppresses the *clk-1* mutant delay in egg production. These observations prompted a model in which the level of oxidized LDL-like lipoproteins correlates with the rate of germline development relative to somatic development. Whether this effect is due to a specific metabolic product or is due to more general cellular responses remains to be determined. The ARK-1 kinase was identified as a possible germline-autonomous mediator since it similarly suppressed *clk-1* mutant defects (Shibata et al. 2003). It will be of interest to determine the origin of the "delay" in germline development that is associated with *clk-1*. In any case, these results underscore additional interesting connections between germline development and lipid biology.

5.5.2 The Germ Line Limits Lifespan

Two observations related to lifespan and the germ line have sparked much debate in the context of evolutionary theory: loss of the germ line extends lifespan and animals outlive their reproductive period (see, for example Mukhopadhyay and

Tissenbaum 2007; Hughes et al. 2007). Following on these considerations, intriguing connections have been made between *C. elegans* lifespan, fat metabolism and the germ line (see recent reviews by Branicky et al. 2010; Watts 2009).

The 1999 report that ablation of the germ line with an intact somatic gonad extends lifespan in *C. elegans* (Hsin and Kenyon 1999) opened a new area of investigation into how the presence of the germ line limits lifespan. This effect depends on DAF-16/FOXO, the nuclear hormone receptor DAF-12, the ankyrin-repeat containing protein KRI-1, the lipase LIPL-4, and the FOXA transcription factor PHA-4 (Hsin and Kenyon 1999; Arantes-Oliveira et al. 2002; Berman and Kenyon 2006; Wang et al. 2008; Lapierre et al. 2011). Germline proliferation in both larval and adult stages appears to influence lifespan (Arantes-Oliveira et al. 2002). In the wake of these results, *glp-1* temperature sensitive (ts) loss-of-function alleles have become a proxy for germline ablation. Results obtained with these alleles alone must be interpreted with caution, however, since *glp-1* may have activities in other tissues such as the nervous system (Singh et al. 2011) that may influence lifespan, feeding and processing of environmental cues.

5.5.3 *Taking Ecology and Evolution into Consideration: A Unifying Hypothesis for Fat, Reproduction, and Aging*

So what is the connection between fat, reproduction, and aging? While a detailed discussion of these connections is beyond the scope of this review, some speculation regarding these connections can be made from the vantage point of the germ line and *C. elegans* ecology. Returning to the themes stated in the introduction, we speculate that the successful evolution of *C. elegans* as a rapidly-reproducing, highly fecund species in environments with patchy resources required exquisite co-evolution of sensory, metabolic, and reproductive strategies. This notion could help explain why the germ line exhibits tremendous responsiveness and plasticity with respect to environmental challenges. One question that emerges, and that will be helpful to consider as the field grapples with the organismal integration of these components, is: to what extent did the demands of reproductive success drive *C. elegans* evolution versus the demands of metabolism? We favor the hypothesis that the physiological responses of the worm to its environment have been honed to maximize the production of at least some offspring. Indeed, larger self-progeny broods produced by strains with elevated sperm production extend generation time, suggesting a possible disadvantage (Hodgkin and Barnes 1991). For a short-lived organism, it may be more important to produce a small number of offspring under as many conditions as possible than to produce the large numbers of progeny seen in the "feasting" laboratory conditions. Even extreme brood-limiting outcomes such as the matricide caused by reduced egg-laying in response to low food (Trent et al. 1983) may ultimately benefit the species, since the surviving progeny can undergo larval arrest and disperse to more favorable environments.

Taking these ecological and evolutionary points into consideration, any possible advantage of extending lifespan after loss of the germ line is difficult to reconcile. Loss of the germ line is a highly abnormal and reproductively (hence evolutionary) dead-end scenario. By contrast, rapid and wild fluctuations in resources were likely common challenges to this species over the course of evolution. Thus, one could speculate that the extension of lifespan that occurs with adult dietary restriction may well enable an aging hermaphrodite to encounter a male and resume reproduction later in life.

How might lipid metabolism figure into these considerations? Under "feast" conditions, major fat stores are released from the intestine and are taken up by developing oocytes (Kimble and Sharrock 1983; Grant and Hirsh 1999). In the abnormal scenario in which oocytes are not produced, lipid-rich material accumulates in the pseudocoelomic space. This observation suggests that the intestine does not receive a "don't dump" signal from the "non-oogenic" germ line. Indeed, germline-less worms accumulate certain lipids and lifespan extension in *glp-1* mutants depends on the targets of several transcription factors that alter fat metabolism. For example, lifespan extension in *glp-1* requires the DAF-16/FOXO-regulated lipase, LIPL-4 (Wang et al. 2008; O'Rourke et al. 2009; Lapierre et al. 2011) as well as the NHR-80/HNF4-regulated stearoyl-CoA desaturase FAT-6, which converts stearic acid to oleic acid (Goudeau et al. 2011). Indeed, exogenous oleic acid can restore longevity to *glp-1* mutants lacking stearoyl-CoA desaturase activity (Goudeau et al. 2011).

One perhaps naïve hypothesis to unify these observations and considerations is that the failure to utilize fats and lipids in the production of new germ cells (e.g., as occurs in the absence of the germ line) may cause the buildup of unusual quantities of particular substrates for substrate-dependent metabolic reactions, and hence shift metabolic flux toward conditions reminiscent of age-defying stages such as dauer where lipids are stored and utilized in a controlled manner. If this were the case, the somatic gonad must act as a sensor or responder—likely by way of DAF-12 (Yamawaki et al. 2010).

5.6 Prospects, Challenges, and Open Questions

Physiology impacts many aspects of germline development and function—and hence reproduction—in all organisms. Because physiology is a whole-organism phenomenon that engages many historically separate fields of biological inquiry (including neurobiology, metabolism, growth, and cellular stress), the possibility of understanding and linking them is both fascinating and daunting. The relative anatomical simplicity, experimental accessibility, and depth of knowledge accrued by the community of *C. elegans* researchers makes *C. elegans* an outstanding model organism for these studies. The boom-and-bust aspects of this organism's ecological resources as well as its possible evolutionary prioritization of reproduction make it a particularly sensitive and therefore attractive model.

The very same whole-organism view that makes these connections so interesting also introduces challenges, both conceptual and experimental. Conceptually, dietary changes affect many aspects of worm biology, and conversely, germline defects can influence whole-worm biology. In addition, the question of whether a given dietary manipulation is altering a developmental process due to a simple reduction of a rate-limiting building block or whether it is triggering a more specific change in a signaling cascade needs to be addressed. Given available genetics and genomics tools, finding genetic changes or network fluxes that reverse the effects of specific physiological manipulations should be feasible. This combination of facile environmental manipulation, cell biology and genetics bodes well for the power of this model system to uncover important mechanisms that link different aspects of physiology, germline development, and fertility.

Experimentally, there are additional considerations. Physiological manipulations require extremely tight controls to reduce variability caused by small changes in conditions (e.g., temperature, differences in plates, presence of contamination) or in synchronization of worm populations. These subtle differences in the handling of worms may account for differences in results between laboratories, so in addition to being carefully controlled internally, they must be well documented to facilitate comparisons between laboratories. Other perhaps less obvious considerations and their consequences are, first, that "food-limited" worms often grow more slowly than well-fed worms. Therefore, the relative effects of any dietary alteration on developmental timing, both in the soma and germ line must be examined to ensure proper interpretation of germline phenotypes. A second (and related) consideration is that many germline phenotypes occur as a secondary consequence of earlier and/or more distal defects and therefore must be interpreted with care. For example, a reduction in germline proliferation in the early L3 stage can delay initial meiotic entry later in the L3 stage and also delay the sperm-oocyte switch in the L4/adult (Killian and Hubbard 2004; Korta et al. 2012) In addition, proper oogenesis requires input from more distal parts of the germ line (Nadarajan et al. 2009; Kim et al. 2012, Chap. 10). Third, different forms of dietary restriction may affect different aspects of germline development. This possibility is likely since limiting factors required to promote cellular processes in mitosis versus gametogenesis, for example, will likely differ. Fourth, changes in the nutritional environment of the larva may confer effects that are latent and only revealed in later stages such as oogenesis. Similarly, aging and prior reproductive status must be taken into account. For example, in a study of late-life fertility after mating, Mendenhall et al. (2011) observed that among animals that are aged and then mated, "food-limited" *eat-2* mutant animals produce progeny significantly later than wild-type under similar mating conditions. However, long-lived *age-1* and *daf-2* mutants do not show a similar phenotype, suggesting that this effect cannot be attributed solely to aging. Fifth, feeding is itself a highly regulated aspect of *C. elegans* biology. For example, pumping rate and feeding behaviors are influenced by previous dietary experience, quality of the diet, starvation, and satiety signals. These controls utilize some of the same signaling pathways implicated in the dauer decision and in the control of larval germline expansion (You et al. 2006, 2008). Combinations of food quality and genetics will be revealing. For example, Lemire and colleagues (Reinke et al. 2010) analyzed the phenotypic effects of two strains of *E. coli*, OP50 and HT115, on wild-type worms and worms carrying a mutation in *nuo-1*, which encodes a subunit of mitochondrial complex I that is required for oxidative phosphorylation. While OP50 and HT115 produce similar broods in wild-type animals, *nuo-1* broods are reduced on HT115. Pathogenic *Bacillus* strains also reduce *C. elegans* fecundity compared OP50 (Rae et al. 2010). Thus, it is clear that not only the amount of food, but also the quality of food affects *C. elegans* reproduction. It will be important to determine the underlying causes of brood size differences. Finally, the potential for intergenerational effects of growth conditions and diet (see, for example, Grishok et al. 2000; Greer et al. 2011; Rechavi et al. 2011) on the germ line must be considered.

Many aspects of the germline response to physiology are completely open areas for investigation. In addition to food and pheromone signals, additional physiological changes that were not featured here influence germline development and reproductive success. These include the presence of sperm, various types of stress, and aging (see Rae et al. 2010; Luo et al. 2010; Hughes et al. 2007, 2011; Kim et al. 2012; Chap. 10). The range of genetic pathways known to interact with physiological sensors and effectors to alter the germ line and reproductive program is not nearly saturated. The mode and anatomical focus of action of these pathways that mediate these effects will be important to help understand how the germ line responds to physiological conditions and how these responses are coordinated with other tissues. For example, the *C. elegans* genome encodes many orphan receptors including large families of nuclear hormone receptors and G-protein-coupled receptors. In addition to reporting environmental conditions, these receptors likely facilitate systemic responses to pheromones, hormones, and other bioactive molecules such as biogenic amines, all of which may influence the germ line. Males and hermaphrodites also likely differ in their reproductive responses as they respond differently to certain physiological changes (see for example Tan et al. 2011; LeBoeuf et al. 2011). In addition, the contribution of cellular processes such as autophagy and intra-germline signaling is open. Understanding the mechanisms of germline sensitivity and plasticity, both at the level of physiological conditions and germline responses are very exciting areas for future studies.

5.7 Broader Implications

Many of the processes and pathways important for the germ line to interpret and respond to physiological changes, including metabolic pathways and signal transduction pathways, are highly conserved between worms, flies, and mammals, suggesting that findings in worms will be broadly applicable. For example, in both worms and flies, insulin/IGF signaling affects the cycling of germline stem cells—specifically in the G2, while TOR and S6K signaling affect both cell cycle and differentiation (Drummond-Barbosa 2008; Lafever et al. 2010; Michaelson et al. 2010; Korta et al. 2012). In addition, these investigations may be relevant to our understanding of stem cell regulation and cancer. For example, in the first half of the twentieth century, many studies explored the phenomenological impact of diet on tumor prevalence and growth (see, e.g., Rous 1914; Tannenbaum and Silverstone 1953), but only recently have organismal dietary changes in mammals been revisited and linked to signal transduction pathways known to promote cancer in humans (Kalaany and Sabatini 2009). The worm offers a simple system to explore these effects in mechanistic depth, and offers the benefit of unbiased function-based gene discovery. We look forward to learning more about the intersection of physiology and the germ line as this exciting field expands.

Acknowledgments We especially thank Tim Schedl for discussions that catalyzed many of the ideas presented here. In addition we thank Barth Grant, Jennifer Watts, Kerry Kornfeld, Marie-Anne Félix, and David Greenstein for comments on the manuscript. We also thank members of the Hubbard lab for discussion, and National Institutes of Health grants R01GM061706, R03HD066005, T32GM07308, F30DK089697 for support.

References

Angelo G, Van Gilst MR (2009) Starvation protects germline stem cells and extends reproductive longevity in *C. elegans*. Science 326(5955):954–958

Arantes-Oliveira N, Apfeld J, Dillin A, Kenyon C (2002) Regulation of life-span by germ-line stem cells in *Caenorhabditis elegans*. Science 295(5554):502–505

Austin J, Kimble J (1987) glp-1 is required in the germ line for regulation of the decision between mitosis and meiosis in *C. elegans*. Cell 51(4):589–599

Avery L (1993) The genetics of feeding in *Caenorhabditis elegans*. Genetics 133(4):897–917

Bargmann CI (2006) Chemosensation in *C. elegans*. WormBook: 1–29

Bargmann CI, Horvitz HR (1991) Chemosensory neurons with overlapping functions direct chemotaxis to multiple chemicals in *C. elegans*. Neuron 7(5):729–742

Baugh LR, Sternberg PW (2006) DAF-16/FOXO regulates transcription of cki-1/Cip/Kip and repression of lin-4 during *C. elegans* L1 arrest. Curr Biol 16(8):780–785

Baugh LR, Demodena J, Sternberg PW (2009) RNA Pol II accumulates at promoters of growth genes during developmental arrest. Science 324(5923):92–94

Berman JR, Kenyon C (2006) Germ-cell loss extends *C. elegans* life span through regulation of DAF-16 by kri-1 and lipophilic-hormone signaling. Cell 124(5):1055–1068

Berry LW, Westlund B, Schedl T (1997) Germ-line tumor formation caused by activation of glp-1, a *Caenorhabditis elegans* member of the Notch family of receptors. Development 124(4): 925–936

Blelloch R, Anna-Arriola SS, Gao D, Li Y, Hodgkin J, Kimble J (1999) The gon-1 gene is required for gonadal morphogenesis in *Caenorhabditis elegans*. Dev Biol 216(1):382–393

Braeckman BP, Houthoofd K, Vanfleteren JR (2009) Intermediary metabolism. WormBook: 1–24

Branicky R, Desjardins D, Liu J-L, Hekimi S (2010) Lipid transport and signaling in *Caenorhabditis elegans*. Dev Dyn 239(5):1365–1377

Brock TJ, Browse J, Watts JL (2006) Genetic regulation of unsaturated fatty acid composition in *C. elegans*. PLoS Genet 2(7):e108

Burnell AM, Houthoofd K, O'Hanlon K, Vanfleteren JR (2005) Alternate metabolism during the dauer stage of the nematode *Caenorhabditis elegans*. Exp Gerontol 40(11):850–856

Butcher RA, Fujita M, Schroeder FC, Clardy J (2007) Small-molecule pheromones that control dauer development in *Caenorhabditis elegans*. Nat Chem Biol 3(7):420–422

Butcher RA, Ragains JR, Kim E, Clardy J (2008) A potent dauer pheromone component in *Caenorhabditis elegans* that acts synergistically with other components. Proc Natl Acad Sci USA 105(38):14288–14292

Butcher RA, Ragains JR, Li W, Ruvkun G, Clardy J, Mak HY (2009) Biosynthesis of the *Caenorhabditis elegans* dauer pheromone. Proc Natl Acad Sci USA 106(6):1875–1879

Cassada RC, Russell RL (1975) The dauerlarva, a post-embryonic developmental variant of the nematode *Caenorhabditis elegans*. Dev Biol 46(2):326–342

Curtis R, O'Connor G, DiStefano PS (2006) Aging networks in *Caenorhabditis elegans*: AMP-activated protein kinase (aak-2) links multiple aging and metabolism pathways. Aging cell 5(2):119–126

Dalfó D, Michaelson D, Hubbard EJA (2012) Sensory regulation of the *C. elegans* germline through TGFß-dependent signaling in the niche. Curr Biol 22:712–719

Darby C (2005) Interactions with microbial pathogens. WormBook: 1–15

Dillin A, Crawford DK, Kenyon C (2002) Timing requirements for insulin/IGF-1 signaling in *C. elegans*. Science 298(5594):830–834

Drummond-Barbosa D (2008) Stem cells, their niches and the systemic environment: an aging network. Genetics 180(4):1787–1797

Drummond-Barbosa D, Spradling AC (2001) Stem cells and their progeny respond to nutritional changes during Drosophila oogenesis. Dev Biol 231(1):265–278

Edison AS (2009) *Caenorhabditis elegans* pheromones regulate multiple complex behaviors. Curr Opin Neurobiol 19(4):378–388

Félix M-A, Braendle C (2010) The natural history of *Caenorhabditis elegans*. Curr Biol 20(22):R965–R969

Fielenbach N, Antebi A (2008) *C. elegans* dauer formation and the molecular basis of plasticity. Genes Dev 22(16):2149–2165

Fontana L, Partridge L, Longo VD (2010) Extending healthy life span–from yeast to humans. Science 328(5976):321–326

Fukuyama M, Rougvie AE, Rothman JH (2006) *C. elegans* DAF-18/PTEN mediates nutrient-dependent arrest of cell cycle and growth in the germline. Curr Biol 16(8):773–779

Gartner A, Boag PR, Blackwell TK (2008) Germline survival and apoptosis. WormBook: 1–20

Golden JW, Riddle DL (1982) A pheromone influences larval development in the nematode *Caenorhabditis elegans*. Science 218(4572):578–580

Golden JW, Riddle DL (1984) The *Caenorhabditis elegans* dauer larva: developmental effects of pheromone, food, and temperature. Dev Biol 102(2):368–378

Goudeau J, Bellemin S, Toselli-Mollereau E, Shamalnasab M, Chen Y, Aguilaniu H (2011) Fatty acid desaturation links germ cell loss to longevity through NHR-80/HNF4 in *C. elegans*. PLoS Biol 9(3):e1000599

Grant B, Hirsh D (1999) Receptor-mediated endocytosis in the *Caenorhabditis elegans* oocyte. Mol Biol Cell 10(12):4311–4326

Greer EL, Brunet A (2009) Different dietary restriction regimens extend lifespan by both independent and overlapping genetic pathways in *C. elegans*. Aging cell 8(2):113–127

Greer EL, Dowlatshahi D, Banko MR, Villen J, Hoang K, Blanchard D, Gygi SP, Brunet A (2007) An AMPK-FOXO pathway mediates longevity induced by a novel method of dietary restriction in *C. elegans*. Curr Biol 17(19):1646–1656

Greer ER, Pérez CL, Van Gilst MR, Lee BH, Ashrafi K (2008) Neural and molecular dissection of a *C. elegans* sensory circuit that regulates fat and feeding. Cell metabolism 8(2):118–131

Greer EL, Maures TJ, Ucar D, Hauswirth AG, Mancini E, Lim JP, Benayoun BA, Shi Y, Brunet A (2011) Transgenerational epigenetic inheritance of longevity in *Caenorhabditis elegans*. Nature 479(7373):365–371

Grishok A, Tabara H, Mello CC (2000) Genetic requirements for inheritance of RNAi in *C. elegans*. Science 287(5462):2494–2497

Gumienny TL, Lambie E, Hartwieg E, Horvitz HR, Hengartner MO (1999) Genetic control of programmed cell death in the *Caenorhabditis elegans* hermaphrodite germline. Development 126(5):1011–1022

Hansen D, Schedl T (2012) Stem cell proliferation versus meiotic fate decision in *C. elegans*. Advances in Experimental Medicine and Biology 757:71–99. (Chap. 4, this volume) Springer, New York

Hansen D, Hubbard EJA, Schedl T (2004) Multi-pathway control of the proliferation versus meiotic development decision in the *Caenorhabditis elegans* germline. Dev Biol 268(2): 342–357

Hardie DG (2011) AMP-activated protein kinase: an energy sensor that regulates all aspects of cell function. Genes Dev 25(18):1895–1908

Henderson ST, Gao D, Lambie EJ, Kimble J (1994) lag-2 may encode a signaling ligand for the GLP-1 and LIN-12 receptors of *C. elegans*. Development 120(10):2913–2924

Hirsh D, Oppenheim D, Klass M (1976) Development of the reproductive system of *Caenorhabditis elegans*. Dev Biol 49(1):200–219

Hodgkin J, Barnes TM (1991) More is not better: brood size and population growth in a self-fertilizing nematode. Proc Roy Soc B 246(1315):19–24

Holt SJ, Riddle DL (2003) SAGE surveys *C. elegans* carbohydrate metabolism: evidence for an anaerobic shift in the long-lived dauer larva. Mech Ageing Dev 124(7):779–800

Honjoh S, Yamamoto T, Uno M, Nishida E (2009) Signalling through RHEB-1 mediates intermittent fasting-induced longevity in *C. elegans*. Nature 457(7230):726–730

Hosono R, Nishimoto S, Kuno S (1989) Alterations of life span in the nematode *Caenorhabditis elegans* under monoxenic culture conditions. Exp Gerontol 24(3):251–264

Houthoofd K (2003) Life extension via dietary restriction is independent of the Ins/IGF-1 signalling pathway in *Caenorhabditis elegans*. Exp Gerontol 38(9):947–954

Houthoofd K, Vanfleteren JR (2006) The longevity effect of dietary restriction in *Caenorhabditis elegans*. Exp Gerontol 41(10):1026–1031

Houthoofd K, Braeckman BP, Lenaerts I, Brys K, De Vreese A, Van Eygen S, Vanfleteren JR (2002a) Axenic growth up-regulates mass-specific metabolic rate, stress resistance, and extends life span in *Caenorhabditis elegans*. Exp Gerontol 37(12):1371–1378

Houthoofd K, Braeckman BP, Lenaerts I, Brys K, De Vreese A, Van Eygen S, Vanfleteren JR (2002b) No reduction of metabolic rate in food restricted *Caenorhabditis elegans*. Exp Gerontol 37(12):1359–1369

Hsin H, Kenyon C (1999) Signals from the reproductive system regulate the lifespan of *C. elegans*. Nature 399(6734):362–366

Hsu H-J, LaFever L, Drummond-Barbosa D (2008) Diet controls normal and tumorous germline stem cells via insulin-dependent and -independent mechanisms in Drosophila. Dev Biol 313(2):700–712

Hu PJ (2007) Dauer. WormBook: 1–19

Hubbard EJ (2011) Insulin and germline proliferation in *Caenorhabditis elegans*. Vitam Horm 87:61–77

Hughes SE, Evason K, Xiong C, Kornfeld K (2007) Genetic and pharmacological factors that influence reproductive aging in nematodes. PLoS Genet 3(2):e25

Hughes SE, Huang C, Kornfeld K (2011) Identification of mutations that delay somatic or reproductive aging of *Caenorhabditis elegans*. Genetics 189(1):341–356

Jaramillo-Lambert A, Ellefson M, Villeneuve AM, Engebrecht J (2007) Differential timing of S phases, X chromosome replication, and meiotic prophase in the *C. elegans* germ line. Dev Biol 308(1):206–221

Jeong P-Y, Jung M, Yim Y-H, Kim H, Park M, Hong E, Lee W, Kim YH, Kim K, Paik Y-K (2005) Chemical structure and biological activity of the *Caenorhabditis elegans* dauer-inducing pheromone. Nature 433(7025):541–545

Jia K, Chen D, Riddle DL (2004) The TOR pathway interacts with the insulin signaling pathway to regulate *C. elegans* larval development, metabolism and life span. Development 131(16):3897–3906

Johnson TE, Mitchell DH, Kline S, Kemal R, Foy J (1984) Arresting development arrests aging in the nematode *Caenorhabditis elegans*. Mech Ageing Dev 28(1):23–40

Jones SJ, Riddle DL, Pouzyrev AT, Velculescu VE, Hillier L, Eddy SR, Stricklin SL, Baillie DL, Waterston R, Marra MA (2001) Changes in gene expression associated with developmental arrest and longevity in *Caenorhabditis elegans*. Genome Res 11(8):1346–1352

Kaeberlein TL, Smith ED, Tsuchiya M, Welton KL, Thomas JH, Fields S, Kennedy BK, Kaeberlein M (2006) Lifespan extension in *Caenorhabditis elegans* by complete removal of food. Aging cell 5(6):487–494

Kalaany NY, Sabatini DM (2009) Tumours with PI3K activation are resistant to dietary restriction. Nature 458(7239):725–731

Kang C, Avery L (2009) Systemic regulation of starvation response in *Caenorhabditis elegans*. Genes Dev 23(1):12–17

Kaplan F, Srinivasan J, Mahanti P, Ajredini R, Durak O, Nimalendran R, Sternberg PW, Teal PEA, Schroeder FC, Edison AS, Alborn HT (2011) Ascaroside expression in *Caenorhabditis elegans* is strongly dependent on diet and developmental stage. PLoS One 6(3):e17804

Kenyon CJ (2010) The genetics of ageing. Nature 464(7288):504–512

Killian DJ, Hubbard EJA (2004) *C. elegans* pro-1 activity is required for soma/germline interactions that influence proliferation and differentiation in the germ line. Development 131(6):1267–1278

Killian DJ, Hubbard EJA (2005) *Caenorhabditis elegans* germline patterning requires coordinated development of the somatic gonadal sheath and the germ line. Dev Biol 279(2):322–335

Kim S, Spike CA, Greenstein D (2012) Control of oocyte growth and meiotic maturation in *C. elegans*. Advances in Experimental Medicine and Biology 757:277–320. (Chap. 10, this volume) Springer, New York

Kimble J, Hirsh D (1979) The postembryonic cell lineages of the hermaphrodite and male gonads in *Caenorhabditis elegans*. Dev Biol 70(2):396–417

Kimble J, Sharrock WJ (1983) Tissue-specific synthesis of yolk proteins in *Caenorhabditis elegans*. Dev Biol 96(1):189–196

Kimble JE, White JG (1981) On the control of germ cell development in *Caenorhabditis elegans*. Dev Biol 81(2):208–219

Kiontke K, Sudhaus W (2006) Ecology of Caenorhabditis species. WormBook: 1–14

Klass MR (1977) Aging in the nematode *Caenorhabditis elegans*: major biological and environmental factors influencing life span. Mech Ageing Dev 6(6):413–429

Klass M, Hirsh D (1976) Non-ageing developmental variant of *Caenorhabditis elegans*. Nature 260(5551):523–525

Kleemann GA, Murphy CT (2009) The endocrine regulation of aging in *Caenorhabditis elegans*. Mol Cell Endocrinol 299(1):51–57

Korta DZ, Hubbard EJA (2010) Soma-germline interactions that influence germline proliferation in *Caenorhabditis elegans*. Dev Dyn 239(5):1449–1459

Korta DZ, Tuck S, Hubbard EJA (2012) S6K links cell fate, cell cycle and nutrient response in *C. elegans* germline stem/progenitor cells. Development 139:859–870

LaFever L, Drummond-Barbosa D (2005) Direct control of germline stem cell division and cyst growth by neural insulin in Drosophila. Science 309(5737):1071–1073

LaFever L, Feoktistov A, Hsu H-J, Drummond-Barbosa D (2010) Specific roles of target of rapamycin in the control of stem cells and their progeny in the Drosophila ovary. Development: 1–10

Lakowski B, Hekimi S (1996) Determination of life-span in *Caenorhabditis elegans* by four clock genes. Science 272(5264):1010–1013

Lakowski B, Hekimi S (1998) The genetics of caloric restriction in *Caenorhabditis elegans*. Proc Natl Acad Sci USA 95(22):13091–13096

Landis JN, Murphy CT (2010) Integration of diverse inputs in the regulation of *Caenorhabditis elegans* DAF-16/FOXO. Dev Dyn 239(5):1405–1412

Lapierre LR, Gelino S, Meléndez A, Hansen M (2011) Autophagy and lipid metabolism coordinately modulate life span in germline-less *C. elegans*. Curr Biol 21(18):1507–1514

LeBoeuf B, Guo X, Garcia LR (2011) The effects of transient starvation persist through direct interactions between CaMKII and ether-a-go-go K + channels in *C. elegans* males. Neuroscience 175:1–17

Lee GD, Wilson MA, Zhu M, Wolkow CA, de Cabo R, Ingram DK, Zou S (2006) Dietary deprivation extends lifespan in *Caenorhabditis elegans*. Aging cell 5(6):515–524

Liu F, Thatcher JD, Epstein HF (1997) Induction of glyoxylate cycle expression in *Caenorhabditis elegans*: a fasting response throughout larval development. Biochemistry 36(1):255–260

Long X, Spycher C, Han ZS, Rose AM, Müller F, Avruch J (2002) TOR deficiency in *C. elegans* causes developmental arrest and intestinal atrophy by inhibition of mRNA translation. Curr Biol 12(17):1448–1461

Luo S, Kleemann GA, Ashraf JM, Shaw WM, Murphy CT (2010) TGF-β and insulin signaling regulate reproductive aging via oocyte and germline quality maintenance. Cell 143(2):299–312

Maciejowski J, Ugel N, Mishra B, Isopi M, Hubbard EJA (2006) Quantitative analysis of germline mitosis in adult *C. elegans*. Dev Biol 292(1):142–151

Macosko EZ, Pokala N, Feinberg EH, Chalasani SH, Butcher RA, Clardy J, Bargmann CI (2009) A hub-and-spoke circuit drives pheromone attraction and social behaviour in *C. elegans*. Nature 458(7242):1171–1175

MacRae TH (2010) Gene expression, metabolic regulation and stress tolerance during diapause. Cell Mol Life Sci 67(14):2405–2424

McCarter J, Bartlett B, Dang T, Schedl T (1999) On the control of oocyte meiotic maturation and ovulation in *Caenorhabditis elegans*. Dev Biol 205(1):111–128

McGovern M, Voutev R, Maciejowski J, Corsi AK, Hubbard EJA (2009) A "latent niche" mechanism for tumor initiation. Proc Natl Acad Sci USA 106(28):11617–11622

McGrath PT, Xu Y, Ailion M, Garrison JL, Butcher RA, Bargmann CI (2011) Parallel evolution of domesticated Caenorhabditis species targets pheromone receptor genes. Nature 477(7364):321–325

McKay RM, McKay JP, Avery L, Graff JM (2003) *C. elegans*: a model for exploring the genetics of fat storage. Dev Cell 4(1):131–142

Meissner B, Boll M, Daniel H, Baumeister R (2004) Deletion of the intestinal peptide transporter affects insulin and TOR signaling in *Caenorhabditis elegans*. J Biol Chem 279(35): 36739–36745

Mendenhall AR, Wu D, Park S-K, Cypser JR, Tedesco PM, Link CD, Phillips PC, Johnson TE (2011) Genetic dissection of late-life fertility in *Caenorhabditis elegans*. J Gerontol A Biol Sci Med Sci 66(8):842–854

Michaelson D, Korta DZ, Capua Y, Hubbard EJA (2010) Insulin signaling promotes germline proliferation in *C. elegans*. Development 137(4):671–680

Mukhopadhyay A, Tissenbaum HA (2007) Reproduction and longevity: secrets revealed by *C. elegans*. Trends Cell Biol 17(2):65–71

Murakami M, Koga M, Ohshima Y (2001) DAF-7/TGF-beta expression required for the normal larval development in *C. elegans* is controlled by a presumed guanylyl cyclase DAF-11. Mech Dev 109(1):27–35

Nadarajan S, Govindan JA, McGovern M, Hubbard EJA, Greenstein D (2009) MSP and GLP-1/ Notch signaling coordinately regulate actomyosin-dependent cytoplasmic streaming and oocyte growth in *C. elegans*. Development 136(13):2223–2234

Narbonne P, Roy R (2006) Inhibition of germline proliferation during *C. elegans* dauer development requires PTEN, LKB1 and AMPK signalling. Development 133(4):611–619

Narbonne P, Roy R (2009) *Caenorhabditis elegans* dauers need LKB1/AMPK to ration lipid reserves and ensure long-term survival. Nature 457(7226):210–214

O'Rourke EJ, Soukas AA, Carr CE, Ruvkun G (2009) *C. elegans* major fats are stored in vesicles distinct from lysosome-related organelles. Cell Metab 10(5):430–435

Onken B, Driscoll M (2010) Metformin induces a dietary restriction-like state and the oxidative stress response to extend *C. elegans* Healthspan via AMPK, LKB1, and SKN-1. PLoS One 5(1):e8758

Panowski SH, Wolff S, Aguilaniu H, Durieux J, Dillin A (2007) PHA-4/Foxa mediates dietrestriction-induced longevity of *C. elegans*. Nature 447(7144):550–555

Pazdernik N, Schedl T (2012) Introduction to germ cell development in *C. elegans*. Advances in Experimental Medicine and Biology 757:1–16. (Chap. 1, this volume) Springer, New York

Pinkston JM, Garigan D, Hansen M, Kenyon C (2006) Mutations that increase the life span of *C. elegans* inhibit tumor growth. Science 313(5789):971–975

Pungaliya C, Srinivasan J, Fox BW, Malik RU, Ludewig AH, Sternberg PW, Schroeder FC (2009) A shortcut to identifying small molecule signals that regulate behavior and development in *Caenorhabditis elegans*. Proc Natl Acad Sci USA 106(19):7708–7713

Rae R, Iatsenko I, Witte H, Sommer RJ (2010) A subset of naturally isolated Bacillus strains show extreme virulence to the free-living nematodes *Caenorhabditis elegans* and Pristionchus pacificus. Environ Microbiol 12(11):3007–3021

Rechavi O, Minevich G, Hobert O (2011) Transgenerational inheritance of an acquired small RNA-based antiviral response in *C. elegans*. Cell 147(6):1248–1256

Reinke SN, Hu X, Sykes BD, Lemire BD (2010) *Caenorhabditis elegans* diet significantly affects metabolic profile, mitochondrial DNA levels, lifespan and brood size. Mol Genet Metab 100(3):274–282

Ren P, Lim CS, Johnsen R, Albert PS, Pilgrim D, Riddle DL (1996) Control of *C. elegans* larval development by neuronal expression of a TGF-beta homolog. Science 274(5291):1389–1391

Riddle DL, Albert PS (1997) Genetic and environmental regulation of Dauer larva development. In: Riddle DL, Blumenthal T, Meyer BJ, Priess JR (eds) *C. elegans* II, 2nd edn. Cold Spring Harbor Laboratory Press, Cold Spring Harbor (NY)

Rous P (1914) The influence of diet on transplanted and spontaneous mouse tumors. J Exp Med 20(5):433–451

Savage-Dunn C (2005) TGF-beta signaling. WormBook: 1–12

Schackwitz WS, Inoue T, Thomas JH (1996) Chemosensory neurons function in parallel to mediate a pheromone response in *C. elegans*. Neuron 17(4):719–728

Schafer WR (2005) Egg-laying. WormBook: 1–7

Schulz TJ, Zarse K, Voigt A, Urban N, Birringer M, Ristow M (2007) Glucose restriction extends *Caenorhabditis elegans* life span by inducing mitochondrial respiration and increasing oxidative stress. Cell Metab 6(4):280–293

Seidel HS, Kimble J (2011) The oogenic germline starvation response in *C. elegans*. PLoS One 6(12):e28074

Shibata Y, Branicky R, Landaverde IO, Hekimi S (2003) Redox regulation of germline and vulval development in *Caenorhabditis elegans*. Science 302(5651):1779–1782

Shtonda BB, Avery L (2006) Dietary choice behavior in *Caenorhabditis elegans*. J Exp Biol 209 (Pt 1):89–102

Singh K, Chao MY, Somers GA, Komatsu H, Corkins ME, Larkins-Ford J, Tucey T, Dionne HM, Walsh MB, Beaumont EK, Hart DP, Lockery SR, Hart AC (2011) *C. elegans* Notch signaling regulates adult chemosensory response and larval molting quiescence. Curr Biol 21(10):825–834

Srinivasan J, Kaplan F, Ajredini R, Zachariah C, Alborn HT, Teal PEA, Malik RU, Edison AS, Sternberg PW, Schroeder FC (2008) A blend of small molecules regulates both mating and development in *Caenorhabditis elegans*. Nature 454(7208):1115–1118

Szewczyk NJ, Udranszky IA, Kozak E, Sunga J, Kim SK, Jacobson LA, Conley CA (2006) Delayed development and lifespan extension as features of metabolic lifestyle alteration in *C. elegans* under dietary restriction. J Exp Biol 209(Pt 20):4129–4139

Taguchi A, White MF (2008) Insulin-like signaling, nutrient homeostasis, and life span. Annu Rev Physiol 70:191–212

Tamai KK, Nishiwaki K (2007) bHLH transcription factors regulate organ morphogenesis via activation of an ADAMTS protease in *C. elegans*. Dev Biol 308(2):562–571

Tan KT, Luo SC, Ho WZ, Lee YH (2011) Insulin/IGF-1 receptor signaling enhances biosynthetic activity and fat mobilization in the initial phase of starvation in adult male *C. elegans*. Cell Metab 14(3):390–402

Tannenbaum A, Silverstone H (1953) Nutrition in relation to cancer. Adv Cancer Res 1:451–501

Trent C, Tsuing N, Horvitz HR (1983) Egg-laying defective mutants of the nematode *Caenorhabditis elegans*. Genetics 104(4):619–647

Van Gilst MR, Hadjivassiliou H, Jolly A, Yamamoto KR (2005) Nuclear hormone receptor NHR-49 controls fat consumption and fatty acid composition in *C. elegans*. PLoS Biol 3(2):e53

Vowels JJ, Thomas JH (1992) Genetic analysis of chemosensory control of dauer formation in *Caenorhabditis elegans*. Genetics 130(1):105–123

Wadsworth WG, Riddle DL (1989) Developmental regulation of energy metabolism in *Caenorhabditis elegans*. Dev Biol 132(1):167–173

Walker AK, Yang F, Jiang K, Ji J-Y, Watts JL, Purushotham A, Boss O, Hirsch ML, Ribich S, Smith JJ, Israelian K, Westphal CH, Rodgers JT, Shioda T, Elson SL, Mulligan P, Najafi-Shoushtari H, Black JC, Thakur JK, Kadyk LC, Whetstine JR, Mostoslavsky R, Puigserver P, Li X, Dyson NJ, Hart AC, Näär AM (2010) Conserved role of SIRT1 orthologs in fasting-dependent inhibition of the lipid/cholesterol regulator SREBP. Genes Dev 24(13):1403–1417

Wang J, Kim SK (2003) Global analysis of dauer gene expression in *Caenorhabditis elegans*. Development 130(8):1621–1634

Wang MC, O'Rourke EJ, Ruvkun G (2008) Fat metabolism links germline stem cells and longevity in *C. elegans*. Science 322(5903):957–960

Watts JL (2009) Fat synthesis and adiposity regulation in *Caenorhabditis elegans*. Trends Endocrinol Metab 20(2):58–65

Wolke U, Jezuit EA, Priess JR (2007) Actin-dependent cytoplasmic streaming in *C. elegans* oogenesis. Development 134(12):2227–2236

Wood WB, Hecht R, Carr S, Vanderslice R, Wolf N, Hirsh D (1980) Parental effects and phenotypic characterization of mutations that affect early development in *Caenorhabditis elegans*. Dev Biol 74(2):446–469

Wood JG, Rogina B, Lavu S, Howitz K, Helfand SL, Tatar M, Sinclair D (2004) Sirtuin activators mimic caloric restriction and delay ageing in metazoans. Nature 430(7000):686–689

Wullschleger S, Loewith R, Hall MN (2006) TOR signaling in growth and metabolism. Cell 124(3):471–484

Yamawaki TM, Berman JR, Suchanek-Kavipurapu M, McCormick M, Maria Gaglia M, Lee S-J, Kenyon C (2010) The somatic reproductive tissues of *C. elegans* promote longevity through steroid hormone signaling. PLoS Biol 8 (8)

You Y-j, Kim J, Cobb M, Avery L (2006) Starvation activates MAP kinase through the muscarinic acetylcholine pathway in *Caenorhabditis elegans* pharynx. Cell Metab 3(4):237–245

You Y-j, Kim J, Raizen DM, Avery L (2008) Insulin, cGMP, and TGF-beta signals regulate food intake and quiescence in *C. elegans*: a model for satiety. Cell Metab 7(3):249–257

Zhang X, Zabinsky R, Teng Y, Cui M, Han M (2011) microRNAs play critical roles in the survival and recovery of *Caenorhabditis elegans* from starvation-induced L1 diapause. Proc Natl Acad Sci USA 108(44):17997–18002

Chapter 6
Meiotic Development in *Caenorhabditis elegans*

Doris Y. Lui and Monica P. Colaiácovo

Abstract *Caenorhabditis elegans* has become a powerful experimental organism with which to study meiotic processes that promote the accurate segregation of chromosomes during the generation of haploid gametes. Haploid reproductive cells are produced through one round of chromosome replication followed by two successive cell divisions. Characteristic meiotic chromosome structure and dynamics are largely conserved in *C. elegans*. Chromosomes adopt a meiosis-specific structure by loading cohesin proteins, assembling axial elements, and acquiring chromatin marks. Homologous chromosomes pair and form physical connections though synapsis and recombination. Synaptonemal complex and crossover formation allow for the homologs to stably associate prior to remodeling that facilitates their segregation. This chapter will cover conserved meiotic processes as well as highlight aspects of meiosis that are unique to *C. elegans*.

Keywords Meiosis • Pairing • Recombination • Synapsis • Cohesion • Germline • *C. elegans*

6.1 Introduction

Meiosis is a specialized cell division process by which sexually reproducing diploid organisms, including humans, produce haploid gametes (i.e., eggs and sperm) to be used for fertilization. This halving in the number of chromosomes is accomplished by following one round of DNA replication with two consecutive rounds of chromosome segregation (meiosis I and meiosis II). Whereas

D.Y. Lui • M.P. Colaiácovo (✉)
Department of Genetics, Harvard Medical School, Boston, MA 02115, USA
e-mail: mcolaiacovo@genetics.med.harvard.edu

T. Schedl (ed.), *Germ Cell Development in C. elegans*, Advances in Experimental Medicine and Biology 757, DOI 10.1007/978-1-4614-4015-4_6,

homologous chromosomes segregate away from each other at meiosis I, sister chromatids segregate to opposite poles of the spindle at meiosis II. The accurate segregation of chromosomes at every cell division is required to prevent aneuploidy, which consists of the formation of cells carrying an incorrect number of chromosomes. This is of tremendous importance given that meiotic chromosome missegregation has a significant impact on human health, as indicated by it being the leading cause of congenital birth defects and miscarriages (Hassold and Hunt 2001). Therefore, there are several mechanisms set in place to ensure proper chromosome segregation, as exemplified by meiosis I, where homologous chromosomes must pair, align, and form physical connections prior to the first division. Several key processes promote the normal progression of these steps during meiosis I. These include the formation of connections between sister chromatids afforded by the establishment of sister chromatid cohesion, the establishment of stable interactions between homologs achieved through the formation of a proteinaceous structure known as the synaptonemal complex (SC), and the repair of programmed meiotic DNA double-strand breaks (DSBs) into crossovers (CO).

Caenorhabditis elegans has several unique features that make it an advantageous model organism to study both meiosis and chromosome dynamics. It has a relatively short generation time and is a tractable genetic system that can be used in both forward and reverse genetic approaches. RNA-mediated interference (RNAi), cosuppression and the generation of transgenic lines with targeted deletions can all be used to examine loss-of-function phenotypes (Frokjaer-Jensen et al. 2010; Dernburg et al. 2000; Timmons and Fire 1998). Chromosome nondisjunction is easily assessed by missegregation of the X chromosome, which determines sex in *C. elegans*. While males have a single sex chromosome (X0), hermaphrodites have two copies (XX). Self-fertilizing hermaphroditic worms lay mostly hermaphroditic progeny and produce males at a very low frequency (<0.2%) (Hodgkin et al. 1979). Mutations affecting meiotic prophase I events result in increased chromosome nondisjunction. Rather than arresting as a result of these defects, aneuploid gametes in *C. elegans* become fertilized but result in inviable offspring as indicated by increased *emb*ryonic lethality (Emb). The Emb phenotype is typically accompanied by a *h*igh *i*ncidence of *m*ales (Him) phenotype.

Distinct morphological changes during its life cycle, as well as its transparency, aid in the developmental staging of worms. Transparency of the entire body of the worm, including the gonad, also facilitates the analysis via whole-mount preparations for immunofluorescence, live or fixed imaging of fluorescent fusions *in vivo*, and *in situ* hybridization (Motohashi et al. 2006; Lee and Schedl 2006; Duerr 2006; Phillips et al. 2009a). This allows for 3D analysis of chromosome dynamics as well as the localization of proteins and specific chromosomal regions in the context of intact meiotic nuclei.

In this chapter, we will focus on the meiosis in hermaphrodite worms as they are producing oocytes. Since the germline contains over half of the total number of nuclei comprising an adult hermaphrodite worm, it provides ample biomass for the study of meiosis (MacQueen et al. 2005; Hirsh et al. 1976). The *C. elegans* gonad is

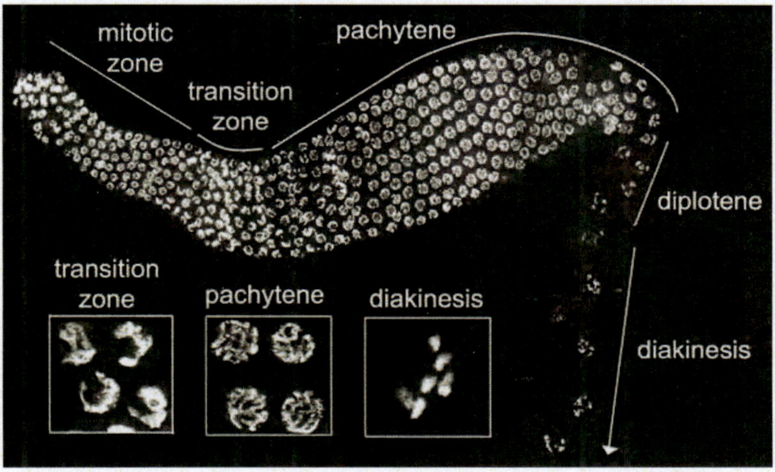

Fig. 6.1 The *C. elegans* gonad. A dissected and DAPI-stained gonad of a hermaphrodite adult worm. Progression from the distal to the proximal end is depicted from left to right. The image is a projection of three-dimensional data stacks of intact nuclei, which were taken approximately halfway through the gonad, to facilitate visualization of nuclear morphology

a bi-lobed structure in which nuclei are ordered in a spatial–temporal gradient such that sequential stages of meiosis are easily visualized (Fig. 6.1) (Crittenden et al. 1994). The most distal end, or proliferative zone, contains germ cell nuclei that are undergoing mitotic divisions and meiotic S-phase (Crittenden et al. 2006; Jaramillo-Lambert et al. 2007; Fox et al. 2011). As nuclei move proximally, they enter the early stages of meiosis (leptotene and zygotene) that correspond to the transition zone, in which chromosomes acquire a polarized organization (Dernburg et al. 1998; Crittenden et al. 1994; Hirsh et al. 1976). In the transition zone, chromosomes pair, initiate recombination, and begin to synapse (Fig. 6.2; Dernburg et al. 1998; MacQueen et al. 2002; Alpi et al. 2003; Colaiacovo et al. 2003). Synapsis is completed by entrance into pachytene, when chromosomes redistribute throughout the nuclear periphery. Crossover recombination is then completed during pachytene within the context of fully synapsed chromosomes. By late pachytene, the synaptonemal complex (SC) starts to disassemble, a process that continues during diplotene. Finally, at diakinesis, six pairs of homologous chromosomes (bivalents) are observed held together by chiasmata, the cytological manifestation resulting from the earlier crossover event between homologous chromosomes underpinned by flanking sister chromatid cohesion (Villeneuve 1994).

The distinct meiotic stages, basic mechanisms, and genes involved in meiotic chromosome dynamics are largely conserved across *taxa* (Tables 6.1 and 6.2). Thus, meiotic studies in *C. elegans*, a model system amenable to a wide range of genetic, molecular, cytological, and biochemical approaches, can provide significant insight into the meiotic processes occurring in other organisms.

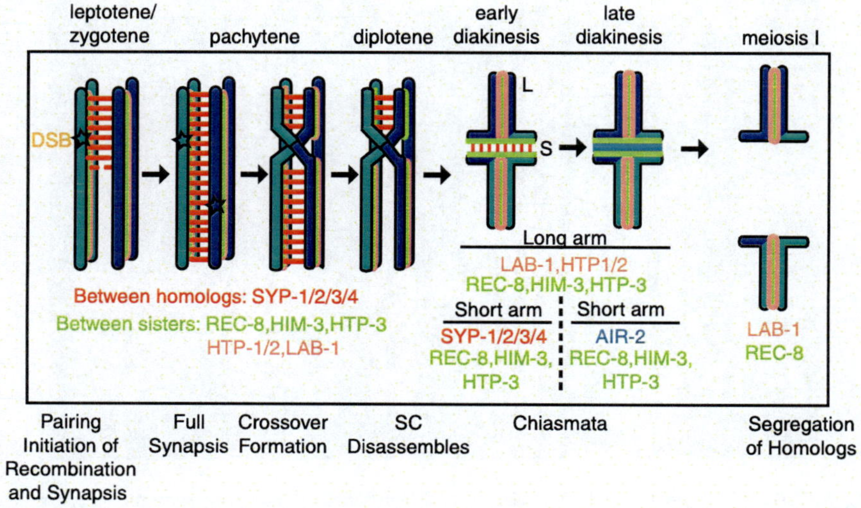

Fig. 6.2 Events during meiotic progression that contribute to the proper segregation of homologs during the first meiotic division

6.2 Sister Chromatid Cohesion

Universally, sister chromatid cohesion joins sister chromatids (Oliveira and Nasmyth 2010). By tethering sister chromatids together at distinct chromosomal loci, the cohesin complex mediates the formation of the loop-axis structure of chromatin (Blat et al. 2002). In the absence of cohesion, sister chromatids segregate prematurely during meiosis I. Specifically, the attachment of chromosomes through cohesion permits cosegregation of sister chromatids at the first meiotic division, whereas it facilitates bi-orientation of sister chromatids on the spindle at the second meiotic division.

The cohesin complex contains four evolutionarily conserved subunits: two *s*tructural *m*aintenance of *c*hromosomes (SMC) subunits, SMC-1/HIM-1 and SMC-3, the non-SMC component SCC-3, and a kleisin subunit (Hagstrom and Meyer 2003). As in plants and mammals, several meiotic α-kleisins exist in *C. elegans* to ensure cohesion between sister chromatids: REC-8, COH-3, and COH-4 (Severson et al. 2009; Pasierbek et al. 2001; Lee and Hirano 2011; Jiang et al. 2007; Bai et al. 1999; Parisi et al. 1999; Lee et al. 2003). The coiled-coil domains of the SMC proteins connect the ATPase-containing "head" domain to the hinge region (Melby et al. 1998). By folding at the hinge region, SMC proteins form intramolecular coiled-coils (Haering et al. 2002). The two SMC proteins, SMC-1 and SMC-3, can then interact at their hinge regions with the N- and C-terminal regions of the kleisin subunits (Haering et al. 2002). Current models suggest that the complex forms a ring structure that acts to embrace the sister chromatids (Haering et al. 2002).

Table 6.1 Conservation of meiotic processes across species

	Homologous pairing	Chromosome movement		
Species	Primary mediator	Primary mediator	SUN and KASH	Polarized chromosomes
A. thaliana	DSB repair	Unknown, but maize uses microtubules (mt) and actin	SUN-1/2	None, but bouquet in rye, maize, and wheat
S. cerevisiae	DSB repair	Actin cables	Mps3 and Csm4	Bouquet; telomere-mediated
S. pombe	Horsetail movement	mt/dynein	Sad1 and Kms1	Horsetail; telomere-mediated
C. elegans	Pairing centers interact via ZIM proteins (autosomes) and HIM-8 (X)	mt/dynein	SUN-1 and ZYG-12	Crescent-shaped organization; PC mediated
D. melanogaster	Females: heterochromatin Males: X–Y pairing uses rDNA repeats, SNM and MNM	mt	–	None
M. musculus	DSB repair	mt	SUN-1/2	Bouquet; telomere-mediated

	Synapsis			Recombination	
Species	Axial/lateral elements	Central region	DSB-dependent?	Meiotic DSBs/nucleus	CO interference?
A. thaliana	ASY1	ZYP1	Yes	153 ± 33 Rad51 foci (Mercier et al. 2005)	Yes
S. cerevisiae	Red1, Hop1, and Mek1	Zip1	Yes	140–170 DSBs (Buhler et al. 2007)	Yes
S. pombe	Rec10, Hop1, and Mek1	None	N/A	Mean: 22.2 Rad51 foci; maximum: 50 Rad51 foci (Lorenz et al. 2006)	No
C. elegans	HIM-3 and HTP-1/2/3	SYP-1/2/3/4	No	23–39 RAD-51 foci (during mid/late pachytene; Nottke et al. 2011; Gao, Saito, and Colaiácovo, personal communication)	Yes
D. melanogaster	CONA	C(3)G and C(2)M	No	Males: none Females: 24.3 γ-HIS2AV foci (Jang et al. 2003)	Females: Yes
M. musculus	SYCP2/3 and HORMAD1/2	SYCP 1, SYCE1/2/3, and TEX12	Yes	230–420 Rad51 foci (reviewed in Baudat and de Massy 2007)	Yes

Table 6.2 Conservation of meiotic recombination across species

	Strand-exchange proteins			
Species	Dmc1, Mnd1/Hop2	Rad51	Pro-CO factors	Meiotic CO intermediate resolution
A. thaliana	Yes	Rad51C	MSH4/5, ZIP3, ZYP1, ZIP4, and RCK (MER3)	MUS81 (9–12% of COs (Berchowitz et al. 2007))
S. cerevisiae	Yes	Rad51	Msh4/5, Zip3, Zip1, Zip2, Spo22 (Zip4), Spo16, and Mer3	Mus81-Mms4 [20% of COs (Argueso et al. 2004; de los Santos et al. 2003)]
S. pombe	Yes	Rhp51	–	Mus81-Eme1 [95–99% of COs (Smith et al. 2003)]
C. elegans	No	RAD-51	MSH-4/5, ZHP-3, SYP-1, SYP-2, SYP-3, SYP-4, and COSA-1	HIM-18 [30% on autosome; 50% on X (Saito et al. 2009)], XPF-1 [21% autosome; and 24% X (Saito et al. 2009)], and MUS-81 (Youds et al. 2010)
D. melanogaster	No	SpnA and SpnB	C(3)G, C(2)M, REC (MCM8), and MEI-218	MEI9-MUS312 [86–92% of COs (Yildiz et al. 2002)] with ERCC1 and HDM; MUS81 [9% of COs (Trowbridge et al. 2007)]
M. musculus	Yes	Rad51	MSH4/5, RNF212, SYCP 1, SYCE1/2/3, TEX12, and TEX11 (ZIP4H)	MUS81 (only required for subset of MLH1-independent COs (Holloway et al. 2008))

	Anti-CO factors				
Species	Sgs1/BLM/RecQ helicases	Disruptors of strand invasion	Meiotic DSBs/nucleus (references in Table 6.1)	Intersister (IS):interhomolog (IH) ratio	IH NCO:CO ratio
A. thaliana	RECQ4A; *recq4A-4* mutants have normal fertility (Mannuss et al. 2010)	SRS2 and AT1G79950? * (Knoll and Puchta 2011)	153 ± 33	–	–
S. cerevisiae	Sgs1; deletion increases IS COs and multi-chromatid joint molecules (Oh et al. 2007)	Srs2 and Hed1	140–170	Total events: 1:5 (Goldfarb and Lichten 2010); JM: 1:2.4 (Schwacha and Kleckner 1994)–1:4.7 (Oh et al. 2007)	1:1 (Allers and Lichten 2001); 2:3 (Mancera et al. 2008); 1:3 (McMahill et al. 2007)

S. pombe	Rqh1; deletion reduces IH CO (Cromie et al. 2008)	Fml1? *(prevents mitotic COs)	22.2	3:1 (Cromie et al. 2006)	1:4 (Cromie et al. 2005)
C. elegans	HIM-6; *him-6* mutants have reduced IH COs (Wicky et al. 2004)	RTEL-1	23–39	–	–
D. melanogaster	Blm/Mus309; *mus309*^{N2} have reduced IH COs (Wicky et al. 2004)	–	24.3	40% of events on ring chromosomes are IS COs (Webber et al. 2004)	4:1 (Hilliker and Chovnick 1981)
M. musculus	BLM; *Blm*$^{-/-}$ spermatocytes have increased chiasmata (Holloway et al. 2010)	RTEL1	230–420	–	–

Species	Number of chromosomes (2N)	CO/meiosis	NCO/meiosis	% COs subject to interference	Gene conversion tract length
A. thaliana	10	9.2 (Mercier et al. 2005)	–	75–84% (Mercier et al. 2005; Higgins et al. 2004)	–
S. cerevisiae	32	90.5 (Mancera et al. 2008)	66.1 (Mancera et al. 2008)	40% (Argueso et al. 2004)	CO: 2 kb (Mancera et al. 2008) NCO: 1.8 kb (Mancera et al. 2008)
S. pombe	6	45 (Munz 1994)	–	None	–
C. elegans	12	6 (genetic map)	–	All (Zalevsky et al. 1999; Kelly et al. 2000)	
D. melanogaster	8	5.6 (genetic map)	–	All (Copenhaver et al. 2002; Baker and Carpenter 1972)	CO: 343 ± 143 bp NCO: 1,208 ± 790 bp (Curtis et al. 1989)
M. musculus	40	Males: 22.6–23.9 Females: 24.1 (Baudat and de Massy 2007)	–	92% (Holloway et al. 2008)	CO: 500 bp NCO: 187 bp (Cole et al. 2010)

* suggested but not fully demonstrated

Sister chromatid cohesion is established during replication (Sherwood et al. 2010). The HEAT/armadillo repeat-containing protein TIM-1, a paralog of the Drosophila circadian clock protein TIMELESS, mediates association of the non-SMC components of cohesin onto chromatin (Chan et al. 2003). HEAT-repeat-containing protein Scc2 and TPR-protein Scc4 are components of the highly conserved cohesin loading complex found in yeast, Xenopus, and humans (Gillespie and Hirano 2004; Watrin et al. 2006; Ciosk et al. 2000; Rollins et al. 2004; Tonkin et al. 2004; Seitan et al. 2006). In *C. elegans*, the ortholog for Scc4 is MAU-2, which likely plays a similar role to its yeast and human counterparts, whereas SCC-2 has been shown to mediate the meiotic loading of cohesin subunits on chromatin (Seitan et al. 2006; Lightfoot et al. 2011).

Maintenance of sister chromatid cohesion requires cohesin subunits and the HEAT-repeat-containing protein PDS-5/EVL-14 (Wang et al. 2003; Nasmyth and Haering 2005). Absence of either PDS-5 or the cohesin subunit SCC-3 results in premature separation of sister chromatids during meiotic prophase in *C. elegans* (Wang et al. 2003). Previously, it was thought that cohesin may act as a scaffold for the loading of additional axis-associated components, which promote chromosome synapsis (Page and Hawley 2004). Recent work, however, suggests that axial components contribute to sister chromatid cohesion, as exemplified by the observation that cohesin components are interdependent with axial/lateral component HTP-3 for stable association on chromosomes (Severson et al. 2009; Kim et al. 2010; Goodyer et al. 2008).

Stepwise removal of sister chromatid cohesion allows for proper segregation of chromosomes at each of the meiotic divisions. Specifically, whereas homologs must segregate away from each other in the first meiotic division, sister chromatids only separate in the second meiotic division. In most eukaryotes, this is accomplished through the incorporation of the meiosis-specific kleisin Rec8 in cohesin complexes. During the meiosis I division, REC-8 cleavage elicits bivalent resolution by releasing the connection holding homologs together (Rogers et al. 2002). However, REC-8 is protected from cleavage at defined regions between sister chromatids to prevent their premature separation at meiosis I (Rogers et al. 2002; Pasierbek et al. 2001). At the second division, the residual REC-8 is cleaved to permit segregation of sister chromatids (Rogers et al. 2002; Pasierbek et al. 2001).

The persistence of sister chromatid cohesion in the absence of REC-8 is due to the presence of the other two meiotic kleisins, COH-3 and COH-4 (Severson et al. 2009). Sister chromatid cohesion is only significantly reduced in the absence of all three kleisins (Severson et al. 2009). However, these paralogs apparently only share partial overlap for other functions during meiosis. This is suggested in part by the observation that *coh-3 coh-4* double mutants are more severely defective for the assembly of the SC than *rec-8* single mutants (Severson et al. 2009). Moreover, REC-8 alone is sufficient to maintain sister chromatid cohesion during the second meiotic division (Severson et al. 2009).

Finally, aside from mediating cohesion between sister chromatids, work in *C. elegans* and other organisms suggests that cohesin may play additional roles. Absence of Rec8 in budding yeast results in a shortened length of meiotic S-phase

(Cha et al. 2000). Since REC-8 localizes to mitotic nuclei in the *C. elegans* gonad (Goodyer et al. 2008; Pasierbek et al. 2001), it has been proposed that REC-8 may also have a function in both the mitotically cycling nuclei and the progression of meiotic S-phase in the germline (Jaramillo-Lambert et al. 2007). Premeiotic nuclei in the germlines of *rec-8* mutants exhibit increased levels of DSBs (Hayashi et al. 2007). Failure to load cohesin subunits also results in an accumulation of meiotic DSBs, which indicates impaired meiotic DSB repair (Pasierbek et al. 2003; Smolikov et al. 2007a; Baudrimont et al. 2011; Lightfoot et al. 2011). This is in line with work in yeast that suggests that cohesin may have a significant role in mediating DNA metabolism. Specifically, absence of cohesin results in altered distribution of meiotic DSBs because cohesin mediates the localization of axial components that recruit factors involved in DSB formation (Kugou et al. 2009; Ellermeier and Smith 2005; Panizza et al. 2011). Moreover, loading of the cohesin complex surrounding DSB sites facilitates the use of the sister chromatid as a template for repair in mitotic cells in budding yeast (Heidinger-Pauli et al. 2008). During meiosis, axial elements promote homolog bias in budding yeast by stimulating the local loss of cohesion in order to release one end of the DSB to undergo repair using the homolog as a template (Kim et al. 2010). However, as recombination progresses, cohesion functions to maintain homolog bias by promoting formation of interhomolog recombination intermediates by inhibiting activity of the other end of the DSB (Kim et al. 2010). *rec8* deletion mutants in yeast and *ord* Drosophila mutants, which fail to localize cohesin on chromosomes, undergo increased recombination between sister chromatids (Webber et al. 2004; Kim et al. 2010). Similarly, DSBs in *rec-8* mutants are likely repaired by intersister recombination given that the univalents observed in diakinesis oocytes in this background have a mostly intact appearance, lacking elevated levels of either chromosome fragments or aggregates (Pasierbek et al. 2001; Colaiacovo et al. 2003; Smolikov et al. 2007a). In *C. elegans*, loading of meiotic cohesin appears to be required for the DNA damage checkpoint to sense and eliminate nuclei with persistent unrepaired DSBs. Specifically, *scc-2* mutants fail to recruit a component of the DNA damage checkpoint to unrepaired meiotic DSBs and activate the apoptotic checkpoint (Lightfoot et al. 2011). Therefore, the regulated establishment, maintenance, and removal of sister chromatid cohesion plays several key roles during meiosis throughout species. Moreover, cohesin also plays important roles in the regulation of DSB distribution and repair.

6.3 Meiotic Pairing of Homologous Chromosomes

In the transition zone of the *C. elegans* germline, chromosomes polarize towards one side of the nuclei, which imparts a crescent-shaped appearance to the DAPI-stained chromatin. This reorganization of the chromosomes is coincident with the pairing of chromosomes as determined by fluorescence in situ hybridization (FISH) (Dernburg et al. 1998). The formation of the crescent-shaped morphology is thought to serve the same purpose as the clustering of chromosomes that is observed in most other organisms during meiotic prophase. In other organisms, both ends of the chromosome localize preferentially near the spindle pole body or the centrosome to form the arrangement known as the "bouquet" (Zickler and Kleckner 1998). Traditionally, the bouquet is thought to assist in the homology search by reducing the volume of searchable space from a three-dimensional to a two-dimensional space in the nucleus, and simplifying the homology search by sorting chromosomes by their lengths (Schlecht et al. 2004; Roeder 1997; Loidl 1990). Since in *C. elegans* only one end of each chromosome attaches to the nuclear envelope, there is not a formation of a classic "bouquet" arrangement, but the reorganization of the chromosomes detected at transition zone is thought to accomplish the same goal by assisting in the pairing of chromosomes (Goldstein 1982).

Defects in crossing over due to either chromosome deletions or the presence of translocations suggested that *C. elegans* chromosomes require *cis*-acting regions to pair their homologs (Rose et al. 1984; McKim et al. 1988; Villeneuve 1994; Rosenbluth and Baillie 1981; Herman et al. 1982). These regions were first defined as homology recognition regions (HRR) and were later given the name "pairing centers" (PCs) (Villeneuve 1994; McKim et al. 1988). PCs were roughly mapped by genetic analysis to the ends of the chromosomes in *C. elegans* (reviewed in Zetka and Rose 1995). PCs contain highly repeated DNA sequence motifs that recruit the C2H2 zinc-finger proteins, ZIM-1, ZIM-2, ZIM-3, and HIM-8 (Phillips et al. 2005, 2009b; Phillips and Dernburg 2006; Sanford and Perry 2001). While HIM-8 binds specifically to the X chromosome, the ZIM proteins mediate the interactions between the autosomes (Phillips and Dernburg 2006; Phillips et al. 2005). Interestingly, there are only three ZIM proteins to mediate pairing between the five autosomes (Phillips and Dernburg 2006). ZIM-1 and ZIM-3 each mediate pairing of two autosomes (Phillips and Dernburg 2006). This suggests that the stable pairing between chromosomes must require a yet unknown additional mechanism to prevent interactions between the nonhomologous chromosomes that share the same ZIM proteins.

It is important to note that pairing centers are not unique to *C. elegans*. Both *Drosophila* males and budding yeast have *cis*-acting regions that promote chromosome pairing as well. Repetitive sequences in the rDNA are required for sex chromosome pairing in *Drosophila* males (McKee 1996). The interaction between the rDNA repeats is mediated by the cohesin Scc3 homolog SNM and the BTB domain-containing protein MNM, which also facilitate the pairing of autosomes in *Drosophila* males (Thomas and McKee 2007; Thomas et al. 2005). While the pairing centers in *C. elegans* and *Drosophila* males promote homologous pairing of chromosomes, centromere coupling, the pairing of centromeric regions during meiosis in budding yeast, occurs between nonhomologous chromosomes (Tsubouchi and Roeder 2005). Centromere pairing of nonhomologous chromosomes precedes homologous pairing and can occur in the absence of recombination (Tsubouchi and Roeder 2005; Obeso and Dawson 2010). This centromere pairing requires both the cohesin component Rec8 and Zip1, a structural component of the central region of the SC in yeast (Tsubouchi and Roeder 2005; Bardhan et al. 2010). Therefore, studies on the mechanism by which pairing centers allow for homolog recognition in *C. elegans* can be informative for the understanding of how *cis*-acting regions promote chromosome associations in other organisms.

X chromosome pairing is particularly robust compared to the autosomes in *C. elegans*. When dynein is depleted, X chromosome pairing is not affected, but the autosomes fail to pair (Sato et al. 2009). The X chromosome pairing is also more resistant to defects in axis morphogenesis and SC formation. Specifically, pairing of the X chromosome is less affected than the pairing of the autosomes in *htp-1* mutants, *him-3* hypomorphs, and *cra-1* mutants (Nabeshima et al. 2004; Martinez-Perez and Villeneuve 2005; Couteau et al. 2004; Smolikov et al. 2008). This efficient pairing of the X chromosome in the *him-3* hypomorphs and *cra-1* mutants translates into less severe synapsis and recombination defects of X chromosomes compared to the autosomes that fail to pair (Smolikov et al. 2008; Couteau et al. 2004). The mechanism by which the X chromosomes pair more efficiently is unknown, but the answer may lie in the chromatin state of the X chromosome. In *C. elegans*, the X chromosome adopts a heterochromatic state and is silenced during meiotic prophase (Kelly et al. 2002). Since transcriptionally silenced regions are often compartmentalized within the nucleus (Cremer et al. 2006), the silenced X chromosomes may preferentially be associated by recruitment to that subnuclear position. On the other hand, it is possible that the heterochromatin of the X may promote pairing more directly as it does in *Drosophila* females (Hawley et al. 1992; Karpen et al. 1996; Dernburg et al. 1996).

Gross changes in chromosome morphology suggest that chromosome movement is highly dynamic during meiotic prophase. In *C. elegans*, meiotic chromosome movement facilitates the timely pairing of chromosomes and is dependent on microtubules (Sato et al. 2009). Depletion of dynein, the microtubule motor, results in pairing delays and failure to synapse chromosomes in this system (Sato et al. 2009). This is analogous to the dynein-mediated horsetail chromosome movement observed in fission yeast that promotes chromosome pairing (Ding et al. 2004; Hiraoka et al. 2000; Miki et al. 2002). However, chromosome movement and pairing is largely independent of dynein in budding yeast and instead requires actin (Koszul et al. 2008; Lui et al. 2006; Trelles-Sticken et al. 2005). Universally, the SUN and KASH proteins that bridge the nuclear membrane in worms, flies, yeast, and humans allow for the transduction of force from the cytoplasm to the nucleus (Starr and Fridolfsson 2010). In *C. elegans*, the KASH protein ZYG-12 is a transmembrane protein that interacts with dynein and spans the outer nuclear membrane (Malone et al. 2003). Meanwhile, ZYG-12 interacts with SUN-1 in the perinuclear space (Minn et al. 2009). Importantly, nocodazole treatment, which depolymerizes microtubles, results in nonhomologous synapsis in *C. elegans* (Sato et al. 2009). It has therefore been proposed that meiotic chromosome movement not only facilitates pairing but also disrupts nonhomologous interactions (Koszul and Kleckner 2009).

Modification of the phosphorylation state of SUN-1 is required for proper chromosome morphogenesis during meiotic prophase. Specifically, phosphorylation of SUN-1 by the Polo-like kinase PLK-2 allows for its aggregation and polarization of the aggregates on one half of the nuclear periphery (Penkner et al. 2009; Harper et al. 2011). Moreover, CHK-2, the ortholog of the checkpoint protein kinases Cds1 in *S. pombe* and Chk2 in mammals, also mediates the phosphorylation of SUN-1, and *chk-2* mutants fail to cluster their chromosomes and pair (MacQueen and

Villeneuve 2001; Penkner et al. 2009). More recently it has been shown that CHK-2 is required for the association of the PBD-domain of PLK-2 with HIM-8 (Harper et al. 2011). Furthermore, the ZIM proteins, but not HIM-8, require CHK-2 to associate with the nuclear membrane, and the ZIM/HIM-8 proteins associate with the nuclear envelope in distinct clusters rather than forming a single cluster (Phillips and Dernburg 2006). Therefore, the crescent-shaped organization acquired by the chromosomes in the transition zone likely results from the ZIM/HIM-8 proteins associating with the polarized SUN-1 aggregates. Interestingly, although the ZIM/HIM-8 proteins colocalize with the SUN/KASH aggregates and PLK-2 (Sato et al. 2009; Harper et al. 2011; Labella et al. 2011), only HIM-8 was found to interact with PLK-2 by yeast two-hybrid (Harper et al. 2011). PC-associated proteins may recruit PLK-2 to the nuclear envelope to phosphorylate SUN-1, which results in its aggregation and polarization (Penkner et al. 2009; Harper et al. 2011; Labella et al. 2011). Finally, SUN-1 dephosphorylation is required to redisperse chromosomes in the nucleus during pachytene (Penkner et al. 2009).

The interaction of PCs on homologous chromosomes facilitates their alignment and may act to stabilize homologous interactions, thereby allowing synapsis to proceed. Although PC interactions are sufficient to stabilize the local pairing of homologs in the absence of synapsis (MacQueen et al. 2005), stable pairing between the homologs along the full length of chromosomes requires synapsis (MacQueen et al. 2002). Pairing of homologs at the non-PC regions of autosomes is mediated by the chromodomain protein MRG-1, which prevents nonhomologous synapsis along regions away from the PCs (Dombecki et al. 2011). In the absence of either PCs or the zinc-finger proteins ZIMs/HIM-8, synapsis of the corresponding chromosome pair does not occur (Phillips and Dernburg 2006; MacQueen et al. 2005; Phillips et al. 2005). However, failure of a single PC to interact does not affect synapsis of the other chromosomes, but it does result in delays in both DSB repair and release from the polarized configuration for the other chromosomes (Phillips and Dernburg 2006; Phillips et al. 2005). Mutants that have defects in synapsis often exhibit an extended transition zone with nuclei that persist in the clustered configuration throughout what corresponds to the pachytene stage in wild-type germlines (MacQueen et al. 2002; Colaiacovo et al. 2003; Smolikov et al. 2007a). PLK-2 is required for prolonging the transition zone in mutants with defects in pairing and/or synapsis (Harper et al. 2011). Therefore, a checkpoint may perhaps exist to help coordinate pairing with synapsis and PLK-2 appears to be involved with this checkpoint (Martinez-Perez and Villeneuve 2005; Harper et al. 2011; Labella et al. 2011).

6.4 Chromosome Synapsis

Synapsis is the zipping-up of paired homologs along their lengths by the proteinaceous scaffold referred to as the synaptonemal complex or SC. The SC is a tripartite structure comprised of a pair of lateral elements and a central region. Lateral elements are formed

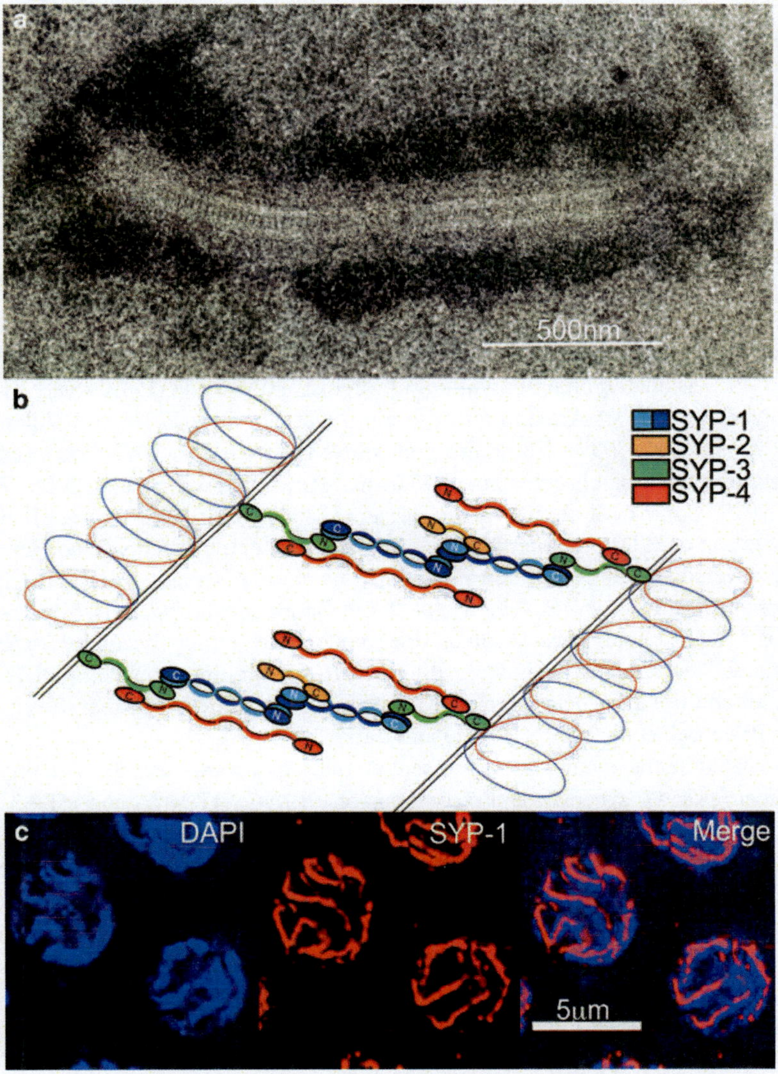

Fig. 6.3 The synaptonemal complex. (**a**) TEM (transmission electron microscopy) image of the structure of the SC between chromosomes in a pachytene nucleus in the *C. elegans* germline. The continuous zipper-like track, comprised of the transverse filaments, is flanked by electron-dense patches of chromatin. (**b**) Schematic of the arrangement of the four central region proteins in the SC of *C. elegans*. (**c**) Immunolocalization of SC proteins in pachytene nuclei. Central region protein SYP-1 (*red*) forms tracks at the interface between DAPI-stained chromosomes (*blue*). The images are projections halfway through three-dimensional data stacks of whole nuclei

via the assembly of proteins along the homologous axes, which in turn are connected via central region components, consisting of transverse filament proteins that bridge the homologous axes. This results in a "ladder-like" organization along the length of paired and aligned homologous chromosomes (Fig. 6.3a; Moses 1968). Within the context of a fully formed SC, homologs are held together with their axes ~100 nm apart throughout most species (Westergaard and von Wettstein 1972). This is also the case in *C. elegans* where axes are separated by an average distance of 118 nm (90–125 nm range; Smolikov et al. 2008). Synapsis begins in the transition zone, where short segments of the central region are observed to associate with chromosomes (MacQueen et al. 2002). Synapsis is completed by pachytene, when thick parallel DAPI-stained tracks are visible in meiotic nuclei, flanking the signal of central region proteins, which localize throughout the full length of the chromosomes (Fig. 6.3c).

Axial element proteins assemble along the lengths of sister chromatids prior to pairing. As axes become closely juxtaposed and homologs start to synapse, the axial elements are referred to as lateral elements. *C. elegans* has multiple HORMA (Hop1, Rev7, and Mad2) domain-containing proteins: HIM-3, HTP-1/2, and HTP-3 (Goodyer et al. 2008; Aravind and Koonin 1998; Martinez-Perez and Villeneuve 2005; Zetka et al. 1999). These are non-cohesin proteins that associate along the longitudinal axes of chromosomes (Martinez-Perez and Villeneuve 2005; Zetka et al. 1999; Couteau et al. 2004; Goodyer et al. 2008; Martinez-Perez et al. 2008). Localization of HTP-3 is interdependent with cohesin components (Goodyer et al. 2008; Severson et al. 2009), and HTP-3 mediates the association of HTP-1/2 and HIM-3 on chromosomes (Severson et al. 2009; Goodyer et al. 2008). While association of HTP-1/2 with chromosomes does not require either the meiosis-specific cohesin REC-8 or HIM-3, bulk loading of HIM-3 requires cohesin, and HTP-1 mediates additional association of HIM-3 with chromosomes (Martinez-Perez et al. 2008; Pasierbek et al. 2003; Couteau and Zetka 2005).

The lateral/axial elements play essential roles in meiotic events and serve to coordinate pairing, synapsis, and recombination. HIM-3, HTP-1/2, and HTP-3 have specialized functions in mediating pairing and proper loading of the central region components of the SC (Martinez-Perez and Villeneuve 2005; Zetka et al. 1999; Couteau et al. 2004; Goodyer et al. 2008). *him-3* null mutants completely fail to cluster and pair their chromosomes (Zetka et al. 1999; Couteau et al. 2004). The failure to cluster chromosomes in *him-3* mutants may reflect the reduced association between the SUN-1 and the ZIM proteins in those mutants (Baudrimont et al. 2010). In contrast, *htp-1* mutants have a less severe defect in pairing than *him-3* mutants and exhibit a few polarized nuclei (Martinez-Perez and Villeneuve 2005; Couteau et al. 2004). In budding yeast, the lack of axial element formation impairs chromosome movement and delays pairing (Trelles-Sticken et al. 2005). Similarly, chromosome axis morphogenesis in *C. elegans* could be coupled to regulation of chromosome movement. Absence of HTP-1/2 can result in nonhomologous synapsis of the autosomes (Martinez-Perez and Villeneuve 2005; Couteau et al. 2004; Couteau and Zetka 2005). By inhibiting the polymerization of the SC when pairing has failed, HTP-1/2 coordinates pairing with SC formation. HTP-3 does not only play a role in pairing and synapsis but is also required for meiotic DSB formation (Goodyer et al. 2008). While *him-3* mutants have normal levels of break formation, but delayed DSB turnover (Couteau et al. 2004), *htp-3* mutants do not form breaks, but can repair breaks induced by irradiation. Therefore, the expansion of the HORMA domain protein class in *C. elegans* has allowed for each of the proteins to attain specialized functions during meiosis.

The central region of the SC consists of an ordered array of transverse filaments repeating along the length of the homologs. In budding yeast, where Zip1 is the only central region protein identified to date, it forms homodimers through its coiled-coil domains (Zickler and Kleckner 1999; Dong and Roeder 2000). The homodimers then interact in an antiparallel manner to span the distance between the lateral elements of the homologs (Liu et al. 1996; Schmekel et al. 1996). However, the organization of the central region of the SC is apparently more complex in higher eukaryotes. In fact, the assembly of the SC in *C. elegans* may be more analogous to that in mammals than in yeast. As opposed to the single central region component present in budding yeast, both mice and *C. elegans* have several central region proteins (Costa et al. 2005; Schramm et al. 2011; Bolcun-Filas et al. 2007; Smolikov et al. 2007b, 2009; Colaiacovo et al. 2003; MacQueen et al. 2002; Hamer et al. 2006; de Vries et al. 2005; Sym et al. 1993). In *C. elegans*, the central region proteins SYP-1/2/3/4 localize between the axes of synapsed homologs (Smolikov et al. 2007b, 2009; Colaiacovo et al. 2003; MacQueen et al. 2002). Moreover, the central region components are interdependent upon each other for their assembly on chromosome axes (Smolikov et al. 2007b, 2009; Colaiacovo et al. 2003). The predicted length of each of the central region proteins is not sufficient to span the entire width of the SC (Smolikov et al. 2009). By assembling into a multi-protein complex, the SYP proteins maintain the characteristic organization and conserved dimensions of the SC (Schild-Prufert et al. 2011). Through a series of immunogold labeling and protein interaction studies, the structure of the SC was determined to consist of a homodimer of SYP-1 that can interact with the C-terminus of SYP-2 and the N-terminus of SYP-3, which in turn interacts with SYP-4 and has its C-terminus located at the chromosome axes (Fig. 6.3b; Schild-Prufert et al. 2011).

Synapsis tends to initiate at the PC end of the chromosome and polymerize down the rest of the chromosome (MacQueen et al. 2005). Moreover, SC polymerization is highly processive once it is nucleated, and can accommodate regions of nonhomology located away from the pairing center (MacQueen et al. 2005). The processivity of SC polymerization is mediated, at least in part, by the tetratricopeptide repeat (TPR) domain-containing protein CRA-1 (Smolikov et al. 2008). *cra-1* mutants form uneven, discontinuous stretches of SC (Smolikov et al. 2008). Unlike yeast, plants and mammals, but similar to flies, synapsis in *C. elegans* can occur even in the absence of recombination initiation (Dernburg et al. 1998; McKim et al. 1998; Giroux et al. 1989; Baudat et al. 2000; Grelon et al. 2001). Importantly, the polymerization of SC components along chromosome axes in the absence of recombination is dependent on CRA-1 (Smolikov et al. 2008). Taken together, this suggests that the dependence of SC formation on DSB formation is largely conserved across taxa. However, the evolution of CRA-1 has allowed meiosis in *C. elegans* to bypass the requirement of DSB initiation for synapsis, but not accurate chromosome segregation.

The formation of a fully mature SC between homologs mediates the progression of meiotic events. Although mutants of central region components can pair their chromosomes, chromosomes fail to stabilize pairing interactions along their lengths (Colaiacovo et al. 2003; MacQueen et al. 2002; Smolikov et al. 2007b, 2009).

Moreover, the clustered configuration characteristic of transition zone is prolonged in mutants that lack central region components of the SC (Smolikov et al. 2009; MacQueen et al. 2002). Thus, similar to budding yeast and mice, SC formation appears to be required for maximal levels of pairing between homologs in *C. elegans* (Nag et al. 1995; Peoples-Holst and Burgess 2005; Peoples et al. 2002; Daniel et al. 2011). The SC also promotes recombination between homologous chromosomes. Mutants defective in SC formation accumulate recombination intermediates, which are visualized as persistent foci of the strand invasion protein RAD-51 in late pachytene (MacQueen et al. 2002; Colaiacovo et al. 2003; Smolikov et al. 2007a, 2009). Synapsis is universally important for the formation of crossovers between homologs, as exemplified by the observation that defects in synapsis result in reduced crossing over in plants, yeast, flies, mammals, and *C. elegans* (de Vries et al. 2005; Page and Hawley 2001; Colaiacovo et al. 2003). Late recombination nodules, which are the dark silver staining bodies visible by electron microscopy (EM) on pachytene chromosomes and represent sites of future crossover formation, tend to be associated with the SC in fungi, flies, plants, and humans (Rasmussen and Holm 1984; Maguire 1966; Carpenter 1975; Zickler 1977). Therefore, the SC plays a key and conserved role in promoting the progression of interhomolog recombination.

Differentiation of the bivalent occurs as chromosomes remodel at the pachytene–diplotene transition around the off-center placed single crossover undergone by every pair of homologs (Nabeshima et al. 2005). Chromosomes therefore acquire a cruciform structure comprised of a pair of long and short arms (Fig. 6.2). Later, this configuration results in the long arms facing the poles and short arms being aligned along the metaphase plate. The SC begins to disassemble during late pachytene and is completely absent from the long arms of the bivalent by diakinesis (MacQueen et al. 2002; Nabeshima et al. 2005). The asymmetric disassembly of the SC requires components involved in crossover formation: ZHP-3, MSH-4, and MSH-5 (Nabeshima et al. 2005; Bhalla et al. 2008). ZHP-3 is a SUMO E3 ligase homologous to yeast Zip3, a protein that plays key roles in crossover formation and promoting the assembly of the central region of the SC (Jantsch et al. 2004; Agarwal and Roeder 2000; Borner et al. 2004). In *C. elegans*, ZHP-3 is not required for SC formation but plays a role in promoting crossover formation (Jantsch et al. 2004; Bhalla et al. 2008). ZHP-3 initially localizes along the length of the chromosome axes during pachytene and then relocalizes during late pachytene to the boundary between the short and long arms of the bivalent to mark the site of crossover formation (Bhalla et al. 2008; Jantsch et al. 2004). Both transgenic worms expressing *zhp-3::gfp* at the restrictive temperature and mutants lacking SUMO polypeptide expression (*smo-1*) fail to remove the SC from the long arms of the bivalent (Bhalla et al. 2008). Finally, proper bivalent maturation also requires proper maintenance of sister chromatid cohesion at the long arms of the bivalent during diakinesis via restriction of the *C. elegans* Aurora B kinase homolog AIR-2 (de Carvalho et al. 2008). This is mediated by LAB-1, a proposed functional analog of the mammalian Shugoshin protein implicated in the regulated two-step removal of sister chromatid cohesion (Kitajima et al. 2004; Marston et al. 2004; Rabitsch et al. 2004; Lee et al. 2008; Llano et al. 2008; de Carvalho et al. 2008).

When AIR-2 is no longer limited to the short arms in the absence of LAB-1, bivalents have aberrant localization of SC components and SC disassembly is delayed until the last oocyte in diakinesis (de Carvalho et al. 2008). Therefore, there are several different elements regulating the timely disassembly of the SC.

Defects in SC formation are detected by the synaptic checkpoint that operates during pachytene to cull nuclei by apoptosis (Bhalla and Dernburg 2005). Moreover, the synapsis checkpoint, which can detect a single pair of asynaptic chromosomes, requires PLK-2 and the AAA–adenosine triphosphatase PCH-2 (Harper et al. 2011; Bhalla and Dernburg 2005). However, in addition to monitoring chromosome asynapsis, the PCH-2 homologs in yeast and Drosophila have been implicated in axis morphogenesis and mediating the outcome of recombination by enforcing homolog bias and crossover interference (COI; Joyce and McKim 2009; Joshi et al. 2009; Wu and Burgess 2006). Taken together, several mechanisms are set in place across species to regulate SC assembly and disassembly, as well as the "quality" of the SC formed throughout meiotic prophase.

6.5 Meiotic Recombination

Crossovers provide sufficient tension to align chromosomes on the spindle (Östergren 1951; Nicklas 1974). Therefore, failure to from crossovers results in chromosomes that segregate randomly during the first meiotic division and subsequent aneuploidy. Chiasmata are the physical manifestation of crossovers between homologs, and crossovers are the product of homologous recombination. Recombination events between homologs allow for exchange of genetic information, and by shuffling the genetic information distributed to gametes that will be used in reproduction, this exchange promotes genetic diversity.

Each *C. elegans* chromosome undergoes multiple DSBs, but only one crossover (CO) forms between each chromosome pair (Mets and Meyer 2009; Barnes et al. 1995; Nottke et al. 2011; Gao, Saito and Colaiacovo, personal communication). Meiotic DSBs that do not become crossovers are repaired as noncrossovers (NCO). There are at least three layers of regulation determining the frequency and distribution of COs: crossover assurance, crossover interference, and crossover homeostasis.

Crossover assurance ensures that each chromosome receives at least one crossover, which is the obligate crossover (Jones 1984). Crossover interference ensures that crossovers are distributed nonrandomly and places them further apart from each other than would be expected by chance (Muller 1916). Therefore, the formation of a CO in a given location is proposed to "interfere" with or inhibit the formation of additional COs nearby. Since *C. elegans* chromosomes only undergo one CO per chromosome, it is an organism that exemplifies strong crossover interference. In other organisms, there are two types of COs that occur in wild-type meiosis: COs either subject or not to interference. In budding yeast, COs subject to interference are mediated by the ZMM proteins [Zip1/2/3, Spo16 (Zip4), Mer3, Msh4, and Msh5], while interference-independent COs are dependent on Mms4 and Mus81 (Borner et al. 2004; Chen et al. 2008; de los

Santos et al. 2003). In *C. elegans*, all COs are normally dependent on HIM-14/MSH-4 and MSH-5, which are the homologs of the yeast ZMM proteins Msh4 and Msh5, respectively (Kelly et al. 2000; Zalevsky et al. 1999). MUS-81-dependent CO formation occurs only in aberrant situations, in which DSBs are in excess or cannot be repaired as NCOs. A subset of crossovers is MUS-81-dependent after X-ray induction and in the absence of the helicase RTEL-1, which is the homolog of Srs2 in budding yeast (Youds et al. 2010).

Crossover interference may require continuous stretches of SC along the chromosomes in *C. elegans*. A three-chromosome fusion consisting of homologous autosomes flanking an unpaired X chromosome that disrupts SC formation between the autosomal segments results in each autosomal segment forming crossovers (Hillers and Villeneuve 2003). Moreover, crossover interference is reduced and double COs occur more frequently in *him-3* hypomorphs that form short SC stretches (Nabeshima et al. 2004). However, continuous SC polymerization along chromosomes does not seem to be required for crossover interference in yeast and mice (Shinohara et al. 2008; de Boer et al. 2006). The decision to make COs as opposed to NCOs is thought to occur very early in budding yeast (Bishop and Zickler 2004; Allers and Lichten 2001; Borner et al. 2004). While not much is known about NCO regulation in *C. elegans*, there is evidence suggesting that an early control/decision regarding COs is exerted at the level of DSB formation. Specifically, CO analysis in condensin I mutants, which exhibit increased DSB formation, revealed both elevated levels and an altered distribution of COs (Mets and Meyer 2009; Tsai et al. 2008). It has been hypothesized that the extended chromatin axes in condensin I mutants result in a higher density of smaller chromatin loops (Mets and Meyer 2009). Moreover, DSBs have been proposed to occur preferentially at chromatin loops and then be recruited to non-sister homologous axes for repair (Blat et al. 2002; Maleki et al. 2007). Since HTP-3 interacts with the nuclease MRE-11, which facilitates resection of DSBs (Goodyer et al. 2008), recruitment of resected DSBs to the axis may occur via this direct interaction. Therefore, by having a higher density of chromatin loops, the potential for DSB formation may be increased in condensin I mutants. More recent work in *C. elegans* suggests that the CO decision is also made during later stages of meiosis. In one study, it was shown that DSBs created after completion of SC assembly (using heat-shock inducible transposon excision) are both subject to and confer interference, competing with endogenous DSBs to become the sole CO (Rosu et al. 2011). In a second study, the introduction of exogenous DSBs by irradiation resulted in the separation of chromosome axes during late pachytene (Couteau and Zetka 2011). Given that axis separation was observed in the CO-deficient mutant *msh-5*, this study inferred that axis separation allowed for DSBs to be repaired as NCO events (Couteau and Zetka 2011). Interestingly, the separation of axes is correlated with a reduction in histone 2A lysine 5 acetylation (H2AK5Ac), which in turn is dependent on HTP-3 (Couteau and Zetka 2011). Taken together, these data suggest that CO control may be exerted by various factors affecting axis morphogenesis at different points during meiotic progression, namely early leptotene/zygotene, when the first meiotic DSBs occur and synapsis is initiating, and late pachytene, when chromosomes are fully synapsed.

Crossover homeostasis maintains CO levels. As a result of crossover homeostasis, the number of COs does not scale to the number of DSBs formed. Specifically, the number of COs is maintained despite a reduction in the number of DSBs formed and probably at the expense of non-crossovers (NCOs). Crossover homeostasis has been most clearly demonstrated in budding yeast by using hypomorhic alleles of the conserved Spo11 endonuclease, which reduce the number of DSBs to varying degrees. Despite a reduction in the number of recombination initiation events, the *spo11* hypomorphs still exhibit normal CO levels (Martini et al. 2006). The existence of crossover homeostasis in *C. elegans* remains to be determined.

Meiotic recombination initiates with the formation of programmed DSBs, which begin in the transition zone (Mets and Meyer 2009). SPO-11 is a topoisomerase II-like protein that forms the DSBs (Dernburg et al. 1998; Keeney 2001) (Fig. 6.4). The DSBs are then resected primarily by MRE-11, RAD-50, and COM-1 to reveal long, 3′ single-stranded DNA tails (Chin and Villeneuve 2001; Hayashi et al. 2007; Penkner et al. 2007; Sun et al. 1991). The single-stranded tails are coated with the RecA homolog RAD-51, whose nucleation on ssDNA and stabilization of the nucleoprotein filament is promoted by BRC-2, the BRCA2 homolog (Petalcorin et al. 2007). Human and yeast homologs of RAD-54, a SWI2/SNF2 chromatin remodeling protein, promote strand invasion of homologous duplex DNA by the nucleoprotein filament (Alexeev et al. 2003; Mets and Meyer 2009; Mazin et al. 2000; Mazina and Mazin 2004). In the absence of RAD-54, DSBs are not repaired and accumulate (Mets and Meyer 2009). Interestingly, strand exchange, which is visualized by the presence of Rad51/Dmc1 foci, typically peaks during leptotene and zygotene in yeast, plants, and mice (Tarsounas et al. 1999; Terasawa et al. 1995; Bishop 1994). In contrast, RAD-51 foci peak during pachytene in *C. elegans* (Alpi et al. 2003; Colaiacovo et al. 2003). DSBs in budding yeast mostly disappear at the transition between zygotene and pachytene as they are converted into either NCO or CO intermediates (Allers and Lichten 2001; Hunter and Kleckner 2001). However, whereas in budding yeast DSB initiation is essential for homolog pairing and synapsis, these processes are primarily mediated by PCs in *C. elegans* (Weiner and Kleckner 1994; Giroux et al. 1989; Peoples et al. 2002; Phillips et al. 2005; MacQueen et al. 2005). Therefore, the earlier kinetics of recombination may not be necessary to promote the progression of those meiotic events in *C. elegans*.

After strand exchange by formation of the D-loop structure, the DSB can be processed into COs or NCOs. NCO formation is promoted by the ortholog of Srs2, RTEL-1, which assists in the ejection of the single-stranded DNA undergoing strand invasion and DNA synthesis (Youds et al. 2010; Barber et al. 2008). DSBs that are destined to become COs are processed into single-end invasions (SEI) and then double Holliday junctions (dHJ) (Allers and Lichten 2001; Hunter and Kleckner 2001). ZHP-3, the homolog of the crossover promoting protein Zip3 in budding yeast, localizes to the sites of obligate crossover formation in late pachytene/diplotene (Youds et al. 2010; Bhalla et al. 2008; Jantsch et al. 2004; Borner et al. 2004). Following DNA synthesis, the second end of the DSB anneals to form a dHJ (Sun et al. 1991; Szostak et al. 1983). Yeast Rad52 has been shown to promote second end capture through its N-terminal region, which promotes single strand annealing (Lao et al. 2008; Sugiyama et al. 1998; Krejci et al. 2002). This property is also

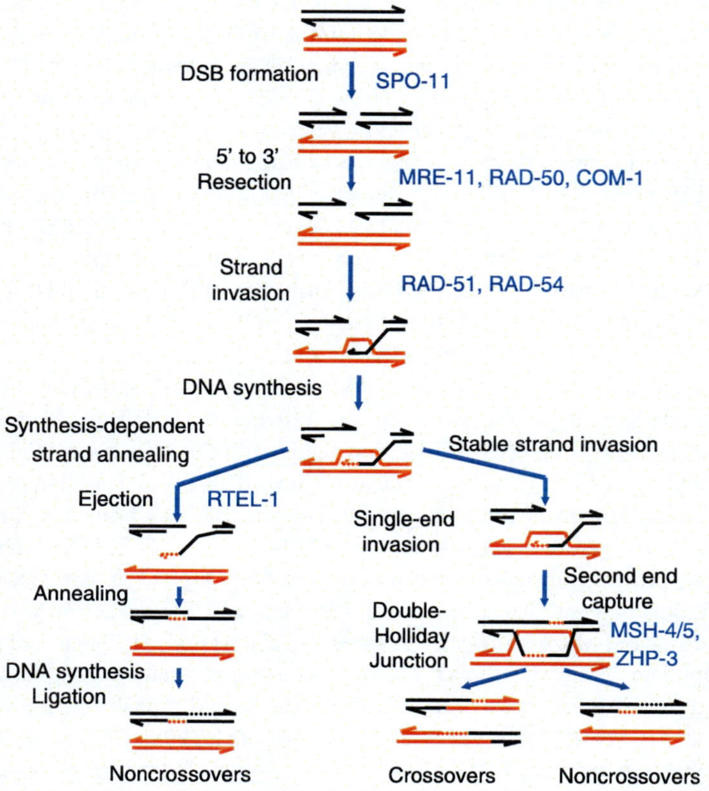

Fig. 6.4 The meiotic recombination pathway in *C. elegans*. Here one homolog is depicted in *black* and the other is shown in *red*. Meiotic recombination is initiated by formation of DSBs. The topoisomerase-like enzyme SPO-11 catalyzes the cleavage of the double-stranded DNA of one sister chromatid. Both 5′ ends of the DSB are rapidly resected by MRE-11, RAD-50, and COM-1 to reveal 3′ single-stranded tails, on which RAD-51 forms a filament. RAD-54 promotes invasion by one end of the DSB into the DNA duplex of the homolog to form the nascent D-loop structure. As DNA synthesis occurs, the D-loop expands. The D-loop structure is processed by two major pathways to yield COs and NCOs. COs, which hold bivalents together, arise from the formation of stable single-end invasions, followed by second end capture and then the formation of double Holliday junctions, which are cleaved by a currently unknown resolvase. NCOs arise from either processing of double Holliday junctions or the synthesis-dependent strand-annealing pathway, through which the invading end of the DSB is ejected so that it can anneal with its sister chromatid. Following annealing, DNA synthesis and ligation occur to complete the formation of NCOs

conserved in the Rad52 ortholog BRC-2, which may therefore serve the same function and assist in second end capture during CO formation in *C. elegans* (Petalcorin et al. 2006). Finally, MSH-4/HIM-14 and MSH-5, which are proposed to form a heterodimer at dHJs in humans, are required for CO formation and assist in the timely processing of NCO products (Winand et al. 1998; Zalevsky et al. 1999; Colaiacovo et al. 2003; Kelly et al. 2000; Snowden et al. 2004).

HJ resolution to yield COs requires cleavage of the structure across the junction followed by ligation. Symmetric cleavage of the HJ allows for ligation immediately following cleavage, but asymmetric cleavage of the HJ requires further processing prior to religation. The main meiotic HJ resolvase in *C. elegans* and most other eukaryotes has yet to be identified (Schwartz and Heyer 2011). Fission yeast utilizes Mus81 as its primary HJ resolvase (Boddy et al. 2001). Although *mus-81* single mutants in *C. elegans* have increased embryonic lethality, recombination intermediates do not persist into late meiotic prophase (Saito et al. 2009). However, MUS-81 is required to process COs stemming from the repair of exposure to gamma-irradiation and those that arise in a *rtel-1* mutant background (Youds et al. 2010). A mutation in *mus-81* is synthetic lethal with a mutation in *him-6*, which encodes for the homolog of the yeast Sgs1 helicase in *C. elegans* (Saito et al. 2009). In budding yeast, Sgs1 prevents multichromatid joint molecules, which involves invasion of the sister chromatid (Oh et al. 2008). Thus, it is likely that MUS-81 is required to process the toxic intermediates that form in *him-6* mutants. The majority of CO products in *Drosophila* is dependent on the MEI-9(XPF-1)/MUS312(HIM-18)/ERCC1 complex (Yildiz et al. 2002; Boyd et al. 1976; Radford et al. 2007). HIM-18/SLX-4 interacts with XPF-1 and SLX-1, but an interaction with ERCC-1 could not be detected by yeast two-hybrid (Saito et al. 2009). HIM-18 appears to be responsible for resolution of only a subset of Holliday junctions during normal meiosis in *C. elegans* because chromatin bridges are visible in some of the bivalents of prometaphase I in *him-18* oocytes and COs are reduced to ~51–70% of wild-type levels (Saito et al. 2009). Human GEN1 and yeast Yen1 have HJ resolvase activity in vitro most similar to bacterial RuvC, which cleaves HJ symmetrically (Bennett et al. 1993; Ip et al. 2008). Ectopically expressed human GEN1 in fission yeast can rescue *mus81* mutants (Lorenz et al. 2010). However, *gen-1* mutants in *C. elegans* have normal brood size and do not exhibit a Him phenotype (Bailly et al. 2010). Since *gen-1* mutants do not show chromosome segregation defects, GEN-1 does not appear to be involved in the processing of HJs during meiosis in *C. elegans*. Alternatively, there could be redundant mechanisms that preclude the analysis of the role of GEN-1 during meiosis in *C. elegans*. Therefore, its role in meiotic recombination remains to be further examined in *C. elegans* as well as in other systems.

Typically, the repair of meiotic DSBs is biased towards HR such that repair will favor CO production, which in turn are required for accurate chromosome segregation. Therefore, alternative forms of DSB repair are suppressed during meiosis to prevent either the use of the sister chromatid as a template for repair or nonhomologous end joining (NHEJ). However, when progression of HR is impaired, alternative forms of DSB repair are engaged. During normal meiosis, NHEJ does not play significant role in the repair of meiotic DSBs (Martin et al. 2005; Clejan et al. 2006). However, either in the absence of synapsis or in the presence of defective sister chromatid cohesion, NHEJ is used to repair DSB breaks (Smolikov et al. 2009; Colaiacovo et al. 2003; Couteau et al. 2004). Specifically, while *syp-3* C-terminal truncation mutants fail to synapse, DSBs eventually turn over to yield a few viable progeny in those mutants (Smolikov et al. 2007a). However, DSB turnover is further impaired in the *syp-3* C-terminal truncation mutant when REC-8 is depleted by

RNAi and in the absence of LIG-4, a component of the NHEJ machinery (Smolikov et al. 2007a). There is also no evidence of chromosome fragmentation in *rad-51* and *brc-1* mutants when HTP-3 is depleted by RNAi suggesting that DSB repair is completed in these mutants when barriers against alternative modes of DSB repair imposed by axis-associated components are no longer in place (Goodyer et al. 2008). BRC-1 is not required for repair of DSBs by HR but functions to repair DSBs in *syp-2* mutants (Adamo et al. 2008). Additionally, breaks form at normal levels and get repaired in *him-3* mutants despite the failure to pair and synapse chromosomes (Couteau et al. 2004). The DSBs in *him-3* mutants may be repaired using the intersister recombination or NHEJ pathways (Couteau et al. 2004). Thus, both the axes and the central region actively inhibit both NHEJ and intersister recombination. Notably, orthologs for proteins involved in homolog bias such as the meiosis-specific strand invasion proteins Dmc1, Hop2, and Mnd1 have not been identified in *C. elegans* (MacQueen et al. 2002; Villeneuve and Hillers 2001). This suggests that *C. elegans* may rely more extensively on just the axes to enforce homolog bias.

Homolog bias may be enforced by axes mediating processing of DSBs at the level of resection prior to late pachytene. RAD-50-dependent loading of RAD-51 in early meiotic prophase is correlated with the ability to form crossovers between homologs (Hayashi et al. 2007). However, the loading of RAD-51 no longer requires RAD-50 in early meiotic prophase when *htp-1* and *him-3* are depleted (Hayashi et al. 2007). Since HTP-3 interacts with MRE-11 (Goodyer et al. 2008), axial elements may promote homolog bias perhaps through direct interaction with components of the DSB repair machinery. Given that homolog bias appears to be restricted to mid-pachytene (Hayashi et al. 2007), the axis may impose a "kinetic constraint" as opposed to a permanent barrier to nonhomologous forms of repair (Goldfarb and Lichten 2010).

Repair of a subset of DSBs has been proposed to help align homologous chromosomes to prevent nonhomologous interactions (Smithies and Powers 1986; Getz et al. 2008). However, *C. elegans* may rely less on DSBs to align chromosomes because they have evolved PCs. The requirement of CO formation to promote homolog alignment is also not present in *Drosophila,* which relies on *cis*-acting regions to pair their chromosomes, and fission yeast, which uses the horsetail movement to promote homolog pairing (discussed above). Importantly, in *C. elegans* COs are nonrandomly distributed on chromosomes and are five times more likely to form at the terminal third of chromosomes (Hillers and Villeneuve 2003). Finally, these hotspots for recombination tend to occur in gene-poor regions of the chromosomes (Barnes et al. 1995). Similarly, in budding yeast, DSB hotspots mostly occur at promoters, but at interstitial regions of chromosomes (Petes 2001; Pan et al. 2011). In contrast, almost half of mouse recombination hotspots occur within genes (Smagulova et al. 2011).

The nonrandom distribution of crossovers reflects the nonrandom distribution of DSBs and DSB formation is influenced by the state of meiotic chromatin. An open chromatin state, modulated in part by post-translationally derived histone modifications, can make the DNA more competent for DSB formation by increasing the accessibility of the recombination machinery (Petes 2001). In support of this model, hotspots for recombination are typically devoid of nucleosomes (Pan et al.

2011; Fan and Petes 1996). However, histone H3 methylation on lysine 9 (H3K9me), which is a modification associated with closed chromatin, is also required for DSB formation in *C. elegans* (Reddy and Villeneuve 2004; Lachner and Jenuwein 2002). *him-17* mutants, which fail to accumulate the H3K9me chromatin mark, phenocopy *spo-11* mutants and lack meiotic DSBs (Reddy and Villeneuve 2004). Thus, compaction of chromatin also plays a role in DSB formation, perhaps by facilitating proper loop-axis conformation. Furthermore, H3K9me patterns in a somatic background also correlate with recombination rates along chromosomes (Liu et al. 2011). One may extrapolate from data on somatic cells and suspect that a similar pattern of H3K9me exists on meiotic chromatin. Specifically, H3K9 methylation is enriched on chromosome arms compared to the central region of chromosomes, where crossover formation tends to be excluded (Liu et al. 2011). Regulation of histone acetylation is apparently also important as suggested by the analysis of XND-1, an AT hook motif-containing protein, which is enriched on autosomes and required for formation of DSBs on the X chromosome (Wagner et al. 2010; Harris et al. 2010). The germlines of *xnd-1* mutants have accumulated histone H2A lysine 5 acetylation (H2AK5Ac) and crossover distribution is altered in this background (Wagner et al. 2010). In *xnd-1* mutants, COs occur more frequently in the middle of chromosomes, which are gene-rich, rather than at the arms of the chromosomes (Wagner et al. 2010). The histone acetyltransferase Tip60 mediates accumulation of H2AK5Ac to allow for exchange of phosphorylated H2Av for unmodified H2Av following induction of DSBs in somatic cells in *Drosophila* and removal of γ-H2AX after DNA damage in human cell culture (Kusch et al. 2004; Ikura et al. 2007). MYS-1, the homolog of Tip60, maybe required for accumulation of H2AK5Ac in *C. elegans* (Wagner et al. 2010); however, more recent data suggests that this may not be the case, as H2K5Ac levels are not affected by RNAi depletion of MYS-1 (Couteau and Zetka 2011). Moreover, CRA-1 localizes to autosomes and is also required for accumulation of H2AK5Ac (J. Gao and M.P. Colaiácovo, personal communication). Therefore, while XND-1 acts to remove H2AK5Ac, CRA-1 and MYS-1 may both act to promote H2AK5Ac (J. Gao and M. Colaiacovo, personal communication). However, further studies will be required to uncover how the interplay between these factors promotes the dynamic regulation of histone modifications in the germline ultimately impacting meiotic DSB formation and repair.

6.6 Late Prophase: Setting the Stage for the Metaphase I—Anaphase I Transition

Chromosomes become highly condensed and remodeled as they approach late prophase to form the characteristic "cruciform bivalent." This remodeling revolves around the site of crossover formation (Nabeshima et al. 2005). Since crossovers form at the terminal third of chromosomes, crossovers asymmetrically divide chromosomes into a long and short arm. During remodeling, different components of the SC adopt distinct localizations along the arms of the bivalent. HTP-1 and

HTP-2 are retained along the long arms of the bivalent (Martinez-Perez et al. 2008), while SYP-1 only localizes to the short arm of the bivalent (Fig. 6.2; Martinez-Perez et al. 2008). However, HIM-3 and HTP-3 localize to both arms of the bivalent with REC-8 (Goodyer et al. 2008; Zetka et al. 1999).

Cleavage of REC-8 along the short arms has been proposed to allow the homologs to segregate to opposite poles and is dependent on its phosphorylation state (Rogers et al. 2002; Schumacher et al. 1998). In yeast, Casein kinase or Dbf4-dependent Cdc7 kinase (DDK)-dependent phosphorylation is responsible for Rec8 cleavage (Ishiguro et al. 2010; Katis et al. 2010). In monocentric organisms, Rec8 is protected at centromeric/pericentromeric regions resulting in a defined region of preservation of cohesion. Shugoshin-mediated dephosphorylation by phosphatase PP2A allows for evasion of Rec8 cleavage in monocentric organisms (Riedel et al. 2006; Kitajima et al. 2006).

As an organism with holocentric chromosomes, *C. elegans* has developed an alternative strategy to designate sites of cohesion retention. The chromosomal passenger complex (CPC), which includes AIR-2 (Aurora B kinase), CSC-1 (Borealin), BIR-1 (Survivin), and ICP-1 (Incenp), localizes to the short arms of the bivalents (Kaitna et al. 2002; Romano et al. 2003; Speliotes et al. 2000). AIR-2 localization is restricted to the short arms by LAB-1 and HTP-1/2 (de Carvalho et al. 2008; Martinez-Perez et al. 2008). Moreover, phosphorylation of REC-8 by AIR-2 licenses cleavage of REC-8 (Rogers et al. 2002). Histone H3 along the short arms is an additional phosphorylation target of AIR-2 (Kaitna et al. 2002; Hsu et al. 2000). Interestingly, the *C. elegans* Shugoshin homolog SGO-1 is dispensable for the retention of REC-8 along the long arms and does not appear to localize to the long arms (de Carvalho et al. 2008). Instead, *sgo-1* mutants have chromosome segregation defects that become evident during meiosis II divisions; *sgo-1* mutants exhibit lagging chromosomes and polar body extrusion (de Carvalho et al. 2008). Therefore, LAB-1 is proposed to antagonize AIR-2-mediated phosphorylation of REC-8 along the long arms of the bivalent by recruiting the *C. elegans* PP1 phosphatases GSP-1 (GLC-7α) and GSP-2 (GLC-7β) (de Carvalho et al. 2008; Rogers et al. 2002), which dephoshorylate REC-8, thereby preventing its cleavage by separase at the long arm (Rogers et al. 2002).

The holocentric nature of *C. elegans* chromosomes requires a unique assembly of the segregation-promoting machinery during meiosis compared to monocentric organisms. This is due in part to the fact that the first meiotic division presents an additional challenge because homologs must segregate to opposite poles. Therefore, kinetochore assembly during the first meiotic division is different from that during mitotic divisions. During mitotic divisions, the kinetochore assembles along the full length of the sister chromatids with microtubules attaching along the length of the chromosomes (Albertson and Thomson 1982). During meiosis I, the holocentric kinetochore forms cup-like structures that surround the long arms of bivalents (Dumont et al. 2010; Monen et al. 2005). Microtubules form bundles that run parallel to the long axis of the bivalents (Wignall and Villeneuve 2009; Howe et al. 2001). Since only a few microtubules attach to the ends of the bivalents, orientation of the bivalents on the spindle is largely driven by interaction of the kinetochores along the length of the long arms with the lateral microtubule bundles (Wignall and Villeneuve 2009; Dumont et al. 2010). This allows one pair of sisters to face one spindle pole and the other pair of sisters belonging to the homolog to face the opposite pole. The short arms of the bivalents align in the center of the meiotic spindle during congression. Although kinetochores promote proper orientation of chromosomes on the spindle axis, they are dispensable for chromosome segregation during anaphase I (Dumont et al. 2010). Instead, microtubules form bundles between the short arms during anaphase that may push outwards to segregate the homologs (Dumont et al. 2010). At the short arms, AIR-2 recruits the spindle checkpoint kinase BUB-1, motor proteins KLP-19 and CEMP-F, and the microtubule-stabilizing protein CLS-1(CLASP) (Dumont et al. 2010). Thus, AIR-2 plays multiple roles to promote the segregation of the homologs: AIR-2 marks REC-8 for cleavage and recruits machinery to actively separate the homologs (Kaitna et al. 2002).

6.7 Summary and Future Perspectives

Multiple processes contribute to ensuring the accurate segregation of homologous chromosomes during meiosis I in *C. elegans*. (1) Chromosomes undergo structural changes via the loading of cohesin protein complexes, assembly of axial elements, and dynamic changes in chromatin marks. (2) Pairing centers facilitate homolog recognition. (3) The SC forms to stabilize pairing interactions and promote crossover formation between homologs. (4) A subset of programmed DSBs are repaired to form COs. (5) The bivalent is differentiated by regulated retention and loss of various proteins along chromosome arms to allow for proper chromosome alignment at the metaphase plate, followed by homolog separation at anaphase I. Although we have gained some understanding of how these meiotic processes function to properly segregate chromosomes in *C. elegans*, a more thorough molecular understanding of the mechanisms promoting accurate segregation remains to be achieved, and we still do not fully understand how these processes are regulated to act coordinately.

C. elegans shares a high degree of gene conservation with higher eukaryotes. While 20–77% of genes in *C. elegans* have a mouse ortholog, approximately 60–80% of all genes in *C. elegans* have a human ortholog and most physiological and stress signaling pathways are also conserved (Lai et al. 2000; Leung et al. 2008; Kaletta and Hengartner 2006). This is a powerful model system that allows for studies of complex biological processes in the context of a multicellular organism. Investigations using *C. elegans* have already fostered a better understanding of meiotic processes that are universal to other organisms (Colaiacovo 2006; Garcia-Muse and Boulton 2007). *C. elegans* is a relevant model organism for the study of reproductive biology that can be utilized to study the impact of environmental factors on meiotic progression. For example, exposure to Bisphenol A, a plasticizer highly prevalent in our environment and which is linked to meiotic chromosome segregation defects in mammals, was demonstrated to alter expression of DSB repair factors resulting in impaired DSB repair in *C. elegans* (Allard and Colaiacovo 2010; Hunt et al. 2003).

Many of the remaining significant questions centered on meiotic processes can be addressed in this model system to further advance our knowledge of what happens in higher eukaryotes. Additional studies can reveal how axis-associated components coordinate both synapsis and recombination to determine the outcome of recombination events. It is still unclear how pairing and synapsis are monitored to avoid nonhomologous interactions. Although there have been extensive studies on both the mechanisms and the proteins involved in recombination, this remains a complex process whose regulation is still poorly understood. Therefore, more studies will be required to tease out both the direct and indirect roles of cohesin, axial elements, and the SC in mediating recombination. Moreover, further studies addressing the dynamic regulation of chromatin marks throughout the germline will enhance our understanding of how sites of recombination are determined and how it can affect downstream repair events. Furthermore, it remains to be determined how bivalent asymmetry is assessed to allow for its differentiation. Importantly, much of our understanding of the meiotic processes in *C. elegans* can shed light on how these processes occur in other organisms including mammals, therefore significantly contributing to our understanding of the mechanisms promoting human reproductive health.

Acknowledgments We thank Dr. Anne Villeneuve and members of the Colaiácovo lab for early access to unpublished results. We also thank Tim Schedl, Sara Beese-Sims, Takamune Saito, and Patrick Allard for critical reading of this manuscript. This work was supported by National Institutes of Health Grant R01GM072551 (to M.P.C.).

References

Adamo A, Montemauri P, Silva N, Ward JD, Boulton SJ, La Volpe A (2008) BRC-1 acts in the inter-sister pathway of meiotic double-strand break repair. EMBO Rep 9(3):287–292

Agarwal S, Roeder GS (2000) Zip3 provides a link between recombination enzymes and synaptonemal complex proteins. Cell 102(2):245–255

Albertson DG, Thomson JN (1982) The kinetochores of *Caenorhabditis elegans*. Chromosoma 86(3):409–428

Alexeev A, Mazin A, Kowalczykowski SC (2003) Rad54 protein possesses chromatin-remodeling activity stimulated by the Rad51-ssDNA nucleoprotein filament. Nat Struct Biol 10(3):182–186

Allard P, Colaiacovo MP (2010) Bisphenol A impairs the double-strand break repair machinery in the germline and causes chromosome abnormalities. Proc Natl Acad Sci USA 107(47):20405–20410

Allers T, Lichten M (2001) Differential timing and control of noncrossover and crossover recombination during meiosis. Cell 106(1):47–57

Alpi A, Pasierbek P, Gartner A, Loidl J (2003) Genetic and cytological characterization of the recombination protein RAD-51 in *Caenorhabditis elegans*. Chromosoma 112(1):6–16

Aravind L, Koonin EV (1998) The HORMA domain: a common structural denominator in mitotic checkpoints, chromosome synapsis and DNA repair. Trends Biochem Sci 23(8):284–286

Argueso JL, Wanat J, Gemici Z, Alani E (2004) Competing crossover pathways act during meiosis in *Saccharomyces cerevisiae*. Genetics 168(4):1805–1816

Bai X, Peirson BN, Dong F, Xue C, Makaroff CA (1999) Isolation and characterization of SYN1, a RAD21-like gene essential for meiosis in *Arabidopsis*. Plant Cell 11(3):417–430

Bailly AP, Freeman A, Hall J, Declais AC, Alpi A, Lilley DM, Ahmed S, Gartner A (2010) The *Caenorhabditis elegans* homolog of Gen1/Yen1 resolvases links DNA damage signaling to DNA double-strand break repair. PLoS Genet 6(7):e1001025

Baker BS, Carpenter AT (1972) Genetic analysis of sex chromosomal meiotic mutants in *Drosophilia melanogaster*. Genetics 71(2):255–286

Barber LJ, Youds JL, Ward JD, McIlwraith MJ, O'Neil NJ, Petalcorin MI, Martin JS, Collis SJ, Cantor SB, Auclair M, Tissenbaum H, West SC, Rose AM, Boulton SJ (2008) RTEL1 maintains genomic stability by suppressing homologous recombination. Cell 135(2):261–271

Bardhan A, Chuong H, Dawson DS (2010) Meiotic cohesin promotes pairing of non-homologous centromeres in early meiotic prophase. Mol Biol Cell 21(11):1799–1809

Barnes TM, Kohara Y, Coulson A, Hekimi S (1995) Meiotic recombination, noncoding DNA and genomic organization in *Caenorhabditis elegans*. Genetics 141(1):159–179

Baudat F, de Massy B (2007) Regulating double-stranded DNA break repair towards crossover or non-crossover during mammalian meiosis. Chromosome Res 15(5):565–577

Baudat F, Manova K, Yuen JP, Jasin M, Keeney S (2000) Chromosome synapsis defects and sexually dimorphic meiotic progression in mice lacking Spo11. Mol Cell 6(5):989–998

Baudrimont A, Penkner A, Woglar A, Machacek T, Wegrostek C, Gloggnitzer J, Fridkin A, Klein F, Gruenbaum Y, Pasierbek P, Jantsch V (2010) Leptotene/zygotene chromosome movement via the SUN/KASH protein bridge in *Caenorhabditis elegans*. PLoS Genet 6(11):e1001219

Baudrimont A, Penkner A, Woglar A, Mamnun YM, Huliek M, Struck C, Schnabel R, Loidl J, Jantsch V (2011) A new thermosensitive smc-3 allele reveals involvement of cohesin in homologous recombination in *C. elegans*. PLoS One 6(9):e24799

Bennett RJ, Dunderdale HJ, West SC (1993) Resolution of Holliday junctions by RuvC resolvase: cleavage specificity and DNA distortion. Cell 74(6):1021–1031

Berchowitz LE, Francis KE, Bey AL, Copenhaver GP (2007) The role of AtMUS81 in interference-insensitive crossovers in *A. thaliana*. PLoS Genet 3(8):e132

Bhalla N, Dernburg AF (2005) A conserved checkpoint monitors meiotic chromosome synapsis in *Caenorhabditis elegans*. Science 310(5754):1683–1686

Bhalla N, Wynne DJ, Jantsch V, Dernburg AF (2008) ZHP-3 acts at crossovers to couple meiotic recombination with synaptonemal complex disassembly and bivalent formation in *C. elegans*. PLoS Genet 4(10):e1000235

Bishop DK (1994) RecA homologs Dmc1 and Rad51 interact to form multiple nuclear complexes prior to meiotic chromosome synapsis. Cell 79(6):1081–1092

Bishop DK, Zickler D (2004) Early decision: meiotic crossover interference prior to stable strand exchange and synapsis. Cell 117(1):9–15

Blat Y, Protacio RU, Hunter N, Kleckner N (2002) Physical and functional interactions among basic chromosome organizational features govern early steps of meiotic chiasma formation. Cell 111(6):791–802

Boddy MN, Gaillard PH, McDonald WH, Shanahan P, Yates JR 3rd, Russell P (2001) Mus81-Eme1 are essential components of a Holliday junction resolvase. Cell 107(4):537–548

Bolcun-Filas E, Costa Y, Speed R, Taggart M, Benavente R, De Rooij DG, Cooke HJ (2007) SYCE2 is required for synaptonemal complex assembly, double strand break repair, and homologous recombination. J Cell Biol 176(6):741–747

Borner GV, Kleckner N, Hunter N (2004) Crossover/noncrossover differentiation, synaptonemal complex formation, and regulatory surveillance at the leptotene/zygotene transition of meiosis. Cell 117(1):29–45

Boyd JB, Golino MD, Setlow RB (1976) The mei-9 alpha mutant of *Drosophila melanogaster* increases mutagen sensitivity and decreases excision repair. Genetics 84(3):527–544

Buhler C, Borde V, Lichten M (2007) Mapping meiotic single-strand DNA reveals a new landscape of DNA double-strand breaks in *Saccharomyces cerevisiae*. PLoS Biol 5(12):e324

Carpenter AT (1975) Electron microscopy of meiosis in Drosophila melanogaster females: II. The recombination nodule—a recombination-associated structure at pachytene? Proc Natl Acad Sci USA 72(8):3186–3189

Cha RS, Weiner BM, Keeney S, Dekker J, Kleckner N (2000) Progression of meiotic DNA replication is modulated by interchromosomal interaction proteins, negatively by Spo11p and positively by Rec8p. Genes Dev 14(4):493–503

Chan RC, Chan A, Jeon M, Wu TF, Pasqualone D, Rougvie AE, Meyer BJ (2003) Chromosome cohesion is regulated by a clock gene paralogue TIM-1. Nature 423(6943):1002–1009

Chen SY, Tsubouchi T, Rockmill B, Sandler JS, Richards DR, Vader G, Hochwagen A, Roeder GS, Fung JC (2008) Global analysis of the meiotic crossover landscape. Dev Cell 15(3):401–415

Chin GM, Villeneuve AM (2001) *C. elegans mre-11* is required for meiotic recombination and DNA repair but is dispensable for the meiotic G(2) DNA damage checkpoint. Genes Dev 15(5):522–534

Ciosk R, Shirayama M, Shevchenko A, Tanaka T, Toth A, Nasmyth K (2000) Cohesin's binding to chromosomes depends on a separate complex consisting of Scc2 and Scc4 proteins. Mol Cell 5(2):243–254

Clejan I, Boerckel J, Ahmed S (2006) Developmental modulation of nonhomologous end joining in *Caenorhabditis elegans*. Genetics 173(3):1301–1317

Colaiacovo MP (2006) The many facets of SC function during *C. elegans* meiosis. Chromosoma 115(3):195–211

Colaiacovo MP, MacQueen AJ, Martinez-Perez E, McDonald K, Adamo A, La Volpe A, Villeneuve AM (2003) Synaptonemal complex assembly in *C. elegans* is dispensable for loading strand-exchange proteins but critical for proper completion of recombination. Dev Cell 5(3):463–474

Cole F, Keeney S, Jasin M (2010) Comprehensive, fine-scale dissection of homologous recombination outcomes at a hot spot in mouse meiosis. Mol Cell 39(5):700–710

Copenhaver GP, Housworth EA, Stahl FW (2002) Crossover interference in *Arabidopsis*. Genetics 160(4):1631–1639

Costa Y, Speed R, Ollinger R, Alsheimer M, Semple CA, Gautier P, Maratou K, Novak I, Hoog C, Benavente R, Cooke HJ (2005) Two novel proteins recruited by synaptonemal complex protein 1 (SYCP1) are at the centre of meiosis. J Cell Sci 118(Pt 12):2755–2762

Couteau F, Zetka M (2005) HTP-1 coordinates synaptonemal complex assembly with homolog alignment during meiosis in *C. elegans*. Genes Dev 19(22):2744–2756

Couteau F, Zetka M (2011) DNA damage during meiosis induces chromatin remodeling and synaptonemal complex disassembly. Dev Cell 20(3):353–363

Couteau F, Nabeshima K, Villeneuve A, Zetka M (2004) A component of *C. elegans* meiotic chromosome axes at the interface of homolog alignment, synapsis, nuclear reorganization, and recombination. Curr Biol 14(7):585–592

Cremer T, Cremer M, Dietzel S, Muller S, Solovei I, Fakan S (2006) Chromosome territories–a functional nuclear landscape. Curr Opin Cell Biol 18(3):307–316

Crittenden SL, Troemel ER, Evans TC, Kimble J (1994) GLP-1 is localized to the mitotic region of the *C. elegans* germ line. Development 120(10):2901–2911

Crittenden SL, Leonhard KA, Byrd DT, Kimble J (2006) Cellular analyses of the mitotic region in the *Caenorhabditis elegans* adult germ line. Mol Biol Cell 17(7):3051–3061

Cromie GA, Rubio CA, Hyppa RW, Smith GR (2005) A natural meiotic DNA break site in *Schizosaccharomyces pombe* is a hotspot of gene conversion, highly associated with crossing over. Genetics 169(2):595–605

Cromie GA, Hyppa RW, Taylor AF, Zakharyevich K, Hunter N, Smith GR (2006) Single Holliday junctions are intermediates of meiotic recombination. Cell 127(6):1167–1178

Cromie GA, Hyppa RW, Smith GR (2008) The fission yeast BLM homolog Rqh1 promotes meiotic recombination. Genetics 179(3):1157–1167

Curtis D, Clark SH, Chovnick A, Bender W (1989) Molecular analysis of recombination events in *Drosophila*. Genetics 122(3):653–661

Daniel K, Lange J, Hached K, Fu J, Anastassiadis K, Roig I, Cooke HJ, Stewart AF, Wassmann K, Jasin M, Keeney S, Toth A (2011) Meiotic homologue alignment and its quality surveillance are controlled by mouse HORMAD1. Nat Cell Biol 13(5):599–610

de Boer E, Stam P, Dietrich AJ, Pastink A, Heyting C (2006) Two levels of interference in mouse meiotic recombination. Proc Natl Acad Sci USA 103(25):9607–9612

de Carvalho CE, Zaaijer S, Smolikov S, Gu Y, Schumacher JM, Colaiacovo MP (2008) LAB-1 antagonizes the Aurora B kinase in *C. elegans*. Genes Dev 22(20):2869–2885 de los Santos T, Hunter N, Lee C, Larkin B, Loidl J, Hollingsworth NM (2003) The Mus81/Mms4 endonuclease acts independently of double-Holliday junction resolution to promote a distinct subset of crossovers during meiosis in budding yeast. Genetics 164(1):81–94

de Vries FA, de Boer E, van den Bosch M, Baarends WM, Ooms M, Yuan L, Liu JG, van Zeeland AA, Heyting C, Pastink A (2005) Mouse Sycp1 functions in synaptonemal complex assembly, meiotic recombination, and XY body formation. Genes Dev 19(11):1376–1389

Dernburg AF, Sedat JW, Hawley RS (1996) Direct evidence of a role for heterochromatin in meiotic chromosome segregation. Cell 86(1):135–146

Dernburg AF, McDonald K, Moulder G, Barstead R, Dresser M, Villeneuve AM (1998) Meiotic recombination in *C. elegans* initiates by a conserved mechanism and is dispensable for homologous chromosome synapsis. Cell 94(3):387–398

Dernburg AF, Zalevsky J, Colaiacovo MP, Villeneuve AM (2000) Transgene-mediated cosuppression in the *C. elegans* germ line. Genes Dev 14(13):1578–1583

Ding DQ, Yamamoto A, Haraguchi T, Hiraoka Y (2004) Dynamics of homologous chromosome pairing during meiotic prophase in fission yeast. Dev Cell 6(3):329–341

Dombecki CR, Chiang AC, Kang HJ, Bilgir C, Stefanski NA, Neva BJ, Klerkx EP, Nabeshima K (2011) The chromodomain protein MRG-1 facilitates the SC-independent homologous pairing during meiosis in *Caenorhabditis elegans*. Dev Cell 21(6):1092–1103

Dong H, Roeder GS (2000) Organization of the yeast Zip1 protein within the central region of the synaptonemal complex. J Cell Biol 148(3):417–426

Duerr JS (2006) Immunohistochemistry. WormBook:1–61

Dumont J, Oegema K, Desai A (2010) A kinetochore-independent mechanism drives anaphase chromosome separation during acentrosomal meiosis. Nat Cell Biol 12(9):894–901

Ellermeier C, Smith GR (2005) Cohesins are required for meiotic DNA breakage and recombination in *Schizosaccharomyces pombe*. Proc Natl Acad Sci USA 102(31):10952–10957

Fan QQ, Petes TD (1996) Relationship between nuclease-hypersensitive sites and meiotic recombination hot spot activity at the HIS4 locus of *Saccharomyces cerevisiae*. Mol Cell Biol 16(5):2037–2043

Fox PM, Vought VE, Hanazawa M, Lee MH, Maine EM, Schedl T (2011) Cyclin E and CDK-2 regulate proliferative cell fate and cell cycle progression in the *C. elegans* germline. Development 138(11):2223–2234

Frokjaer-Jensen C, Davis MW, Hollopeter G, Taylor J, Harris TW, Nix P, Lofgren R, Prestgard-Duke M, Bastiani M, Moerman DG, Jorgensen EM (2010) Targeted gene deletions in *C. elegans* using transposon excision. Nat Methods 7(6):451–453

Garcia-Muse T, Boulton SJ (2007) Meiotic recombination in *Caenorhabditis elegans*. Chromosome Res 15(5):607–621

Getz TJ, Banse SA, Young LS, Banse AV, Swanson J, Wang GM, Browne BL, Foss HM, Stahl FW (2008) Reduced mismatch repair of heteroduplexes reveals "non"-interfering crossing over in wild-type *Saccharomyces cerevisiae*. Genetics 178(3):1251–1269

Gillespie PJ, Hirano T (2004) Scc2 couples replication licensing to sister chromatid cohesion in *Xenopus* egg extracts. Curr Biol 14(17):1598–1603

Giroux CN, Dresser ME, Tiano HF (1989) Genetic-control of chromosome synapsis in yeast meiosis. Genome 31(1):88–94

Goldfarb T, Lichten M (2010) Frequent and efficient use of the sister chromatid for DNA double-strand break repair during budding yeast meiosis. PLoS Biol 8(10):e1000520

Goldstein P (1982) The synaptonemal complexes of *Caenorhabditis elegans*: pachytene karyotype analysis of male and hermaphrodite wild-type and him mutants. Chromosoma 86(4):577–593

Goodyer W, Kaitna S, Couteau F, Ward JD, Boulton SJ, Zetka M (2008) HTP-3 links DSB formation with homolog pairing and crossing over during *C. elegans* meiosis. Dev Cell 14(2):263–274

Grelon M, Vezon D, Gendrot G, Pelletier G (2001) AtSPO11-1 is necessary for efficient meiotic recombination in plants. EMBO J 20(3):589–600

Haering CH, Lowe J, Hochwagen A, Nasmyth K (2002) Molecular architecture of SMC proteins and the yeast cohesin complex. Mol Cell 9(4):773–788

Hagstrom KA, Meyer BJ (2003) Condensin and cohesin: more than chromosome compactor and glue. Nat Rev Genet 4(7):520–534

Hamer G, Gell K, Kouznetsova A, Novak I, Benavente R, Hoog C (2006) Characterization of a novel meiosis-specific protein within the central element of the synaptonemal complex. J Cell Sci 119(Pt 19):4025–4032

Harper NC, Rillo R, Jover-Gil S, Assaf ZJ, Bhalla N, Dernburg AF (2011) Pairing centers recruit a Polo-like kinase to orchestrate meiotic chromosome dynamics in *C. elegans*. Dev Cell 21(5):934–947

Harris TW, Antoshechkin I, Bieri T, Blasiar D, Chan J, Chen WJ, De La Cruz N, Davis P, Duesbury M, Fang R, Fernandes J, Han M, Kishore R, Lee R, Muller HM, Nakamura C, Ozersky P, Petcherski A, Rangarajan A, Rogers A, Schindelman G, Schwarz EM, Tuli MA, Van Auken K, Wang D, Wang X, Williams G, Yook K, Durbin R, Stein LD, Spieth J, Sternberg PW (2010) WormBase: a comprehensive resource for nematode research. Nucleic Acids Res 38(Database issue):D463–D467

Hassold T, Hunt P (2001) To err (meiotically) is human: the genesis of human aneuploidy. Nat Rev Genet 2(4):280–291

Hawley RS, Irick H, Zitron AE, Haddox DA, Lohe A, New C, Whitley MD, Arbel T, Jang J, McKim K et al (1992) There are two mechanisms of achiasmate segregation in *Drosophila* females, one of which requires heterochromatic homology. Dev Genet 13(6):440–467

Hayashi M, Chin GM, Villeneuve AM (2007) *C. elegans* germ cells switch between distinct modes of double-strand break repair during meiotic prophase progression. PLoS Genet 3(11):e191

Heidinger-Pauli JM, Unal E, Guacci V, Koshland D (2008) The kleisin subunit of cohesin dictates damage-induced cohesion. Mol Cell 31(1):47–56

Herman RK, Kari CK, Hartman PS (1982) Dominant X-chromosome nondisjunction mutants of *Caenorhabditis elegans*. Genetics 102(3):379–400

Higgins JD, Armstrong SJ, Franklin FC, Jones GH (2004) The *Arabidopsis* MutS homolog AtMSH4 functions at an early step in recombination: evidence for two classes of recombination in *Arabidopsis*. Genes Dev 18(20):2557–2570

Hillers KJ, Villeneuve AM (2003) Chromosome-wide control of meiotic crossing over in *C. elegans*. Curr Biol 13(18):1641–1647

Hilliker AJ, Chovnick A (1981) Further observations on intragenic recombination in *Drosophila melanogaster*. Genet Res 38(3):281–296

Hiraoka Y, Ding DQ, Yamamoto A, Tsutsumi C, Chikashige Y (2000) Characterization of fission yeast meiotic mutants based on live observation of meiotic prophase nuclear movement. Chromosoma 109(1–2):103–109

Hirsh D, Oppenheim D, Klass M (1976) Development of the reproductive system of *Caenorhabditis elegans*. Dev Biol 49(1):200–219

Hodgkin J, Horvitz HR, Brenner S (1979) Nondisjunction mutants of the nematode *Caenorhabditis elegans*. Genetics 91(1):67–94

Holloway JK, Booth J, Edelmann W, McGowan CH, Cohen PE (2008) MUS81 generates a subset of MLH1-MLH3-independent crossovers in mammalian meiosis. PLoS Genet 4(9):e1000186

Holloway JK, Morelli MA, Borst PL, Cohen PE (2010) Mammalian BLM helicase is critical for integrating multiple pathways of meiotic recombination. J Cell Biol 188(6):779–789

Howe M, McDonald KL, Albertson DG, Meyer BJ (2001) HIM-10 is required for kinetochore structure and function on *Caenorhabditis elegans* holocentric chromosomes. J Cell Biol 153(6):1227–1238

Hsu JY, Sun ZW, Li X, Reuben M, Tatchell K, Bishop DK, Grushcow JM, Brame CJ, Caldwell JA, Hunt DF, Lin R, Smith MM, Allis CD (2000) Mitotic phosphorylation of histone H3 is governed by Ipl1/aurora kinase and Glc7/PP1 phosphatase in budding yeast and nematodes. Cell 102(3):279–291

Hunt PA, Koehler KE, Susiarjo M, Hodges CA, Ilagan A, Voigt RC, Thomas S, Thomas BF, Hassold TJ (2003) Bisphenol a exposure causes meiotic aneuploidy in the female mouse. Curr Biol 13(7):546–553

Hunter N, Kleckner N (2001) The single-end invasion: an asymmetric intermediate at the double-strand break to double-Holliday junction transition of meiotic recombination. Cell 106(1):59–70

Ikura T, Tashiro S, Kakino A, Shima H, Jacob N, Amunugama R, Yoder K, Izumi S, Kuraoka I, Tanaka K, Kimura H, Ikura M, Nishikubo S, Ito T, Muto A, Miyagawa K, Takeda S, Fishel R,

Igarashi K, Kamiya K (2007) DNA damage-dependent acetylation and ubiquitination of H2AX enhances chromatin dynamics. Mol Cell Biol 27(20):7028–7040

Ip SC, Rass U, Blanco MG, Flynn HR, Skehel JM, West SC (2008) Identification of Holliday junction resolvases from humans and yeast. Nature 456(7220):357–361

Ishiguro T, Tanaka K, Sakuno T, Watanabe Y (2010) Shugoshin-PP2A counteracts casein-kinase-1-dependent cleavage of Rec8 by separase. Nat Cell Biol 12(5):500–506

Jang JK, Sherizen DE, Bhagat R, Manheim EA, McKim KS (2003) Relationship of DNA double-strand breaks to synapsis in *Drosophila*. J Cell Sci 116(Pt 15):3069–3077

Jantsch V, Pasierbek P, Mueller MM, Schweizer D, Jantsch M, Loidl J (2004) Targeted gene knockout reveals a role in meiotic recombination for ZHP-3, a Zip3-related protein in *Caenorhabditis elegans*. Mol Cell Biol 24(18):7998–8006

Jaramillo-Lambert A, Ellefson M, Villeneuve AM, Engebrecht J (2007) Differential timing of S phases, X chromosome replication, and meiotic prophase in the *C. elegans* germ line. Dev Biol 308(1):206–221

Jiang L, Xia M, Strittmatter LI, Makaroff CA (2007) The *Arabidopsis* cohesin protein SYN3 localizes to the nucleolus and is essential for gametogenesis. Plant J 50(6):1020–1034

Jones GH (1984) The control of chiasma distribution. Symp Soc Exp Biol 38:293–320

Joshi N, Barot A, Jamison C, Borner GV (2009) Pch2 links chromosome axis remodeling at future crossover sites and crossover distribution during yeast meiosis. PLoS Genet 5(7):e1000557

Joyce EF, McKim KS (2009) *Drosophila* PCH2 is required for a pachytene checkpoint that monitors double-strand-break-independent events leading to meiotic crossover formation. Genetics 181(1):39–51

Kaitna S, Pasierbek P, Jantsch M, Loidl J, Glotzer M (2002) The aurora B kinase AIR-2 regulates kinetochores during mitosis and is required for separation of homologous chromosomes during meiosis. Curr Biol 12(10):798–812

Kaletta T, Hengartner MO (2006) Finding function in novel targets: *C. elegans* as a model organism. Nat Rev Drug Discov 5(5):387–398

Karpen GH, Le MH, Le H (1996) Centric heterochromatin and the efficiency of achiasmate disjunction in *Drosophila* female meiosis. Science 273(5271):118–122

Katis VL, Lipp JJ, Imre R, Bogdanova A, Okaz E, Habermann B, Mechtler K, Nasmyth K, Zachariae W (2010) Rec8 phosphorylation by casein kinase 1 and Cdc7-Dbf4 kinase regulates cohesin cleavage by separase during meiosis. Dev Cell 18(3):397–409

Keeney S (2001) The mechanism and control of meiotic recombination initiation. Curr Top Dev Biol 52:1–53

Kelly KO, Dernburg AF, Stanfield GM, Villeneuve AM (2000) *Caenorhabditis elegans* msh-5 is required for both normal and radiation-induced meiotic crossing over but not for completion of meiosis. Genetics 156(2):617–630

Kelly WG, Schaner CE, Dernburg AF, Lee MH, Kim SK, Villeneuve AM, Reinke V (2002) X-chromosome silencing in the germline of *C. elegans*. Development 129(2):479–492

Kim KP, Weiner BM, Zhang L, Jordan A, Dekker J, Kleckner N (2010) Sister cohesion and structural axis components mediate homolog bias of meiotic recombination. Cell 143(6):924–937

Kitajima TS, Kawashima SA, Watanabe Y (2004) The conserved kinetochore protein shugoshin protects centromeric cohesion during meiosis. Nature 427(6974):510–517

Kitajima TS, Sakuno T, Ishiguro K, Iemura S, Natsume T, Kawashima SA, Watanabe Y (2006) Shugoshin collaborates with protein phosphatase 2A to protect cohesin. Nature 441(7089):46–52

Knoll A, Puchta H (2011) The role of DNA helicases and their interaction partners in genome stability and meiotic recombination in plants. J Exp Bot 62(5):1565–1579

Koszul R, Kleckner N (2009) Dynamic chromosome movements during meiosis: a way to eliminate unwanted connections? Trends Cell Biol 19(12):716–724

Koszul R, Kim KP, Prentiss M, Kleckner N, Kameoka S (2008) Meiotic chromosomes move by linkage to dynamic actin cables with transduction of force through the nuclear envelope. Cell 133(7):1188–1201

Krejci L, Song B, Bussen W, Rothstein R, Mortensen UH, Sung P (2002) Interaction with Rad51 is indispensable for recombination mediator function of Rad52. J Biol Chem 277(42):40132–40141
Kugou K, Fukuda T, Yamada S, Ito M, Sasanuma H, Mori S, Katou Y, Itoh T, Matsumoto K, Shibata T, Shirahige K, Ohta K (2009) Rec8 guides canonical Spo11 distribution along yeast meiotic chromosomes. Mol Biol Cell 20(13):3064–3076
Kusch T, Florens L, Macdonald WH, Swanson SK, Glaser RL, Yates JR 3rd, Abmayr SM, Washburn MP, Workman JL (2004) Acetylation by Tip60 is required for selective histone variant exchange at DNA lesions. Science 306(5704):2084–2087
Labella S, Woglar A, Jantsch V, Zetka M (2011) Polo kinases establish links between meiotic chromosomes and cytoskeletal forces essential for homolog pairing. Dev Cell 21(5):948–958
Lachner M, Jenuwein T (2002) The many faces of histone lysine methylation. Curr Opin Cell Biol 14(3):286–298
Lai CH, Chou CY, Ch'ang LY, Liu CS, Lin W (2000) Identification of novel human genes evolutionarily conserved in *Caenorhabditis elegans* by comparative proteomics. Genome Res 10(5):703–713
Lao JP, Oh SD, Shinohara M, Shinohara A, Hunter N (2008) Rad52 promotes postinvasion steps of meiotic double-strand-break repair. Mol Cell 29(4):517–524
Lee J, Hirano T (2011) RAD21L, a novel cohesin subunit implicated in linking homologous chromosomes in mammalian meiosis. J Cell Biol 192(2):263–276
Lee MH, Schedl T (2006) RNA in situ hybridization of dissected gonads. WormBook:1–7
Lee J, Iwai T, Yokota T, Yamashita M (2003) Temporally and spatially selective loss of Rec8 protein from meiotic chromosomes during mammalian meiosis. J Cell Sci 116(Pt 13):2781–2790
Lee J, Kitajima TS, Tanno Y, Yoshida K, Morita T, Miyano T, Miyake M, Watanabe Y (2008) Unified mode of centromeric protection by shugoshin in mammalian oocytes and somatic cells. Nat Cell Biol 10(1):42–52
Leung MC, Williams PL, Benedetto A, Au C, Helmcke KJ, Aschner M, Meyer JN (2008) *Caenorhabditis elegans*: an emerging model in biomedical and environmental toxicology. Toxicol Sci 106(1):5–28
Lightfoot J, Testori S, Barroso C, Martinez-Perez E (2011) Loading of meiotic cohesin by SCC-2 is required for processing of DSBs and for the DNA damage checkpoint. Curr Biol 32(17):1421–1430
Liu JG, Yuan L, Brundell E, Bjorkroth B, Daneholt B, Hoog C (1996) Localization of the N-terminus of SCP1 to the central element of the synaptonemal complex and evidence for direct interactions between the N-termini of SCP1 molecules organized head-to-head. Exp Cell Res 226(1):11–19
Liu T, Rechtsteiner A, Egelhofer TA, Vielle A, Latorre I, Cheung MS, Ercan S, Ikegami K, Jensen M, Kolasinska-Zwierz P, Rosenbaum H, Shin H, Taing S, Takasaki T, Iniguez AL, Desai A, Dernburg AF, Kimura H, Lieb JD, Ahringer J, Strome S, Liu XS (2011) Broad chromosomal domains of histone modification patterns in *C. elegans*. Genome Res 21(2):227–236
Llano E, Gomez R, Gutierrez-Caballero C, Herran Y, Sanchez-Martin M, Vazquez-Quinones L, Hernandez T, de Alava E, Cuadrado A, Barbero JL, Suja JA, Pendas AM (2008) Shugoshin-2 is essential for the completion of meiosis but not for mitotic cell division in mice. Genes Dev 22(17):2400–2413
Loidl J (1990) The initiation of meiotic chromosome pairing: the cytological view. Genome 33(6):759–778
Lorenz A, Estreicher A, Kohli J, Loidl J (2006) Meiotic recombination proteins localize to linear elements in *Schizosaccharomyces pombe*. Chromosoma 115(4):330–340
Lorenz A, West SC, Whitby MC (2010) The human Holliday junction resolvase GEN1 rescues the meiotic phenotype of a *Schizosaccharomyces pombe mus81* mutant. Nucleic Acids Res 38(6):1866–1873
Lui DY, Peoples-Holst TL, Mell JC, Wu HY, Dean EW, Burgess SM (2006) Analysis of close stable homolog juxtaposition during meiosis in mutants of *Saccharomyces cerevisiae*. Genetics 173(3):1207–1222
MacQueen AJ, Villeneuve AM (2001) Nuclear reorganization and homologous chromosome pairing during meiotic prophase require *C. elegans* chk-2. Genes Dev 15(13):1674–1687
MacQueen AJ, Colaiacovo MP, McDonald K, Villeneuve AM (2002) Synapsis-dependent and -independent mechanisms stabilize homolog pairing during meiotic prophase in *C. elegans*. Genes Dev 16(18):2428–2442

MacQueen AJ, Phillips CM, Bhalla N, Weiser P, Villeneuve AM, Dernburg AF (2005) Chromosome sites play dual roles to establish homologous synapsis during meiosis in *C. elegans*. Cell 123(6):1037–1050

Maguire MP (1966) The relationship of crossing over to chromosome synapsis in a short paracentric inversion. Genetics 53(6):1071–1077

Maleki S, Neale MJ, Arora C, Henderson KA, Keeney S (2007) Interactions between Mei4, Rec114, and other proteins required for meiotic DNA double-strand break formation in *Saccharomyces cerevisiae*. Chromosoma 116(5):471–486

Malone CJ, Misner L, Le Bot N, Tsai MC, Campbell JM, Ahringer J, White JG (2003) The *C. elegans* hook protein, ZYG-12, mediates the essential attachment between the centrosome and nucleus. Cell 115(7):825–836

Mancera E, Bourgon R, Brozzi A, Huber W, Steinmetz LM (2008) High-resolution mapping of meiotic crossovers and non-crossovers in yeast. Nature 454(7203):479–485

Mannuss A, Dukowic-Schulze S, Suer S, Hartung F, Pacher M, Puchta H (2010) RAD5A, RECQ4A, and MUS81 have specific functions in homologous recombination and define different pathways of DNA repair in *Arabidopsis thaliana*. Plant Cell 22(10):3318–3330

Marston AL, Tham WH, Shah H, Amon A (2004) A genome-wide screen identifies genes required for centromeric cohesion. Science 303(5662):1367–1370

Martin JS, Winkelmann N, Petalcorin MI, McIlwraith MJ, Boulton SJ (2005) RAD-51-dependent and -independent roles of a *Caenorhabditis elegans* BRCA2-related protein during DNA double-strand break repair. Mol Cell Biol 25(8):3127–3139

Martinez-Perez E, Villeneuve AM (2005) HTP-1-dependent constraints coordinate homolog pairing and synapsis and promote chiasma formation during *C. elegans* meiosis. Genes Dev 19(22):2727–2743

Martinez-Perez E, Schvarzstein M, Barroso C, Lightfoot J, Dernburg AF, Villeneuve AM (2008) Crossovers trigger a remodeling of meiotic chromosome axis composition that is linked to two-step loss of sister chromatid cohesion. Genes Dev 22(20):2886–2901

Martini E, Diaz RL, Hunter N, Keeney S (2006) Crossover homeostasis in yeast meiosis. Cell 126(2):285–295

Mazin AV, Bornarth CJ, Solinger JA, Heyer WD, Kowalczykowski SC (2000) Rad54 protein is targeted to pairing loci by the Rad51 nucleoprotein filament. Mol Cell 6(3):583–592

Mazina OM, Mazin AV (2004) Human Rad54 protein stimulates DNA strand exchange activity of hRad51 protein in the presence of Ca2+. J Biol Chem 279(50):52042–52051

McKee BD (1996) The license to pair: identification of meiotic pairing sites in *Drosophila*. Chromosoma 105(3):135–141

McKim KS, Howell AM, Rose AM (1988) The effects of translocations on recombination frequency in *Caenorhabditis elegans*. Genetics 120(4):987–1001

McKim KS, Green-Marroquin BL, Sekelsky JJ, Chin G, Steinberg C, Khodosh R, Hawley RS (1998) Meiotic synapsis in the absence of recombination. Science 279(5352):876–878

McMahill MS, Sham CW, Bishop DK (2007) Synthesis-dependent strand annealing in meiosis. PLoS Biol 5(11):e299

Melby TE, Ciampaglio CN, Briscoe G, Erickson HP (1998) The symmetrical structure of structural maintenance of chromosomes (SMC) and MukB proteins: long, antiparallel coiled coils, folded at a flexible hinge. J Cell Biol 142(6):1595–1604

Mercier R, Jolivet S, Vezon D, Huppe E, Chelysheva L, Giovanni M, Nogue F, Doutriaux MP, Horlow C, Grelon M, Mezard C (2005) Two meiotic crossover classes cohabit in Arabidopsis: one is dependent on MER3, whereas the other one is not. Curr Biol 15(8):692–701

Mets DG, Meyer BJ (2009) Condensins regulate meiotic DNA break distribution, thus crossover frequency, by controlling chromosome structure. Cell 139(1):73–86

Miki F, Okazaki K, Shimanuki M, Yamamoto A, Hiraoka Y, Niwa O (2002) The 14-kDa dynein light chain-family protein Dlc1 is required for regular oscillatory nuclear movement and efficient recombination during meiotic prophase in fission yeast. Mol Biol Cell 13(3):930–946

Minn IL, Rolls MM, Hanna-Rose W, Malone CJ (2009) SUN-1 and ZYG-12, mediators of centrosome-nucleus attachment, are a functional SUN/KASH pair in *Caenorhabditis elegans*. Mol Biol Cell 20(21):4586–4595

Monen J, Maddox PS, Hyndman F, Oegema K, Desai A (2005) Differential role of CENP-A in the segregation of holocentric *C. elegans* chromosomes during meiosis and mitosis. Nat Cell Biol 7(12):1248–1255

Moses MJ (1968) Synaptinemal complex. Annu Rev Genet 2:363–412

Motohashi T, Tabara H, Kohara Y (2006) Protocols for large scale in situ hybridization on *C. elegans* larvae. WormBook:1–8

Muller HJ (1916) The mechanism of crossing over. Am Nat 50:193–221

Munz P (1994) An analysis of interference in the fission yeast *Schizosaccharomyces pombe*. Genetics 137(3):701–707

Nabeshima K, Villeneuve AM, Hillers KJ (2004) Chromosome-wide regulation of meiotic crossover formation in *Caenorhabditis elegans* requires properly assembled chromosome axes. Genetics 168(3):1275–1292

Nabeshima K, Villeneuve AM, Colaiacovo MP (2005) Crossing over is coupled to late meiotic prophase bivalent differentiation through asymmetric disassembly of the SC. J Cell Biol 168(5):683–689

Nag DK, Scherthan H, Rockmill B, Bhargava J, Roeder GS (1995) Heteroduplex DNA formation and homolog pairing in yeast meiotic mutants. Genetics 141(1):75–86

Nasmyth K, Haering CH (2005) The structure and function of SMC and kleisin complexes. Annu Rev Biochem 74:595–648

Nicklas RB (1974) Chromosome segregation mechanisms. Genetics 78(1):205–213

Nottke AC, Beese-Sims SE, Pantelena LF, Reinke V, Shi Y, Colaiacovo MP (2011) SPR-5 is a histone H3K4 demethylase with a role in meiotic double-strand break repair. Proc Natl Acad Sci USA 108(31):12805–12810

Obeso D, Dawson DS (2010) Temporal characterization of homology-independent centromere coupling in meiotic prophase. PLoS One 5(4):e10336

Oh SD, Lao JP, Hwang PY, Taylor AF, Smith GR, Hunter N (2007) BLM ortholog, Sgs1, prevents aberrant crossing-over by suppressing formation of multichromatid joint molecules. Cell 130(2):259–272

Oh SD, Lao JP, Taylor AF, Smith GR, Hunter N (2008) RecQ helicase, Sgs1, and XPF family endonuclease, Mus81-Mms4, resolve aberrant joint molecules during meiotic recombination. Mol Cell 31(3):324–336

Oliveira RA, Nasmyth K (2010) Getting through anaphase: splitting the sisters and beyond. Biochem Soc Trans 38(6):1639–1644

Östergren G (1951) The mechanism of co-orientation in bivalents, and multivalents. Berlingska boktr, Lund

Page SL, Hawley RS (2001) c(3)G encodes a *Drosophila* synaptonemal complex protein. Genes Dev 15(23):3130–3143

Page SL, Hawley RS (2004) The genetics and molecular biology of the synaptonemal complex. Annu Rev Cell Dev Biol 20:525–558

Pan J, Sasaki M, Kniewel R, Murakami H, Blitzblau HG, Tischfield SE, Zhu X, Neale MJ, Jasin M, Socci ND, Hochwagen A, Keeney S (2011) A hierarchical combination of factors shapes the genome-wide topography of yeast meiotic recombination initiation. Cell 144(5):719–731

Panizza S, Mendoza MA, Berlinger M, Huang L, Nicolas A, Shirahige K, Klein F (2011) Spo11-accessory proteins link double-strand break sites to the chromosome axis in early meiotic recombination. Cell 146(3):372–383

Parisi S, McKay MJ, Molnar M, Thompson MA, van der Spek PJ, van Drunen-Schoenmaker E, Kanaar R, Lehmann E, Hoeijmakers JH, Kohli J (1999) Rec8p, a meiotic recombination and sister chromatid cohesion phosphoprotein of the Rad21p family conserved from fission yeast to humans. Mol Cell Biol 19(5):3515–3528

Pasierbek P, Jantsch M, Melcher M, Schleiffer A, Schweizer D, Loidl J (2001) A *Caenorhabditis elegans* cohesion protein with functions in meiotic chromosome pairing and disjunction. Genes Dev 15(11):1349–1360

Pasierbek P, Fodermayr M, Jantsch V, Jantsch M, Schweizer D, Loidl J (2003) The *Caenorhabditis elegans* SCC-3 homologue is required for meiotic synapsis and for proper chromosome disjunction in mitosis and meiosis. Exp Cell Res 289(2):245–255

Penkner A, Portik-Dobos Z, Tang L, Schnabel R, Novatchkova M, Jantsch V, Loidl J (2007) A conserved function for a *Caenorhabditis elegans* Com1/Sae2/CtIP protein homolog in meiotic recombination. EMBO J 26(24):5071–5082

Penkner AM, Fridkin A, Gloggnitzer J, Baudrimont A, Machacek T, Woglar A, Csaszar E, Pasierbek P, Ammerer G, Gruenbaum Y, Jantsch V (2009) Meiotic chromosome homology search involves modifications of the nuclear envelope protein Matefin/SUN-1. Cell 139(5):920–933

Peoples TL, Dean E, Gonzalez O, Lambourne L, Burgess SM (2002) Close, stable homolog juxtaposition during meiosis in budding yeast is dependent on meiotic recombination, occurs independently of synapsis, and is distinct from DSB-independent pairing contacts. Genes Dev 16(13):1682–1695

Peoples-Holst TL, Burgess SM (2005) Multiple branches of the meiotic recombination pathway contribute independently to homolog pairing and stable juxtaposition during meiosis in budding yeast. Genes Dev 19(7):863–874

Petalcorin MI, Sandall J, Wigley DB, Boulton SJ (2006) CeBRC-2 stimulates D-loop formation by RAD-51 and promotes DNA single-strand annealing. J Mol Biol 361(2):231–242

Petalcorin MI, Galkin VE, Yu X, Egelman EH, Boulton SJ (2007) Stabilization of RAD-51-DNA filaments via an interaction domain in *Caenorhabditis elegans* BRCA2. Proc Natl Acad Sci USA 104(20):8299–8304

Petes TD (2001) Meiotic recombination hot spots and cold spots. Nat Rev Genet 2(5):360–369

Phillips CM, Dernburg AF (2006) A family of zinc-finger proteins is required for chromosome-specific pairing and synapsis during meiosis in *C. elegans*. Dev Cell 11(6):817–829

Phillips CM, Wong C, Bhalla N, Carlton PM, Weiser P, Meneely PM, Dernburg AF (2005) HIM-8 binds to the X chromosome pairing center and mediates chromosome-specific meiotic synapsis. Cell 123(6):1051–1063

Phillips CM, McDonald KL, Dernburg AF (2009a) Cytological analysis of meiosis in *Caenorhabditis elegans*. Methods Mol Biol 558:171–195

Phillips CM, Meng X, Zhang L, Chretien JH, Urnov FD, Dernburg AF (2009b) Identification of chromosome sequence motifs that mediate meiotic pairing and synapsis in *C. elegans*. Nat Cell Biol 11(8):934–942

Rabitsch KP, Gregan J, Schleiffer A, Javerzat JP, Eisenhaber F, Nasmyth K (2004) Two fission yeast homologs of *Drosophila* Mei-S332 are required for chromosome segregation during meiosis I and II. Curr Biol 14(4):287–301

Radford SJ, McMahan S, Blanton HL, Sekelsky J (2007) Heteroduplex DNA in meiotic recombination in *Drosophila* mei-9 mutants. Genetics 176(1):63–72

Rasmussen SW, Holm PB (1984) The synaptonemal complex, recombination nodules and chiasmata in human spermatocytes. Symp Soc Exp Biol 38:271–292

Reddy KC, Villeneuve AM (2004) *C. elegans* HIM-17 links chromatin modification and competence for initiation of meiotic recombination. Cell 118(4):439–452

Riedel CG, Katis VL, Katou Y, Mori S, Itoh T, Helmhart W, Galova M, Petronczki M, Gregan J, Cetin B, Mudrak I, Ogris E, Mechtler K, Pelletier L, Buchholz F, Shirahige K, Nasmyth K (2006) Protein phosphatase 2A protects centromeric sister chromatid cohesion during meiosis I. Nature 441(7089):53–61

Roeder GS (1997) Meiotic chromosomes: it takes two to tango. Genes Dev 11(20):2600–2621

Rogers E, Bishop JD, Waddle JA, Schumacher JM, Lin R (2002) The aurora kinase AIR-2 functions in the release of chromosome cohesion in *Caenorhabditis elegans* meiosis. J Cell Biol 157(2):219–229

Rollins RA, Korom M, Aulner N, Martens A, Dorsett D (2004) *Drosophila* Nipped-B protein supports sister chromatid cohesion and opposes the stromalin/Scc3 cohesion factor to facilitate long-range activation of the cut gene. Mol Cell Biol 24(8):3100–3111

Romano A, Guse A, Krascenicova I, Schnabel H, Schnabel R, Glotzer M (2003) CSC-1: a subunit of the Aurora B kinase complex that binds to the Survivin-like protein BIR-1 and the Incenp-like protein ICP-1. J Cell Biol 161(2):229–236

Rose AM, Baillie DL, Curran J (1984) Meiotic pairing behavior of two free duplications of linkage group I in *Caenorhabditis elegans*. Mol Gen Genet 195(1–2):52–56

Rosenbluth RE, Baillie DL (1981) The genetic analysis of a reciprocal translocation, eT1(III; V), in *Caenorhabditis elegans*. Genetics 99(3–4):415–428

Rosu S, Libuda DE, Villeneuve AM (2011) Robust crossover assurance and regulated intehomolog access maintain meiotic crossover number. Science 334(6060):1286–1289

Saito TT, Youds JL, Boulton SJ, Colaiacovo MP (2009) *Caenorhabditis elegans* HIM-18/SLX-4 interacts with SLX-1 and XPF-1 and maintains genomic integrity in the germline by processing recombination intermediates. PLoS Genet 5(11):e1000735

Sanford C, Perry MD (2001) Asymmetrically distributed oligonucleotide repeats in the *Caenorhabditis elegans* genome sequence that map to regions important for meiotic chromosome segregation. Nucleic Acids Res 29(14):2920–2926

Sato A, Isaac B, Phillips CM, Rillo R, Carlton PM, Wynne DJ, Kasad RA, Dernburg AF (2009) Cytoskeletal forces span the nuclear envelope to coordinate meiotic chromosome pairing and synapsis. Cell 139(5):907–919

Schild-Prufert K, Saito TT, Smolikov S, Gu Y, Hincapie M, Hill DE, Vidal M, McDonald K, Colaiacovo MP (2011) Organization of the synaptonemal complex during meiosis in *Caenorhabditis elegans*. Genetics 189:411–421

Schlecht HB, Lichten M, Goldman ASH (2004) Compartmentalization of the yeast meiotic nucleus revealed by analysis of ectopic recombination. Genetics 168(3):1189–1203

Schmekel K, Meuwissen RL, Dietrich AJ, Vink AC, van Marle J, van Veen H, Heyting C (1996) Organization of SCP1 protein molecules within synaptonemal complexes of the rat. Exp Cell Res 226(1):20–30

Schramm S, Fraune J, Naumann R, Hernandez-Hernandez A, Hoog C, Cooke HJ, Alsheimer M, Benavente R (2011) A novel mouse synaptonemal complex protein is essential for loading of central element proteins, recombination, and fertility. PLoS Genet 7(5):e1002088

Schumacher JM, Golden A, Donovan PJ (1998) AIR-2: An Aurora/Ipl1-related protein kinase associated with chromosomes and midbody microtubules is required for polar body extrusion and cytokinesis in *Caenorhabditis elegans* embryos. J Cell Biol 143(6):1635–1646

Schwacha A, Kleckner N (1994) Identification of joint molecules that form frequently between homologs but rarely between sister chromatids during yeast meiosis. Cell 76(1):51–63

Schwartz EK, Heyer WD (2011) Processing of joint molecule intermediates by structure-selective endonucleases during homologous recombination in eukaryotes. Chromosoma 120(2):109–127

Seitan VC, Banks P, Laval S, Majid NA, Dorsett D, Rana A, Smith J, Bateman A, Krpic S, Hostert A, Rollins RA, Erdjument-Bromage H, Tempst P, Benard CY, Hekimi S, Newbury SF, Strachan T (2006) Metazoan Scc4 homologs link sister chromatid cohesion to cell and axon migration guidance. PLoS Biol 4(8):e242

Severson AF, Ling L, van Zuylen V, Meyer BJ (2009) The axial element protein HTP-3 promotes cohesin loading and meiotic axis assembly in *C. elegans* to implement the meiotic program of chromosome segregation. Genes Dev 23(15):1763–1778

Sherwood R, Takahashi TS, Jallepalli PV (2010) Sister acts: coordinating DNA replication and cohesion establishment. Genes Dev 24(24):2723–2731

Shinohara M, Oh SD, Hunter N, Shinohara A (2008) Crossover assurance and crossover interference are distinctly regulated by the ZMM proteins during yeast meiosis. Nat Genet 40(3):299–309

Smagulova F, Gregoretti IV, Brick K, Khil P, Camerini-Otero RD, Petukhova GV (2011) Genome-wide analysis reveals novel molecular features of mouse recombination hotspots. Nature 472(7343):375–378

Smith GR, Boddy MN, Shanahan P, Russell P (2003) Fission yeast Mus81.Eme1 Holliday junction resolvase is required for meiotic crossing over but not for gene conversion. Genetics 165(4):2289–2293

Smithies O, Powers PA (1986) Gene conversions and their relation to homologous chromosome pairing. Philos Trans R Soc Lond B Biol Sci 312(1154):291–302

Smolikov S, Eizinger A, Hurlburt A, Rogers E, Villeneuve AM, Colaiacovo MP (2007a) Synapsis-defective mutants reveal a correlation between chromosome conformation and the mode of double-strand break repair during *Caenorhabditis elegans* meiosis. Genetics 176(4):2027–2033

Smolikov S, Eizinger A, Schild-Prufert K, Hurlburt A, McDonald K, Engebrecht J, Villeneuve AM, Colaiacovo MP (2007b) SYP-3 restricts synaptonemal complex assembly to bridge paired chromosome axes during meiosis in *Caenorhabditis elegans*. Genetics 176(4):2015–2025

Smolikov S, Schild-Prufert K, Colaiacovo MP (2008) CRA-1 uncovers a double-strand break-dependent pathway promoting the assembly of central region proteins on chromosome axes during *C. elegans* meiosis. PLoS Genet 4(6):e1000088

Smolikov S, Schild-Prufert K, Colaiacovo MP (2009) A yeast two-hybrid screen for SYP-3 interactors identifies SYP-4, a component required for synaptonemal complex assembly and chiasma formation in *Caenorhabditis elegans* meiosis. PLoS Genet 5(10):e1000669

Snowden T, Acharya S, Butz C, Berardini M, Fishel R (2004) hMSH4-hMSH5 recognizes Holliday Junctions and forms a meiosis-specific sliding clamp that embraces homologous chromosomes. Mol Cell 15(3):437–451

Speliotes EK, Uren A, Vaux D, Horvitz HR (2000) The survivin-like *C. elegans* BIR-1 protein acts with the Aurora-like kinase AIR-2 to affect chromosomes and the spindle midzone. Mol Cell 6(2):211–223

Starr DA, Fridolfsson HN (2010) Interactions between nuclei and the cytoskeleton are mediated by SUN-KASH nuclear-envelope bridges. Annu Rev Cell Dev Biol 26:421–444

Sugiyama T, New JH, Kowalczykowski SC (1998) DNA annealing by RAD52 protein is stimulated by specific interaction with the complex of replication protein A and single-stranded DNA. Proc Natl Acad Sci USA 95(11):6049–6054

Sun H, Treco D, Szostak JW (1991) Extensive 3′-overhanging, single-stranded DNA associated with the meiosis-specific double-strand breaks at the *ARG4* recombination initiation site. Cell 64(6):1155–1161

Sym M, Engebrecht JA, Roeder GS (1993) ZIP1 is a synaptonemal complex protein required for meiotic chromosome synapsis. Cell 72(3):365–378

Szostak JW, Orr-Weaver TL, Rothstein RJ, Stahl FW (1983) The double-strand-break repair model for recombination. Cell 33(1):25–35

Tarsounas M, Morita T, Pearlman RE, Moens PB (1999) RAD51 and DMC1 form mixed complexes associated with mouse meiotic chromosome cores and synaptonemal complexes. J Cell Biol 147(2):207–220

Terasawa M, Shinohara A, Hotta Y, Ogawa H, Ogawa T (1995) Localization of RecA-like recombination proteins on chromosomes of the lily at various meiotic stages. Genes Dev 9(8):925–934

Thomas SE, McKee BD (2007) Meiotic pairing and disjunction of mini-X chromosomes in *Drosophila* is mediated by 240-bp rDNA repeats and the homolog conjunction proteins SNM and MNM. Genetics 177(2):785–799

Thomas SE, Soltani-Bejnood M, Roth P, Dorn R, Logsdon JM Jr, McKee BD (2005) Identification of two proteins required for conjunction and regular segregation of achiasmate homologs in *Drosophila* male meiosis. Cell 123(4):555–568

Timmons L, Fire A (1998) Specific interference by ingested dsRNA. Nature 395(6705):854

Tonkin ET, Wang TJ, Lisgo S, Bamshad MJ, Strachan T (2004) NIPBL, encoding a homolog of fungal Scc2-type sister chromatid cohesion proteins and fly Nipped-B, is mutated in Cornelia de Lange syndrome. Nat Genet 36(6):636–641

Trelles-Sticken E, Adelfalk C, Loidl J, Scherthan H (2005) Meiotic telomere clustering requires actin for its formation and cohesin for its resolution. J Cell Biol 170(2):213–223

Trowbridge K, McKim K, Brill SJ, Sekelsky J (2007) Synthetic lethality of *Drosophila* in the absence of the MUS81 endonuclease and the DmBlm helicase is associated with elevated apoptosis. Genetics 176(4):1993–2001

Tsai CJ, Mets DG, Albrecht MR, Nix P, Chan A, Meyer BJ (2008) Meiotic crossover number and distribution are regulated by a dosage compensation protein that resembles a condensin subunit. Genes Dev 22(2):194–211

Tsubouchi T, Roeder GS (2005) A synaptonemal complex protein promotes homology-independent centromere coupling. Science 308(5723):870–873

Villeneuve AM (1994) A cis-acting locus that promotes crossing over between X chromosomes in *Caenorhabditis elegans*. Genetics 136(3):887–902

Villeneuve AM, Hillers KJ (2001) Whence meiosis? Cell 106(6):647–650

Wagner CR, Kuervers L, Baillie DL, Yanowitz JL (2010) xnd-1 regulates the global recombination landscape in *Caenorhabditis elegans*. Nature 467(7317):839–843

Wang F, Yoder J, Antoshechkin I, Han M (2003) *Caenorhabditis elegans* EVL-14/PDS-5 and SCC-3 are essential for sister chromatid cohesion in meiosis and mitosis. Mol Cell Biol 23(21):7698–7707

Watrin E, Schleiffer A, Tanaka K, Eisenhaber F, Nasmyth K, Peters JM (2006) Human Scc4 is required for cohesin binding to chromatin, sister-chromatid cohesion, and mitotic progression. Curr Biol 16(9):863–874

Webber HA, Howard L, Bickel SE (2004) The cohesion protein ORD is required for homologue bias during meiotic recombination. J Cell Biol 164(6):819–829

Weiner BM, Kleckner N (1994) Chromosome pairing via multiple interstitial interactions before and during meiosis in yeast. Cell 77(7):977–991

Westergaard M, von Wettstein D (1972) The synaptinemal complex. Annu Rev Genet 6:71–110

Wicky C, Alpi A, Passannante M, Rose A, Gartner A, Muller F (2004) Multiple genetic pathways involving the *Caenorhabditis elegans* Bloom's syndrome genes *him-6*, *rad-51*, and *top-3* are needed to maintain genome stability in the germ line. Mol Cell Biol 24(11):5016–5027

Wignall SM, Villeneuve AM (2009) Lateral microtubule bundles promote chromosome alignment during acentrosomal oocyte meiosis. Nat Cell Biol 11(7):839–844

Winand NJ, Panzer JA, Kolodner RD (1998) Cloning and characterization of the human and *Caenorhabditis elegans* homologs of the *Saccharomyces cerevisiae* MSH5 gene. Genomics 53(1):69–80

Wu HY, Burgess SM (2006) Two distinct surveillance mechanisms monitor meiotic chromosome metabolism in budding yeast. Curr Biol 16(24):2473–2479

Yildiz O, Majumder S, Kramer B, Sekelsky JJ (2002) *Drosophila* MUS312 interacts with the nucleotide excision repair endonuclease MEI-9 to generate meiotic crossovers. Mol Cell 10(6):1503–1509

Youds JL, Mets DG, McIlwraith MJ, Martin JS, Ward JD, NJ ON, Rose AM, West SC, Meyer BJ, Boulton SJ (2010) RTEL-1 enforces meiotic crossover interference and homeostasis. Science 327(5970):1254–1258

Zalevsky J, MacQueen AJ, Duffy JB, Kemphues KJ, Villeneuve AM (1999) Crossing over during *Caenorhabditis elegans* meiosis requires a conserved MutS-based pathway that is partially dispensable in budding yeast. Genetics 153(3):1271–1283

Zetka M, Rose A (1995) The genetics of meiosis in *Caenorhabditis elegans*. Trends Genet 11(1):27–31

Zetka MC, Kawasaki I, Strome S, Muller F (1999) Synapsis and chiasma formation in *Caenorhabditis elegans* require HIM-3, a meiotic chromosome core component that functions in chromosome segregation. Genes Dev 13(17):2258–2270

Zickler D (1977) Development of the synaptonemal complex and the "recombination nodules" during meiotic prophase in the seven bivalents of the fungus *Sordaria macrospora* Auersw. Chromosoma 61(4):289–316

Zickler D, Kleckner N (1998) The leptotene-zygotene transition of meiosis. Annu Rev Genet 32:619–697

Zickler D, Kleckner N (1999) Meiotic chromosomes: integrating structure and function. Annu Rev Genet 33:603–754

Chapter 7
Spermatogenesis

Diana S. Chu and Diane C. Shakes

Abstract During spermatogenesis, pluripotent germ cells differentiate to become efficient delivery vehicles to the oocyte of paternal DNA. Though male and female germ cells both undergo meiosis to produce haploid complements of DNA, at the same time they also each undergo distinct differentiation processes that result in either sperm or oocytes. This review will discuss our current understanding of mechanisms of sperm formation and differentiation in *Caenorhabditis elegans* gained from studies that employ a combination of molecular, transcriptomic, and cell biological approaches. Many of these processes also occur during spermatogenesis in other organisms but with differences in timing, molecular machinery, and morphology. In *C. elegans*, sperm differentiation is implemented by varied modes of gene regulation, including the genomic organization of genes important for sperm formation, the generation of sperm-specific small RNAs, and the interplay of specific transcriptional activators. As sperm formation progresses, chromatin is systematically remodeled to allow first for the implementation of differentiation programs, then for sperm-specific DNA packaging required for transit of paternal genetic and epigenetic information. Sperm also exhibit distinctive features of meiotic progression, including the formation of a unique karyosome state and the centrosomal-based segregation of chromosomes during symmetric meiotic divisions. Sperm-specific organelles are also assembled and remodeled as cells complete meiosis and individualize in preparation for activation, morphogenesis, and the acquisition of motility. Finally, in addition to DNA, sperm contribute specific cellular factors that contribute to successful embryogenesis.

D.S. Chu (✉)
Department of Biology, San Francisco State University,
1600 Holloway Avenue, San Francisco, CA, USA
e-mail: chud@sfsu.edu

D.C. Shakes (✉)
Department of Biology, College of William and Mary, Williamsburg, VA, USA
e-mail: dcshak@wm.edu

T. Schedl (ed.), *Germ Cell Development in C. elegans*, Advances in Experimental Medicine and Biology 757, DOI 10.1007/978-1-4614-4015-4_7,

Keywords Differentiation • Chromosome segregation • Paternal epigenetics • Meiosis • Organelle morphogenesis • Spermiogenesis • Sperm maturation • Transcriptional regulation

7.1 Introduction

Although sperm and oocytes share a mission to contribute a haploid genome to the new embryo, each has specialized functions. Small sperm are streamlined to efficiently navigate the female reproductive tract whereas large oocytes are packed with maternal materials necessary for embryonic development. Thus, the production of functionally and morphologically distinct gametes requires a precise interplay between shared and distinctive aspects of their differentiation programs. In this review we will discuss recent advances that inform our current understanding of processes required for sperm formation, particularly the spermatogenesis-specific features of germ cell formation. Our discussion will generally follow the temporal progression of spermatogenesis, while highlighting how individual processes overlap and influence one another. In particular, this chapter will emphasize the varied modes of gene regulation that underlie sperm differentiation, the systematic remodeling of chromatin required for differentiation and delivery of paternal epigenetic information, the sperm-specific features of meiotic progression, the assembly and remodeling of sperm-specific organelles, the regulated process of sperm activation, and the acquisition of cell motility. Many of these processes also occur during spermatogenesis in other organisms but with differences in timing, molecular machinery, and morphology that will be discussed as relevant.

The rapidly expanding toolkit for studying *Caenorhabditis elegans* has uncovered a broad range of molecular mechanisms important for sperm formation. Early ultrastructural studies of the male gonad and isolation of spermatogenesis-defective genetic mutants helped define the stages of sperm formation and laid a foundation for recent genomic and transcriptomic approaches. Technological advances in RNA interference (RNAi) and reverse genetic deletion mutant screening methods are now facilitating the analysis of candidate gene function in various stages of sperm formation and fertilization. Likewise, cytological studies of gamete formation have benefited from advances in imaging techniques and the ability to make transgenic animals that express fluorescently labeled fusion proteins. Unlike previous reviews of *C. elegans* spermatogenesis that have focused on insights gained from genetic mutants (L'Hernault 2006; Nelson et al. 1982), this review will integrate what has been learned about molecular mechanisms of *C. elegans* spermatogenesis using a myriad of experimental approaches.

7.2 Brief Overview of Sperm Formation in *C. elegans*

C. elegans has many features that make it an ideal model for studying spermatogenesis. Approximately half the adult body mass of both males and hermaphrodites is devoted to germ cell formation and both sexes make sperm. Hermaphrodites generate approximately 300 sperm during the last larval stage before switching to exclusively producing oocytes. Hermaphrodites are self-fertile; sperm that they make during their last larval stage are stored within one of two spermathecae, where they fertilize passing oocytes during each ovulation event. In contrast, males continue to make sperm, which they transfer to hermaphrodites through mating (Ward and Carrel 1979). Relative to hermaphrodite sperm, male sperm are both larger and more competitive (LaMunyon and Ward 1995, 1998); however, the information provided in this review will apply to sperm formation in both sexes unless otherwise noted.

Germ cell formation can be easily observed through the transparent cuticle of both sexes in fixed and live samples (L'Hernault 2006; McCarter et al. 1999; Shakes et al. 2009). Because gamete formation occurs in a linear progression along the length of the gonad (Seydoux and Schedl 2001), germ cells can be staged cytologically by their location within the gonad arm, their nuclear morphology, and the presence of distinct marker proteins, which can be assessed by immunostaining (Figs. 7.1a and 7.2). In the subsections below we will give a brief overview of the key events that occur within each of these zones shown in Figs. 7.1a and 7.2. These subsections are intended to place the events of spermatogenesis in context of other chapters in this book and to introduce readers to the key events of spermatogenesis before devoting the remainder of the chapter to an in-depth discussion of many of these sperm-specific features.

7.2.1 Mitotic Region

Somatic cells, called distal tip cells, define the most distal end of the gonad (Fig. 7.1a). In hermaphrodites, one distal tip cell is positioned at each end of the two gonad arms while males have both distal tip cells positioned at the end of the single-armed gonad (Kimble and Crittenden 2007; Byrd and Kimble 2009). Adjacent to the distal tip cell(s), sexually uncommitted germ cells undergo repeated rounds of mitotic duplication. The molecular mechanisms responsible for specifying the size and extent of this mitotic proliferation zone are covered in Chap. 4 (Hansen and Schedl 2012). Surprisingly, recent studies reveal that mitotically proliferating germ cells divide more rapidly in male gonads than in hermaphrodite gonads, an observation that suggests that the germ cells are sexually dimorphic even before they fully commit to an oocyte or sperm fate (Morgan et al. 2010).

7.2.2 Meiotic Entry, the Sperm/Oocyte Switch, and Early Meiotic Prophase

Exit from the mitotic region (Fig. 7.1a) is tightly coordinated with two events that are covered in other chapters of this series: transition to meiosis in Chap. 4 (Hansen and Schedl 2012) and sex determination in Chap. 3 (Zanetti and Puoti 2012).

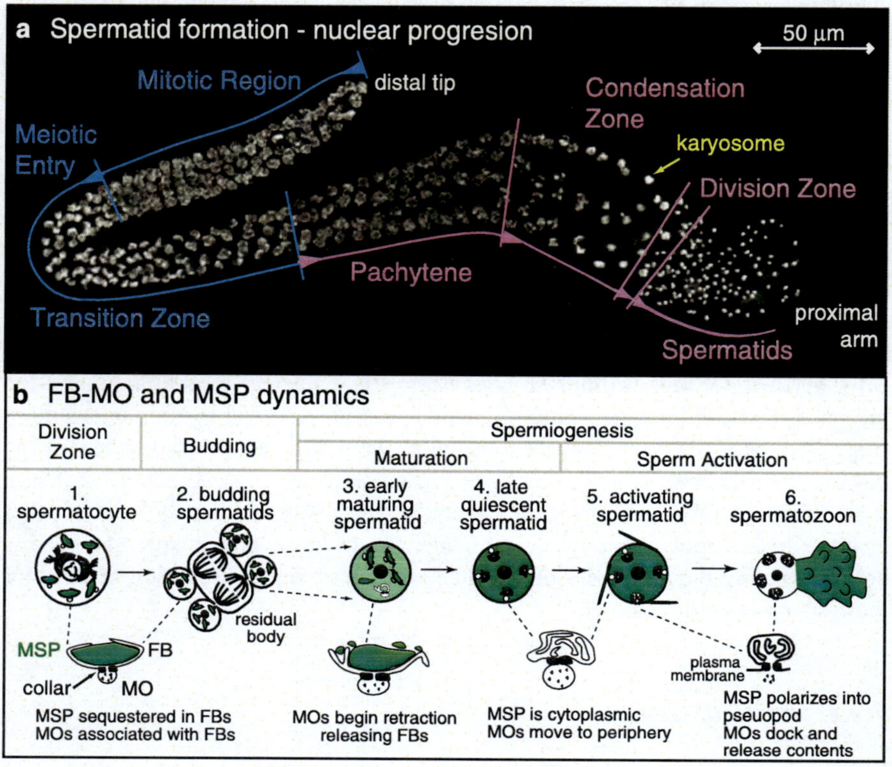

Fig. 7.1 The progression of spermatid formation and pseudopod assembly. (**a**) Changes in nuclear morphology during spermatid formation. A single-armed wild-type male gonad visualized using DAPI and fluorescence microscopy. Regions of the gonad are labeled: mitotic, meiotic entry, and the transition zone (*blue*) and pachytene, the condensation zone, meiotic division zone, and haploid spermatids (*purple*). A nucleus exhibiting the karysome morphology is indicated in *yellow*. *Scale bar* represent 50 µm. (**b**) Key stages in FB-MO and MSP dynamics. The process of spermiogenesis includes budding, maturation, and activation. (1) A spermatocyte in diakinesis (before nuclear envelope breakdown) contains multiple, fully mature fibrous body (FB)–membranous organelle (MO) complexes. The major sperm protein (MSP) (*green*) is assembled into FBs that are enveloped by the arms of the MO. The MO head is the vesicle below the electron-dense collar (two *dark bars*). (2) After the MI and MII divisions the late-stage budding spermatid is fully polarized with the FB-MOs and chromatin masses partitioned to the spermatids and the intact spindle microtubules partitioned to the central residual body. (3) In an early maturing spermatid, the arms of the MO retract as the FBs are released into the cytoplasm and begin to disassemble. (4) A late-stage quiescent spermatid in which the MOs are docked and MSP is cytosolic. (5) Upon exposure to an activator, spermatids initially form microspikes as the MOs begin to fuse at the collar with the plasma membrane. (6) Motile spermatozoon with a distinct cell body containing fused MOs and MSP-filled pseudopod

The molecular events that drive the early stages of meiotic development, including chromosome pairing, synapsis, and recombination are thought to occur similarly in both males and hermaphrodites and are described in Chap. 6 (Lui and Colaiácovo 2012). Conversely, the programs of *C. elegans* oogenesis and spermatogenesis differ

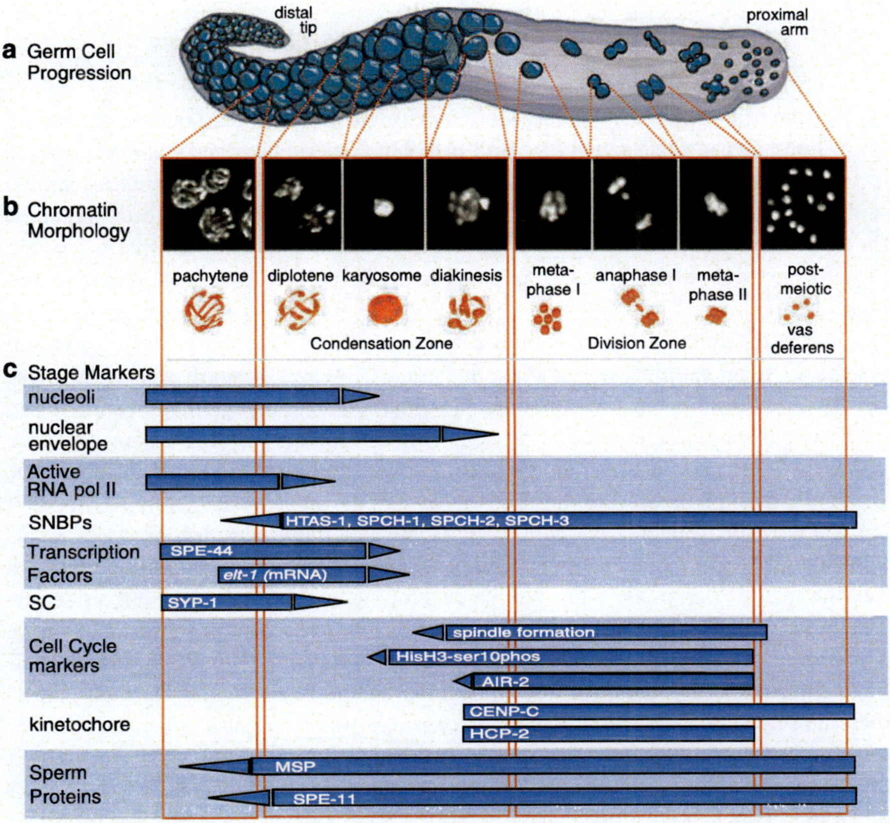

Fig. 7.2 Overview of the overlapping events that occur during late meiotic prophase of sperm cell formation. (**a**) Schematic of the progression of male germline cells (*blue*) during spermatogenesis. Cells are attached to the rachis through karyosome formation, then bud off of the rachis to undergo meiotic divisions. After anaphase II, haploid cells bud from residual bodies to form spermatids. (**b**) The corresponding chromatin morphology of cells highlighted in (**a**). DAPI-stained and schematic drawings (*red*) of the nuclear morphology of cells in the stages of late meiotic prophase indicated. (**c**) Staging of sperm cells can also be monitored by the presence of specific cell structures, organelles, and macromolecules, which are represented as *blue bars*

in that spermatocytes progress through meiotic prophase at a faster rate (20–24 h rather than 54–60 h) (Jaramillo-Lambert et al. 2007) and, unlike oocytes, are not subjected to a checkpoint for DNA damage and meiotic recombination errors that results in damaged cells being removal by apoptosis (Gartner et al. 2000; Jaramillo-Lambert et al. 2010). However, to date, the earliest marker of spermatogenesis is the recently discovered transcriptional regulator, SPE-44, which can first be observed on the chromatin of sperm but not oocyte producing germ lines during early pachytene (Figs. 7.2c and 7.4) and will be discussed in Sect. 7.3.3.

7.2.3 Late Meiotic Prophase and the Meiotic Divisions

By late meiotic prophase, several features distinguish sperm- and oocyte-producing germ lines. During late pachytene and diplotene, spermatogenesis-enriched genes are being robustly transcribed and sperm-specific structures are being assembled within cells (Figs. 7.1a and 7.2). Then, immediately prior to the meiotic divisions, the spermatocytes become transcriptionally inactive and enter a sperm-specific karyosome stage, which is described in more detail in Sect. 7.4.1 (Fig. 7.1a) (Shakes et al. 2009).

During this time, the chromosomes of both oocytes and spermatocytes resolve and condense in preparation for meiotic divisions. However several aspects of these events differ in oocytes and spermatocytes, particularly in regard to events in late pachytene through the condensation zone (Figs. 7.1a and 7.2). First, shortly before initiation of the meiotic divisions, the spermatocyte chromatin becomes differentially structured through the incorporation of sperm-specific nuclear basic proteins (SNBPs) (Chu et al. 2006). Second, in contrast to oocytes that mature in response to extracellular signals (Kim et al. 2012), spermatocytes proceed directly from meiotic prophase into the meiotic divisions. Third, unlike developing oocytes, which lose their centrioles during meiotic prophase and subsequently segregate their chromosomes on an acentriolar spindle (Albertson and Thomson 1993; Kim and Roy 2006; Peters et al. 2010; Wignall and Villeneuve 2009), developing spermatocytes retain their centrioles and segregate their chromosomes on centriole-based spindles. Since the spermatocyte spindles are nucleated by centrosomes rather than through a chromatin-mediated mechanism, spindle assembly can initiate earlier in spermatocytes, before the breakdown of the nuclear envelope (Shakes et al. 2009). Lastly, in oocytes, these divisions are asymmetric and yield a single gamete whereas in spermatocytes, the divisions are symmetric and yield four equally sized gametes (L'Hernault 2006).

7.2.4 Spermiogenesis and Motility

After the completion of the meiotic divisions, spermatids develop into motile spermatozoa, a process referred to as spermiogenesis (Fig. 7.1b). We define this process as beginning immediately after anaphase II and involving three major phases. In an initial budding phase, a combination of polarization and spermatid budding mechanisms result in the separation of individual spermatids from a central residual body, in which materials that are unnecessary for the subsequent development and function of the sperm are discarded. In the subsequent, poorly characterized maturation phase, the sperm chromatin becomes surrounded by an electron-dense, RNA-enriched perinuclear halo (Ward et al. 1981). Sperm-specific complexes and organelles, which house components required for sperm motility and fertilization, also remodel in preparation for an extended quiescent state. In a final activation phase, quiescent, spherical sperm transform into bipolar and motile spermatozoa upon exposure to sex-specific extracellular signaling molecules. Motility of the crawling spermatozoa is driven by the regulated assembly and disassembly of a nematode-specific cytoskeletal protein known as the Major Sperm Protein (MSP).

Exciting advances in our understanding of some of the molecular mechanisms required for the specification of sperm formation during each of these phases are summarized in Fig. 7.2.

7.3 Genomics and Transcriptomics: Programming the Genome for Sperm Differentiation

Our understanding of how germline stem cells differentiate into motile spermatozoa has benefited from diverse experimental approaches. "Systems-wide" approaches have not only identified a comprehensive list of genes whose expression is "spermatogenesis-enriched," but are also enabling the exploration of how their global transcriptional profiles are modified under different experiment conditions or between mutant backgrounds. In one key approach (Reinke et al. 2000, 2004), factors that contribute to sperm formation were identified through pair-wise comparisons of the expression profiles of mutant hermaphrodites that produce only oocytes (*fem-1* lf), only sperm (*fem-3* gf), or completely lack a germ line (*glp-4(bn2)*) (Barton et al. 1987; Beanan and Strome 1992; Nelson et al. 1978). In another, sperm-enriched factors were identified by comparing protein or RNA profiles from mutants that produce only sperm or only oocytes (Chu et al. 2006; Han et al. 2009). These studies are revealing the variety of molecular mechanisms that regulate expression of spermatogenesis-enriched genes. In this section, we will discuss four levels of regulation: (1) the organization and genetic structure of spermatogenesis-enriched genes within the genome, (2) sperm-specific histones and other basic nuclear proteins that modify the structure of the sperm chromatin, (3) regulatory transcription factors that govern the large-scale implementation of the sperm differentiation program, and (4) sperm-specific small RNAs at work during sperm differentiation.

7.3.1 Organization and Genetic Structure of Spermatogenesis-Enriched Genes

Several genome-wide expression studies have identified genes whose transcription is differentially regulated during spermatogenesis (Reinke et al. 2000, 2004; Bamps and Hope 2008; Maeda et al. 2001; Wang et al. 2009). These microarray studies distinguished genes with sperm-enriched expression from others, including germ line-intrinsic, oocyte-enriched, and somatic-enriched. Though all germ cell-specific classes have commonalities, sperm genes exhibit several distinct features.

For example, the X chromosome contains relatively few genes that are expressed in either the germ line or the developing gametes, chromosome IV contains an abundance of sperm-enriched genes, and chromosome I contains an over-representation of germline-enriched genes. These finding suggests that genes may exhibit large- and small-scale clustering patterns within the genome that aid in their coordinated regulation. Supporting this hypothesis, an integrated analysis of germline microarray and *in situ* expression data revealed that many germline and oocyte genes are tightly clustered in small groups within operons (Reinke and Cutter 2009). In contrast, genes expressed during spermatogenesis were largely excluded from operons, indicating they do not exhibit small-scale clustering. Interestingly, germ line-intrinsic and oocyte-enriched genes exhibit numerous similarities in temporal regulation, response to RNAi, and the functional classes of their protein products, whereas sperm enriched genes have distinct temporal expression profiles, are generally refractory to RNAi, and encode many novel protein products. Taken together, these studies suggest that the gene expression program necessary for the execution of the sperm fate involves large-scale changes in chromatin that are aided, in part, by the genomic organization of spermatogenesis-enriched genes.

7.3.2 Chromatin Organization and Global Transcription Activation Are Coupled

Throughout spermatogenesis, the processes of meiosis, sperm differentiation, and chromatin remodeling are intimately intertwined. One distinctive feature of chromatin remodeling during spermatogenesis is the introduction of SNBPs that regulate transcription (Bettegowda and Wilkinson 2011). For example, in mammals, high levels of transcription are promoted in meiotic cells by histone acetylation and incorporation of histone variants (Lewis et al. 2003; Sassone-Corsi 2002). During this time, sperm-specific transcriptional regulatory factors must access DNA in order to implement specific differentiation programs (Kimmins et al. 2004). Once meiosis is completed, somatic histones are replaced in a stepwise fashion: first by other sperm-specific histone variants, then by transition proteins, and ultimately by protamine proteins (Kimmins and Sassone-Corsi 2005; Govin et al. 2004; Braun 2001). As a result, transcription becomes globally repressed as DNA becomes increasingly compacted. The repackaging of sperm DNA is hypothesized to streamline the sperm DNA cargo for efficient mobility, protect the DNA package from potential environmental damage, and carry potential paternal epigenetic information to the nascent embryo (Caron et al. 2005; Miller et al. 2010; Wu and Chu 2008).

In *C. elegans*, candidate SNBPs have been identified through comparisons of the proteomic profiles of sperm and oocyte chromatin (Chu et al. 2006). These include a sperm-specific histone H2A variant called HTAS-1 (H Two A Sperm) and three putative protamines called SPCH-1, 2, and 3 (SPerm CHromatin), all of which initially incorporate into chromosomes during late pachytene and remain associated with mature sperm chromatin (Chu et al. 2006) (Figs. 7.2 and 7.3a).

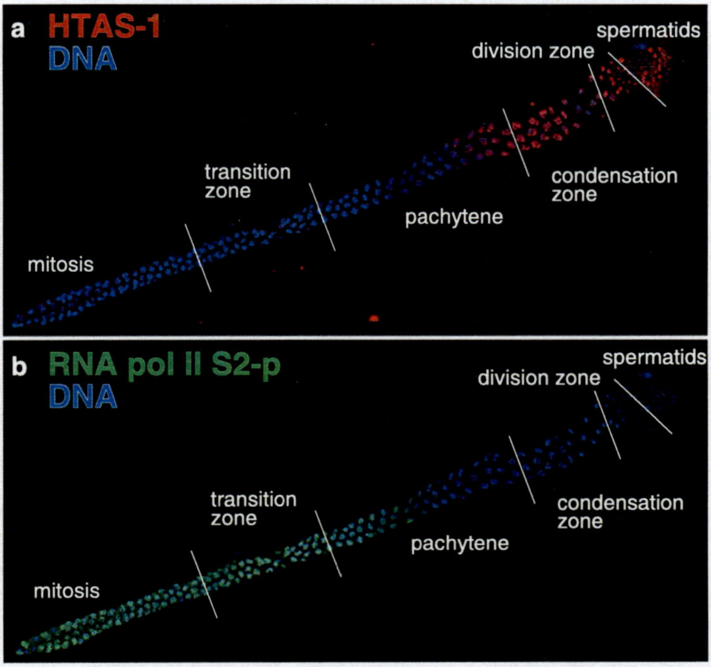

Fig. 7.3 Changes in transcriptional regulation correlate with the alteration of chromatin composition. Immunolocalization of (**a**) the sperm-specific histone H2A variant, HTAS-1 (*red*) and (**b**) elongating RNA polymerase II (phosphorylated on the C-terminal domain on serine 2, detected using Abcam H5 antibody ab24758) (*green*). Regions of the male germ line are indicated. DNA is shown in *blue*. (**a**) HTAS-1 incorporates into sperm chromatin as cells condense for meiotic divisions. (**b**) High levels of actively elongating RNA polymerase (*green*) decrease dramatically as chromosomes condense for meiotic divisions, indicating global transcriptional activation is curtailed by the karyosome stage

Overall, the incorporation of these SNBPs correlates with global transcriptional down-regulation; levels of RNA polymerase II actively engaged in transcriptional elongation remain elevated through mid-pachytene but then decrease significantly as chromosomes condense in preparation for meiotic divisions (Fig. 7.3b) (Shakes et al. 2009). However, it has yet to be determined the extent to which subsets of genes required for sperm formation may escape transcriptional down-regulation. When compared to the process in mammals, the incorporation of SNBPs into spermatocyte chromatin during meiosis may seem precocious. However, we hypothesize that this early incorporation of SNBPs may have evolved to support the comparatively rapid progression of *C. elegans* spermatogenesis. One consequence of early SNBP incorporation is that the meiotic chromosomes of *C. elegans* spermatocytes may be distinct from those of either oocytes or mitotically dividing germ cells. In addition, the shutdown of transcription prior to the meiotic divisions implies that both the progression of sperm morphogenesis and sperm activation must be driven solely by post-transcription regulatory mechanisms.

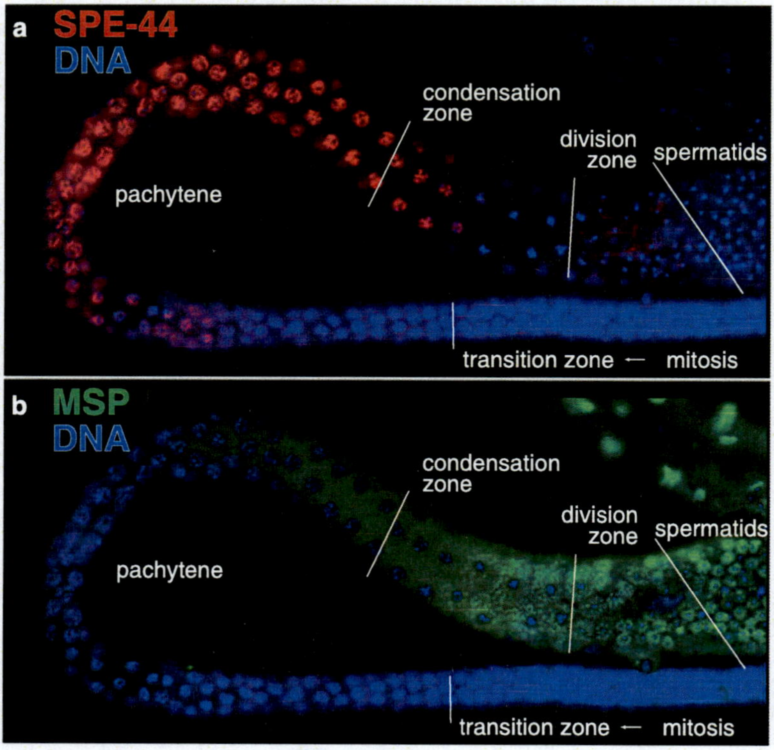

Fig. 7.4 Sperm-specific transcription factors and the expression of sperm proteins. Immunolocalization of (**a**) the SPE-44 transcription factor (*red*) and (**b**) the Major Sperm Protein, MSP (*green*) within an isolated and fixed male gonad. DNA is shown in *blue*. Regions of the male germ line are indicated. (**a**) SPE-44 (*green*) is expressed early in pachytene until chromosomes condense for meiotic divisions. (**b**) MSP is synthesized beginning in pachytene and subsequently localizes to distinct FBs, which localize as oblong stripes,within the condensation zone. MSP partitions to the spermatids in FBs but in mature, quiescent spermatids (*far right*) it disassembles and fills the cytoplasm

7.3.3 Transcription Factors That Coordinate Sperm Differentiation

The role of individual transcription factors identified from the list of spermatogenesis-enriched genes has also revealed coordinated regulation of sperm formation. For example, the transcription factor SPE-44 was identified from the list of 1,343 spermatogenesis-enriched genes (Reinke 2002; Reinke et al. 2000, 2004; Kulkarni et al. 2012). A homozygous mutant of *spe-44* is spermatogenesis defective (*spe*). Evidence suggests that SPE-44 functions as an early global transcriptional activator required for spermatocyte differentiation (Kulkarni et al. 2012). First, SPE-44 is expressed exclusively during spermatogenesis in nuclei of undifferentiated spermatocytes, well before levels of the sperm-specific protein MSP accumulate (Fig. 7.4). During pachytene, SPE-44 localizes strongly on autosomes but not the X chromosome, which is largely transcriptionally silent during spermatogenesis (Kelly et al. 2002; Reinke et al. 2000). Microarray analysis also reveals that two-thirds of the more than 500 genes whose expression is down-regulated in *spe-44* mutants were previously classified as spermatogenesis-enriched genes (Reinke 2002; Reinke et al. 2000, 2004; Kulkarni et al. 2012). As a result, *spe-44* mutants produce arrested spermatocytes with defects in multiple sperm-specific processes including the assembly of FBs within spermatocytes and the spermatid budding process that follows anaphase II. These results suggest that SPE-44 may function, presumably in conjunction with other transcription factors, as a key, sperm-specific transcriptional regulator of sperm differentiation.

In another example, the promoters of spermatogenesis-enriched genes were compared with those of all other *C. elegans* genes in order to identify enriched DNA sequence motifs that could act as transcription factor binding sites (del Castillo-Olivares et al. 2009). A subset of 45 sperm-enriched genes was found to share a single bipartite consensus that was then used as a binding target in a yeast 1-hybrid screen for potential sperm-specific transcription factors (del Castillo-Olivares et al. 2009; Klass et al. 1988; Shim et al. 1995). This screen identified a single GATA-type transcription factor, ELT-1, that regulates hypodermal expression in the embryo but is also highly expressed in the male germ line (Spieth et al. 1991). In sperm producing germ lines *elt-1* mRNA is present beginning from mid-pachytene; in oocyte producing germ lines, it is undetectable. Reduction of ELT-1 in either RNAi mediated knockdown or knockout mosaics resulted in defects in sperm production, including post-meiotic morphological and motility defects (del Castillo-Olivares et al. 2009). ELT-1 is also amongst the genes that are down-regulated in absence of SPE-44 (Kulkarni et al. 2012). Importantly, no phenotypes were observed in oocytes, though affected worms exhibited additional somatic defects, consistent with the role of ELT-1 in other somatic tissues (del Castillo-Olivares et al. 2009). This role for ELT-1 in *C. elegans* spermatogenesis is consistent with roles of GATA transcription factor family members in the regulation of mammalian sperm development (Ketola et al. 1999, 2000).

7.3.4 Regulation of Sperm Formation by Sperm-Specific Small RNAs

Studies in *C. elegans* of small non-coding RNAs, important regulators of mRNA degradation, translational repression, and chromatin structure, have revealed they contribute to the proper regulation of large groups of sperm genes. Three classes of small RNAs are known: microRNAs (miRNAs), Piwi-interacting RNAs (piRNAs), and endogeneous small interfering RNAs (endo-siRNAs) (Kim et al. 2009). Each class is generated by a distinct mechanism and each employs a distinct effector pathway to regulate gene expression (Suh and Blelloch 2011). Temperature-sensitive roles in *C. elegans* spermatogenesis have been found for specific subclasses of endo-siRNAs and piRNAs.

7.3.4.1 piRNAs

piRNAs have evolutionarily conserved functions in male fertility (Aravin et al. 2006; Houwing et al. 2007). In mammals, piRNAs are expressed in both pachytene spermatocytes and round spermatids—stages that are blocked by mutations in Piwi proteins, a distinct family of Argonaut proteins (Deng and Lin 2002; Kuramochi-Miyagawa et al. 2004). piRNAs, which do not require Dicer for their generation, are predominantly expressed in the germ line and are complementary to genomic repeat sequences (Aravin et al. 2006; Grivna et al. 2006). piRNAs in *Drosophila* are associated with the repression of transposable elements during spermatogenesis (Malone and Hannon 2009). Thus, a primary function *ascribed* to piRNAs is to guard the integrity of the genome in the germ line from foreign invaders like transposons.

In *C. elegans*, piRNAs function in both sperm and oocyte-producing germ lines. Slightly shorter than the typical 25–33 nucleotide piRNAs in other organisms, *C. elegans* piRNAs are 21 base pairs in length, begin with U, and physically associate with the Piwi protein PRG-1 (Han et al. 2009; Kato et al. 2009; Ruby et al. 2006; Batista et al. 2008; Das et al. 2008). The expression of these 21U piRNAs is restricted to the germ line, and strains with null mutations in either of the two Piwi homologs, *prg-1* and *prg-2*, result in temperature-sensitive mitotic and meiotic germline defects. Thus the Piwi protein PRG-1 is important for robust thermotolerance during spermatogenesis (Batista et al. 2008; Wang and Reinke 2008). At 25°C, *prg-1* males produce spermatocytes but very few spermatids, and the spermatids that are produced are defective in sperm activation (Wang and Reinke 2008). In the male germ line, PRG-1 localizes in a perinuclear fashion to the germline P-granules in developing spermatocytes but disappears as the chromatin condenses following pachytene (Batista et al. 2008). This localization to P-granules is intriguing as other P-granule associated factors have been reported to exhibit analogous, maternal-specific, temperature-sensitive, null-phenotypes (Smith et al. 2002; Spike et al. 2008).

Microarray analysis of *prg-1* single mutants shows that a subset of sperm-enriched transcripts are affected by lack of PRG-1; though studies conflict as to whether PRG-1 upregulates or represses these genes (Batista et al. 2008; Wang and Reinke 2008). Targets of piRNAs include the Tc3 transposon, whose transposition is elevated at least a 100-fold in *prg-1; prg-2* double mutants (Batista et al. 2008; Das et al. 2008). This repression of transposon activity suggests piRNAs may function in maintaining the genomic stability of germ cells, a role consistent with that characterized for piRNAs in other organisms. The mechanism for how piRNAs inhibit transposition in any organism is not yet understood.

C. elegans piRNAs have distinct features from those in other organisms. Besides being shorter, they have an upstream eight nucleotide core sequence, suggesting they are individually transcribed, not processed from a transcript of clustered piRNAs genes as seen in other organisms (Das et al. 2008). *C. elegans* piRNAs also lack any overlap, suggesting a different mode of amplification than that proposed in other organisms (Das et al. 2008; Aravin et al. 2007; Brennecke et al. 2007; Gunawardane et al. 2007). Thus mechanisms of small RNA regulation for germline function may have both conserved and adaptive features.

7.3.4.2 Endo-siRNAs

Characterization of siRNA function in the *C. elegans* germ line revealed an important role for small RNAs not only in response to foreign DNA but also in endogeneous gene regulation. Endo-siRNAs are generated from long dsRNAs that either base pair with complementary dsRNAs or form long hairpins (Okamura and Lai 2008; Golden et al. 2008). In *C. elegans*, response to long double-stranded RNAs involves two phases. First primary siRNAs are generated via cleavage by the endonuclease Dicer (Bernstein et al. 2001; Grishok et al. 2001; Ketting et al. 2001; Knight and Bass 2001). Then secondary siRNAs are produced by RNA-directed RNA polymerases (RdRPs), which amplify the silencing response (Aoki et al. 2007; Sijen et al. 2001). Post-transcriptional silencing is achieved as siRNAs partner with specific Argonaute proteins as part of the RISC complex, which degrades target mRNAs. The mechanism of silencing depends on the complementarity between the endo-siRNA and target mRNA: perfect base-pairing to the target mRNA results in mRNA cleavage while imperfect base-pairing causes silencing by an unknown mechanism (Zeng and Cullen 2003).

Sperm and oocytes have distinct classes of endo-siRNAs (Han et al. 2009). Deep sequencing of purified sperm, oocytes, and embryos in *C. elegans* revealed sperm and oocyte-specific classes of 26 nucleotide long small RNAs with a strong bias for a G as the first nucleotide (Han et al. 2009). Both classes of 26G RNAs are exact complements to target mRNAs and their generation requires members of the *C. elegans* RISC complex, including Dicer (DCR-1), an RdRP called RRF-3, and Dicer associated factors called ERI-1 and ERI-3 (Pavelec et al. 2009; Gent et al. 2009; Han et al. 2009). Oocyte-specific 26G RNAs are maternally inherited and silence gene expression during zygotic development (Han et al. 2009). In contrast, targets of the sperm-specific 26G RNAs include the sperm-enriched transcripts identified by previous germline microarray data (Reinke et al. 2000, 2004; Gent et al. 2009; Han et al. 2009). This suggests that sperm-specific 26G RNAs function to down-regulate the levels of sperm-enriched genes (Han et al. 2009). Indeed, these particular sperm genes are significantly upregulated in *eri-1* and *rrf-3* mutants (Gent et al. 2009; Han et al. 2009).

In an analogous fashion to the Piwi protein PRG-1, the RISC complex members *eri-1, eri-3, rrf-3, and dcr-1* are required for spermatogenesis at elevated temperatures (Pavelec et al. 2009; Gent et al. 2009). At 25°C, mutant males undergo improper cytokinesis and chromosome segregation and produce misshapen spermatids that contain abnormal chromatin masses and excess tubulin. Although most of these mutant spermatids fail to activate, the motile spermatozoa that are formed can fertilize oocytes. However, the resulting embryos are nonviable (Gent et al. 2009). During oogenesis, Dicer is also required for the proper organization of the hermaphrodite germ line during the pachytene stage, and *dcr-1* null hermaphrodites produce irregularly shaped, nonfunctional, endomitotic oocytes (Knight and Bass 2001). As such, unlike *eri-1* and *rrf-3*, Dicer functions more broadly in both siRNA and miRNA pathways (Knight and Bass 2001) and has recently been shown to regulate ribonucleoprotein assembly in the hermaphrodite germ line (Beshore et al. 2011). In contrast, although *eri-1* and *rrf-3* are required for 26G RNA production in both spermatocytes and oocytes, loss of either *eri-1* or *rrf-3* causes no discernable detects in either oocytes or the resulting embryos (Pavelec et al. 2009; Gent et al. 2009). Why the loss of sperm-specific 26G RNAs results in male infertility while loss of oocyte 26Gs does not adversely affect oogenesis or embryonic development is yet unknown.

How does the RISC complex generate different sperm and oocyte targets? Distinct Argonaute protein components confer specificity to the RISC complex (Han et al. 2009). The Argonaut ERGO-1 is required for expression of oocyte but not sperm 26G RNAs; however, depletion of ERGO-1 results in near wild-type fertility. In contrast, depletion of two Argonaut proteins, T22B3.2 and ZK757.3, specifically decreases the expression of 26G RNAs in sperm but not in oocytes (Han et al. 2009; Conine et al. 2010) and results in thermosensitive male infertility. Future studies that investigate either the role of specific sperm targets or alternative functions for RISC complex members in male fertility promise to provide mechanistic insights regarding how 26G RNAs function specifically in the context of sperm development and fertility.

7.4 Meiotic Progression: Sex-Specific Features of Preparing for and Undergoing Meiotic Divisions

As summarized in the overview (Sect. 7.2.3), oocytes and spermatocytes undergo meiotic divisions that differ in the timing of production and the number and size of their end products. Underlying these difference are gamete-specific differences in chromatin condensation, cell cycle progression, and kinetochore structure.

7.4.1 Sex-Specific Differences in Preparing Chromosomes for Segregation

Though the mechanisms required to facilitate recombination appear similar in both sexes, the resolution of chromosomes in preparation for meiotic divisions differs. Following the pachytene stage, SYP-1, a component of the synaptonemal complex required for pairing, synapsis, and recombination, is removed earlier from spermatocyte chromosomes than from oocyte chromosomes (Chap. 6, Lui and Colaiácovo 2012; Shakes et al. 2009). Chromatin composition may also influence meiotic events. In other organisms, sperm chromatin is tightly compacted through the incorporation of SNBPs during a prolonged post-meiotic differentiation stage. However, in *C. elegans,* SNBPs are incorporated as spermatocytes exit pachytene (Chu et al. 2006). This pre-loading of chromatin structural elements may allow for the efficient compaction of mature *C. elegans* sperm chromatin even in the absence of a prolonged post-meiotic processing period (Shakes et al. 2009).

Another striking feature is the formation of a karyosome, in which spermatocyte chromosomes retain their structural organization but come together to form a single, constricted mass (Shakes et al. 2009). In contrast, oocyte chromosomes remain as individual entities as they condense before meiotic divisions. Karyosome formation occurs in a broad range of animals; however, it more typically occurs during oogenesis with the hypothesized purpose of facilitating chromosome remodeling prior to the meiotic divisions (Gruzova and Parfenov 1993; Orr-Weaver et al. 1995; Sanyal et al. 1976). In *C. elegans* spermatocytes, karyosome formation occurs after desynapsis and the global down-regulation of transcription but before nuclear envelope breakdown (Shakes et al. 2009). The function of karyosome formation remains poorly understood in any organism, but is hypothesized to facilitate rapid progression to metaphase I after breakdown of the nuclear envelope.

C. elegans spermatocytes and oocytes also differ in how they transition through the cell cycle in preparation for chromosome segregation. In both, the chromatin is phosphorylated on histone H3 (HisH3-ser10) as the chromosomes condense prior to meiotic divisions. This phosphorylation is mediated by the aurora kinase (AIR-2), which regulates both kinetochores and release of chromosome cohesion during meiosis (Rogers et al. 2002; Hsu et al. 2000; McCarter et al. 1999; Burrows et al. 2006; Schumacher et al. 1998). However, during late prophase of oogenesis, AIR-2 and phosphosphorylated HisH3-ser10 can only be detected on the chromosomes of diakinetic oocytes that have received an MSP-based signal from sperm to undergo oocyte maturation. In contrast, during spermatogenesis, HisH3-ser10 phosphorylation occurs earlier, during the diplotene stage and before AIR-2 is present (Shakes et al. 2009). In addition, AIR-2 and phosphorylated HisH3-ser10 exhibit gamete-specific localization patterns during the diplotene, karyosome, and diakinesis stages (Shakes et al. 2009). These results suggest that, in diplotene spermatocytes, an additional kinase may be necessary to phosphorylate HisH3-ser10. In summary, when compared to the process during oogenesis, chromosome condensation during spermatogenesis initiates both more rapidly and, potentially, independently of an extracellular signal.

7.4.2 Sex-Specific Similarities and Differences in Kinetochore Structure

The presence of centrioles in sperm but not oocyte meiosis necessitates alterations in the mechanics of chromosome segregation. During oocyte meiosis, kinetochores assemble as cup-shaped structures, which are thought to enable the sides of tightly compacted rounded chromosomes to interface between organized microtubule bundles (Albertson and Thomson 1993; Maddox et al. 2004; Monen et al. 2005; Wignall and Villeneuve 2009; Schvarzstein et al. 2010). However, it is equally plausible that this cup-shaped morphology addresses the shared challenge of segregating homologs rather than sister chromatids during the first meiotic division. After all, despite dramatic differences in their interactions with microtubules, the kinetochores of spermatocytes are likewise cup-shaped and thus more similar in overall structure to those in oocytes (Albertson and Thomson 1993; Howe et al. 2001; Shakes et al. 2009) than to the mitotic kinetochores, which localize in a poleward fashion along the length of mitotic chromosomes (Dernburg 2001; Kitagawa 2009; Maddox et al. 2004). Importantly, oocytes and spermatocyte kinetochores do exhibit notable differences in the specific arrangement of individual inner and outer kinetochore components, including CENP-C^{HCP-4}, HCP-1, and HCP-2 (Shakes et al. 2009). These differences may facilitate attachment to structurally distinct microtubule spindles or may be a consequence of the incorporation of SNBPs in sperm chromatin (Shakes et al. 2009). Overall, many aspects of the mechanics of chromosome segregation during sperm meiosis remain to be resolved.

7.5 The Assembly and Function of Sperm-Specific Organelles: Packing for the Trip

In all species, sperm cells change morphology during the later stages of spermatogenesis. In mammals, this differentiation process is supported by a post-meiotic burst of sperm-specific transcription and translation. In contrast, in *C. elegans,* global transcriptional activation ceases shortly before the meiotic divisions (Shakes et al. 2009). Furthermore, translation largely halts after the meiotic divisions when translational machinery is discarded in the residual body (Ward et al. 1981). As a result, *C. elegans* sperm assemble modular, "pre-fabricated" sperm-specific organelles prior to the meiotic divisions, which are subsequently modified and regulated during the course of differentiation by post-translational modifications.

Striking features of developing *C. elegans* spermatocytes are the sperm-specific FB–MO complexes (Roberts et al. 1986; Ward et al. 1981; Wolf et al. 1978). FB–MO complexes can first be observed during late pachytene, expanding in size through the meiotic divisions (Figs. 7.1b, 7.4b and 7.5b). The FBs sequester newly synthesized filaments of the MSP in a paracrystaline-like state. The MO component is a Golgi-derived organelle with three regions: a glycoprotein-filled "head" portion, a body with a highly convoluted membrane morphology that presumably maximizes its surface area, and an electron-dense collar that separates these two domains (Fig. 7.1b). In developing spermatocytes, the FBs and MOs are intimately associated; the microvillus-like arms of the MO body envelope the growing FBs. Following anaphase II, FB–MO complexes then partition to the budding spermatids. After the spermatids detach from the residual body, the MO membranes retract from and release the FB. At that point the detached MOs dock with the plasma membrane and the FBs disassemble.

Our understanding of the FBs is expanding with the discovery of genes required for FB assembly/disassembly dynamics. To date, the assembly of MSP into FBs is known to require both a member of the casein kinase I superfamily, *spe-6* (Muhlrad and Ward 2002; Varkey et al. 1993) and SPE-7, a spermatogenesis-enriched cytosolic protein which seems to function as an essential structural and scaffolding component (M. Presler, K. Messina, and D. Shakes, unpublished data). In terms of

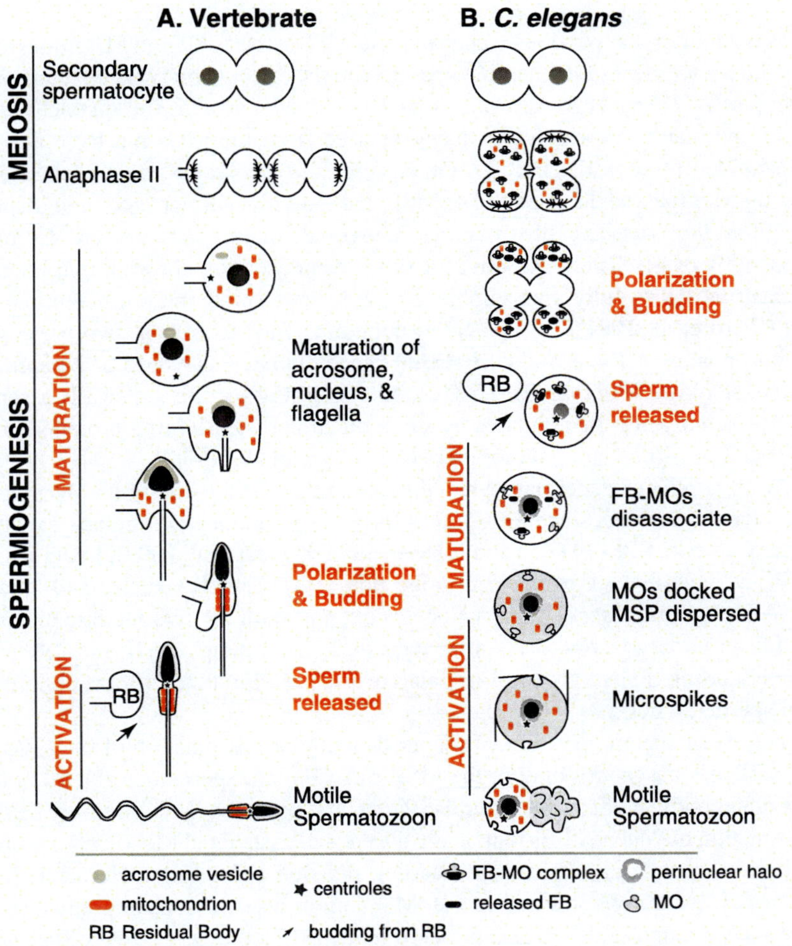

Fig. 7.5 Comparison of spermiogenesis in (**a**) vertebrates and (**b**) *C. elegans,* highlighting analogous events. (**a**) In vertebrates, following anaphase II, spermatocytes undergo incomplete cytokinesis to generate four, interconnected haploid spermatids. These spermatids then undergo a multi-week maturation process of spermiogenesis that includes the following events: a burst of sperm-specific transcription and translation, the formation of a mature acrosome, the mature flagellum, and the compaction and reshaping of the nucleus. Materials unneeded by the spermatozoon are then partitioned into a residual body (RB) as the spermatozoon completes cellularization. Sperm activation causes a spermatozoon to become fully motile. (**b**) In *C. elegans*, following anaphase II, spermatocytes initiate a cleavage furrow that regresses and morphs into a polarization and budding process during which time unneeded materials are partitioned away from the differentiating sperm and left in a central residual body as spermatids detach. During a short (minutes-long) maturation step the MOs mature and dock, the FBs disassociate and subsequently disassemble, and an RNA-enriched perinuclear halo forms around the compact chromatin mass. Male spermatids are stored in this quiescent state until stimulated by extracellular signals to active and form bipolar, motile spermatozoa. Both hermaphrodite and male sperm activation occurs in less than 10 minutes

FB disassembly, it has previously been thought MSP is released from FBs immediately upon the detachment of the haploid sperm from the residual body (L'Hernault 2006; Roberts et al. 1986). However, others have observed a sizable and distinct population of spermatids in which FBs have released from the MOs but have not fully disassembled (Wu et al. 2012, D. Greenstein, and D. Shakes, unpublished data). This suggests that the disassembly of MSP into individual dimers and their dispersal throughout the cytoplasm involves a more gradual "maturation" process. Complete disassembly of FBs is not essential for subsequent pseudopod formation; however, mutants that fail to fully disassemble their FBs have small and inefficient pseudopods (Ward et al. 1981). In a potential counterbalance to SPE-6, two paralogs of PP1 phosphatase (GSP-3 and GSP-4) have recently been shown to be essential for FB disassembly (Wu et al. 2012). Thus, the emerging picture is that FB assembly and disassembly may be regulated by a combination of scaffolding proteins such as SPE-7 and post-translational modifiers like SPE-6 and GSP-3/4.

Within *C. elegans* spermatocytes, the assembly of MOs likely involves homotypic vacuolar fusions, since MOs fail to form in spermatocytes that lack the HOPS complex protein, SPE-39 (Zhu and L'Hernault 2003; Zhu et al. 2009). In the absence of SPE-39, spermatocytes fill with large numbers of small vesicles and "naked" FBs. Although most *spe-39* spermatocytes attempt both the two meiotic divisions, they fail to undergo cytokinesis after meiosis I. After anaphase II, they attempt a normal budding division but fail to polarize and are only able to form abnormally small spermatid buds.

Analysis of specific *spe* mutants suggests that close association of the MOs and FBs facilitates the partitioning of the MSP-rich FBs into spermatids and away from the residual body. For example, in *spe-39* mutants, defects in MO assembly result in the formation of "naked" FBs that, in the absence of associated MOs, fail to properly partition to spermatids during the budding division (Zhu and L'Hernault 2003). Similar defects are observed when FB–MO complexes disassociate prematurely as in *spe-4* (Arduengo et al. 1998) and *spe-10* mutants (Gleason et al. 2006; Shakes and Ward 1989). SPE-4 encodes a member of the presenilin family and is thought to function in the regulated cleavage of adjacent integral membrane proteins, as presenilin does for the amyloid precursor protein, APP. In *spe-4* mutants, the cleavage of specific integral membrane proteins is presumably dysregulated. As a result, *spe-4* spermatocytes can undergo normal meiotic chromosome segregation but are unable to either polarize or initiate the budding division following anaphase II (Arduengo et al. 1998; L'Hernault and Arduengo 1992). In the absence of the MO-localized, transmembrane palmitoyl transferase protein SPE-10, the MOs effectively partition to spermatids; however, the "naked" FBs are left behind in the residual body where they either remain or bud off directly as small cytoplasts (Gleason et al. 2006; Shakes and Ward 1989). Studies of sperm from other nematode species may also differentiate the functions of FB from MOs. For example, ultrastructural studies (Justine 2002; Justine and Jamieson 2000) reveal that spermatocytes in certain other nematode species either exhibit morphologically distinct MO-structures (Turpeenniemi 1998; Shepherd and Clark 1983) or lack MOs altogether (Lee and Anya 1967; Shepherd and Clark 1983; Yushin and Commans 2005).

Further studies in both *C. elegans* and other nematode species are needed to better understand the intricacies of FB assembly and disassembly, as well as the seemingly parallel process of MSP assembly and disassembly within the pseudopod of crawling spermatozoa. In particular, there is evidence in *C. elegans* that genes required for FB assembly and disassembly, like *spe-6* and the PP1 phosphatases *gsp-3* and *gsp-4*, function again later in both sperm activation and MSP-based pseudopod motility (Muhlrad and Ward 2002; Wu et al. 2012). Furthermore, independent biochemical studies of *Ascaris* spermatozoa have convincingly shown that the rapid assembly of MSP complexes at the leading edge of the pseudopod is mediated by a combination of integral membrane phosphoproteins, cytosolic proteins, and the casein I kinase-related protein called MSP polymerization-activating kinase (MPAK) (Chap. 11, Marcello et al. 2012; Buttery et al. 2003; Italiano et al. 1996; LeClaire et al. 2003; Aitken and De Iuliis 2007), while the disassembly of MSP at the base of the pseudopod is mediated by a PP2 phosphatase (Yi et al. 2007).

7.6 Spermiogenesis

Spermiogenesis refers to the post-meiotic processes that convert sessile haploid spermatids into motile spermatozoa. In both insects and vertebrates, spermiogenesis involves the formation of an acrosome, hypercondensation of the chromatin, construction of the flagella, elimination of excess cytoplasm, and separation of individual sperm from the larger syncytium. In most organisms, these processes take days to weeks to complete and are supported by a final post-meiotic burst of sperm-specific transcription and an extended period of protein synthesis. In *C. elegans*, the term "spermiogenesis" has previously been used in reference to the 10-min transformation of fully cellularized spherical spermatids into bipolar crawling spermatozoa (L'Hernault 2006; Shakes and Ward 1989; Ward et al. 1983). However, in order to draw more informative parallels between the developmental events in *C. elegans* and other organisms, including vertebrates, here we define spermiogenesis as including three phases (Fig. 7.5). After anaphase II, spermatids first individualize by budding and forming a residual body. The second phase, which we term "spermatid maturation," directly follows anaphase II and includes polarization of cellular components, spermatid budding, and maturation of nuclear and cytoplasmic components. The later phase, which we refer to as "sperm activation," includes the conversion of quiescent, spherical spermatids into motile, bipolar spermatozoa.

7.6.1 Polarization and Budding

At the completion of anaphase II, a shallow cleavage furrow initiates at the midpoint between the two haploid nuclei that have segregated to opposite poles via centrosome-organized spindles (Fig. 7.5b) (Shakes et al. 2009, 2011; Ward et al. 1981).

However this transient furrow rapidly regresses as the cells bud from a central residual body. Cellular components are then differentially partitioned into either the bud or residual body. Actin, endoplasmic reticulum, and ribosomes accumulate within the residual body while the FB–MO complexes and mitochondria partition along with the nuclei into the budding spermatids (Ward 1986). Towards the end of this polarization process, the microtubules detach from the centrioles and move into the residual body. We propose that this polarization is analogous to the cytoplasmic shedding events that typically occur at the end of spermiogenesis as individual spermatids detach from the syncytium in other organisms (Fig. 7.5a, b).

The molecular mechanisms that underlie the spermatid budding division and the associated polarization events remain poorly understood. Analysis of *spe* mutants suggests that the two events are regulated by distinct actin-mediated processes. Both are disrupted in spermatocytes deficient in the actin-binding, kelch-related protein SPE-26 (Varkey et al. 1995), whereas spermatocytes that lack SPE-15 (myosin VI) can bud but are unable to differentially partition their cellular components (Kelleher et al. 2000). The constriction forces associated with spermatid bud formation remain unknown; ultrastructural studies indicate that the constriction points between the buds and residual body are devoid of either an actin/myosin ring or microtubules (Ward et al. 1983). Centrosomes and microtubules may set up the initial patterning of the budding division as the number of spermatid buds correlates with the number of centrosomes in spermatocytes that over-replicate their centrosomes (Peters et al. 2010). Interestingly, both polarization and spermatid budding can be uncoupled from cell cycle progression beyond metaphase I (Golden et al. 2000). Analysis of mutants with temperature-sensitive defects in the anaphase-promoting complex (APC/C) revealed that budding can proceed despite the presence of a stabilized arrested metaphase I spindle (Golden et al. 2000). Such mutant budding yield residual bodies that contain an intact metaphase spindle (Golden et al. 2000), and spermatids that lack chromatin but can nevertheless activate, crawl, and fertilize oocytes (Sadler and Shakes 2000). It has yet to be determined whether either spermatid budding or polarization can occur in the absence of assembled microtubules. Although much remains to be learned about both the polarization and budding events, their striking physical and functional similarities to the process of "sperm individualization" in other organisms suggest that comparative studies between the two will be informative for both (Fig. 7.5a, b).

7.6.2 Spermatid Maturation

By analogy to the events of spermiogenesis in other organisms, the "maturation" events of *C. elegans* spermiogenesis include the remodeling and docking of the MOs, the disassembly of MSP from the FBs, the final remodeling of chromatin, and the formation of an RNA-enriched perinuclear halo (Fig. 7.5b). Although spermatid maturation has not been a widely used term in the *C. elegans* literature, we suggest that it provides a useful term to describe the numerous cellular changes that occur in spermatids from the time that they separate from the residual body until they are either stored in a quiescent state within the male seminal vesicle or are signaled by sperm activators to undergo morphogenesis.

7.6.3 Sperm Activation and Cellular Morphogenesis

Unlike other events in spermatogenesis, sperm activation is triggered by extracellular signals, which have yet to be identified. When activated *in vitro,* spermatids can be observed to first develop several long, thin microspikes around the spherical cell before fully polarizing to form a distinct pseudopod and cell body (Shakes and Ward 1989; Ward et al. 1983; Fig. 7.1b). During this time, the docked MOs fuse with the plasma membrane in the cell body in a manner that releases soluble glycoproteins from the head while the electron-dense collar of the MO establishes a permanent fusion pore with the plasma membrane that opens into the invaginated, pocket-like membrane of the MO body (Fig. 7.1b) (Nelson and Ward 1980; Wolf et al. 1978).

To date, the extracellular activators and underlying the cellular response machinery remain incompletely understood. Early studies revealed that spermatids isolated from *C. elegans* males could be activated *in vitro* by the addition of either the protease Pronase or substances that elevated the intracellular pH (triethanolamine or the N + K + ionophore monesin) (Nelson and Ward 1980; Shakes and Ward 1989). More recently, wortmannin was identified as another potent activator, suggesting a potential role for PI(3,4,5)P3 signaling in the process (Bae et al. 2009). Although all of these *in vitro* activators, particularly Pronase, continue to be used in an experimental context, their in vivo molecular targets remain undefined.

Genetic screens have revealed sex-specific activation mechanisms (L'Hernault et al. 1988; Minniti et al. 1996; Nance et al. 1999, 2000; Shakes and Ward 1989; Geldziler et al. 2005). For example, mutations in the so-called "*spe-8* class" genes disrupt sperm activation in affected hermaphrodites but not in their male siblings. These mutants share complex phenotypic traits with two features. First, mutant hermaphrodites are self-sterile because the sperm they produce do not activate unless exposed to and activated "in trans" by male seminal fluid from either wild-type or *spe-8* class mutant males. Second, the mutant males are fertile and their sperm activate normally both *in vivo* and in response to activators that raise the intracellular pH; however, their sperm arrest with microspikes when exposed to Pronase. The "*spe-8* class" of genes encodes diverse proteins: SPE-8 is a non-receptor tyrosine kinase with an SH2 domain, SPE-27 is a soluble protein and SPE-12, SPE-19, and SPE-29 are all transmembrane proteins. These initial studies suggested that male and hermaphrodite sperm were normally activated by distinct sex-specific activators but that both male and hermaphrodite sperm redundantly express the cellular machinery required to respond to either activator (Fig. 7.6a).

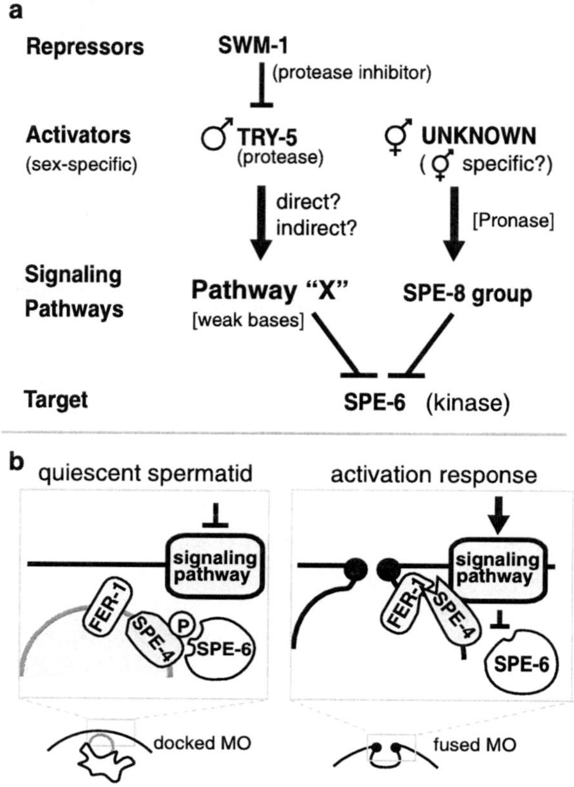

Fig. 7.6 Signaling and response elements involved in sperm activation. (**a**) Diagram of a two-pathway model for sperm activation. The SWM-1 protease inhibitor is present in both males and hermaphrodites, but its only known target to date is the male-specific protease TRY-5. The action of TRY-5 as a component of a male-specific activation pathway may be direct or indirect. The cellular response pathway downstream of TRY-5 (Pathway "X") is present in both male and hermaphrodite sperm. The SPE-8 group components comprise a second cellular response pathway, which is also present in both male and hermaphrodite sperm. The *in vivo* activator of this SPE-8 pathway has yet to be either molecularly or mutationally identified. This unknown activator is definitely present in hermaphrodites, and it may or may not be redundantly present in males. The *in vitro* activator Pronase activates sperm via the SPE-8 pathway. *In vitro* activation by weak bases bypasses the absence of either TRY-5 or SPE-8 group components. The two cellular response pathways converge to inhibit the SPE-6 kinase. (**b**) Hypothetical model for sperm activation events (adopted from Gosney et al. 2008). Close-up views of the region indicated by the *gray boxes* are shown on the schematic of docked or fused MOs below. In quiescent spermatids, cytosolic SPE-6 actively phosphorylates and thus inhibits the MO membrane protein SPE-4. Upon sperm activation, signaling components activate to inhibit SPE-6. As a result, SPE-4 becomes active and can cleave FER-1. The proteolytically processed form of FER-1 promotes MO fusion with the plasma membrane

In an independent study, Stanfield and Villeneuve identified a sperm activation mutant in which only the males were infertile (Stanfield and Villeneuve 2006). In this case, the males were infertile because their sperm activated precociously within the seminal vesicle, a situation that proved to be incompatible with successful sperm transfer. Notably, *swm-1* encodes an extracellular protease inhibitor, implying that a protease target of SWM-1 either serves directly as an *in vivo* activator or indirectly as part of an activation cascade. Conversely since *swm-1* is normally expressed in both males and hermaphrodites, either hermaphrodite fertility is unaffected by precocious sperm activation or hermaphrodites lack one or more downstream elements of this SWM-1 regulated activation cascade.

In fact, one key target of SWM-1 is a male-specific, seminal fluid protease called TRY-5 (Smith and Stanfield 2011). Analysis of GFP fusion constructs revealed that TRY-5 is expressed and stored in the valve and vas deferens cells of the male gonad until it is secreted and transferred to the hermaphrodite during ejaculation. Notably, Smith and Stanfield found that males that only lack TRY-5 are fertile, whereas males that lack both TRY-5 and SPE-8 are infertile. This observation is consistent with a model in which male sperm can be activated through either of two distinct and redundant pathways. However, it leaves open the question of whether the sperm of *try-5* males is being activated "in trans" by a hermaphrodite-specific activator within the uterus or whether males redundantly express activators for both the *spe-8* and pathway "X" response elements (Fig. 7.6a). A key distinction between these two models is whether the activator of the *spe-8* pathway will prove to be hermaphrodite specific in its expression.

Intriguingly, these activation pathways also link back to SPE-6 (casein kinase I) and SPE-4 (presenilin), which were discussed earlier in the context of FB assembly and the regulated disassociation the FB from the MO. Specific non-null alleles of both *spe-6* and *spe-4* both genetically suppress the self-sterility of *spe-8* class hermaphrodites and cause precocious sperm activation in males (Gosney et al. 2008; Muhlrad and Ward 2002). These results suggest that SPE-6 and SPE-4 may function as shared elements in the both the TRY-5 and SPE-8 activation pathways (Gosney et al. 2008; Smith and Stanfield 2011). In one scenario (Fig. 7.6b), soluble SPE-6 within unactivated sperm phosphorylates and thus inhibits the function of SPE-4 within the membrane of unfused MOs (Gosney et al. 2008). However sperm activation by either sperm activation pathway inhibits SPE-6, enabling SPE-4 to become active and cleave adjacent transmembrane proteins (Gosney et al. 2008).

One proposed target of SPE-4 is FER-1 (Fig. 7.6b). FER-1 is a multi-pass transmembrane protein that localizes to the MOs of unactivated spermatids but, upon MO fusion, distributes to both MO body and the plasma membrane. Multiple forms of FER-1 are produced from proteolytically processing during sperm activation (Achanzar and Ward 1997; Washington and Ward 2006). When sperm from *fer-1* mutants are exposed to activators, they fail in MO fusion. *fer-1* sperm are able to form pseudopods; however, they are abnormally short and fertilization incompetent. This is because proteins that are essential for the sperm–oocyte interactions, like the tetraspanin protein SPE-38, likely remain sequestered in the unfused MOs (Chatterjee et al. 2005) (Chap. 11, Marcello et al. 2012). FER-1 is suspected to be a direct participant in the fusion event since other members of the ferlin superfamily are also associated with membrane fusion events.

7.7 The Deliverables

The ultimate goal of spermatogenesis is to generate haploid spermatozoa that can find and fertilize oocytes to contribute genetically to the subsequent generation. While the details of MSP-based motility and fertilization will be covered in Chap. 11 (Marcello et al. 2012), we briefly consider here the contributions that sperm make to the embryo.

1. *A haploid complement of DNA with epigenetic information.* A conserved feature of spermatogenesis in all organisms involves the large-scale but incomplete replacement of histones by protamines within sperm chromatin. This change accommodates a tighter packing state and protects the chromatin from genotoxic factors (Miller et al. 2009). Recent studies suggest that the remaining histones, modified in the context of sex-specific differences in gene transcription within the germ line, retain epigenetic information that continues to pattern gene transcription in both the gametes and the early embryo (Sha and Fire 2005; Arico et al. 2011).
2. *A centriole pair that is required to generate the first mitotic spindle.* Whereas the centrosomes of oocytes are lost during oogenesis (Kim and Roy 2006), each haploid sperm inherits a single pair of centrioles (Albertson 1984; Albertson and Thomson 1993). After fertilization, this centriole pair duplicates and combines with maternal components to generate the two active centrosomes that establish the two poles of the embryo's first mitotic spindle (Dammermann et al. 2008; Pelletier et al. 2006).
3. *A cue that specifies the anterior–posterior axis of the embryo.* In *C. elegans*, the sperm cues the anterior–posterior axis. Interestingly, this cue does not reflect either the point of sperm entry (Goldstein and Hird 1996) or the paternally contributed DNA (Sadler and Shakes 2000). Instead, multiple lines of evidence suggest that polarity is specified by the paternally contributed centrosome and their nucleation of a microtubule aster on one side of the embryo (O'Connell et al. 2000; Cowan and Hyman 2004; Hamill et al. 2002; Wallenfang and Seydoux 2000). In addition, the sperm has been implicated breaking the symmetry of the acto-myosin network and initiating a Rho-mediated cortical flow by delivering a localized bolus of CYK-4, a sperm-enriched Rho-GAP (Jenkins et al. 2006), while also contributing to the localized depletion of the Rho-GEF ECT-2 in the immediate vicinity of the centrosome (Motegi and Sugimoto 2006).
4. *SPE-11, a sperm-supplied factor that is required for egg activation.* In *C. elegans*, the oocyte chromosomes complete their meiotic divisions and form an impermeable eggshell only after fertilization. When wild-type oocytes are fertilized by sperm that lack SPE-11, the resulting zygotes fail to either produce polar bodies or construct a functional eggshell (Browning and Strome 1996; Hill et al. 1989; McNally and McNally 2005; Johnston et al. 2010). Within spermatozoa, the SPE-11 protein localizes to the perinuclear halo (Browning and Strome 1996), and within *spe-11* mutant sperm, the perinuclear halo is structurally aberrant (Hill et al. 1989). Despite these defects, oocytes fertilized by *spe-11* mutant sperm produce viable embryos if functional SPE-11 is expressed in the oocytes (Browning and Strome 1996).

5. *Paternal RNAs*. Early cytological studies suggest that the perinuclear halo may be enriched in RNA (Ward et al. 1981). Small RNAs are part of the paternal cargo, as sperm-specific small RNAs, like the 26G endo-siRNAs described in Sect. 7.3.4.2, were identified from purified sperm (Han et al. 2009). It is yet unclear if a specific cadre of mRNAs is also carried over to the new embryo upon fertilization. Future studies are necessary to reveal the extent to which the paternal RNA component influences zygotic development.

7.8 Summary and Future Perspectives

Overall, a big picture view of what it takes to generate motile spermatozoa reveals numerous mechanisms that overlap and intertwine after cells commit to spermatogenesis.

At the level of gene expression, implementation of the spermatocyte fate likely requires a team of transcriptional regulators, including both spermatogenesis-specific genes like *spe-44* and non-cell specific factors like *elt-1,* acting on genes whose genomic organization is distinct from that of either germline-specific or oocyte-specific genes. Equally important, the spermatogenesis program uses mechanisms to subsequently turn off global gene expression as the chromatin is packaged for long-term protection. These changes are mediated by a combination of post-translational histone modifications and the incorporation of various sperm basic nuclear proteins (SBNPs). Such changes place epigenetic marks on the chromatin that can affect gene expression of paternally inherited chromatin within in the embryo. In addition, both piRNAs and endo-siRNAs appear to play a back-up role in facilitating the appropriate repression of gene expression, particularly under conditions of stress.

In parallel, spermatocytes must accomplish meiotic chromosome segregation. A better understanding of the mechanistic differences between the meiotic divisions of spermatocytes and oocytes will help us distinguish the features that specifically facilitate the segregation of homologs in meiosis I from those that accommodate the distinct microtubule structures of meiotically dividing oocytes and spermatocytes. For example, in spermatocytes, core machinery, like centrosomes and kinetochores, are utilized differentially from oocytes. Progression of spermatocytes from meiotic prophase into M-phase is likewise distinct as it features a unique karyosome state and distinctions in the timing of desynapsis.

A fresh perspective on the events following anaphase II suggests interesting and informative analogies between the spermiogenesis programs of *C. elegans* amoeboid sperm and the flagellated spermatozoa of vertebrates and *Drosophila*. Although differentially ordered, both types include distinct phases of polarization/budding, spermatid maturation, and sperm activation that may involve a subset of analogous molecular mechanisms (Fig. 7.5). One distinction to note is the fate of centrosomes. During nematode spermiogenesis, the centrosome is specifically turned off and the microtubules are discarded whereas in vertebrates and *Drosophila*, the centrosome is converted into a basal body that facilitates assembly of the flagella (Li et al. 1998). Continuing investigations of centrosome dynamics and fate should provide important new insights regarding their function.

Like sperm, many other cell types, including spores and pluripotent stems cells, undergo extended periods of quiescence. Recent studies of *C. elegans* sperm activation suggest that a quiescent state for *C. elegans* sperm is essential for efficient sperm transfer and appropriate activation within hermaphrodites. Future investigations of underlying molecular mechanisms may provide important insights into the spatial and temporal regulation of differentiation and morphogenesis.

Although sperm from different organisms can be morphologically distinct, studies defining both conserved and adapted features between organisms can help identify key components required for spermatogenesis. As a result, the application of *C. elegans* as a model system has great potential to make further significant contributions to our understanding of the processes of chromosome segregation, differentiation, morphogenesis, and motility acquisition.

Acknowledgments We thank David Greenstein and Harold Smith for sharing data prior to publication. We thank Meghann Shorrock for assistance with figures. We also thank Dana Byrd, Margaret Jow, and Kari Price for critical reading of this manuscript. This work was supported grants from the National Science Foundation to D.S.C. (MCB-0747515) and the National Institutes of Health to D.S.C. (R15 HD068996) and D.C.S. (R15 GM096309).

References

Achanzar WE, Ward S (1997) A nematode gene required for sperm vesicle fusion. J Cell Sci 110(Pt 9):1073–1081

Aitken RJ, De Iuliis GN (2007) Origins and consequences of DNA damage in male germ cells. Reprod Biomed Online 14(6):727–733

Albertson DG (1984) Formation of the first cleavage spindle in nematode embryos. Dev Biol 101(1):61–72

Albertson DG, Thomson JN (1993) Segregation of holocentric chromosomes at meiosis in the nematode, *Caenorhabditis elegans*. Chromosome Res 1(1):15–26

Aoki K, Moriguchi H, Yoshioka T, Okawa K, Tabara H (2007) In vitro analyses of the production and activity of secondary small interfering RNAs in *C. elegans*. EMBO J 26(24):5007–5019

Aravin A, Gaidatzis D, Pfeffer S, Lagos-Quintana M, Landgraf P, Iovino N, Morris P, Brownstein MJ, Kuramochi-Miyagawa S, Nakano T, Chien M, Russo JJ, Ju J, Sheridan R, Sander C, Zavolan M, Tuschl T (2006) A novel class of small RNAs bind to MILI protein in mouse testes. Nature 442(7099):203–207

Aravin AA, Hannon GJ, Brennecke J (2007) The Piwi-piRNA pathway provides an adaptive defense in the transposon arms race. Science 318(5851):761–764

Arduengo PM, Appleberry OK, Chuang P, L'Hernault SW (1998) The presenilin protein family member SPE-4 localizes to an ER/Golgi derived organelle and is required for proper cytoplasmic partitioning during *Caenorhabditis elegans* spermatogenesis. J Cell Sci 111(Pt 24):3645–3654

Arico JK, Katz DJ, van der Vlag J, Kelly WG (2011) Epigenetic patterns maintained in early *Caenorhabditis elegans* embryos can be established by gene activity in the parental germ cells. PLoS Genet 7(6):e1001391

Bae YK, Kim E, L'Hernault SW, Barr MM (2009) The CIL-1 PI 5-phosphatase localizes TRP Polycystins to cilia and activates sperm in *C. elegans*. Curr Biol 19(19):1599–1607

Bamps S, Hope IA (2008) Large-scale gene expression pattern analysis, *in situ*, in *Caenorhabditis elegans*. Brief Funct Genomic Proteomic 7(3):175–183

Barton MK, Schedl TB, Kimble J (1987) Gain-of-function mutations of *fem-3*, a sex-determination gene in *Caenorhabditis elegans*. Genetics 115(1):107–119

Batista PJ, Ruby JG, Claycomb JM, Chiang R, Fahlgren N, Kasschau KD, Chaves DA, Gu W, Vasale JJ, Duan S, Conte D Jr, Luo S, Schroth GP, Carrington JC, Bartel DP, Mello CC (2008) PRG-1 and 21U-RNAs interact to form the piRNA complex required for fertility in *C. elegans*. Mol Cell 31(1):67–78

Beanan MJ, Strome S (1992) Characterization of a germ-line proliferation mutation in *C. elegans*. Development 116(3):755–766

Bernstein E, Caudy AA, Hammond SM, Hannon GJ (2001) Role for a bidentate ribonuclease in the initiation step of RNA interference. Nature 409(6818):363–366

Beshore EL, McEwen TJ, Jud MC, Marshall JK, Schisa JA, Bennett KL (2011) *C. elegans* dicer interacts with the P-granule component GLH-1 and both regulate germline RNPs. Dev Biol 350(2):370–381

Bettegowda A, Wilkinson MF (2011) Transcription and post-transcriptional regulation of spermatogenesis. Philos Trans R Soc Lond B Biol Sci 365(1546):1637–1651

Braun RE (2001) Packaging paternal chromosomes with protamine. Nat Genet 28(1):10–12

Brennecke J, Aravin AA, Stark A, Dus M, Kellis M, Sachidanandam R, Hannon GJ (2007) Discrete small RNA-generating loci as master regulators of transposon activity in *Drosophila*. Cell 128(6):1089–1103

Browning H, Strome S (1996) A sperm-supplied factor required for embryogenesis in *C. elegans*. Development 122(1):391–404

Burrows AE, Sceurman BK, Kosinski ME, Richie CT, Sadler PL, Schumacher JM, Golden A (2006) The *C. elegans* Myt1 ortholog is required for the proper timing of oocyte maturation. Development 133(4):697–709

Buttery SM, Ekman GC, Seavy M, Stewart M, Roberts TM (2003) Dissection of the *Ascaris* sperm motility machinery identifies key proteins involved in major sperm protein-based amoeboid locomotion. Mol Biol Cell 14(12):5082–5088

Byrd DT, Kimble J (2009) Scratching the niche that controls *Caenorhabditis elegans* germline stem cells. Semin Cell Dev Biol 20(9):1107–1113

Caron C, Govin J, Rousseaux S, Khochbin S (2005) How to pack the genome for a safe trip. Prog Mol Subcell Biol 38:65–89

Chatterjee I, Richmond A, Putiri E, Shakes DC, Singson A (2005) The *Caenorhabditis elegans spe-38* gene encodes a novel four-pass integral membrane protein required for sperm function at fertilization. Development 132(12):2795–2808

Chu D, Liu H, Nix P, Wu T, Ralston E, Yates J, Meyer B (2006) Sperm chromatin proteomics identifies evolutionarily conserved fertility factors. Nature 443(7107):101–105

Conine CC, Batista PJ, Gu W, Claycomb JM, Chaves DA, Shirayama M, Mello CC (2010) Argonautes ALG-3 and ALG-4 are required for spermatogenesis-specific 26 G-RNAs and thermotolerant sperm in *Caenorhabditis elegans*. Proc Natl Acad Sci USA 107(8):3588–3593

Cowan CR, Hyman AA (2004) Centrosomes direct cell polarity independently of microtubule assembly in *C. elegans* embryos. Nature 431(7004):92–96

Dammermann A, Maddox PS, Desai A, Oegema K (2008) SAS-4 is recruited to a dynamic structure in newly forming centrioles that is stabilized by the gamma-tubulin-mediated addition of centriolar microtubules. J Cell Biol 180(4):771–785

Das PP, Bagijn MP, Goldstein LD, Woolford JR, Lehrbach NJ, Sapetschnig A, Buhecha HR, Gilchrist MJ, Howe KL, Stark R, Matthews N, Berezikov E, Ketting RF, Tavare S, Miska EA (2008) Piwi and piRNAs act upstream of an endogenous siRNA pathway to suppress Tc3 transposon mobility in the *Caenorhabditis elegans* germline. Mol Cell 31(1):79–90

del Castillo-Olivares A, Kulkarni M, Smith HE (2009) Regulation of sperm gene expression by the GATA factor ELT-1. Dev Biol 333(2):397–408

Deng W, Lin H (2002) miwi, a murine homolog of piwi, encodes a cytoplasmic protein essential for spermatogenesis. Dev Cell 2(6):819–830

Dernburg AF (2001) Here, there, and everywhere: kinetochore function on holocentric chromosomes. J Cell Biol 153(6):F33–F38

Gartner A, Milstein S, Ahmed S, Hodgkin J, Hengartner MO (2000) A conserved checkpoint pathway mediates DNA damage—induced apoptosis and cell cycle arrest in *C. elegans*. Mol Cell 5(3):435–443

Geldziler B, Chatterjee I, Singson A (2005) The genetic and molecular analysis of *spe-19*, a gene required for sperm activation in *Caenorhabditis elegans*. Dev Biol 283(2):424–436

Gent JI, Schvarzstein M, Villeneuve AM, Gu SG, Jantsch V, Fire AZ, Baudrimont A (2009) A *Caenorhabditis elegans* RNA-directed RNA polymerase in sperm development and endogenous RNAi. Genetics 183(4):1297–1314

Gleason EJ, Lindsey WC, Kroft TL, Singson AW, L'Hernault SW (2006) *spe-10* encodes a DHHC-CRD zinc-finger membrane protein required for endoplasmic reticulum/Golgi membrane morphogenesis during *Caenorhabditis elegans* spermatogenesis. Genetics 172(1):145–158

Golden A, Sadler PL, Wallenfang MR, Schumacher JM, Hamill DR, Bates G, Bowerman B, Seydoux G, Shakes DC (2000) Metaphase to anaphase (*mat*) transition-defective mutants in *Caenorhabditis elegans*. J Cell Biol 151(7):1469–1482

Golden DE, Gerbasi VR, Sontheimer EJ (2008) An inside job for siRNAs. Mol Cell 31(3):309–312

Goldstein B, Hird SN (1996) Specification of the anteroposterior axis in *Caenorhabditis elegans*. Development 122(5):1467–1474

Gosney R, Liau WS, Lamunyon CW (2008) A novel function for the presenilin family member spe-4: inhibition of spermatid activation in *Caenorhabditis elegans*. BMC Dev Biol 8:44

Govin J, Caron C, Lestrat C, Rousseaux S, Khochbin S (2004) The role of histones in chromatin remodelling during mammalian spermiogenesis. Eur J Biochem 271(17):3459–3469

Grishok A, Pasquinelli AE, Conte D, Li N, Parrish S, Ha I, Baillie DL, Fire A, Ruvkun G, Mello CC (2001) Genes and mechanisms related to RNA interference regulate expression of the small temporal RNAs that control *C. elegans* developmental timing. Cell 106(1):23–34

Grivna ST, Beyret E, Wang Z, Lin H (2006) A novel class of small RNAs in mouse spermatogenic cells. Genes Dev 20(13):1709–1714

Gruzova MN, Parfenov VN (1993) Karyosphere in oogenesis and intranuclear morphogenesis. Int Rev Cytol 144:1–52

Gunawardane LS, Saito K, Nishida KM, Miyoshi K, Kawamura Y, Nagami T, Siomi H, Siomi MC (2007) A slicer-mediated mechanism for repeat-associated siRNA 5′ end formation in *Drosophila*. Science 315(5818):1587–1590

Hamill DR, Severson AF, Carter JC, Bowerman B (2002) Centrosome maturation and mitotic spindle assembly in *C. elegans* require SPD-5, a protein with multiple coiled-coil domains. Dev Cell 3(5):673–684

Han T, Manoharan AP, Harkins TT, Bouffard P, Fitzpatrick C, Chu DS, Thierry-Mieg D, Thierry-Mieg J, Kim JK (2009) 26G endo-siRNAs regulate spermatogenic and zygotic gene expression in *Caenorhabditis elegans*. Proc Natl Acad Sci USA 106(44):18674–18679

Hansen D, Schedl T (2012) Stem cell proliferation versus meiotic fate decision in *C. elegans*. Advances in Experimental Medicine and Biology 757:71–99. (Chap. 4, this volume) Springer, New York

Hill DP, Shakes DC, Ward S, Strome S (1989) A sperm-supplied product essential for initiation of normal embryogenesis in *Caenorhabditis elegans* is encoded by the paternal-effect embryonic-lethal gene, spe-11. Dev Biol 136(1):154–166

Houwing S, Kamminga LM, Berezikov E, Cronembold D, Girard A, van den Elst H, Filippov DV, Blaser H, Raz E, Moens CB, Plasterk RH, Hannon GJ, Draper BW, Ketting RF (2007) A role for Piwi and piRNAs in germ cell maintenance and transposon silencing in Zebrafish. Cell 129(1):69–82

Howe M, McDonald KL, Albertson DG, Meyer BJ (2001) HIM-10 is required for kinetochore structure and function on *Caenorhabditis elegans* holocentric chromosomes. J Cell Biol 153(6):1227–1238

Hsu J, Sun Z, Li X, Reuben M, Tatchell K, Bishop D, Grushcow J, Brame C, Caldwell J, Hunt D, Lin R, Smith M, Allis C (2000) Mitotic phosphorylation of histone H3 is governed by Ipl1/aurora kinase and Glc7/PP1 phosphatase in budding yeast and nematodes. Cell 102(3):279–291

Italiano JE Jr, Roberts TM, Stewart M, Fontana CA (1996) Reconstitution in vitro of the motile apparatus from the amoeboid sperm of *Ascaris* shows that filament assembly and bundling move membranes. Cell 84(1):105–114

Jaramillo-Lambert A, Ellefson M, Villeneuve AM, Engebrecht J (2007) Differential timing of S phases, X chromosome replication, and meiotic prophase in the *C. elegans* germ line. Dev Biol 308(1):206–221

Jaramillo-Lambert A, Harigaya Y, Vitt J, Villeneuve A, Engebrecht J (2010) Meiotic errors activate checkpoints that improve gamete quality without triggering apoptosis in male germ cells. Curr Biol 20(23):2078–2089

Jenkins N, Saam JR, Mango SE (2006) CYK-4/GAP provides a localized cue to initiate anteroposterior polarity upon fertilization. Science 313(5791):1298–1301

Johnston WL, Krizus A, Dennis JW (2010) Eggshell chitin and chitin-interacting proteins prevent polyspermy in *C. elegans*. Curr Biol 20(21):1932–1937

Justine JL (2002) Male and female gametes and fertilization. In: Biology of nematodes. Taylor & Francis, London

Justine JL, Jamieson BGM (2000) Nematode, vol IX, part B. Progress in male gamete ultrastructure and phylogeny. Reproductive biology of invertebrates, vol IX, part B. Wiley, Chichester

Kato M, de Lencastre A, Pincus Z, Slack FJ (2009) Dynamic expression of small non-coding RNAs, including novel microRNAs and piRNAs/21U-RNAs, during *Caenorhabditis elegans* development. Genome Biol 10(5):R54

Kelleher JF, Mandell MA, Moulder G, Hill KL, L'Hernault SW, Barstead R, Titus MA (2000) Myosin VI is required for asymmetric segregation of cellular components during *C. elegans* spermatogenesis. Curr Biol 10(23):1489–1496

Kelly WG, Schaner CE, Dernburg AF, Lee M-H, Kim SK, Villeneuve AM, Reinke V (2002) X chromosome silencing in the germline of *C. elegans*. Development 129(2):479–492

Ketola I, Rahman N, Toppari J, Bielinska M, Porter-Tinge SB, Tapanainen JS, Huhtaniemi IT, Wilson DB, Heikinheimo M (1999) Expression and regulation of transcription factors GATA-4 and GATA-6 in developing mouse testis. Endocrinology 140(3):1470–1480

Ketola I, Pentikainen V, Vaskivuo T, Ilvesmaki V, Herva R, Dunkel L, Tapanainen JS, Toppari J, Heikinheimo M (2000) Expression of transcription factor GATA-4 during human testicular development and disease. J Clin Endocrinol Metab 85(10):3925–3931

Ketting RF, Fischer SE, Bernstein E, Sijen T, Hannon GJ, Plasterk RH (2001) Dicer functions in RNA interference and in synthesis of small RNA involved in developmental timing in *C. elegans*. Genes Dev 15(20):2654–2659

Kim DY, Roy R (2006) Cell cycle regulators control centrosome elimination during oogenesis in *Caenorhabditis elegans*. J Cell Biol 174(6):751–757

Kim VN, Han J, Siomi MC (2009) Biogenesis of small RNAs in animals. Nat Rev Mol Cell Biol 10(2):126–139

Kim S, Spike CA, Greenstein D (2012) Control of oocyte growth and meiotic maturation in *C. elegans*. Advances in Experimental Medicine and Biology 757:277–320. (Chap. 10, this volume) Springer, New York

Kimble J, Crittenden SL (2007) Control of germline stem cells, entry into meiosis, and the sperm/oocyte decision in *C. elegans*. Annu Rev Cell Dev Biol 23:405–433

Kimmins S, Sassone-Corsi P (2005) Chromatin remodelling and epigenetic features of germ cells. Nature 434(7033):583–589

Kimmins S, Kotaja N, Davidson I, Sassone-Corsi P (2004) Testis-specific transcription mechanisms promoting male germ-cell differentiation. Reproduction 128(1):5–12

Kitagawa R (2009) Key players in chromosome segregation in *Caenorhabditis elegans*. Front Biosci 14:1529–1557

Klass M, Ammons D, Ward S (1988) Conservation in the 5′ flanking sequences of transcribed members of the *Caenorhabditis elegans* major sperm protein gene family. J Mol Biol 199(1):15–22

Knight SW, Bass BL (2001) A role for the RNase III enzyme DCR-1 in RNA interference and germ line development in *Caenorhabditis elegans*. Science 293(5538):2269–2271
Kulkarni M, Shakes DC, Guevel K, Smith HE (2012) SPE-44 implements sperm cell fate. PLoS Genet. 8(4):e1002678
Kuramochi-Miyagawa S, Kimura T, Ijiri TW, Isobe T, Asada N, Fujita Y, Ikawa M, Iwai N, Okabe M, Deng W, Lin H, Matsuda Y, Nakano T (2004) Mili, a mammalian member of piwi family gene, is essential for spermatogenesis. Development 131(4):839–849
LaMunyon CW, Ward S (1995) Sperm precedence in a hermaphroditic nematode (*Caenorhabditis elegans*) is due to competitive superiority of male sperm. Experientia 51(8):817–823
LaMunyon CW, Ward S (1998) Larger sperm outcompete smaller sperm in the nematode *Caenorhabditis elegans*. Proc R Soc Lond B Biol Sci 265(1409):1997–2002
LeClaire LL 3rd, Stewart M, Roberts TM (2003) A 48 kDa integral membrane phosphoprotein orchestrates the cytoskeletal dynamics that generate amoeboid cell motility in *Ascaris* sperm. J Cell Sci 116(Pt 13):2655–2663
Lee DL, Anya AO (1967) The structure and development of the spermatozoon of Aspiculuris tetraptera (Nematoda). J Cell Sci 2(4):537–544
Lewis JD, Abbott DW, Ausio J (2003) A haploid affair: core histone transitions during spermatogenesis. Biochem Cell Biol 81(3):131–140
L'Hernault SW (2006) Spermatogenesis. WormBook:1–14
L'Hernault SW, Arduengo PM (1992) Mutation of a putative sperm membrane protein in *Caenorhabditis elegans* prevents sperm differentiation but not its associated meiotic divisions. J Cell Biol 119(1):55–68
L'Hernault SW, Shakes DC, Ward S (1988) Developmental genetics of chromosome I spermatogenesis-defective mutants in the nematode *Caenorhabditis elegans*. Genetics 120(2):435–452
Li K, Xu EY, Cecil JK, Turner FR, Megraw TL, Kaufman TC (1998) *Drosophila* centrosomin protein is required for male meiosis and assembly of the flagellar axoneme. J Cell Biol 141(2):455–467
Lui DY, Colaiácovo MP (2012) Meiotic development in *C. elegans*. In: Schedl T (ed) Advances in experimental medicine and biology, Chap. 6. Springer, Boston
Maddox PS, Oegema K, Desai A, Cheeseman IM (2004) "Holo"er than thou: chromosome segregation and kinetochore function in *C. elegans*. Chromosome Res 12(6):641–653
Maeda I, Kohara Y, Yamamoto M, Sugimoto A (2001) Large-scale analysis of gene function in *Caenorhabditis elegans* by high-throughput RNAi. Curr Biol 11(3):171–176
Malone CD, Hannon GJ (2009) Small RNAs as guardians of the genome. Cell 136(4):656–668
Marcello MR, Singaravelu G, Singson A (2012) Fertilization. Advances in Experimental Medicine and Biology 757:321–350. (Chap. 11, this volume) Springer, New York
McCarter J, Bartlett B, Dang T, Schedl T (1999) On the control of oocyte meiotic maturation and ovulation in *Caenorhabditis elegans*. Dev Biol 205(1):111–128
McNally KL, McNally FJ (2005) Fertilization initiates the transition from anaphase I to metaphase II during female meiosis in *C. elegans*. Dev Biol 282(1):218–230
Miller D, Brinkworth M, Iles D (2009) Paternal DNA packaging in spermatozoa: more than the sum of its parts? DNA, histones, protamines and epigenetics. Reproduction 139(2):287–301
Miller D, Brinkworth M, Iles D (2010) Paternal DNA packaging in spermatozoa: more than the sum of its parts? DNA, histones, protamines and epigenetics. Reproduction 139(2):287–301
Minniti AN, Sadler C, Ward S (1996) Genetic and molecular analysis of spe-27, a gene required for spermiogenesis in *Caenorhabditis elegans* hermaphrodites. Genetics 143(1):213–223
Monen J, Maddox PS, Hyndman F, Oegema K, Desai A (2005) Differential role of CENP-A in the segregation of holocentric *C. elegans* chromosomes during meiosis and mitosis. Nat Cell Biol 7(12):1248–1255
Morgan DE, Crittenden SL, Kimble J (2010) The *C. elegans* adult male germline: stem cells and sexual dimorphism. Dev Biol 346(2):204–214
Motegi F, Sugimoto A (2006) Sequential functioning of the ECT-2 RhoGEF, RHO-1 and CDC-42 establishes cell polarity in *Caenorhabditis elegans* embryos. Nat Cell Biol 8(9):978–985
Muhlrad PJ, Ward S (2002) Spermiogenesis initiation in *Caenorhabditis elegans* involves a casein kinase 1 encoded by the spe-6 gene. Genetics 161(1):143–155

Nance J, Minniti AN, Sadler C, Ward S (1999) spe-12 encodes a sperm cell surface protein that promotes spermiogenesis in *Caenorhabditis elegans*. Genetics 152(1):209–220

Nance J, Davis EB, Ward S (2000) spe-29 encodes a small predicted membrane protein required for the initiation of sperm activation in *Caenorhabditis elegans*. Genetics 156(4):1623–1633

Nelson GA, Ward S (1980) Vesicle fusion, pseudopod extension and amoeboid motility are induced in nematode spermatids by the ionophore monensin. Cell 19(2):457–464

Nelson GA, Lew KK, Ward S (1978) Intersex, a temperature-sensitive mutant of the nematode *Caenorhabditis elegans*. Dev Biol 66(2):386–409

Nelson GA, Roberts TM, Ward S (1982) *Caenorhabditis elegans* spermatozoan locomotion: amoeboid movement with almost no actin. J Cell Biol 92(1):121–131

O'Connell KF, Maxwell KN, White JG (2000) The *spd-2* gene is required for polarization of the anteroposterior axis and formation of the sperm asters in the *Caenorhabditis elegans* zygote. Dev Biol 222(1):55–70

Okamura K, Lai EC (2008) Endogenous small interfering RNAs in animals. Nat Rev Mol Cell Biol 9(9):673–678

Orr-Weaver TL, Parfenov VN, Dudina LM, Kostiuchek DF, Gruzova MN, Parfenov V, Potchukalina G, Dudina L, Kostyuchek D, Gruzova M, Sanyal MK, Taymor ML, Berger MJ (1995) Meiosis in *Drosophila*: seeing is believing. Proc Natl Acad Sci USA 92 (23):10443–10449

Pavelec DM, Lachowiec J, Duchaine TF, Smith HE, Kennedy S (2009) Requirement for ERI/DICER complex in endogenous RNAi and sperm development in *Caenorhabditis elegans*. Genetics 183(4):1283–1295

Pelletier L, O'Toole E, Schwager A, Hyman AA, Muller-Reichert T (2006) Centriole assembly in *Caenorhabditis elegans*. Nature 444(7119):619–623

Peters N, Perez DE, Song MH, Liu Y, Muller-Reichert T, Caron C, Kemphues KJ, O'Connell KF (2010) Control of mitotic and meiotic centriole duplication by the Plk4-related kinase ZYG-1. J Cell Sci 123(Pt 5):795–805

Reinke V (2002) Functional exploration of the *C. elegans* genome using DNA microarrays. Nat Genet 32(Suppl):541–546

Reinke V, Cutter AD (2009) Germline expression influences operon organization in the *Caenorhabditis elegans* genome. Genetics 181(4):1219–1228

Reinke V, Smith HE, Nance J, Wang J, Van Doren C, Begley R, Jones SJ, Davis EB, Scherer S, Ward S, Kim SK (2000) A global profile of germline gene expression in *C. elegans*. Mol Cell 6(3):605–616

Reinke V, Gil IS, Ward S, Kazmer K (2004) Genome wide germline enriched and sex biased expression profiles in *Caenorhabditis elegans*. Development 131(2):311–323

Roberts TM, Pavalko FM, Ward S (1986) Membrane and cytoplasmic proteins are transported in the same organelle complex during nematode spermatogenesis. J Cell Biol 102(5):1787–1796

Rogers E, Bishop JD, Waddle JA, Schumacher JM, Lin R (2002) The aurora kinase AIR-2 functions in the release of chromosome cohesion in *Caenorhabditis elegans* meiosis. J Cell Biol 157(2):219–229

Ruby JG, Jan C, Player C, Axtell MJ, Lee W, Nusbaum C, Ge H, Bartel DP (2006) Large-scale sequencing reveals 21U-RNAs and additional microRNAs and endogenous siRNAs in *C. elegans*. Cell 127(6):1193–1207

Sadler PL, Shakes DC (2000) Anucleate *Caenorhabditis elegans* sperm can crawl, fertilize oocytes and direct anterior-posterior polarization of the 1-cell embryo. Development 127(2):355–366

Sanyal MK, Taymor ML, Berger MJ (1976) Cytologic features of oocytes in the adult human ovary. Fertil Steril 27(5):501–510

Sassone-Corsi P (2002) Unique chromatin remodeling and transcriptional regulation in spermatogenesis. Science 296(5576):2176–2178

Schumacher JM, Golden A, Donovan PJ (1998) AIR-2: an Aurora/Ipl1-related protein kinase associated with chromosomes and midbody microtubules is required for polar body extrusion and cytokinesis in *Caenorhabditis elegans* embryos. J Cell Biol 143(6):1635–1646

Schvarzstein M, Wignall SM, Villeneuve AM (2010) Coordinating cohesion, co-orientation, and congression during meiosis: lessons from holocentric chromosomes. Genes Dev 24(3):219–228

Seydoux G, Schedl T (2001) The germline in *C. elegans*: origins, proliferation, and silencing. Int Rev Cytol 203:139–185

Sha K, Fire A (2005) Imprinting capacity of gamete lineages in *Caenorhabditis elegans*. Genetics 170(4):1633–1652

Shakes DC, Ward S (1989) Mutations that disrupt the morphogenesis and localization of a sperm-specific organelle in *Caenorhabditis elegans*. Dev Biol 134(2):307–316

Shakes DC, Wu JC, Sadler PL, Laprade K, Moore LL, Noritake A, Chu DS (2009) Spermatogenesis-specific features of the meiotic program in *Caenorhabditis elegans*. PLoS Genet 5(8):e1000611

Shakes DC, Allen AK, Albert KM, Golden A (2011) *emb-1* encodes the APC16 subunit of the *Caenorhabditis elegans* anaphase-promoting complex. Genetics 189(2):549–560

Shepherd AM, Clark SA (1983) The structure and development of the spermatozoon of *Aspicularis tetraptera* (Nematoda). J Cell Sci 2:537–544

Shim YH, Bonner JJ, Blumenthal T (1995) Activity of a *C. elegans* GATA transcription factor, ELT-1, expressed in yeast. J Mol Biol 253(5):665–676

Sijen T, Fleenor J, Simmer F, Thijssen KL, Parrish S, Timmons L, Plasterk RH, Fire A (2001) On the role of RNA amplification in dsRNA-triggered gene silencing. Cell 107(4):465–476

Smith JR, Stanfield GM (2011) TRY-5 is a sperm-activating protease in *Caenorhabditis elegans* seminal fluid. PLoS Genet 7(11):e1002375

Smith P, Leung-Chiu WM, Montgomery R, Orsborn A, Kuznicki K, Gressman-Coberly E, Mutapcic L, Bennett K (2002) The GLH proteins, *Caenorhabditis elegans* P granule components, associate with CSN-5 and KGB-1, proteins necessary for fertility, and with ZYX-1, a predicted cytoskeletal protein. Dev Biol 251(2):333–347

Spieth J, Shim YH, Lea K, Conrad R, Blumenthal T (1991) elt-1, an embryonically expressed *Caenorhabditis elegans* gene homologous to the GATA transcription factor family. Mol Cell Biol 11(9):4651–4659

Spike C, Meyer N, Racen E, Orsborn A, Kirchner J, Kuznicki K, Yee C, Bennett K, Strome S (2008) Genetic analysis of the *Caenorhabditis elegans* GLH family of P-granule proteins. Genetics 178(4):1973–1987

Stanfield GM, Villeneuve AM (2006) Regulation of sperm activation by SWM-1 is required for reproductive success of *C. elegans* males. Curr Biol 16(3):252–263

Suh N, Blelloch R (2011) Small RNAs in early mammalian development: from gametes to gastrulation. Development 138(9):1653–1661

Turpeenniemi TA (1998) Ultrastructure of spermatozoa in the nematode *Halalaimus dimorphus* (Nemata: Oxystominidae). J Nematol 30(4):391–403

Varkey JP, Jansma PL, Minniti AN, Ward S (1993) The *Caenorhabditis elegans* spe-6 gene is required for major sperm protein assembly and shows second site non-complementation with an unlinked deficiency. Genetics 133(1):79–86

Varkey JP, Muhlrad PJ, Minniti AN, Do B, Ward S (1995) The *Caenorhabditis elegans spe-26* gene is necessary to form spermatids and encodes a protein similar to the actin-associated proteins kelch and scruin. Genes Dev 9(9):1074–1086

Wallenfang MR, Seydoux G (2000) Polarization of the anterior-posterior axis of *C. elegans* is a microtubule-directed process. Nature 408(6808):89–92

Wang G, Reinke V (2008) A *C. elegans* Piwi, PRG-1, regulates 21U-RNAs during spermatogenesis. Curr Biol 18(12):861–867

Wang X, Zhao Y, Wong K, Ehlers P, Kohara Y, Jones SJ, Marra MA, Holt RA, Moerman DG, Hansen D (2009) Identification of genes expressed in the hermaphrodite germ line of *C. elegans* using SAGE. BMC Genomics 10:213

Ward S (1986) The asymmetric localization of gene products during the development of *Caenorhabditis elegans* spermatozoa. Gametogenesis and the early embryo. A.R. Liss, New York, pp. 55–75

Ward S, Carrel JS (1979) Fertilization and sperm competition in the nematode *Caenorhabditis elegans*. Dev Biol 73(2):304–321

Ward S, Argon Y, Nelson GA (1981) Sperm morphogenesis in wild-type and fertilization-defective mutants of *Caenorhabditis elegans*. J Cell Biol 91(1):26–44

Ward S, Hogan E, Nelson GA (1983) The initiation of spermiogenesis in the nematode *Caenorhabditis elegans*. Dev Biol 98(1):70–79

Washington NL, Ward S (2006) FER-1 regulates Ca2+–mediated membrane fusion during *C. elegans* spermatogenesis. J Cell Sci 119 (Pt 12):2552–2562

Wignall SM, Villeneuve AM (2009) Lateral microtubule bundles promote chromosome alignment during acentrosomal oocyte meiosis. Nat Cell Biol 11(7):839–844

Wolf N, Hirsh D, McIntosh JR (1978) Spermatogenesis in males of the free-living nematode, *Caenorhabditis elegans*. J Ultrastruct Res 63(2):155–169

Wu TF, Chu DS (2008) Epigenetic processes implemented during spermatogenesis distinguish the paternal pronucleus in the embryo. Reprod Biomed Online 16(1):13–22

Wu JC, Go AC, Samson M, Cintra T, Mirsoian S, Wu TF, Jow MM, Routman EJ, Chu DS (2012) Sperm development and motility are regulated by PP1 phosphatases in *Caenorhabditis elegans*. Genetics 190(1):143–157

Yi K, Buttery SM, Stewart M, Roberts TM (2007) A Ser/Thr kinase required for membrane-associated assembly of the major sperm protein motility apparatus in the amoeboid sperm of *Ascaris*. Mol Biol Cell 18(5):1816–1825

Yushin VV, Commans A (2005) ltrastructure of sperm development in the free-living marine nematode *Metachromadora itoi* (Chromadoria, Desmodorida). Acta Zoologica 86(4):255–265

Zanetti S, Puoti A (2012) Sex determination in the *C. elegans* germline. Advances in Experimental Medicine and Biology 757:41–69. (Chap. 3, this volume) Springer, New York

Zeng Y, Cullen BR (2003) Sequence requirements for micro RNA processing and function in human cells. RNA 9(1):112–123

Zhu GD, L'Hernault SW (2003) The *Caenorhabditis elegans spe-39* gene is required for intracellular membrane reorganization during spermatogenesis. Genetics 165(1):145–157

Zhu GD, Salazar G, Zlatic SA, Fiza B, Doucette MM, Heilman CJ, Levey AI, Faundez V, L'Hernault SW (2009) SPE-39 family proteins interact with the HOPS complex and function in lysosomal delivery. Mol Biol Cell 20(4):1223–1240

Chapter 8
Translational Control in the *Caenorhabditis elegans* Germ Line

Marco Nousch and Christian R. Eckmann

Abstract Translational control is a prevalent form of gene expression regulation in the *Caenorhabditis elegans* germ line. Linking the amount of protein synthesis to mRNA quantity and translational accessibility in the cell cytoplasm provides unique advantages over DNA-based controls for developing germ cells. This mode of gene expression is especially exploited in germ cell fate decisions and during oogenesis, when the developing oocytes stockpile hundreds of different mRNAs required for early embryogenesis. Consequently, a dense web of RNA regulators, consisting of diverse RNA-binding proteins and RNA-modifying enzymes, control the translatability of entire mRNA expression programs. These RNA regulatory networks are tightly coupled to germ cell developmental progression and are themselves under translational control. The underlying molecular mechanisms and RNA codes embedded in the mRNA molecules are beginning to be understood. Hence, the *C. elegans* germ line offers fertile grounds for discovering post-transcriptional mRNA regulatory mechanisms and emerges as great model for a systems level understanding of translational control during development.

Keywords RNA regulatory network • Post-transcriptional RNA regulation • Translational control • RNA-binding protein • Polyadenylation • Deadenylation • RNA decay • miRNA

M. Nousch • C.R. Eckmann (✉)
Max Planck Institute of Molecular Cell Biology and Genetics,
Pfotenhauerstrasse 108, 01307 Dresden, Germany
e-mail: mnousch@mpi-cbg.de; eckmann@mpi-cbg.de

T. Schedl (ed.), *Germ Cell Development in C. elegans*, Advances in Experimental Medicine and Biology 757, DOI 10.1007/978-1-4614-4015-4_8,

8.1 Introduction

Development depends on the coordinated execution of gene expression programs in a spatial and temporal manner. These programs are regulated at the transcriptional and post-transcriptional level. The *Caenorhabditis elegans* germ line has emerged as a model for both the requirement and complexity of post-transcriptional gene expression control. The importance of post-transcriptional control is reflected at the simplest level by the sheer number of RNA regulatory proteins identified that play important roles in the development and function of the germ line. By contrast, transcription seems to be the primary mechanism for regulating gene expression in most somatic tissues.

Possible explanations for the dominance of translational control are the germ line's special organization, its unique functionality and the short time frame available for gametogenesis and early embryogenesis in *C. elegans*. The adult germ line is a syncytial tube-like organ, which harbors germ cells of two different cell cycles and many stages of differentiation. A pool of mitotically dividing cells is located at its distal most end. Germ cells undergoing the meiotic program are spatially arranged in the remaining tissue in a distal to proximal fashion (see Lui and Colaiácovo 2012). In adult hermaphrodites diakinesis-stage oocytes remain associated with the germline syncytium until they fully cellularize. Differentiated oocytes at the very proximal end complete meiotic progression with oocyte maturation are ovulated, fertilized and undergo the meiotic divisions (see Kim et al. 2012). While sperm production happens in the hermaphrodite prior to oogenesis, spermatogenesis is a continuous process in the male with spermatocytes cellularizing before diakinesis (see Chu and Shakes 2012). Consequently, germ cells have to undergo two different types of cell divisions, mitosis of undifferentiated cells and meiosis of two sex-specific differentiation programs, which occur in a spatially and temporally distinct manner while sharing a common cytoplasm. These constraints demand a system that supports cytoplasmic gene regulation, such as translational control.

In addition to the germ line's spatial character, a number of functional criteria argue for the importance of post-transcriptional regulation. From diakinesis until after the completion of the first mitotic division, all chromosomes are highly condensed. Consequently, the transcriptional machinery has restricted access to the genome, therefore limiting the reach of transcriptional control in these stages (see Robertson and Lin 2012). Also, female meiosis is completed after oocyte–sperm fusion and maternally donated mRNA and protein factors continue to direct early embryogenesis, as zygotic gene transcription does not begin in somatic blastomeres until the four-cell stage (see Robertson and Lin 2012). Furthermore, a fundamental characteristic of germ cells is their totipotency, allowing the zygote to differentiate into all somatic tissues after fertilization. Ectopic germline expression of master transcription factors necessary for specific somatic differentiation programs, or the elimination of certain post-transcriptional regulators leads to a loss of germ cell fate identity and the formation of neuronal, muscle, or gut cells in the germ line (Ciosk et al. 2006; Tursun et al. 2011). It is conceivable that general transcriptional activity is tightly controlled to avoid unwanted cell fate commitment. Lastly, post-transcriptional control may offer a speed advantage. In transcriptional control an input signal has to be transmitted into the nucleus, the appropriate mRNA has to be at first synthesized, matured, quality controlled, and exported to the cytoplasm. In contrast, an input signal received by the cytoplasmic post-transcriptional control machinery accesses directly a pool of pre-made cytoplasmically localized mRNAs, awaiting translation.

8.2 Fundamentals of Translational Control

8.2.1 Life of an mRNA and the Concept of mRNPs

From its birth until its death, an mRNA passes through many different activity states and subcellular territories. One can globally divide the life of an mRNA into nuclear and cytoplasmic phases. In the nucleus the pre-mRNA is transcribed, co-transcriptionally modified at each end and spliced, before the mature mRNA is exported through the nuclear pores. In the cytoplasm, the mRNA's lifetime is marked by its translation and degradation. Over the past two decades a picture has emerged in which cytoplasmic RNA regulatory events are important gatekeepers that control the amount of protein synthesis in nearly all stages of germ cells development.

In the different subcellular territories, diverse proteins associate with any given mRNA, forming biochemically definable entities termed mRNA-protein particles (mRNPs). In general, activity state-specific mRNPs are considered as functional units and accompany the mRNA throughout its life. In these units mRNAs are structurally modified or transported within the cell or a syncitial tissue. More importantly, mRNPs represent integration platforms for developmental control. The nature of an mRNP is defined by its protein components and its subcellular location (discussed in Sect. 8.4.3). Proteins of an mRNP contact the RNA either indirectly or directly as designated RNA-binding proteins. Different families of RNA-binding proteins are encoded in the worm genome, of which many show either germline-specific expression or germline-enriched expression. Some of the best-studied examples are discussed in Sect. 8.4.1.

The mRNA itself provides the *cis*-regulatory information that is decoded by RNA-binding proteins recognizing structural and/or sequence-specific elements. A typical mature mRNA consists of unique parts encoded in the DNA sequence and non-encoded generic parts. DNA-encoded parts are the *O*pen *R*eading *F*rame (ORF), which serves as the protein synthesis template, and the 5′ and 3′ *u*n*t*ranslated *r*egions (UTRs), which flank the ORF (Wilkie et al. 2003). Both UTRs can possess valuable information that influences the mRNA's capacity to serve as a protein synthesis template. Regulatory features of the 5′UTR that have negative effects on protein production are short upstream open reading frames (uORFs) and stem-loop structures or binding sites that are recognized by RNA-binding proteins. 5′UTR structures that can have positive effects on protein production are internal ribosome entry sites (IRESes) (Jackson 2005). Although a number of cases underlining the importance of these features have been described in yeast, *Drosophila* and mammals, examples in the worm remain to be identified. Features of the 3′UTR include interaction sites for sequence-specific RNA-binding proteins and microRNA-containing RNA silencing complexes. Numerous examples of 3′UTR-mediated controls exist in the *C. elegans* germ line and are discussed in Sect. 8.5. Most importantly, two structures that are not encoded in the DNA but are present on every mRNA are the 5′cap structure and the 3′ poly(A) tail. Both structures are added to the mRNA in the nucleus as part of the mRNA maturation process and are essential for stability and translatability (discussed in Sects. 8.3.1 and 8.3.2). It is the combination of all these mRNA-intrinsic features that influence the capability and strength of post-transcriptional regulation. The synergy or antagonism of multiple features and the availability of the *trans*-acting RNA regulators form the basis for mRNA-specific translational effector networks, organized in larger mRNP units, which is the topic of Sects. 8.4.2 and 8.4.3.

8.2.2 *Events in the Cytoplasm: mRNA Quantity Versus Translatability*

In the cytoplasm, an mRNA encounters three different fate choices. (1) It may enter directly the translating pool of mRNAs, giving rise to a protein. (2) It may enter the mRNA decay pathway, being removed permanently from the translating pool. (3) It may be subject to translational repression, withholding the mRNA from the translational pool without degradation. Although mechanistically different, mRNA degradation and stable repression may lead to similar amounts of protein produced from a single mRNA. While the balance of general RNA decay and active translation establishes a steady-state concentration of bulk protein produced, post-transcriptional regulation offsets this balance, allowing for larger differences between mRNA and protein amounts. For example, regulated mRNA degradation can lead to a local change in mRNA abundance and consequently establishes a subcellular gradient of mRNA available for translation. The effect of this quantitative difference can be magnified by additionally regulating the translational accessibility of an mRNA via qualitative features, such as the length of the 3′ poly(A) tail. Stable repression of an mRNA without degradation is a delicate task and is employed for protein production in a temporal and/or spatial manner. In a syncytial tissue mRNA repression is often a prerequisite for intracellular mRNA localization or transport. Once the mRNA reaches its destination the repression has to be reversed and the mRNA is activated. In complex tissues, such as the germ line, it is the interplay of mRNA translation, storage/transport and degradation that is the basis for the cytoplasmic phase of post-transcriptional gene regulation; this interplay dictates when, where, how, and to which extent synthesis of a particular protein is carried out.

8.2.3 Concept of Translational Control: Basic Properties and Control Possibilities

Strictly speaking, translational control is the regulation of ribosomal activity and, therefore, is not identical to post-transcriptional mRNA regulation. However, the operational definition of translational control includes any activity that affects the amount of protein output for an mRNA by either stimulating or reducing ribosome association. The majority of known regulatory mechanisms are repressive ones, arguing that the default state of most mRNAs is likely geared towards optimal translation. The process of translation is divided into three different phases: initiation, elongation, and termination. The initiation phase comprises all steps required for the assembly of a translationally competent 80S (S, Svedberg) ribosome on the start codon, i.e., 40S ribosome subunit recruitment, scanning and 60S subunit joining. During the elongation phase the 80S ribosome moves along the mRNA and synthesizes the polypeptide chain. When the ribosome encounters the stop codon, the termination phase is initiated, leading to a disassembly of the 80S ribosome and liberation of the newly synthesized protein (for more detail, see Mathews et al. 2007).

Although translation can be regulated at any of the three phases, the majority of regulatory events described so far target the initiation phase (Fig. 8.1). Translation initiation is a multistep act, reflected in the large number of factors taking part in this process. Across species, the six translation initiation factors are formed from many more individual components than the elongation or release factors. In *C. elegans*, more than 40 putative translation initiation factor components are encoded in the genome, in contrast to only six elongation factor components and two release factors. The individual initiation factors form large protein complexes, each fulfilling a distinct function during the initiation phase. Hence, the full complement of eIFs and their availability provides the basis for a high protein synthesis capacity. Two structural components of every mRNA contribute to efficient translation, the cap-structure and the poly(A) tail. Most importantly, their synergy for translational initiation is much more than the sum of each individual feature (Tarun and Sachs 1995; Wells et al. 1998). The basis for this observation is the proposed formation of the closed loop connection between cap- and tail-structures (Fig. 8.1a). Consequently, controlling the accessibility to, and the quality of the 5′cap and the 3′tail represent key mechanistic entry points in translational control. Each entry point is utilized in the *C. elegans* germ line and specific examples are discussed in Sects. 8.3.1 and 8.3.2.

8.2.4 Two Aspects of Translational Control: Global Versus Gene-Specific

Translational control can be classified into two different modes, global and gene-specific. (1) Global regulation affects all cap-carrying mRNAs in a cell or tissue by targeting the functionality of eIFs or ribosomal subunits, mostly via post-translational

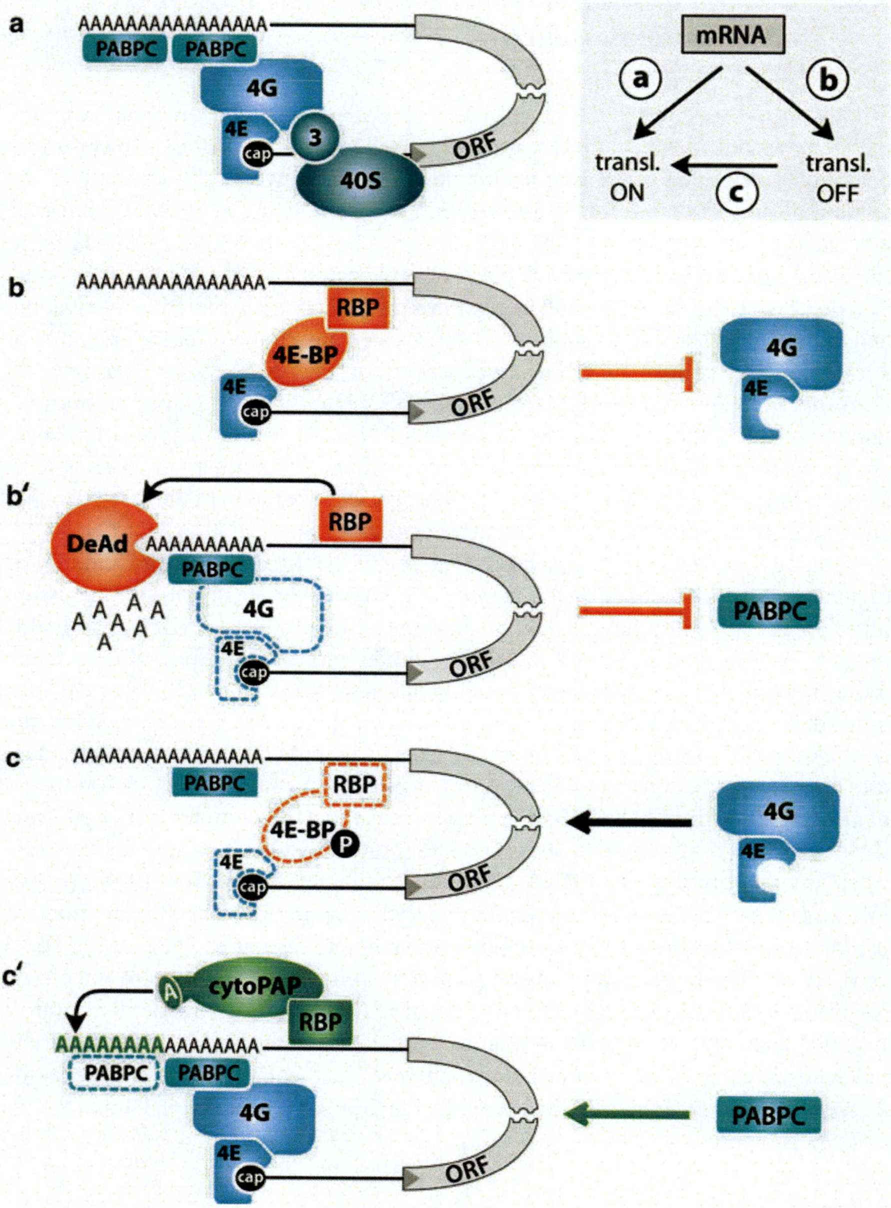

Fig. 8.1 Schematic representation of translation initiation, repression, and re-activation. **(a)** Cap-mediated translation initiation. Translation initiation factor eIF4G (4G) binds to eIF4E (4E) and the cytoplasmic poly(A)-binding protein (PABPC), facilitating circularization of the mRNA's tails. The central position of eIF4G in the closed loop aids translational initiation by indirectly recruiting the small ribosomal subunit (40S) to the mRNA via eIF3 (3). ORF, open reading frame; arrowhead in ORF indicates start codon. **(b)** Interfering with the placement of functional eIF4G results in translational repression. Specifically recruited RNA-bound proteins

modifications (Gebauer and Hentze 2004). This profound type of regulation is used in stress situations, such as toxic exposure or starvation, to quickly shut down new protein synthesis in order to dictate an appropriate stress response. Detailed mechanistic studies regarding this type of regulation have not been conducted in *C. elegans*, but indications for its existence can be found. For example, in response to heat stress, key eIFs are phosphorylated, which correlates with a general reduction in the abundance of translating ribosomes (Nousch and Eckmann, unpublished data). A different and more indirect indication is the presence of large cytoplasmic RNP structures in response to environmental stresses (Jud et al. 2008). The protein composition of these structures resemble in their protein components and dynamics stress granules of yeast or mammalian cells, which are thought to contain repressed translation initiation complexes (Buchan et al. 2011; Kedersha et al. 2005). (2) Gene-specific regulation affects only individual or a defined group of mRNAs. Here, not a canonical translation factor, but rather the mRNA itself is the direct target of regulation, utilizing sequence-specific RNA-binding proteins to recruit activating or repressive protein complexes.

8.3 Mechanistic Aspects of Translational Control in the Germ Line

8.3.1 Cap-Mediated Regulation

The 5′cap structure is a modified guanine nucleotide that protects the mRNA from 5′ to 3′ exonucleolytic decay (Furuichi et al. 1977; Shimotohno et al. 1977). In the majority of eukaryotes, mRNAs carry a guanosine that is mono-methylated at position seven (m7G), whereas a class of small nuclear RNAs carries a tri-methylated guanosine (TMG) with two methyl-groups at position two and one at position seven (Reddy et al. 1992). Either 5′cap is present on mRNAs in *C. elegans*, which depends on the nuclear history of RNA synthesis. mRNA produced by the canonical pre-mRNA maturation processes carries an m7G cap, while *trans*-spliced mRNAs to splice-leader sequences carries a TMG cap (Lasda and Blumenthal 2011). Importantly, *trans*-splicing is prevalent in worms and generates stereotyped 5′end sequences that replace, in the most extreme cases such as the *gld-3* mRNA (Eckmann et al. 2002),

Fig. 8.1 (continued) (RBP) guide eIF4E-binding proteins (4E-BP) to exclude eIF4G, forming alternative eIF4E/4E-BP complexes. (**b′**) Alternatively, RBPs may enhance the activity of deadenylases (DeAd), which indirectly limit the number of PABPCs associated with the poly(A) tail, therefore lowering the probability of eIF4G binding. (**c**) Translational re-activation of repressed mRNA. 4E-BP repressive complexes are destabilized and displaced from the mRNA by phosphorylation. (**c′**) Shortened poly(A) tails are re-elongated by cytoplasmic poly(A) polymerases (cytoPAP), which are stimulated upon mRNP remodeling. The translational initiation machinery efficiently recognizes a translationally re-activated mRNA, which is able to attract more PABPC the entire gene-encoded 5′UTR. This leaves limited or no room for gene-specific 5′UTR translational control sequences, and perhaps explains why the 3′UTR-mediated translational control mechanisms are more prevalent (Merritt et al. 2008).

The cap structure is the docking point for a unique group of proteins, the cap-binding family of proteins (Rhoads 2009). In the cytoplasm, eIF4E recognizes specifically the 5′cap and assists translation initiation as part of eIF4F, a larger protein complex recruiting the small 40S ribosomal subunit to the mRNA (Mathews et al. 2007) (Fig. 8.1a). Besides eIF4E, two additional proteins build the core of eIF4F: the scaffolding protein eIF4G, which provides the structural basis for the complex and is the essential bridging factor to form a closed loop structure; and eIF4A, an RNA helicase that unwinds the 5′UTR, aiding the 40S ribosomal scanning process to locate the initial start codon (Fig. 8.1a). Homologs for all three factors are encoded in the *C. elegans* genome, with five paralogs of eIF4E, two of eIF4A and one for eIF4G (Table 8.1). The latter exists as several isoforms presumably due to alternative splicing events, adding to further modes of regulation (Contreras et al. 2008). Although any eIF4 complex member represents in principle a target for regulation, it is the availability or functionality of eIF4E that is modulated most commonly across organisms. Thus, it is not surprising that the best-studied translation factors in *C. elegans* are the five eIF4E paralogs, *ife-1* to *ife-5*, possessing diverse cap selectivity. IFE-3 and IFE-4 bind preferentially to the m7G cap, whereas IFE-2, IFE-3, and IFE-5 bind to both, m7G and TMG caps (Jankowska-Anyszka et al. 1998; Keiper et al. 2000; Miyoshi et al. 2002). All paralogs are expressed in the germ line, with the exception of IFE-4, which is produced in muscles and neurons (Dinkova et al. 2005). Only one of the IFEs is absolutely essential; loss of IFE-3 activity results in embryonic lethality (Keiper et al. 2000). This suggests that IFE-1, IFE-2, and IFE-5 may compensate for each other and that IFE-3 is either used for bulk or essential embryonic mRNAs. Consistent with this notion, more subtle and specific defects have been reported for the other paralogs. IFE-1 and IFE-2 have roles limited to germline development. Loss of *ife-1* activity leads to defective gametogenesis in males and hermaphrodites (Henderson et al. 2009), suggesting that IFE-1 functions to guide general differentiation programs. IFE-2 has an even more defined role. It regulates meiosis in hermaphrodites, because *ife-2* mutants display severe chromosome segregation defects at elevated temperature (Song et al. 2010). Both factors have in common that they are important for efficient translation of specific mRNA subsets. For example, in the adult hermaphrodite, *oma-1* and *mex-1* mRNAs are not efficiently translated in the *ife-1* mutant and *msh-5* mRNA in the *ife-2* mutant. This suggests that *C. elegans* utilizes different cap-binding isoforms to positively regulate small groups of mRNAs, which appear functionally connected. It remains to be determined how the different IFEs define their target mRNAs and if additional RNA-binding factors help to select the mRNA.

Translational repression via the 5′cap prevents the assembly of a functional eIF4F complex (Fig. 8.1). This is achieved by controlling the availability of a single eIF4F component or by blocking essential interactions among eIF4F complex members. For example, the same peptide motif in eIF4E that physically contacts eIF4G is also recognized by a group of regulatory 4E-binding proteins

Table 8.1 Components of the eIF4F complex and a selection of general translation factors implicated in germline development

Translation factors and regulators	Protein	*C. elegans* name	WB gene ID	Remarks
eIF4F complex	eIF4E	IFE-1	F53A2.6	Affinity to m7G, TMG (Jankowska-Anyszka et al. 1998; Keiper et al. 2000)
		IFE-2	R04A9.4	Affinity to m7G, TMG (Jankowska-Anyszka et al. 1998; Keiper et al. 2000)
		IFE-3	B0348.6	Affinity to m7G (Jankowska-Anyszka et al. 1998; Keiper et al. 2000)
		IFE-4	C05D9.5	Affinity to m7G, expressed in muscle and neurons (Keiper et al. 2000; Dinkova et al. 2005)
		IFE-5	Y57A10A.30	Affinity to m7G, TMG (Keiper et al. 2000)
	eIF4A	INF-1	F57B9.6	Embryonic lethal by RNAi (Simmer et al. 2003)
			F57B9.3	Larval arrest by RNAi (Simmer et al. 2003)
	eIF4G	IFG-1	M110.4	Prevents increase of GL apoptosis (Contreras et al. 2008)
General translation factors	eIF5A	IFF-1	T05G5.10	Promotes GL proliferation (Hanazawa et al. 2004)
	eIF5B	IFFB-1	Y54F10BM.2	Promotes GL proliferation and differentiation (Yu et al. 2006)
	eEF1A	EEF-1A.1 (EFT-3, GLP-3)	F31E3.5	Required for all germ cell mitosis (Kadyk et al. 1997)
	PABPC	PAB-1	Y106G6H.2	Promotes GL proliferation (Ciosk et al. 2004; Maciejowski et al. 2005; Ko et al. 2010)
		PAB-2	F18H3.3	Promotes fertility (Ceron et al. 2007)
Translational control regulator	4E-BP	SPN-2 (PQN-45)	F56F3.1	Promotes embryogenesis (Li et al. 2009), oogenesis Guven-Ozkan et al. 2010
		PGL-1	ZK381.4	Promotes GL proliferation and gametogenesis (Amiri et al. 2001
	PAIP	ATX-2	D2045.1	Required for sp–oo switch (Ciosk et al. 2004)

For the translation initiation factor eIF4F all *C. elegans* factors are shown. General translation factors are included, provided a specific germline function is known
GL germ line, *WB* Wormbase

(4E-BPs) (Gebauer and Hentze 2004). Hence, 4E-BPs compete with eIF4G for eIF4E and act as global translational repressors, as they do not assist 40S ribosome recruitment. Stable and mRNA-specific repression is further assisted by 3′UTR-associated RNA-binding proteins. Thereby the 5′cap is stably bound by an eIF4E/4E-BP complex, which forms an alternative closed loop structure with the 3′UTR-anchored RNA-binding protein, leading to a decrease in protein production and the formation of a 4E-BP-poisoned mRNP (Rhoads 2009) (Fig. 8.1b). Although examples exist from other organisms, no general 4E-BP translation regulator has yet been described in *C. elegans*. However, SPN-2, a 4E-BP regulator involved in gene-specific regulation, is active in the germ line as a maternally expressed factor (Li et al. 2009). SPN-2, also known as PQN-52, interacts with all IFEs, except IFE-4, and with the RNA-binding proteins OMA-1/2, which are highly abundant in oocytes. Two mRNA targets of an IFE/SPN-2/OMA complex have been identified (Li et al. 2009; Guven-Ozkan et al. 2010). *zif-1* mRNA, encoding a substrate-binding subunit of an E3 ubiquitin ligase, and *mei-1* mRNA, encoding a subunit of the microtubule-severing enzyme katanin. Oocyte repression of *zif-1* mRNA ensures high protein levels of PIE-1, a P lineage-specific transcriptional repressor donated maternally to the early embryo and a protein target of ubiquitin-mediated decay in somatic blastomeres (Strome 2005) (also see Wang and Seydoux 2012). Robust and fast elimination of the katanin subunit MEI-1 is important for the fertilized embryo to switch from an oogenic meiotic spindle to a first mitotic spindle, which is accomplished in two parallel steps in the early embryo (Clark-Maguire and Mains 1994a, b; Srayko et al. 2000) (also see Kim et al. 2012, Chap. 10). While MEI-1 is degraded via ubiquitin-mediated proteolysis (Bowerman and Kurz 2006), *mei-1* mRNA is translationally repressed to prevent new protein synthesis (Li et al. 2009). The effectiveness of this mechanism is underscored by numerous 4E-BPs in other organisms, such as Maskin or Cup. Maskin is important for the maturation of *Xenopus* oocytes (Stebbins-Boaz et al. 1999), whereas Cup has a role in axis formation in *Drosophila* embryos (Nelson et al. 2004). A reversal of 4E-BP-mediated repression is envisioned to depend on its phosphorylation status (Barnard et al. 2005) (Fig. 8.1c). In conclusion, 5′cap-mediated translational repression is a frequent mechanism during oocyte maturation and early embryogenesis, however, less often observed during post-embryonic *C. elegans* germline development.

mRNA decapping leads to immediate 5′–3′-directed RNA degradation. *C. elegans* homologs of the yeast enzymes and associated factors involved in decapping and 5′–3′ degradation are summarized in Table 8.2. Although the major decapping enzyme DCAP-2 is present in germ cells, no specific role in germline development has been identified (Lall et al. 2005). Interestingly, the important decapping enhancer PATR-1 is only weakly expressed during *C. elegans* germline development (Boag et al. 2008). In the absence of an additional decapping enhancer, it appears that specific mRNA-degradation has a minor role in post-transcriptional mRNA regulation in *C. elegans*.

Table 8.2 Factors of the mRNA deadenylation and decay machinery

RNA processing	Factor or complex	*C. elegans* name	WB gene ID
Deadenylation	CCR4-NOT	CCR-4	ZC518.3
		CCF-1	Y56A3A.20
		LET-711 (NTL-1)	F57B9.2
		NTL-2	B0286.4
		NTL-3	Y56A3A.1
		NTL-4	C49H3.5
		NTL-9	C26E6.3
	PAN2/PAN3	PANL-2	F31E3.4
		PANL-3	ZK632.7
	PARN	K10C8.1	K10C8.1
5′–3′ decay	Decapping enzyme	DCAP-2	F52G2.1
	Dhh1p	CGH-1	C07H6.5
	Pat1p	PATR-1	F43G6.9
	Lsm complex	LSM-1	F40F8.9
		GUT-2 (LSM-2)	T10G3.6
		LSM-3	Y62E10A.12
		LSM-4	F32A5.7
		LSM-5	F28F8.3
		LSM-6	Y71G12B.14
		LSM-7	ZK593.7
	Xrn1p	XRN-1	Y39G8C.1
		XRN-2	Y48B6A.3

Deadenylation and 5′–3′ decay proteins with sequence similarity to biochemically defined yeast or human factors
WB Wormbase

8.3.2 Poly(A)-Tail Length Control

The most dynamic structure of an mRNA is its poly(A) tail. Nuclear polyadenylation is a co-transcriptional default process that liberates the RNA from its site of transcription and assists mRNA export (Sachs and Wahle 1993). In the cytoplasm, the homopolymeric A-tail is removed as part of the natural mRNA decay pathway (Decker and Parker 1993). However, the poly(A) tail is also a platform for regulatory translational control mechanisms that exploit its two cytoplasmic functions, enhancing mRNA stability and translatability (Mathews et al. 2007). In particular, the length of the poly(A) tail is an indicator of the mRNA's fate. A short tail makes an mRNA less attractive for translation initiation and renders it rather unstable, whereas a long tail stimulates translation initiation and stabilizes an mRNA (Munroe and Jacobson 1990; Decker and Parker 1993). The molecular basis for this phenomena is a sequence-specific RNA-binding protein that decorates the mRNA's tail, cytoplasmic poly(A)-binding protein (PABPC) (Otero et al. 1999). A single PABPC molecule binds approximately 20 adenosines and interacts directly with the eIF4F complex, via eIF4G, assisting closed-loop formation (Wells et al. 1998; Baer and Kornberg 1980, 1983) (Fig. 8.1a). Hence, a long poly(A) tail is able to recruit more PABP molecules than a short poly(A) tail, enhancing the probability of frequent translational initiation. Two PABPC proteins are present in *C. elegans*, PAB-1 and PAB-2 (Table 8.1). Only *pab-1,* not *pab-2,* seems essential for germline development. Loss of *pab-1* results in sterile animals with a strong germline proliferation defect, indicating that PAB-1-related poly(A)-tail metabolism is critical for germ cell divisions (Ciosk et al. 2004; Maciejowski et al. 2005; Ko et al. 2010).

The process of poly(A)-tail shortening is termed deadenylation and is carried out by deadenylases (Garneau et al. 2007) (Fig. 8.1b′). Three major deadenylation complexes have been described so far: the CCR4-NOT complex, the PAN2/PAN3 complex and PARN. All complexes show conservation from yeast to humans and, with the exception of the PAN2/3 complex, have been linked to mRNA-specific translational control (Garneau et al. 2007). Homologs of all three complex components can be found in the *C. elegans* genome, of which the majority remains uncharacterized (Table 8.2). Although the CCR4-NOT complex is the major cytoplasmic deadenylase in *Drosophila* and yeast (Garneau et al. 2007), it is also involved in mRNA-specific translational regulation, consistent with reported functions of its individual components during *C. elegans* germline development. The absence of CCF-1, the *C. elegans* homolog of POP2, which is one of the two catalytic subunits from the presumed worm CCR4-NOT complex, results in pachytene-arrested germ cells without differentiation into sperm or oocytes (Molin and Puisieux 2005). Upon RNAi knockdown of CCR-4, the second deadenylase of the presumed complex, more subtle defects are observed that are limited to the stability of specific mRNAs (Schmid et al. 2009). While a partial reduction of function of the *C. elegans* NOT1 homolog, LET-711, affects the first mitotic division of the one-cell embryo, a strong loss of function results in sterile germ lines (DeBella et al. 2006). Together these suggest that CCF-1/CCR-4/LET-711-mediated deadenylation is important for meiotic progression, germ cell differentiation and early embryogenesis. Given the unique phenotypic defects of each complex member, it is also likely that some mRNAs that are keys for a specific process in germ cell development are differentially affected by the activity of individual deadenylase complex components. Yet, a formal biochemical demonstration of the existence of a CCR4-NOT complex is still missing in *C. elegans.*

While shortening of the poly(A) tail is the initial step of canonical mRNA degradation, the removal of an mRNA is not an obligatory fate. mRNAs with a shortened tail can escape degradation and persist as stable yet translationally silenced molecules. The underlying molecular mechanisms are less clear; nevertheless, it is most likely assisted by specific mRNA-associated proteins that package the mRNA into translationally dormant mRNPs. mRNA-associated factors that recruit the presumed CCF-1/CCR-4/LET-711 deadenylase to specific mRNAs and enhancing poly(A)-tail shortening are slowly emerging and discussed in Sect. 8.5.

Translational activation of deadenylated mRNAs requires remodeling of the repressed mRNP and the resynthesis of a longer poly(A) tail. Cytoplasmic poly(A) polymerases (cytoPAPs) elongate the homopolymeric A-tail and reintroduce the mRNA into the translating pool (Fig. 8.1c′). A crucial aspect of cytoPAPs is that, contrary to nuclear PAP, the enzyme lacks a predictable RNA-binding domain (Eckmann et al. 2011). They associate with mRNA targets via other mRNP components and, therefore, are also referred to as noncanonical poly(A) polymerases (Wang et al. 2002). cytoPAPs are conserved across species with two representatives described in *C. elegans*, GLD-2 and GLD-4 (Wang et al. 2002; Schmid et al. 2009). As their gene name indicates—*gld* stands for *g*erm*l*ine *d*evelopment defective—both proteins are implicated in many germline functions, ranging from germ cell fate decisions to meiotic progression, gametogenesis and early embryogenesis. The molecular composition of both cytoPAP complexes is discussed in detail in Sect. 8.4.2.

Interestingly, developmentally regulated mRNAs show complex poly(A) tail dynamics in the germ line, which are further exploited in the early embryo, illustrating a tight connection between cytoplasmic deadenylation and polyadenylation (see Sect. 8.5). The effectiveness, versatility, and the ability to fine-tune protein synthesis, rather than just establishing an all-or-non situation, make poly(A)-tail length control probably the most widely used mechanism in germline development to control the exact amount of protein synthesis at each stage.

8.3.3 miRNA-Mediated Regulation

Gene expression regulation by microRNAs (miRNA) has emerged as a widespread mechanism. miRNAs belong to the group of noncoding RNAs, are small in size (20–25 nt), and are generated from local hairpin structures by the action of two RNA endonucleases, Drosha and Dicer (Kim et al. 2009b). A mature miRNA is ultimately loaded into the RNA-induced silencing complex (RISC), which contains Argonaute-family proteins. As part of the RISC complex, miRNAs serve as RNA-recognition devices. They base-pair with complementary mRNA sequences preferentially located in the 3′UTR, leading to the subsequent translational repression or degradation of the mRNA (Fabian et al. 2010). Although miRNAs were first identified in *C. elegans* as regulators of developmental timing (Lee et al. 1993; Wightman et al. 1993), a direct involvement in germline development is still lacking. However, some indications exist: The absence of either miRNA processing factors, *drsh-1* (Drosha) or *drc-1* (Dicer), leads to sterile animals (Denli et al. 2004; Knight and Bass 2001) (Table 8.3). RNAi-mediated down-regulation of ALG-1 and ALG-2, two of 24 worm Argonaute (Ago) proteins, results in weak germ cell differentiation defects (Grishok et al. 2001), and a specific micro-RNA family is required for DNA damage response in germ cells (Kato et al. 2009).

Another group of noncoding RNAs with germline functions are the endogenously produced small interfering RNAs, called 26G-RNAs (Han et al. 2009; Conine et al. 2010). This group of endo-siRNAs is 26 nt in length, preferentially starts with guanine, is exclusively present in the germ line, and shows perfect complementarity to their target transcripts. Two non-overlapping subsets of 26G RNAs are expressed during spermatogenesis (class I) and oogenesis (class II). Mapping of the 26G RNAs to the genome shows that they preferentially target protein-coding genes that are

Table 8.3 Factors involved in miRNA biogenesis and miRNA-mediated gene silencing

miRNA factor	Protein	WB gene ID	Germline phenotypes
Drosha	DRSH-1	F26E4.10	Reduced fertility (Denli et al. 2004)
Pasha	PASH-1	T22A3.5	Reduced brood size by RNAi
Dicer	DCR-1	K12H4.8	Abnormal oocytes (Knight and Bass 2001)
Dicer-related helicases	DRH-1	F15B10.2	Required redundantly with DRH-2 for GL and somatic RNAi (Tabara et al. 2002; Lu et al. 2009)
	DRH-2	C01B10.1	Required redundantly with DRH-1 for GL and somatic RNAi (Tabara et al. 2002; Lu et al. 2009)
	DRH-3	D2005.5	Abnormal chromosome arrangements in pachytene (Nakamura et al. 2007; She et al. 2009)
Argonaute	ALG-1	F48F7.1	Redundant with ALG-2 shows GC differentiation defects (Grishok et al. 2001)
	ALG-2	T07D3.7	Redundant with ALG-1 shows GC differentiation defects (Grishok et al. 2001)
GW182	AIN-1	C06G1.4	No apparent GL defect in mutants or by RNAi
	AIN-2	B0041.2	No apparent GL defect in mutants or by RNAi

Germline development-associated miRNA processing factors and miRNA-mediated translational silencing complexes are given. Factors of other non-coding RNAs, such as endo-siRNAs or piRNAs, are not included
GL germ line, *WB* Wormbase, *GC* germ cell

expressed during spermatogenesis or oogenesis/early embryogenesis, respectively. In their absence the abundance of putative target mRNAs increases, suggesting that 26G RNAs specifically repress mRNAs by enhancing their degradation. The two classes of 26G RNAs are suggested to be sorted into distinct RISC complexes, based on the observation that different Ago-proteins are essential for the presence of the different classes of 26G-RNAs in the worm. Whereas ERGO-1 is important for class II, the abundance of class I depends highly on the two paralogs ALG-3 and ALG-4 (Han et al. 2009; Conine et al. 2010). The functional relevance of class II 26G-RNA remains to be demonstrated, but class I 26G-RNAs are linked to gene regulation during spermatogenesis. This is based on the observation that the absence of *alg-3* and *alg-4* results in the production of defective sperm at elevated temperatures (Han et al. 2009; Conine et al. 2010).

8.4 Connections Between Global and Specific Regulators of Translational Control

8.4.1 RNA-Binding Protein Families in the Germ Line

Genome-wide gene expression analysis shows that in *C. elegans* RNA metabolism-associated proteins are significantly more abundant in the germ line than in the soma (Wang et al. 2009a). Not surprisingly, most RNA-associated protein families are involved in numerous aspects of germline development. This is further underscored by evolutionary conservation of the relevant RNA-binding protein families and an expansion in number of individual protein family members in *C. elegans*, which is consistent with a further diversification of their biological roles. While some protein families are quite closely related in their overall protein architecture and function (e.g., PUF or STAR proteins), others show poor sequence similarities outside their family-defining structural domains and have diverse RNA regulatory functions (e.g., KH or Nanos-like proteins).

Although most RNA-binding proteins use a particular protein domain(s) to interact with an mRNA, they do so with varying degrees of specificity and affinity. One can roughly divide them into binders with high or low selectivity, which reflects on their molecular roles in mRNP complex formation. A high-selectivity binder has high RNA-affinity, close to the single-digit nanomolar range, and recognizes a well-defined target sequence motif (e.g., PUF proteins and the STAR protein GLD-1) (Bernstein et al. 2005; Stumpf et al. 2008; Ryder et al. 2004). The group of low-selectivity binders is quite broad. Proteins belonging to this group display RNA-affinities in an upper two-digit to a three-digit nanomolar range and can, but not always do, display some general interaction preferences to a loosely definable consensus sequence, such as AU-rich sequences bound by *Xenopus* CPEBs (Hake et al. 1998). Whereas high selectivity RNA-binding proteins, defined here as RNA selectors, are envisioned as the primary targeting unit in an mRNP complex, the low selectivity ones may help to narrow down the mRNA target pools further. Additionally, low affinity binder may stabilize the mRNP structure or provide further regulatory capacity to the mRNP (e.g., GLD-3).

In the following section some of the best-studied RNA-binding protein families are described in more detail. We have limited our discussion to four protein families. Other RNA-binding proteins and their characteristics are summarized in Table 8.4.

8.4.1.1 PUF Proteins

One of the most distinguished RNA selector family in the *C. elegans* germ line is composed of the PUF (*Pu*milio and *F*BF) proteins (Wickens et al. 2002), FBF-1 and FBF-2 (collectively referred to as FBF), and PUF-3 to PUF-12, of which PUF-9 is exclusively somatic (Nolde et al. 2007). All PUF proteins possess eight, ~40 aa long, consecutively arranged PUF-repeats, which form a single RNA-recognition platform. Extensive structural analysis revealed the global domain architecture of the staged PUF-repeats into an arched superhelix with an inner RNA-binding surface and an outer protein interaction surface (Edwards et al. 2001). Each individual PUF-repeat contacts a single RNA nucleotide, providing the molecular basis for its defined RNA target motif, the FBF-binding element (FBE) (Bernstein et al. 2005). Although the UGU core sequence of an FBE is essential for all tested PUF proteins to bind RNAs with high affinity, additional flanking nucleotides add to their selectivity and mode of recognition. For example, PUF-8 preferentially binds to an eight nucleotide long motif with the consensus sequence of 5′-**UGU**ANAUA-3′, whereas FBF proteins prefer a nine nucleotide long 5′-**UGU**RNNAUA-3′ (R, purine; N, any base)

Table 8.4 RNA-binding proteins involved in germ cell development

Protein family	Protein	WB gene ID	5′/3′ regulator[a]	GL protein expression[b]	GL function
PUF proteins	PUF-1 (FBF-1)	H12I13.4	3′	A	sp–oo; mit–mei
	PUF-2 (FBF-2)	F21H12.5	3′	A	sp–oo; mit–mei
	PUF-3	Y45F10A.2	3′		emb
	PUF-4	M4.2	3′		
	PUF-5	F54C9.8	3′	D	oo
	PUF-6	F18A11.1	3′		oo
	PUF-7	B0273.2	3′		oo
	PUF-8	C30G12.7	3′	A	sp–oo; sp mei
	PUF-9[c]	W06B11.2	3′		
	PUF-11	Y73B6BL.38	3′		
	PUF-12	ZK945.3	3′		
K homology (KH) proteins	GLD-1	T23G11.3	5′ and 3′	B	mit–mei; oo mei; GC identity
	GLD-3	T07F8.3	3′	C	sp–oo; mit–mei
	MEX-3	F53G12.5		C	mei; GC identity; prolif
CPEB proteins	CPB-1	C40H1.1	3′		sp mei
	CPB-2	C30B5.3	3′		
	CPB-3	B0414.5	3′	B	sp–oo; mit–mei; oo
	FOG-1	Y54E10A.4	3′	B	sp–oo; mit–mei
Nanos proteins	NOS-1	R03D7.7	3′		GC viability
	NOS-2	ZK1127.1	3′		GC viability
	NOS-3	Y53C12B.3	3′	U/D[d]	sp–oo; mit–mei
DEAD-box ATP-dependent RNA helicases	GLH-1	T21G5.3		U	prolif; gametogenesis
	GLH-2	C55B7.1		U	prolif; gametogenesis
	GLH-3	B0414.6		U	
	GLH-4	T12F5.3		U	prolif; gametogenesis
	CGH-1	C07H6.5		U	oo
	LAF-1	Y71H2AM.19			sp–oo
	VBH-1	Y54E10A.9		U	sp–oo; emb
Y-box proteins	CEY-1	F33A8.3			
	CEY-2	F46F11.2			
	CEY-3	M01E11.5			
	CEY-4	Y39A1C.3			
Other RNA-binding motif containing proteins	LARP-1	R144.7		U	sp–oo; oo
	OMA-1	C09G9.6	3′	D	oo
	OMA-2	ZC513.6	3′	D	oo
	RNP-8	R119.7	3′	D	sp–oo; oo
	CAR-1	Y18D10A.17		U	emb
	DAZ-1	F56D1.7		A	sp–oo; oo mei

RNA regulator protein families with evolutionary defined RNA-binding domains are listed. Most proteins are referred to in the main text. Y-box proteins (Boag et al. 2005) are also known as cold-shock domain proteins. The proteins LARP-1(Nykamp et al. 2008; Zanin et al. 2010) and CAR-1(Squirrell et al. 2006; Audhya et al. 2005; Boag et al. 2005) contain an La-type and LSM-like RNA-binding motif, respectively

GC germ cell. *WB* Wormbase. Reported germline(GL) functions: *sp–oo* sperm-to-oocyte switch, *mit–mei* mitosis-to-meiosis switch, *sp* spermatogenesis, *oo* oogenesis, *emb* early embryogenesis, *mei* meiotic progression, *prolif* proliferation

[a]5′ or 3′ end-mediated translational control regulation

[b]Protein distribution in the germ line. Type A to D and U are according to Fig. 8.2

[c]Expressed in the soma

[d]phosphorylated NOS-3

sequence (Wang et al. 2009b). To comply with the one nucleotide per PUF-repeat logic, the central nucleotide in the RNA-binding motif of FBF is flipped away from the protein and remains exposed, diversifying the binding repertoire of PUF proteins (Opperman et al. 2005). Functionally the PUF and FBF proteins are also quite diverse, which correlates with their distinct protein expression profile in the germ line (see Sect. 8.4.2). Both FBF proteins are essential for actively dividing germ stem cells (Crittenden et al. 2002), meiotic entry (Lamont et al. 2004), and the sperm-to-oocyte switch (Zhang et al. 1997). PUF-8 is important for germline proliferation (Ariz et al. 2009), the sperm-to-oocyte switch (Bachorik and Kimble 2005), and male meiotic progression (Subramaniam and Seydoux 2003). While PUF-5, -6, -7 assist oocyte differentiation and maturation (Lublin and Evans 2007), PUF-3 is essential for early embryogenesis (Sonnichsen et al. 2005). For the most part, PUF proteins are translational repressors (discussed in detail in Sect. 8.5.3); however, in certain circumstances they may also act as translational activators (further discussed in Sects. 8.4.2 and 8.4.3). We term such dual regulators here as translational effectors.

8.4.1.2 Nanos Proteins

A protein family closely connected to PUF proteins consists of three *Drosophila* Nanos-related proteins, NOS-1, NOS-2, and NOS-3. The defining criterion of these germ cell-enriched RNA-binding proteins is the presence of two consecutive C-terminal CCHC zinc fingers assumed to be important for RNA association. In analogy to *Drosophila* Nanos, they are considered as high affinity binders with little sequence specificity (Curtis et al. 1997). While together all three worm Nanos proteins assist postembryonic germ cell proliferation and germ cell survival in the later larval stages, individual family members have also additional functions. Maternal NOS-2 protein is important for efficient primordial germ cell incorporation into the somatic gonadal primordium (Jadhav et al. 2008). Furthermore, NOS-1 and NOS-2 are redundantly required to block primordial germ cell proliferation in the first larval stage, upon starvation (Jadhav et al. 2008). Essential roles of NOS-3 in germ cells adopting the meiotic or oogenic fate are revealed when other redundantly acting RNA regulators are eliminated (Hansen et al. 2004; Eckmann et al. 2004).

8.4.1.3 KH Proteins

The hnRNP K homology (KH) domain represents a versatile protein fold of ~70–100 aa in length. Depending on its detailed amino-acid composition, KH domains bind ssDNA, RNA and/or proteins with varying binding affinities. Structural analysis revealed that a single KH domain accommodates at a maximum four nucleotides (Valverde et al. 2008). Thus, additional flanking protein sequences are necessary to extend the nucleic acid-binding surface and to increase its RNA-binding affinity and selectivity. Alternatively, multiple KH domains are combined in one protein to expand its RNA-binding potential. Proteins that carry either one or multiple KH domains are assigned to essentially all aspects of RNA metabolism. In *C. elegans*, three diverse KH domain proteins, GLD-1, GLD-3, and MEX-3, direct germline development.

The STAR protein GLD-1 is to date the most comprehensively characterized RNA selector of the *C. elegans* germ line. Its single KH domain is flanked by N-terminal and C-terminal sequences that form a unique structural arrangement, known as the *s*ignal *t*ransduction and *a*ctivation of *R*NA (STAR) domain (Vernet and Artzt 1997). This conserved maxi-KH domain can form functional homodimers (Chen et al. 1997) and contacts RNA in a sequence-specific manner with high affinity in the low nanomolar range (Ryder et al. 2004). Originally a six nucleotide-long recognition sequence was determined in vitro, which was subsequently extended to a 7-mer motif (UACUAAC) based on a much larger number of in vivo associated mRNA targets (Wright et al. 2011). GLD-1-binding motifs (GBMs) are present in both the 5′UTR or the 3′UTR of target mRNA (Lee and Schedl 2004). All verified mRNA targets are translationally repressed upon GLD-1 association, which in some cases also protects uORF-containing mRNAs from non-sense-mediated decay (Lee and Schedl 2004). Although detailed molecular mechanisms remain to be determined, the latter finding suggests that GLD-1 may inhibit translational initiation, 80S ribosome assembly or translation elongation (Mootz et al. 2004). It is also conceivable that the mechanism of GLD-1-mediated translational repression depends on the target mRNA and its associated factors. This is further underscored by mutations in GLD-1 that affect the regulation of a few mRNA targets but not others (Schumacher et al. 2005). Given that a large number of verified GLD-1 target mRNAs encode proteins involved in diverse biological roles (see Table 8.5), the importance of GLD-1 for germline development is easy to comprehend. It regulates the balance of proliferation vs. meiotic entry (Hansen et al. 2004), female meiotic progression (Francis et al. 1995a, b), physiological apoptosis (Schumacher et al. 2005), the sperm-to-oocyte switch (Jan et al. 1999), and maintenance of germ cell totipotency (Ciosk et al. 2006).

The multi-KH domain protein GLD-3, together with its paralog BCC-1, are the Bicaudal-C (BicC) protein family RNA regulators in *C. elegans* (Eckmann et al. 2002). They contain 5 KH domains arranged in tandem connected via very short amino acid linkers, whereby each individual domain adopts a classic KH fold (Eckmann et al. 2002; Nakel et al. 2010). Three-dimensional structural analysis has revealed that all five KH domains of GLD-3 have extensive contacts with each other, forming a tightly packaged protein core, which is in contrast to a previously assumed "beads on a string" organization. Consistent with GLD-3 having an extremely low affinity for homopolymeric guanidine RNA (Jedamzik and Eckmann, unpublished results), the typical GxxG RNA-contacting loops (G, glycine) of known RNA-binding KH domains are missing either one or both of the two glycine residues (Nakel et al. 2010). GLD-3 binds GLD-2, and GLS-1 (see below). The amino-terminal KH domain region of GLD-3 is the binding site for GLD-2 (Eckmann et al. 2004) and GLS-1 (Rybarska et al. 2009), thus the multi-KH domain arrangement in GLD-3 likely serves as a large protein interaction platform, rather than providing an RNA selector function. The inferred scaffolding and regulatory functions of GLD-3 in

Table 8.5 mRNAs targets of 3′ end-mediated translational control

mRNA	GL function	Repressor	Activator	*cis*-regulatory motif in 3′UTR
glp-1	GL proliferation	GLD-1	GLD-2[a]	TCCTAAC; ATCTCAC; GACTAAT (Wright et al. 2011)
fbf-1	Maintenance of GL stem cells, sp–oo switch	FBF-x		UGUAAUAUU; UGUGCCAUC (Lamont et al. 2004)
fbf-2	Maintenance of GL stem cells, sp–oo switch	FBF-x		UGUAAUAUU (Lamont et al. 2004)
cye-1	Promotion of mitosis	GLD-1		TACTTAC; AATTAAC; TACTCAT; ATCTCAC (Wright et al. 2011)
cki-2	Inhibition of mitosis	FBF-x	GLD-2[a]	UGUGAAUUU; UGUCCAUUU[b]; UGUGUUCUA[b]; TGTUUUUUU[a] (Kalchhauser et al. 2011)
gld-1	Entry into meiosis, mit–mei, meiotic progression, GL totipotency	FBF-x	GLD-2	UGUGCCAUA; UGUGCCAUA (Crittenden et al. 2002)
him-3	Synaptonemal complex component	FBF-x		UGUGCAAUG (Merritt and Seydoux 2010)
htp-1	Synaptonemal complex component	FBF-x		UGUAAAAUG (Merritt and Seydoux 2010)
htp-2	Synaptonemal complex component	FBF-x		UGUAACAUG; UGUACAAUG(Merritt and Seydoux 2010)
syp-2	Synaptonemal complex component	FBF-x		UGUAUCAUU (Merritt and Seydoux 2010)
syp-3	Synaptonemal complex component	FBF-x		UGUCGAAUG; UGUAAUAUU (Merritt and Seydoux 2010)
cep-1	GL apoptosis	GLD-1		n.d.
rme-2	Yolk receptor in oo	GLD-1		TACTAAA (Wright et al. 2011)
oma-1	oo maturation	GLD-1	GLD-2	TACTAAC (Wright et al. 2011)
oma-2	oo maturation	GLD-1	GLD-2	TACTAAC; CACTAAC (Wright et al. 2011)
lip-1	Germline proliferation	FBF-x		UGUAAAAUC; UGUGCCAUC (Lee et al. 2006)
mpk-1	oo maturation, GL apoptosis	FBF-x		UGUAUCAUA; UGUAAUAUA (Lee et al. 2007a)
fem-3	sp–oo switch	FBF-x	GLD-2[a]	UGUGUCAUU (Zhang et al. 1997)
fog-1	sp–oo switch	FBF-x	GLD-2[a]	UGUAAAAUC; UGUAUCAUG; UGUAUCAUU (Thompson et al. 2005)

(continued)

Table 8.5 (continued)

mRNA	GL function	Repressor	Activator	*cis*-regulatory motif in 3′UTR
tra-2	sp–oo switch	GLD-1	GLD-2[a]	4 × UACUCA (Ryder et al. 2004)
mex-3	GL proliferation, GL totipotency, establishment of embryonic cell fates	GLD-1		n.d.
pal-1	Specification of cell fates in embryo (maternally provided)	GLD-1		n.d.
		MEX-3		AUAGAGCUUCUUUAUUUA; UUAGGAAAAAGUUUA (Pagano et al. 2009)
		PUF-8		UGUACAAAA (Mainpal et al. 2011)

Translationally controlled mRNAs that are confirmed mRNA targets of indicated translational effectors (indicated as Repressor or Activator) by either in vitro binding assays, gel shift assays, transgenic reporter construct expression studies, or by a combination of these methods. FBF-x experiments did not distinguish between FBF-1 and FBF-2. If no particular motif was described yet (n.d.), translational regulation of the mRNA was at least demonstrated to dependent on the entire 3′UTR

GL germ line, *sp–oo* sperm-to-oocyte, *mit–mei* mitosis-to-meiosis decision, *oo* oocyte

[a]Proposed target mRNAs based on their association with GLD-2 in genome-wide RNA target analysis via co-immunoprecipitation experiments

[b]Have only weak binding affinities for FBF-x

mRNP complexes are consistent with its elaborate expression pattern and its global involvement in germline development. Germ cells require GLD-3 function to initiate the meiotic fate, progress beyond meiotic pachytene, and to achieve proper meiotic chromosome segregation (Eckmann et al. 2002, 2004; Rybarska et al. 2009). Additionally, GLD-3 assists in specification of the male germline fate in hermaphrodites and males (Eckmann et al. 2002, 2004). Maternal GLD-3 regulates early embryogenesis and maintains germline survival in the post-embryonic germline (Rybarska et al. 2009; Eckmann et al. 2002). The functions of GLD-3 are carried out by two prevalent protein isoforms of GLD-3, GLD-3S, and GLD-3L, which are distinguished by unique C-termini. BCC-1 is most similar to GLD-3 in its amino-terminal KH domain arrangement but significantly differs in its C-terminus from GLD-3. Functional roles of BCC-1 remain unclear.

A different multi-KH domain protein is the evolutionarily conserved RNA-binding protein MEX-3 (Buchet-Poyau et al. 2007). Originally discovered as a maternal-effect lethal gene that regulates early embryonic cell fates, *mex-3* is also active in the postembryonic germ line (Draper et al. 1996; Mootz et al. 2004; Ciosk et al. 2004, 2006; Ariz et al. 2009). MEX-3 is a 3′UTR-associated translational repressor and prevents the premature accumulation of the maternally donated embryonic cell fate determinant PAL-1 in the growing oocytes (Draper et al. 1996; Mootz et al. 2004; Hunter and Kenyon 1996; Jadhav et al. 2008). MEX-3 contains two prototypical KH domains and a bipartite RNA sequence motif was defined as a consensus MEX-3 RNA-binding site, using an in vitro reiterative RNA selection approach. This MEX-3 recognition element (MRE) consists of two four-nucleotide-binding motifs, which are separated by 0–8 nucleotides: DKAGN$_{(0–8)}$UHUA (D, everything except a C; K, is a G or U; H, everything except G) (Pagano et al. 2009). Consistent with the notion that each individual KH domain of MEX-3 may contact one motif to achieve overall high RNA selectivity and affinity, the elimination of both motifs is more detrimental to MEX-3 RNA binding than compromising an individual motif of the MRE (Pagano et al. 2009). Although numerous candidate MEX-3 target mRNAs were suggested (Pagano et al. 2009), little is known about how MEX-3 functions with GLD-1 to maintain totipotency of germ cells (Ciosk et al. 2006) and functions with PUF-8 in germ cell proliferation (Ariz et al. 2009).

8.4.1.4 CPEB Proteins

*C*ytoplasmic *p*olyadenylation *e*lement-*b*inding (CPEB) proteins represent an evolutionarily conserved protein family with a stereotypic multi-domain organization (Mendez and Richter 2001). Two centrally placed RNA recognition motifs (RRMs) are followed by two consecutive zinc finger motifs of the CCCC- and CCHH-types, respectively. All four domains are required for RNA binding in *Xenopus* CPEB1, the best-characterized protein family member (Hake et al. 1998; Richter 2007). CPEB1 recognizes a 3′UTR located sequence motif, termed the cytoplasmic polyadenylation element (CPE), of the AU-rich consensus sequence $U_4A_{1–2}U$ (Stebbins-Boaz et al. 1996; de Moor and Richter 1997). In *Xenopus* oocytes, CPE-mediated translational control depends on the phosphorylation status of CPEB1: hypo-phosphorylated CPEB1 acts as a translational repressor, while hyper-phosphorylated CPEB1 acts as a translational activator, thereby CPEB1 represents a molecular switch of CPE-containing maternal mRNAs (Richter 2007). In *C. elegans*, the CPEB family is comprised of four proteins of yet undefined RNA-binding capacity, CPB-1 to CPB-4. As *C. elegans* 3′UTRs are in general quite AU-rich it is unclear how many mRNAs would serve as specific CPB targets, and if CPBs serve as a translational regulatory molecular switch. In contrast to the other RNA regulators discussed in this section, the mRNAs of most CPB proteins are either expressed abundantly or even exclusively in male germ cells (Luitjens et al. 2000). Consistent with this CPB-4/FOG-1 is essential for germ cells to adopt the male fate in hermaphrodites (Barton and Kimble 1990). Moreover, FOG-1 promotes germ cell proliferation in a dose-dependent manner (Thompson et al. 2005). CPB-1 aids meiotic progression of spermatogenic germ cells (Luitjens et al. 2000). Similar to its homologs in *Xenopus* and *Drosophila*, CPB-3 functions in early oogenesis, preventing excessive physiological germline apoptosis (Boag et al. 2005). Additional functions may include the regulation of the sperm-to-oocyte switch and the mitosis-to-meiosis decision (Hasegawa et al. 2006). No roles of CPB-2 have yet been reported.

8.4.2 Expression and Activity Domains of RNA Regulators

The overall organization of the adult hermaphrodite germ line is perfectly suited for correlating protein expression levels with germ cell fates. While mRNA and protein gradients can be analyzed in a spatially stretched out distal to proximal axis, this arrangement reflects in reality a gradient of high temporal resolution, with undifferentiated mitotic germ cells near the distal end and fully differentiated gametes at the proximal end. This cell biological advantage compensates for the downside of being unable to isolate developmentally staged germ cells for biochemical experiments. For ease of description we refer to the mitotic region and the early stages of meiotic prophase I (leptotene, zygotene, and pachytene) as the distal part of the adult germ line. Germ cells undergoing diplotene are confined to the loop region, a morphological hallmark of the germline tube. The proximal part of the germ line contains differentiated germ cells, which in the case of oocytes are in diakinesis or in the case of sperm have completed the meiotic divisions.

Little data on de novo RNA synthesis activity of germ cells is available (Sheth et al. 2010; Schisa et al. 2001; Starck 1977; Starck et al. 1983). However, a comparison of the steady-state levels of mRNAs, deduced from the in situ hybridizations of many specific mRNAs generated by several laboratories, reveals the following general picture. Bulk transcriptional activity of the adult female germ line appears to be confined to the distal arm, especially to the more proximal mitotic region and early stages of prophase I (Schisa et al. 2001; Starck 1977), giving rise to essentially two prevalent mRNA expression patterns, a ubiquitous

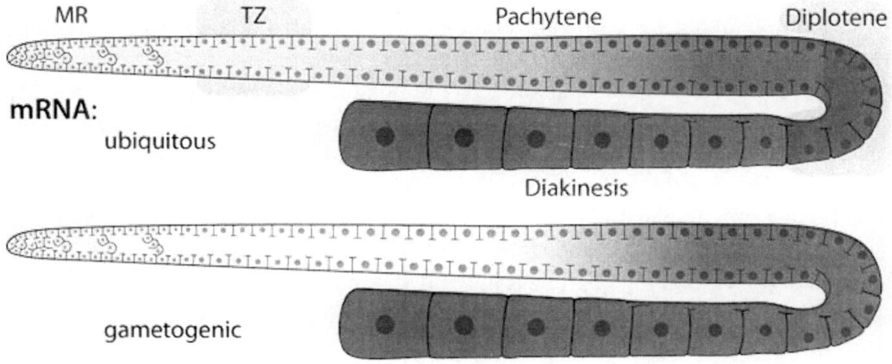

Fig. 8.2 Distribution of mRNAs in the adult germ line. The relative mRNA abundance in female germ cells is illustrated in magenta as observed by in situ hybridization experiments. A ubiquitous and a gametogenic expression pattern are depicted. Distal is top left and differentiated oocytes are most proximal. Previously made sperm is not shown. MR, mitotic region. For further details, see text, Sect. 8.4.2

and a gametogenic pattern, each sharing the common feature of abundant mRNA accumulation in oocytes (Fig. 8.2). Ubiquitously expressed mRNAs are present at lower levels in the more distal germ line when compared to their amounts in oocytes. Often the level increases before the loop region, which may correlate with an additional transcriptional burst in the late pachytene stage. By contrast, the gametogenic expression pattern is exclusively dominated by transcriptional activity in the pachytene region with no detectable mRNA in earlier developmental stages. In both cases, mRNAs are stockpiled in oocytes as maternal load for the early embryo, suggesting that such mRNAs are either important for early embryogenesis or they indirectly escape mRNA decay, as RNA degradation mechanisms may not be active in growing oocytes and their clearance is initiated upon fertilization (Seydoux and Fire 1994). A similar situation is present in the male germ line; mRNA is either produced in the mitotic region and/or in a second wave during late pachytene (Klass et al. 1982). However, little to no mRNA remains detectable in mature sperm, presumably due to exclusion of most cytoplasmic components in the last maturation steps of spermatogenesis or due to active mRNA degradation in the late stages of meiosis. Nevertheless, some RNAs may be in sperm and paternally donated to the zygote, like in other organisms (Bourc'his and Voinnet 2010).

In strong contrast to these simple and generic mRNA expression patterns, the derived protein expression patterns are far more diverse and complex, exemplifying the importance of translational control. A direct correlation between RNA and protein levels rarely exists. A better correlation is observed between protein amounts and their activities, although additional layers of regulation can occur. In addition, all protein patterns are presumably further shaped by the intrinsic stability of the encoded protein. However, little is known yet about regulated protein degradation during postembryonic germline development.

General and specific post-transcriptional RNA regulators have distinct protein expression patterns, often nicely paralleling their requirements for germline development (Table 8.4). While general RNA regulators are expressed in most germ cells, giving rise to a ubiquitous protein expression pattern across the germ line, specific RNA regulators can be selectively expressed, occupying distinct germline territories. This is especially true for RNA-binding proteins. In the case of translational repressors, five distinct protein expression patterns are prevalent in the adult hermaphrodite (Types A–D and U, Fig. 8.3): The PUF proteins FBF-1, FBF-2, and PUF-8 are predominantly expressed in the distal most part of the germ line, corresponding to the mitotic region and the initial stages of meiosis (Type A) (Crittenden et al. 2002; Lamont et al. 2004; Ariz et al. 2009). The expression pattern of the maxi-KH protein GLD-1 covers a larger region of the distal part extending to the germline loop, overlapping with the mitotic region and the early meiotic stages until diplotene (Type B) (Jones et al. 1996). MEX-3 expression is inverse to Type B, being abundant in the mitotic region, very low in early prophase, but abundant again in late prophase (Type C) (Mootz et al. 2004; Ciosk et al. 2006). PUF-5 and the zinc finger proteins OMA-1 and OMA-2 are restricted to meiotic stages beyond the germline loop, corresponding to the last two meiotic prophase stages, diplotene and diakinesis (Type D) (Detwiler et al. 2001; Lublin and Evans 2007).

NOS-3, a presumed translational co-repressor of FBF, is an example of a ubiquitously expressed germline protein (Type U) (Kraemer et al. 1999; Arur et al. 2011). Nevertheless, its temporal activity during female germline development is modulated at the post-translational level by MAP kinase (MPK-1) (Arur et al. 2011). Non-phosporylated NOS-3 is restricted to the distal part of the germ line, corresponding to a Type A pattern, whereas phosporylated NOS-3 (phospho-NOS-3) accumulates during pachytene and persists until diakinesis, similar to a Type D pattern (Fig. 8.3). The resulting overlap between non-phosporylated NOS-3 and FBF has functional consequences with respect to controlling the expression of the sex determining protein FEM-3 (Arur et al. 2011). At the molecular level, the interaction of FBF and NOS-3 is sensitive to the phosphorylation status of NOS-3, as FBF binds in vitro with higher affinity to non-phospho-NOS-3 than to phospho-NOS-3, suggesting that MPK-1-phosphorylated NOS-3 can no longer engage in the translational repression of *fem-3* (Arur et al. 2011). Although no other example of activity is currently available that demonstrates post-translational RNP-activity changes, it is conceivable that this type of regulation is prevalent. Especially, MPK-1 may appear as a master regulatory kinase for controlling the activity of RNA regulators as it is ubiquitously expressed and its activated form is abundant prior to the loop region and in the proximal part of the germ line, underscoring its many roles during germline development (Lee et al. 2007b; Arur et al. 2009, 2011; Lee et al. 2007a). Other kinases and different post-translational modifications likely also add to regulate mRNP activities.

Subtle deviations from the five dominant expression patterns exist and the relative amounts of a given RNA regulator may vary in detail among distinct germ cell stages. This may reflect dose-dependent requirements of germ cell fate regulation whereby protein amounts correlate directly with the activity of an RNA regulator.

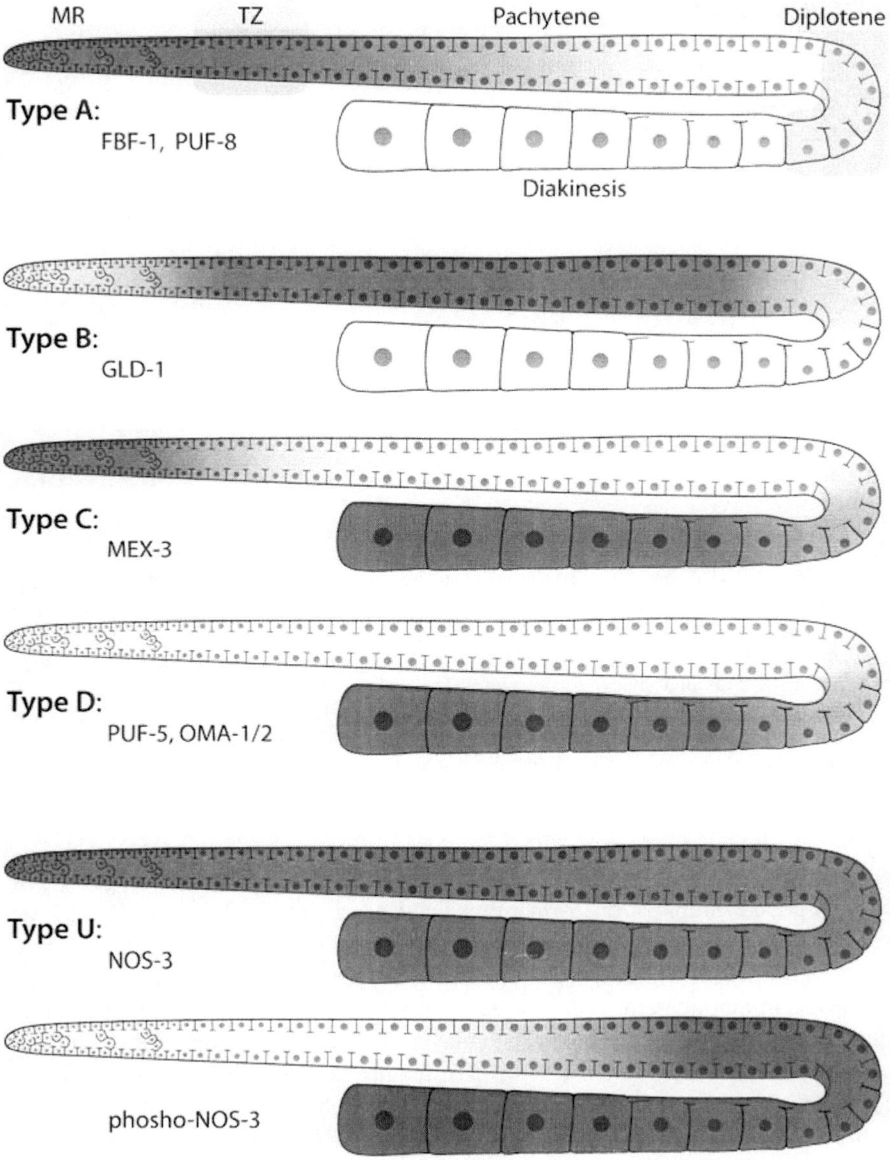

Fig. 8.3 Predominant distribution of translational repressors in the adult germ line. The relative protein abundance in female germ cells is illustrated in red. Orientation and labels as in Fig. 8.2. Five prevailing expression patterns can be distinguished: Type A—mainly restricted to the mitotic region, B—mainly restricted to very early prophase and the pachytene region, C—is an example of a complex expression pattern, which demonstrates that also combinations of Type A and Type D are possible, D—mainly limited to developing oocytes in diplotene and diakinesis, U—ubiquitous expression, which can be further limited by post-translational modifications to restrict protein activity such as phosphorylation (phospho). Deviations and other combination of these categories are possible. The sharpness of the boundaries needs to be adjusted for each RNA regulator in detail

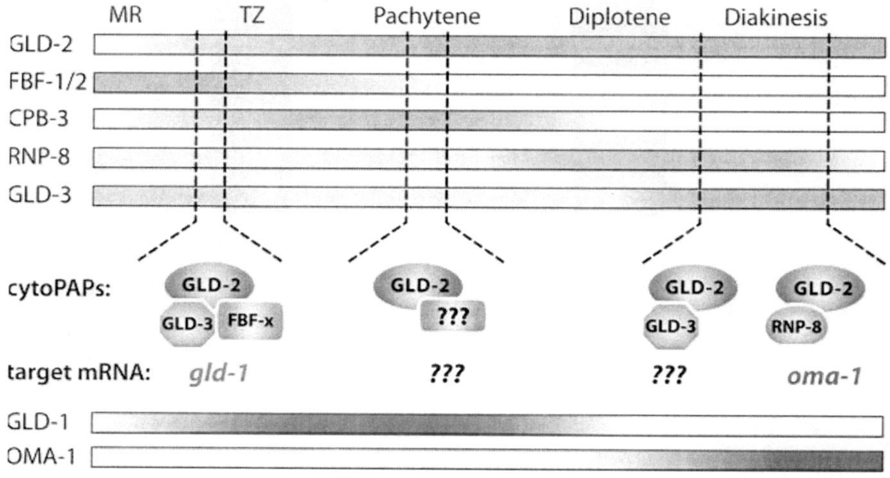

Fig. 8.4 Different translational activator complexes dominate specific regions in the adult germ line. The relative protein distribution across the female germ line for GLD-2 cytoPAP components, FBF-1, FBF-2, RNP-8, and GLD-3 is depicted in shades of orange. The corresponding developmental germ cell stage is indicated. Based on protein abundance an assembly of GLD-2/GLD-3/FBF-x is likely to be dominant in the mitotic region (MR) and very early meiosis (TZ, transition zone), and of GLD-2/GLD-3 and GLD-2/RNP-8 during diakinesis. Examples of established target mRNAs of the individual cytoPAPs are given. Translational activator complexes are contrasted with the expression domain of GLD-1 and OMA-1, two translational repressors. *gld-1* mRNA is translationally activated by GLD-2 cytoPAP complexes. *oma-1* mRNA is a target of GLD-1 repression and GLD-2 activation. For further details, see Sects. 8.4.3 and 8.5.3

Alternatively, the expression of the RNA regulator follows indirectly germ cell fates. A good example of the former type is the expression profile of GLD-1/STAR in the distal most part of the female germ line (Fig. 8.4), where a two-step increase of GLD-1 expression is observed: GLD-1 levels rise from virtually undetectable to low levels in mitotic germ cells closest to the transition zone. GLD-1 expression is further boosted to its highest expression level in the transition zone and plateaus in pachytene (Type B, Fig. 8.3). This level difference is critical: low levels of GLD-1 correlate with its known role in meiotic entry and high GLD-1 levels are consistent with its essential role in female meiotic progression. By contrast, raising the usually low amounts of GLD-1 during the development of the mitotic region, all distal germ cells enter meiosis at the expense of further mitotic activity (Crittenden et al. 2002; Hansen et al. 2004).

An example of a Type A variation is FBF-1, as its protein expression in the mitotic region is not uniform. FBF-1 abundance is increased in the more centrally positioned mitotic cells compared to the distal and proximal flanking germ cells (Crittenden et al. 2002). Conversely, FBF-2 may be even more enriched in the distal-most germ cells of the mitotic region and less prevalent in the proximal part of the mitotic region (Lamont et al. 2004). The differences between both expression patterns were attributed to a positive transcriptional response of *fbf-2* mRNA expression, stimulated by the *glp-1*/Notch signaling pathway (Lamont et al. 2004). Their different expression levels were also correlated with a distinct influence of each FBF protein on the size of the mitotic region (Lamont et al. 2004).

The protein expression patterns of many translational activators resemble those of the translational repressors (Table 8.4 and Fig. 8.4): DAZ-1, a conserved RRM-containing protein, follows a type A pattern, CPB-3 fits a type B pattern, and the RRM protein RNP-8 fits an extended modified type D pattern, as it is abundant in pachytene (Hasegawa et al. 2006; Maruyama et al. 2005; Kim et al. 2009a). The multi-KH domain protein GLD-3 expression is similar to a type C pattern (Eckmann et al. 2002). GLD-3, and RNP-8 are protein interactors of the ubiquitously expressed cytoPAP GLD-2 (Kim et al. 2009a; Wang et al. 2002) (Fig. 8.4). Further, GLD-3, RNP-8, and FBF stimulate GLD-2 poly(A) polymerase activity in vitro (Wang et al. 2002; Kim et al. 2009a). Thus, given their spatially distinct expression patterns it is likely that different cytoPAP complexes are formed, activating translation of specific mRNA targets at specific germline stages (Fig. 8.4). For example, it is proposed that distally expressed FBF and GLD-3 may control GLD-2 activity for promoting meiotic entry (Eckmann et al. 2004). A GLD-2/GLD-3 complex may be important for meiotic progression. GLD-3 and RNP-8 may control GLD-2 activity separately to promote the sperm or oocyte fate, respectively (Kim et al. 2009a), but together they may regulate maternal mRNA storage during oocyte maturation (Kim et al. 2010). A peculiarity of the GLD-2 expression pattern is its low abundance in the distal-most mitotic region, which may reflect a lower demand for GLD-2 in mitotic cells and/or that mitotic cells regulate GLD-2 in a specific manner. Certainly high levels of GLD-2 correlate with its essential roles in meiotic progression. GLD-4 and GLS-1, two components of the second poly(A) polymerase complex, are ubiquitously expressed (Schmid et al. 2009; Rybarska et al. 2009).

Interestingly, some RNA regulators also display sexually dimorphic expression patterns, suggesting fundamental differences in the requirement of the individual factors for male and female gametogenesis. An obvious example is the translational repressor GLD-1. Low levels of GLD-1 promote entry into meiosis and high levels promote female meiotic progression. Consistent with these functions, GLD-1 is weakly expressed in pre-meiotic male germ cells and remains low during very early stages of meiosis (Jones et al. 1996). Further, meiosis commitment defects are restricted to female *gld-1* mutant germ lines (Francis et al. 1995a). A second example is the protein expression profile of the RNA regulator GLD-3. Its Type C pattern appears inverse to the Type B pattern of GLD-1 in females (Eckmann et al. 2004) (Fig. 8.4). In the male germ line, GLD-3 accumulates steadily during all stages of male gametogenesis and maintains highest expression levels in metaphase of meiosis I and II (Eckmann et al. 2002). The lack of an apparent down-regulation of GLD-3 in male pachytene compared to female pachytene remains unclear but an artificial elevation of GLD-3 in female pachytene causes germ cells to arrest in meiosis (Jedamzik and Eckmann, unpublished results). Therefore, this sexual dimorphic expression pattern reflects presumably dose-dependent differences for RNA regulators controlling male and female meiotic progression.

8.4.3 Systems Biology of RNA Regulatory Networks

A remarkably and almost universal feature of germ cells is their overt reluctance to employ transcription factor networks to generate gene expression patterns, unlike somatic cells. Especially in *C. elegans* female germ cells, gene regulation appears rather permissively instructed at the DNA level. Transgenic reporter analysis of a dozen germline-enriched genes demonstrated that female protein expression patterns can be recapitulated to a large extent with the gene's cognate 3′UTR, while male germ cell differentiation genes rely largely on promoter-mediated gene regulation (Merritt et al. 2008). These experiments underscore the regulatory power of 3′UTRs and highlight the fact that mRNAs can encode all the information necessary to regulate their expression (Kuersten and Goodwin 2003). Within mRNAs, multiple protein-binding and regulatory sites (e.g., for miRNAs) together facilitate the formation of distinct mRNPs. Here, different RNP protein components may act on a single mRNA and single mRNP protein components may act on multiple mRNA species. The combinatorial coordination of *cis*-acting sites and *trans*-acting factors provide the structural framework of RNA regulatory networks that coordinate gene expression of germ cells. This is further facilitated by the extreme modularity of the protein–RNA interactions, which also provide the basis for evolutionary rewiring within the network to adapt cell fate decisions to developmental or environmental changes.

RNA regulatory networks are built of many different modules. Each module is composed of three integral components, which organize themselves into mRNP units. It is important to keep in mind that not all mRNAs of a given gene are regulated in the same way within the germ line. Rather, mRNPs are flexible operational entities that exchange their constituents, even within cellular territories. The organizational principle of an mRNPs may be viewed as three "layers" on top of the nucleating mRNA itself. At the heart of an mRNP unit is the RNA selector protein, which recognizes a *cis*-regulatory site and thereby flags the RNA target for regulation. The RNA selector represents the physical link between the mRNA and the mRNP, and may act in certain instances even as a seed for mRNP formation. The second principal mRNP components are mRNA-associated proteins that primarily assist the formation of larger and stable mRNP units, by having RNA affinity themselves and/or binding directly to the RNA selector. This second layer of proteins represents regulators of RNA selector capacity by further narrowing the RNA target group and provides the basis for the functional output of the mRNP (i.e., translational repression/activation or RNA stability). Further, they form an extended signaling input platform that integrates developmental controls into the formation, disassembly or remodeling of mRNPs. An example of this class includes RNA helicases. The third layer consists of mRNP-associated components that either enforce or change the activity of the mRNP. In this category belong RNA-modifying enzymes, such as deadenylases or poly(A) polymerases. Nevertheless, it is possible that the functions of two components are combined into a single molecule, like in the case of the RNA selector FBF. In addition to its established role as a translational repressor, FBF may also have the capacity to act as a translational activator, based on the composition of the mRNP unit. Data on *gld-1* mRNA regulation suggest that it forms mRNPs that require FBF for their translational activation and repression (Suh et al. 2009). Hence, FBF-mediated regulatory functions appear to be dictated by the type of mRNP that is assembled on the mRNA target. This concept may even apply for many mRNPs that use RNA selectors for dual functions. Therefore, such RNA selectors are better designated as translational effectors.

The establishment of RNA regulatory networks in the germ line is strictly correlated with the germ cell fate and the developmental stage of the germ cells. Integral to the network is that mRNP protein components are themselves subject to translational control, forming a strongly connected web of regulators with sharp boundaries of expression territories throughout the germ line. Especially the central nodes of the network, the RNA selectors, are precisely controlled. As discussed in Sect. 8.4.2, the expression domains of FBF, GLD-1 and PUF-5 establish an interdependent regulatory system reaching from the distal to the proximal part of the germ line. FBF limits GLD-1 expression (Crittenden et al. 2002), whereas GLD-1 limits OMA-1/2 (Lee and Schedl 2004) and PUF-5 (Lee and Schedl 2001) expression (Fig. 8.4). The initiation of this sequential negative repression cascade is further enforced by negative auto-regulation and cross-regulation, as observed for FBF-1 and FBF-2 (Lamont et al. 2004). These small circuits may be in place to speed up the response time and reduce the cell-to-cell variation in protein levels that are due to fluctuations in their production rate.

A major aspect of RNA regulatory networks in the germ line is the high redundancy of its modules that govern germ cell fate decisions and the strong enforcement level once the decisions have been made. The two best examples are the sperm-to-oocyte decision and meiotic entry, which are both regulated by PUF proteins. Combinations of two out of three different PUF proteins, FBF-1, FBF-2, and PUF-8, translationally repress multiple players of the sex determination pathway, achieving a tight suppression of sperm promoting factors at several different levels of the gene hierarchy (see Zanetti and Puoti 2012). A minimum of two parallel-acting pathways ensures meiotic entry. Two genetically independently acting translational repressors (NOS-3 and GLD-1) form a negative feedback loop to repress mitotic genes, i.e., GLD-1 translationally represses Notch/*glp-1* mRNA (Marin and Evans 2003) and cyclin E/*cye-1* mRNA (Biedermann et al. 2009). Concomitantly, the two translational activators, GLD-2 and GLD-3, enforce spatially regulated meiotic entry by presumably activating yet unknown meiotic genes (Eckmann et al. 2004). The bimodality of this cell fate switch is further assisted by FBF, which translationally represses GLD-1 (Crittenden et al. 2002), cyclin-dependent kinase inhibitor (CKI-1) accumulation (Kalchhauser et al. 2011), and the ectopic expression of meiotic proteins, such as the synaptonemal complex components HIM-3, HTP-1, SYP-1, and SYP-2 (Merritt and Seydoux 2010) (Table 8.5). Once female germ cells have entered meiosis, abundant GLD-1 levels ensure meiotic progression. GLD-1 accumulation is promoted by the redundant action of the two distinct poly(A) polymerases GLD-2 and GLD-4/GLS-1 (Suh et al. 2006; Schmid et al. 2009). This tight connectivity of the RNA regulatory network nodes and the enormous redundancy of the system allowed initially the identification of the key RNA selectors, as they are central to the network's gene expression capacity. Lastly, it becomes apparent that different mRNA species associated with defined RNA-binding protein components appear to encode functionally related proteins, leading to coordinated gene expression patterns.

An obstacle in elucidating the wiring of the RNA regulatory system is that simple forward genetic approaches have reached their limits in identifying mRNA targets of the network. Consequently, much more focused genetic screens are required (Schumacher et al. 2005) and biochemical approaches combined with modern molecular detection systems need to be pursued (Kim et al. 2010; Kershner and Kimble 2010; Wright et al. 2011). Although this is not limited to *C. elegans*, as most RNA-binding selector proteins in all species will have multiple targets, a few target mRNAs were identified in such screens by mutations that clustered in 3′UTRs. These gain-of-function mutations removed a key translational repressor site in *tra-2* (Goodwin et al. 1993) and *fem*-3 (Ahringer and Kimble 1991), which encode two essential sex-determination pathway components (see Zanetti and Puoti 2012). A fortuitous combination of the dose sensitive nature of the sex-determination pathway and an easy to score morphological phenotype enabled their discovery. Further, both genes appear to represent an Achilles' heels in a cell fate decision that presumably evolved recently and therefore is less complex. Together with the dose sensitivity of some RNA selectors and RNA-associated components, these examples also underscore that a quantitative assessment of the system is key in understanding the networks wiring. Although new technologies are available to tackle the RNA-target repertoire of RNA selectors, quantifications of the amounts necessary to build functional distinct mRNPs, its protein components and the number and strength of diverse *cis*-regulatory elements, remain a challenge for the future in understanding germline gene regulation.

8.5 Regulation of Specific Germ Cell Fate Decisions Via mRNP Activities

The emerging concepts of the detailed molecular mechanisms of translational control in the *C. elegans* germ line are derived from the studies on mRNAs that encode key germ cell fate determinants. The first mRNAs identified, *tra-2* and *fem-3* (Ahringer and Kimble 1991; Goodwin et al. 1993), turned out to be controlled by two distinct RNA regulatory machineries and became paradigms of GLD-1- and FBF-mediated translational repression (Zhang et al. 1997; Jan et al. 1999). Recently, a few additional targets were added to this list of well-studied mRNAs, which include *gld-1* and *glp-1* mRNA. Many more mRNA targets are currently being discovered in genome-wide studies using large-scale RNP immunopurifications, coupled to microarray discovery or next-generation RNA-sequencing techniques. Here, we will focus on a few select examples that will serve as paradigms of translational control mechanisms. Other mRNAs subject to translational control with defined and validated *cis*-regulatory elements are listed in Table 8.5.

8.5.1 Multidimensional Translational Control of tra-2 mRNA

The sex determining gene *tra-2* is required for female cell fates (see Zanetti and Puoti 2012, Chap. 3). As *tra-2* activity must be lower than *fem-3* activity during spermatogenesis in hermaphrodites, *tra-2* mRNA is translationally repressed in a GLD-1-containing mRNP (Jan et al. 1999). Gain-of-function 3′UTR mutations in *tra-2* affect one or both of two direct repeat elements (DREs), which contain RNA-binding sites for the maxi-KH domain protein GLD-1 (Goodwin et al. 1993; Jan et al. 1999). At least two functional GLD-1 binding motifs (GBMs) are present in the *tra-2* 3′UTR, which are bound by GLD-1 from worm extracts (Clifford et al. 2000). A comparison of wild-type and GBM-lacking 3′UTR sequences shows that the poly(A) tail of wild-type *tra-2* mRNA is shorter than that of the non-repressed mRNA (Jan et al. 1997). Also, reporter RNA constructs that carry GBMs are rapidly deadenylated in *Xenopus* embryos (Thompson et al. 2000). Taken together, this suggests that GLD-1 may repress *tra-2* mRNAs via a poly(A) tail-mediated mechanism, e.g., by recruiting mRNA deadenylases. However, experimental tests of this mechanistic aspect of *tra-2* mRNA regulation in *C. elegans* germ cells have not yet been performed.

Certainly, additional factors are needed for *tra-2* translational regulation. Animals heterozygote for loss-of-function mutations in the RNA-dependent DEAD-box RNA helicase LAF-1 are sperm-to-oocyte switch defective and contain higher levels of TRA-2 protein (Goodwin et al. 1997). As the isolated mutations presumably affect LAF-1's ATPase activity or RNA-binding capacity, it has been suggested that LAF-1-mediated remodeling of GLD-1-containing mRNPs may be important for efficient *tra-2* mRNA translational repression. In addition, GLD-1/*tra-2* mRNPs contain FOG-2, a special type of F-box protein (Clifford et al. 2000). Although the *tra-2* 3′UTR can recruit FOG-2 and GLD-1 proteins from worm extracts, the binding of GLD-1 to *tra-2* mRNA is not dependent on FOG-2. Rather, FOG-2 may represent a unique co-factor that influences GLD-1-mediated translational repression, via defined protein interaction sites between FOG-2 and GLD-1. The true molecular function of both RNA-associated components remains unclear.

A second mechanism contributes to *tra-2* mRNA regulation, which illustrates the tight interplay of mRNP history with mRNA fates. This mechanism involves TRA-1, a conserved zinc finger transcription factor. TRA-1 binds to a sequence element in the *tra-2* 3′UTR adjacent to the GBMs and facilitates the nuclear export of *tra-2* mRNA (Graves et al. 1999). The export of TRA-1 protein and *tra-2* mRNA is interdependent and a deletion of the TRA-1-binding element in *tra-2* mRNA results in the nuclear accumulation of both TRA-1 protein and *tra-2* mRNA (Segal et al. 2001). Importantly, *tra-2* mRNA is not exported via the canonical RNA export pathway of polyadenylated mRNAs, but rather uses an alternative pathway mediated by NXF-2, REF-1, and REF-2, which facilitates a more efficient translation regulation (Kuersten et al. 2004). Other conserved general RNA regulators, such as members of the exon junction complex affect the sperm-to-oocyte switch and likely regulate components of the sex determination pathway (Li et al. 2000). Together this suggests that the nuclear history of mRNA may directly influence its cytoplasmic fate.

8.5.2 *Reiterated Translational Control of glp-1 mRNA During Development*

glp-1 is important for maintaining proliferation in the mitotic region and its mRNA is of the ubiquitous type (Fig. 8.2, see Hansen and Schedl 2012). In addition to its post-embryonic germline function, *glp-1* is also essential for anterior cell fates in the early embryo (Austin and Kimble 1987; Priess et al. 1987). GLP-1 protein expression is therefore tightly controlled. *glp-1* mRNA is translationally repressed during multiple stages of germ cell development via different RNA selectors. In undifferentiated oocytes during early meiosis *glp-1* mRNA is responsive to GLD-1-mediated repression (Marin and Evans 2003), as it carries several GBMs in its 3′UTR (Wright et al. 2011). In oocytes, where GLD-1 is absent, members of the PUF-family, PUF-5 and PUF-6/7, suppress efficient GLP-1 protein synthesis (Lublin and Evans 2007). During early stages of embryogenesis, *glp-1* mRNA is subject to POS-1-mediated repression, a maternally provided zinc finger-containing protein (Ogura et al. 2003). This illustrates nicely that an mRNA can be repressed by different translational regulators at different points during germ cell development. However, detailed mechanistic insight into how *glp-1* mRNA is translationally repressed is still lacking.

8.5.3 *Distinct Poly(A)-Tail Length Control Mechanisms in PUF-Mediated mRNA Regulation*

fem-3 mRNA is a target of FBF-mediated translational repression (Zhang et al. 1997). FEM-3 protein expression must be lowered after initial spermatogenesis in hermaphrodites to facilitate oogenesis. The RNA selector FBF and the FBF-associated co-repressor NOS-3 translationally repress ubiquitously expressed *fem-3* mRNA, possibly in the pre-meiotic germ cells (Arur et al. 2011). One FBF-binding element (FBE) is present in the *fem-3* 3′UTR (Zhang et al. 1997), which was initially genetically defined as mutations that conferred a *fem-3* gain-of-function phenotype (Barton et al. 1987). Comparative Northern blot analysis of such *fem-3* mutations with wild-type clearly demonstrated that *fem-3(gf)* mRNAs possess a longer poly(A) tail (Ahringer and Kimble 1991). Together this suggests that FBF-mediated translational regulation involves poly(A)-tail shortening of *fem-3* mRNA to inhibit FEM-3 accumulation in the mitotic region. As mentioned in Sect. 8.4.2, the co-repressor function of NOS-3 depends on its phosphorylation status.

A similar picture of poly(A)tail-mediated translational regulation emerged with *gld-1* mRNA, an FBF target mRNA that contains two FBEs in its 3′UTR (Crittenden et al. 2002). Intriguingly, the *gld-1(oz10)* allele is a deletion affecting the *gld-1* 3′UTR. It eliminates both FBEs and confers also a semi-dominant gain-of-function sperm-to-oocyte phenotype (Francis et al. 1995a; Jones and Schedl 1995). In agreement with this, more GLD-1 protein is expressed in the mitotic region, where FBF is active (Jones et al. 1996). Unfortunately, the interpretation of this result is hampered by a second-site mutation in the GLD-1 protein that causes an amino acid change of unknown consequence (Jones and Schedl 1995). Consistent with FBE-mediated translational repression, the elimination of FBF-1 causes precocious GLD-1 protein synthesis in mitotic germ cells (Crittenden et al. 2002). In vitro studies show that FBF-bound synthetic polyadenylated RNA is sensitive to a Pop2p-containing deadenylase complex purified from yeast (Suh et al. 2009). Compatible with a general ability of PUF proteins to physically associate with Pop2p-type deadenylases, FBF binds CCF-1 in vitro (Suh et al. 2009). Although a formal test of the CCF-1–FBF interaction in vivo is lacking, it is reasonable to envision that FBF-mediated translational repression causes a shortening of the poly(A) tail via the recruitment of a CCF-1/CCR-4/LET-711 complex.

Although evidence for a worm CCF-1/CCR-4/LET-711 deadenylase complex is currently lacking (see Sect. 8.3.2), it is interesting to note that *ccr-4* activity is linked to *gld-1* mRNA stability. This has been observed in the *gld-2 gld-4* double mutant, which lacks polyadenylation activity of GLD-2 and GLD-4 cytoPAPs (Schmid et al. 2009). Both GLD-type cytoPAPs are required for efficient GLD-1 protein synthesis after FBF-mediated translational repression. In the absence of both cytoPAPs, a destabilization of *gld-1* mRNA is observed, which can be prevented either by elimination of FBF or CCR-4; however, no protein accumulation of GLD-1 occurred (Schmid et al. 2009). Hence, translational activation is assumed to be a consequence of poly(A)-tail elongation of *gld-1* mRNA. Consistent with this view is that *gld-1* mRNA poly(A)-tail length is shortened in the absence of GLD-2 (Suh et al. 2006).

In summary, the emerging picture from these two examples is that a single FBE may require FBF and co-repressors for efficient translational repression. Multiple FBEs may recruit more efficiently FBF molecules to the mRNP and depending on the mRNP unit formed, FBF elicits deadenylation of the target mRNA or participates in its polyadenylation-mediated translational activation.

8.5.4 RNA Regulatory Networks in the Male Germ Line

The *C. elegans* XO male germ line produces sperm only. While most RNA regulators are shared between male and female germ cells, some are unique to either sex, largely correlating with the two different gametogenesis programs. Like their female counterparts, spermatogenic cells utilize almost the exact same set of RNA regulatory factors to achieve the mitosis-to-meiosis decision and meiotic progression. However, the RNA regulatory network is not wired identically, for example, GLD-1 is not crucial for male meiotic progression (Francis et al. 1995a). Although the male germ line does not promote the switch to the oocyte fate, RNA regulators such as FOG-1 and GLD-3 are important for initiating and maintaining the sperm fate, respectively (Eckmann et al. 2002; Thompson et al. 2005). To maintain meiotic progression, CPB-1 (Luitjens et al. 2000), PUF-8 (Ariz et al. 2009) and the activity of both cyto-PAPs, GLD-2 (Kadyk and Kimble 1998) and GLD-4 (Schmid et al. 2009), is essential, suggesting that poly(A)-tail metabolism plays also a central role in the male germ line. Furthermore, 5′cap-mediated regulation has been documented to assist male gametogenesis (Amiri et al. 2001). However, much less is known about the individual RNA regulatory components in the male, representing fertile ground for future research.

8.6 Concluding Remarks

The evolutionary conservation of the many germline RNA regulators across metazoans highlights the importance of post-transcriptional gene expression control. Research done on the *C. elegans* germ line allowed us to start grasping its magnitude and complexity in a developmental setting. However, our understanding of translational control at a systems level is far from complete and will require new tools. In vitro reconstitution assays need to be to setup to define the contribution of mRNP components in their respective environment. Also, in vitro translation assays to measure mRNP activities have to be developed. Our current knowledge of individual mRNPs, their composition and the interplay between different mRNPs has to be clearly expanded. Once individual mRNP units are defined and the *cis*-regulatory code of mRNAs is deciphered, a fully integrated view of all mRNP units into a developmentally changing RNA regulatory network will remain a last big challenge.

References

Ahringer J, Kimble J (1991) Control of the sperm-oocyte switch in *Caenorhabditis elegans* hermaphrodites by the fem-3 3′ untranslated region. Nature 349(6307):346–348. doi:10.1038/349346a0

Amiri A, Keiper BD, Kawasaki I, Fan Y, Kohara Y, Rhoads RE, Strome S (2001) An isoform of eIF4E is a component of germ granules and is required for spermatogenesis in *C. elegans*. Development 128(20):3899–3912

Ariz M, Mainpal R, Subramaniam K (2009) *C. elegans* RNA-binding proteins PUF-8 and MEX-3 function redundantly to promote germline stem cell mitosis. Dev Biol 326(2):295–304. doi:10.1016/j.ydbio.2008.11.024

Arur S, Ohmachi M, Nayak S, Hayes M, Miranda A, Hay A, Golden A, Schedl T (2009) Multiple ERK substrates execute single biological processes in *Caenorhabditis elegans* germ-line development. Proc Natl Acad Sci USA 106(12):4776–4781. doi:10.1073/pnas.0812285106

Arur S, Ohmachi M, Berkseth M, Nayak S, Hansen D, Zarkower D, Schedl T (2011) MPK-1 ERK controls membrane organization in *C. elegans* oogenesis via a sex-determination module. Dev Cell 20(5):677–688. doi:10.1016/j.devcel.2011.04.009

Audhya A, Hyndman F, McLeod IX, Maddox AS, Yates JR 3rd, Desai A, Oegema K (2005) A complex containing the Sm protein CAR-1 and the RNA helicase CGH-1 is required for embryonic cytokinesis in *Caenorhabditis elegans*. J Cell Biol 171(2):267–279. doi:10.1083/jcb.200506124

Austin J, Kimble J (1987) glp-1 is required in the germ line for regulation of the decision between mitosis and meiosis in *C. elegans*. Cell 51(4):589–599

Bachorik JL, Kimble J (2005) Redundant control of the *Caenorhabditis elegans* sperm/oocyte switch by PUF-8 and FBF-1, two distinct PUF RNA-binding proteins. Proc Natl Acad Sci USA 102(31):10893–10897. doi:10.1073/pnas.0504593102

Baer BW, Kornberg RD (1980) Repeating structure of cytoplasmic poly(A)-ribonucleoprotein. Proc Natl Acad Sci USA 77(4):1890–1892

Baer BW, Kornberg RD (1983) The protein responsible for the repeating structure of cytoplasmic poly(A)-ribonucleoprotein. J Cell Biol 96(3):717–721

Barnard DC, Cao Q, Richter JD (2005) Differential phosphorylation controls Maskin association with eukaryotic translation initiation factor 4E and localization on the mitotic apparatus. Mol Cell Biol 25(17):7605–7615. doi:10.1128/MCB.25.17.7605-7615.2005

Barton MK, Kimble J (1990) fog-1, a regulatory gene required for specification of spermatogenesis in the germ line of *Caenorhabditis elegans*. Genetics 125(1):29–39

Barton MK, Schedl TB, Kimble J (1987) Gain-of-function mutations of fem-3, a sex-determination gene in *Caenorhabditis elegans*. Genetics 115(1):107–119

Bernstein D, Hook B, Hajarnavis A, Opperman L, Wickens M (2005) Binding specificity and mRNA targets of a *C. elegans* PUF protein, FBF-1. RNA 11(4):447–458. doi:10.1261/rna.7255805

Biedermann B, Wright J, Senften M, Kalchhauser I, Sarathy G, Lee MH, Ciosk R (2009) Translational repression of cyclin E prevents precocious mitosis and embryonic gene activation during *C. elegans* meiosis. Dev Cell 17(3):355–364. doi:10.1016/j.devcel.2009.08.003

Boag PR, Nakamura A, Blackwell TK (2005) A conserved RNA-protein complex component involved in physiological germline apoptosis regulation in *C. elegans*. Development 132(22):4975–4986. doi:10.1242/dev.02060

Boag PR, Atalay A, Robida S, Reinke V, Blackwell TK (2008) Protection of specific maternal messenger RNAs by the P body protein CGH-1 (Dhh1/RCK) during *Caenorhabditis elegans* oogenesis. J Cell Biol 182(3):543–557. doi:10.1083/jcb.200801183

Bourc'his D, Voinnet O (2010) A small-RNA perspective on gametogenesis, fertilization, and early zygotic development. Science 330(6004):617–622. doi:10.1126/science.1194776

Bowerman B, Kurz T (2006) Degrade to create: developmental requirements for ubiquitin-mediated proteolysis during early *C. elegans* embryogenesis. Development 133(5):773–784. doi:10.1242/dev.02276

Buchan JR, Yoon JH, Parker R (2011) Stress-specific composition, assembly and kinetics of stress granules in Saccharomyces cerevisiae. J Cell Sci 124(Pt 2):228–239. doi:10.1242/jcs.078444

Buchet-Poyau K, Courchet J, Le Hir H, Seraphin B, Scoazec JY, Duret L, Domon-Dell C, Freund JN, Billaud M (2007) Identification and characterization of human Mex-3 proteins, a novel family of evolutionarily conserved RNA-binding proteins differentially localized to processing bodies. Nucleic Acids Res 35(4):1289–1300. doi:10.1093/nar/gkm016

Ceron J, Rual JF, Chandra A, Dupuy D, Vidal M, van den Heuvel S (2007) Large-scale RNAi screens identify novel genes that interact with the *C. elegans* retinoblastoma pathway as well as splicing-related components with synMuv B activity. BMC Dev Biol 7:30. doi:10.1186/1471-213X-7-30

Chen T, Damaj BB, Herrera C, Lasko P, Richard S (1997) Self-association of the single-KH-domain family members Sam68, GRP33, GLD-1, and Qk1: role of the KH domain. Mol Cell Biol 17(10):5707–5718

Chu DS, Shakes DC (2012) Spermatogenesis. Advances in Experimental Medicine and Biology 757:171–203. (Chap. 7, this volume) Springer, New York

Ciosk R, DePalma M, Priess JR (2004) ATX-2, the *C. elegans* ortholog of ataxin 2, functions in translational regulation in the germline. Development 131(19):4831–4841. doi:10.1242/dev.01352

Ciosk R, DePalma M, Priess JR (2006) Translational regulators maintain totipotency in the *Caenorhabditis elegans* germline. Science 311(5762):851–853. doi:10.1126/science.1122491

Clark-Maguire S, Mains PE (1994a) Localization of the mei-1 gene product of *Caenorhaditis elegans*, a meiotic-specific spindle component. J Cell Biol 126(1):199–209

Clark-Maguire S, Mains PE (1994b) mei-1, a gene required for meiotic spindle formation in *Caenorhabditis elegans*, is a member of a family of ATPases. Genetics 136(2):533–546

Clifford R, Lee MH, Nayak S, Ohmachi M, Giorgini F, Schedl T (2000) FOG-2, a novel F-box containing protein, associates with the GLD-1 RNA binding protein and directs male sex determination in the *C. elegans* hermaphrodite germline. Development 127(24):5265–5276

Conine CC, Batista PJ, Gu W, Claycomb JM, Chaves DA, Shirayama M, Mello CC (2010) Argonautes ALG-3 and ALG-4 are required for spermatogenesis-specific 26 G-RNAs and thermotolerant sperm in *Caenorhabditis elegans*. Proc Natl Acad Sci USA 107(8):3588–3593. doi:10.1073/pnas.0911685107

Contreras V, Richardson MA, Hao E, Keiper BD (2008) Depletion of the cap-associated isoform of translation factor eIF4G induces germline apoptosis in *C. elegans*. Cell Death Differ 15(8):1232–1242. doi:10.1038/cdd.2008.46

Crittenden SL, Bernstein DS, Bachorik JL, Thompson BE, Gallegos M, Petcherski AG, Moulder G, Barstead R, Wickens M, Kimble J (2002) A conserved RNA-binding protein controls germline stem cells in *Caenorhabditis elegans*. Nature 417(6889):660–663. doi:10.1038/nature754

Curtis D, Treiber DK, Tao F, Zamore PD, Williamson JR, Lehmann R (1997) A CCHC metal-binding domain in Nanos is essential for translational regulation. EMBO J 16(4):834–843. doi:10.1093/emboj/16.4.834

de Moor CH, Richter JD (1997) The Mos pathway regulates cytoplasmic polyadenylation in Xenopus oocytes. Mol Cell Biol 17(11):6419–6426

DeBella LR, Hayashi A, Rose LS (2006) LET-711, the *Caenorhabditis elegans* NOT1 ortholog, is required for spindle positioning and regulation of microtubule length in embryos. Mol Biol Cell 17(11):4911–4924. doi:10.1091/mbc.E06-02-0107

Decker CJ, Parker R (1993) A turnover pathway for both stable and unstable mRNAs in yeast: evidence for a requirement for deadenylation. Genes Dev 7(8):1632–1643

Denli AM, Tops BB, Plasterk RH, Ketting RF, Hannon GJ (2004) Processing of primary microRNAs by the microprocessor complex. Nature 432(7014):231–235. doi:10.1038/nature03049

Detwiler MR, Reuben M, Li X, Rogers E, Lin R (2001) Two zinc finger proteins, OMA-1 and OMA-2, are redundantly required for oocyte maturation in *C. elegans*. Dev Cell 1(2):187–199

Dinkova TD, Keiper BD, Korneeva NL, Aamodt EJ, Rhoads RE (2005) Translation of a small subset of *Caenorhabditis elegans* mRNAs is dependent on a specific eukaryotic translation initiation factor 4E isoform. Mol Cell Biol 25(1):100–113. doi:10.1128/MCB.25.1.100-113.2005

Draper BW, Mello CC, Bowerman B, Hardin J, Priess JR (1996) MEX-3 is a KH domain protein that regulates blastomere identity in early *C. elegans* embryos. Cell 87(2):205–216

Eckmann CR, Kraemer B, Wickens M, Kimble J (2002) GLD-3, a bicaudal-C homolog that inhibits FBF to control germline sex determination in *C. elegans*. Dev Cell 3(5):697–710

Eckmann CR, Crittenden SL, Suh N, Kimble J (2004) GLD-3 and control of the mitosis/meiosis decision in the germline of *Caenorhabditis elegans*. Genetics 168(1):147–160. doi:10.1534/genetics.104.029264

Eckmann CR, Rammelt C, Wahle E (2011) Control of poly(A) tail length. Wiley Interdiscip Rev RNA 2(3):348–361. doi:10.1002/wrna.56

Edwards TA, Pyle SE, Wharton RP, Aggarwal AK (2001) Structure of Pumilio reveals similarity between RNA and peptide binding motifs. Cell 105(2):281–289

Fabian MR, Sonenberg N, Filipowicz W (2010) Regulation of mRNA translation and stability by microRNAs. Annu Rev Biochem 79:351–379. doi:10.1146/annurev-biochem-060308-103103

Francis R, Barton MK, Kimble J, Schedl T (1995a) gld-1, a tumor suppressor gene required for oocyte development in *Caenorhabditis elegans*. Genetics 139(2):579–606

Francis R, Maine E, Schedl T (1995b) Analysis of the multiple roles of gld-1 in germline development: interactions with the sex determination cascade and the glp-1 signaling pathway. Genetics 139(2):607–630

Furuichi Y, LaFiandra A, Shatkin AJ (1977) 5′-Terminal structure and mRNA stability. Nature 266(5599):235–239

Garneau NL, Wilusz J, Wilusz CJ (2007) The highways and byways of mRNA decay. Nat Rev Mol Cell Biol 8(2):113–126. doi:10.1038/nrm2104

Gebauer F, Hentze MW (2004) Molecular mechanisms of translational control. Nat Rev Mol Cell Biol 5(10):827–835. doi:10.1038/nrm1488

Goodwin EB, Okkema PG, Evans TC, Kimble J (1993) Translational regulation of tra-2 by its 3′ untranslated region controls sexual identity in *C. elegans*. Cell 75(2):329–339

Goodwin EB, Hofstra K, Hurney CA, Mango S, Kimble J (1997) A genetic pathway for regulation of tra-2 translation. Development 124(3):749–758

Graves LE, Segal S, Goodwin EB (1999) TRA-1 regulates the cellular distribution of the tra-2 mRNA in *C. elegans*. Nature 399(6738):802–805. doi:10.1038/21682

Grishok A, Pasquinelli AE, Conte D, Li N, Parrish S, Ha I, Baillie DL, Fire A, Ruvkun G, Mello CC (2001) Genes and mechanisms related to RNA interference regulate expression of the small temporal RNAs that control *C. elegans* developmental timing. Cell 106(1):23–34

Guven-Ozkan T, Robertson SM, Nishi Y, Lin R (2010) zif-1 translational repression defines a second, mutually exclusive OMA function in germline transcriptional repression. Development 137(20):3373–3382. doi:10.1242/dev.055327

Hake LE, Mendez R, Richter JD (1998) Specificity of RNA binding by CPEB: requirement for RNA recognition motifs and a novel zinc finger. Mol Cell Biol 18(2):685–693

Han T, Manoharan AP, Harkins TT, Bouffard P, Fitzpatrick C, Chu DS, Thierry-Mieg D, Thierry-Mieg J, Kim JK (2009) 26 G endo-siRNAs regulate spermatogenic and zygotic gene expression in *Caenorhabditis elegans*. Proc Natl Acad Sci USA 106(44):18674–18679. doi:10.1073/pnas.0906378106

Hanazawa M, Kawasaki I, Kunitomo H, Gengyo-Ando K, Bennett KL, Mitani S, Iino Y (2004) The *Caenorhabditis elegans* eukaryotic initiation factor 5A homologue, IFF-1, is required for germ cell proliferation, gametogenesis and localization of the P-granule component PGL-1. Mech Dev 121(3):213–224. doi:10.1016/j.mod.2004.02.001

Hansen D, Schedl T (2012) Stem cell proliferation versus meiotic fate decision in *C. elegans*. Advances in Experimental Medicine and Biology 757:71–99. (Chap. 4, this volume) Springer, New York

Hansen D, Wilson-Berry L, Dang T, Schedl T (2004) Control of the proliferation versus meiotic development decision in the *C. elegans* germline through regulation of GLD-1 protein accumulation. Development 131(1):93–104. doi:10.1242/dev.00916

Hasegawa E, Karashima T, Sumiyoshi E, Yamamoto M (2006) *C. elegans* CPB-3 interacts with DAZ-1 and functions in multiple steps of germline development. Dev Biol 295(2):689–699. doi:10.1016/j.ydbio.2006.04.002

Henderson MA, Cronland E, Dunkelbarger S, Contreras V, Strome S, Keiper BD (2009) A germline-specific isoform of eIF4E (IFE-1) is required for efficient translation of stored mRNAs and maturation of both oocytes and sperm. J Cell Sci 122(Pt 10):1529–1539. doi:10.1242/jcs.046771

Hunter CP, Kenyon C (1996) Spatial and temporal controls target pal-1 blastomere-specification activity to a single blastomere lineage in *C. elegans* embryos. Cell 87(2):217–226

Jackson RJ (2005) Alternative mechanisms of initiating translation of mammalian mRNAs. Biochem Soc Trans 33(Pt 6):1231–1241. doi:10.1042/BST20051231

Jadhav S, Rana M, Subramaniam K (2008) Multiple maternal proteins coordinate to restrict the translation of *C. elegans* nanos-2 to primordial germ cells. Development 135(10):1803–1812. doi:10.1242/dev.013656

Jan E, Yoon JW, Walterhouse D, Iannaccone P, Goodwin EB (1997) Conservation of the *C. elegans* tra-2 3′UTR translational control. EMBO J 16(20):6301–6313. doi:10.1093/emboj/16.20.6301

Jan E, Motzny CK, Graves LE, Goodwin EB (1999) The STAR protein, GLD-1, is a translational regulator of sexual identity in *Caenorhabditis elegans*. EMBO J 18(1):258–269. doi:10.1093/emboj/18.1.258

Jankowska-Anyszka M, Lamphear BJ, Aamodt EJ, Harrington T, Darzynkiewicz E, Stolarski R, Rhoads RE (1998) Multiple isoforms of eukaryotic protein synthesis initiation factor 4E in *Caenorhabditis elegans* can distinguish between mono- and trimethylated mRNA cap structures. J Biol Chem 273(17):10538–10542

Jones AR, Schedl T (1995) Mutations in gld-1, a female germ cell-specific tumor suppressor gene in *Caenorhabditis elegans*, affect a conserved domain also found in Src-associated protein Sam68. Genes Dev 9(12):1491–1504

Jones AR, Francis R, Schedl T (1996) GLD-1, a cytoplasmic protein essential for oocyte differentiation, shows stage- and sex-specific expression during *Caenorhabditis elegans* germline development. Dev Biol 180(1):165–183. doi:10.1006/dbio.1996.0293

Jud MC, Czerwinski MJ, Wood MP, Young RA, Gallo CM, Bickel JS, Petty EL, Mason JM, Little BA, Padilla PA, Schisa JA (2008) Large P body-like RNPs form in *C. elegans* oocytes in response to arrested ovulation, heat shock, osmotic stress, and anoxia and are regulated by the major sperm protein pathway. Dev Biol 318(1):38–51. doi:10.1016/j.ydbio.2008.02.059

Kadyk LC, Kimble J (1998) Genetic regulation of entry into meiosis in *Caenorhabditis elegans*. Development 125(10):1803–1813

Kadyk LC, Lambie EJ, Kimble J (1997) glp-3 is required for mitosis and meiosis in the *Caenorhabditis elegans* germ line. Genetics 145(1):111–121

Kalchhauser I, Farley BM, Pauli S, Ryder SP, Ciosk R (2011) FBF represses the Cip/Kip cell-cycle inhibitor CKI-2 to promote self-renewal of germline stem cells in *C. elegans*. EMBO J. doi:10.1038/emboj.2011.263

Kato M, de Lencastre A, Pincus Z, Slack FJ (2009) Dynamic expression of small non-coding RNAs, including novel microRNAs and piRNAs/21U-RNAs, during *Caenorhabditis elegans* development. Genome Biol 10(5):R54. doi:10.1186/gb-2009-10-5-r54

Kedersha N, Stoecklin G, Ayodele M, Yacono P, Lykke-Andersen J, Fritzler MJ, Scheuner D, Kaufman RJ, Golan DE, Anderson P (2005) Stress granules and processing bodies are dynamically linked sites of mRNP remodeling. J Cell Biol 169(6):871–884. doi:10.1083/jcb.200502088

Keiper BD, Lamphear BJ, Deshpande AM, Jankowska-Anyszka M, Aamodt EJ, Blumenthal T, Rhoads RE (2000) Functional characterization of five eIF4E isoforms in *Caenorhabditis elegans*. J Biol Chem 275(14):10590–10596

Kershner AM, Kimble J (2010) Genome-wide analysis of mRNA targets for *Caenorhabditis elegans* FBF, a conserved stem cell regulator. Proc Natl Acad Sci USA 107(8):3936–3941. doi:10.1073/pnas.1000495107

Kim KW, Nykamp K, Suh N, Bachorik JL, Wang L, Kimble J (2009a) Antagonism between GLD-2 binding partners controls gamete sex. Dev Cell 16(5):723–733. doi:10.1016/j.devcel.2009.04.002

Kim VN, Han J, Siomi MC (2009b) Biogenesis of small RNAs in animals. Nat Rev Mol Cell Biol 10(2):126–139. doi:10.1038/nrm2632

Kim KW, Wilson TL, Kimble J (2010) GLD-2/RNP-8 cytoplasmic poly(A) polymerase is a broad-spectrum regulator of the oogenesis program. Proc Natl Acad Sci USA 107(40):17445–17450. doi:10.1073/pnas.1012611107

Kim S, Spike CA, Greenstein D (2012) Control of oocyte growth and meiotic maturation in *C. elegans*. Advances in Experimental Medicine and Biology 757:277–320. (Chap. 10, this volume) Springer, New York

Klass M, Dow B, Herndon M (1982) Cell-specific transcriptional regulation of the major sperm protein in *Caenorhabditis elegans*. Dev Biol 93(1):152–164

Knight SW, Bass BL (2001) A role for the RNase III enzyme DCR-1 in RNA interference and germ line development in *Caenorhabditis elegans*. Science 293(5538):2269–2271. doi:10.1126/science.1062039

Ko S, Park JH, Lee AR, Kim E, Jiyoung K, Kawasaki I, Shim YH (2010) Two mutations in pab-1 encoding poly(A)-binding protein show similar defects in germline stem cell proliferation but different longevity in *C. elegans*. Mol Cells 30(2):167–172. doi:10.1007/s10059-010-0103-2

Kraemer B, Crittenden S, Gallegos M, Moulder G, Barstead R, Kimble J, Wickens M (1999) NANOS-3 and FBF proteins physically interact to control the sperm-oocyte switch in *Caenorhabditis elegans*. Curr Biol 9(18):1009–1018

Kuersten S, Goodwin EB (2003) The power of the 3′ UTR: translational control and development. Nat Rev Genet 4(8):626–637. doi:10.1038/nrg1125

Kuersten S, Segal SP, Verheyden J, LaMartina SM, Goodwin EB (2004) NXF-2, REF-1, and REF-2 affect the choice of nuclear export pathway for tra-2 mRNA in *C. elegans*. Mol Cell 14(5):599–610. doi:10.1016/j.molcel.2004.05.004

Lall S, Piano F, Davis RE (2005) *Caenorhabditis elegans* decapping proteins: localization and functional analysis of Dcp1, Dcp2, and DcpS during embryogenesis. Mol Biol Cell 16(12):5880–5890. doi:10.1091/mbc.E05-07-0622

Lamont LB, Crittenden SL, Bernstein D, Wickens M, Kimble J (2004) FBF-1 and FBF-2 regulate the size of the mitotic region in the *C. elegans* germline. Dev Cell 7(5):697–707. doi:10.1016/j.devcel.2004.09.013

Lasda EL, Blumenthal T (2011) Trans-splicing. Wiley Interdiscip Rev RNA 2(3):417–434. doi:10.1002/wrna.71

Lee MH, Schedl T (2001) Identification of in vivo mRNA targets of GLD-1, a maxi-KH motif containing protein required for *C. elegans* germ cell development. Genes Dev 15(18):2408–2420. doi:10.1101/gad.915901

Lee MH, Schedl T (2004) Translation repression by GLD-1 protects its mRNA targets from nonsense-mediated mRNA decay in *C. elegans*. Genes Dev 18(9):1047–1059. doi:10.1101/gad.1188404

Lee RC, Feinbaum RL, Ambros V (1993) The *C. elegans* heterochronic gene lin-4 encodes small RNAs with antisense complementarity to lin-14. Cell 75(5):843–854

Lee MH, Hook B, Lamont LB, Wickens M, Kimble J (2006) LIP-1 phosphatase controls the extent of germline proliferation in *Caenorhabditis elegans*. EMBO J 25(1):88–96. doi:10.1038/sj.emboj.7600901

Lee MH, Hook B, Pan G, Kershner AM, Merritt C, Seydoux G, Thomson JA, Wickens M, Kimble J (2007a) Conserved regulation of MAP kinase expression by PUF RNA-binding proteins. PLoS Genet 3(12):e233. doi:10.1371/journal.pgen.0030233

Lee MH, Ohmachi M, Arur S, Nayak S, Francis R, Church D, Lambie E, Schedl T (2007b) Multiple functions and dynamic activation of MPK-1 extracellular signal-regulated kinase signaling in *Caenorhabditis elegans* germline development. Genetics 177(4):2039–2062. doi:10.1534/genetics.107.081356

Li W, Boswell R, Wood WB (2000) mag-1, a homolog of Drosophila mago nashi, regulates hermaphrodite germ-line sex determination in *Caenorhabditis elegans*. Dev Biol 218(2):172–182. doi:10.1006/dbio.1999.9593

Li W, DeBella LR, Guven-Ozkan T, Lin R, Rose LS (2009) An eIF4E-binding protein regulates katanin protein levels in *C. elegans* embryos. J Cell Biol 187(1):33–42. doi:10.1083/jcb.200903003

Lu R, Yigit E, Li WX, Ding SW (2009) An RIG-I-Like RNA helicase mediates antiviral RNAi downstream of viral siRNA biogenesis in *Caenorhabditis elegans*. PLoS Pathog 5(2):e1000286. doi:10.1371/journal.ppat.1000286

Lublin AL, Evans TC (2007) The RNA-binding proteins PUF-5, PUF-6, and PUF-7 reveal multiple systems for maternal mRNA regulation during *C. elegans* oogenesis. Dev Biol 303(2):635–649. doi:10.1016/j.ydbio.2006.12.004

Lui DY, Colaiácovo MP (2012) Meiotic development in *C. elegans*. Advances in Experimental Medicine and Biology 757:133–170. (Chap. 6, this volume) Springer, New York

Luitjens C, Gallegos M, Kraemer B, Kimble J, Wickens M (2000) CPEB proteins control two key steps in spermatogenesis in *C. elegans*. Genes Dev 14(20):2596–2609

Maciejowski J, Ahn JH, Cipriani PG, Killian DJ, Chaudhary AL, Lee JI, Voutev R, Johnsen RC, Baillie DL, Gunsalus KC, Fitch DH, Hubbard EJ (2005) Autosomal genes of autosomal/X-linked duplicated gene pairs and germ-line proliferation in *Caenorhabditis elegans*. Genetics 169(4):1997–2011. doi:10.1534/genetics.104.040121

Mainpal R, Priti A, Subramaniam K (2011) PUF-8 suppresses the somatic transcription factor PAL-1 expression in *C. elegans* germline stem cells. Dev Biol 360(1):195–207. doi:10.1016/j.ydbio.2011.09.021

Marin VA, Evans TC (2003) Translational repression of a *C. elegans* Notch mRNA by the STAR/KH domain protein GLD-1. Development 130(12):2623–2632

Maruyama R, Endo S, Sugimoto A, Yamamoto M (2005) *Caenorhabditis elegans* DAZ-1 is expressed in proliferating germ cells and directs proper nuclear organization and cytoplasmic core formation during oogenesis. Dev Biol 277(1):142–154. doi:10.1016/j.ydbio.2004.08.053

Mathews MB, Sonenberg N, Hershey J (2007) Translational control in biology and medicine. Cold Spring Harbor Laboratory Press, Cold Spring Harbor

Mendez R, Richter JD (2001) Translational control by CPEB: a means to the end. Nat Rev Mol Cell Biol 2(7):521–529. doi:10.1038/35080081

Merritt C, Seydoux G (2010) The Puf RNA-binding proteins FBF-1 and FBF-2 inhibit the expression of synaptonemal complex proteins in germline stem cells. Development 137(11):1787–1798. doi:10.1242/dev.050799

Merritt C, Rasoloson D, Ko D, Seydoux G (2008) 3′ UTRs are the primary regulators of gene expression in the *C. elegans* germline. Curr Biol 18(19):1476–1482. doi:10.1016/j.cub.2008.08.013

Miyoshi H, Dwyer DS, Keiper BD, Jankowska-Anyszka M, Darzynkiewicz E, Rhoads RE (2002) Discrimination between mono- and trimethylated cap structures by two isoforms of *Caenorhabditis elegans* eIF4E. EMBO J 21(17):4680–4690

Molin L, Puisieux A (2005) *C. elegans* homologue of the Caf1 gene, which encodes a subunit of the CCR4-NOT complex, is essential for embryonic and larval development and for meiotic progression. Gene 358:73–81. doi:10.1016/j.gene.2005.05.023

Mootz D, Ho DM, Hunter CP (2004) The STAR/Maxi-KH domain protein GLD-1 mediates a developmental switch in the translational control of *C. elegans* PAL-1. Development 131(14):3263–3272. doi:10.1242/dev.01196

Munroe D, Jacobson A (1990) mRNA poly(A) tail, a 3′ enhancer of translational initiation. Mol Cell Biol 10(7):3441–3455

Nakamura M, Ando R, Nakazawa T, Yudazono T, Tsutsumi N, Hatanaka N, Ohgake T, Hanaoka F, Eki T (2007) Dicer-related drh-3 gene functions in germ-line development by maintenance of chromosomal integrity in *Caenorhabditis elegans*. Genes Cells 12(9):997–1010. doi:10.1111/j.1365-2443.2007.01111.x

Nakel K, Hartung SA, Bonneau F, Eckmann CR, Conti E (2010) Four KH domains of the *C. elegans* Bicaudal-C ortholog GLD-3 form a globular structural platform. RNA 16(11):2058–2067. doi:10.1261/rna.2315010

Nelson MR, Leidal AM, Smibert CA (2004) Drosophila Cup is an eIF4E-binding protein that functions in Smaug-mediated translational repression. EMBO J 23(1):150–159. doi:10.1038/sj.emboj.7600026

Nolde MJ, Saka N, Reinert KL, Slack FJ (2007) The *Caenorhabditis elegans* pumilio homolog, puf-9, is required for the 3′UTR-mediated repression of the let-7 microRNA target gene, hbl-1. Dev Biol 305(2):551–563. doi:10.1016/j.ydbio.2007.02.040

Nykamp K, Lee MH, Kimble J (2008) *C. elegans* La-related protein, LARP-1, localizes to germline P bodies and attenuates Ras-MAPK signaling during oogenesis. RNA 14(7):1378–1389. doi:10.1261/rna.1066008

Ogura K, Kishimoto N, Mitani S, Gengyo-Ando K, Kohara Y (2003) Translational control of maternal glp-1 mRNA by POS-1 and its interacting protein SPN-4 in *Caenorhabditis elegans*. Development 130(11):2495–2503

Opperman L, Hook B, DeFino M, Bernstein DS, Wickens M (2005) A single spacer nucleotide determines the specificities of two mRNA regulatory proteins. Nat Struct Mol Biol 12(11):945–951. doi:10.1038/nsmb1010

Otero LJ, Ashe MP, Sachs AB (1999) The yeast poly(A)-binding protein Pab1p stimulates in vitro poly(A)-dependent and cap-dependent translation by distinct mechanisms. EMBO J 18(11):3153–3163. doi:10.1093/emboj/18.11.3153

Pagano JM, Farley BM, Essien KI, Ryder SP (2009) RNA recognition by the embryonic cell fate determinant and germline totipotency factor MEX-3. Proc Natl Acad Sci USA 106(48):20252–20257. doi:10.1073/pnas.0907916106

Priess JR, Schnabel H, Schnabel R (1987) The glp-1 locus and cellular interactions in early *C. elegans* embryos. Cell 51(4):601–611

Reddy R, Singh R, Shimba S (1992) Methylated cap structures in eukaryotic RNAs: structure, synthesis and functions. Pharmacol Ther 54(3):249–267
Rhoads RE (2009) eIF4E: new family members, new binding partners, new roles. J Biol Chem 284(25):16711–16715. doi:10.1074/jbc.R900002200
Richter JD (2007) CPEB: a life in translation. Trends Biochem Sci 32(6):279–285. doi:10.1016/j.tibs.2007.04.004
Robertson S, Lin R (2012) The oocyte-to-embryo transition. Advances in Experimental Medicine and Biology, 757:351–372. (Chap. 12, this volume) Springer, New York
Rybarska A, Harterink M, Jedamzik B, Kupinski AP, Schmid M, Eckmann CR (2009) GLS-1, a novel P granule component, modulates a network of conserved RNA regulators to influence germ cell fate decisions. PLoS Genet 5(5):e1000494. doi:10.1371/journal.pgen.1000494
Ryder SP, Frater LA, Abramovitz DL, Goodwin EB, Williamson JR (2004) RNA target specificity of the STAR/GSG domain post-transcriptional regulatory protein GLD-1. Nat Struct Mol Biol 11(1):20–28. doi:10.1038/nsmb706
Sachs A, Wahle E (1993) Poly(A) tail metabolism and function in eucaryotes. J Biol Chem 268(31):22955–22958
Sachs AB, Bond MW, Kornberg RD (1986) A single gene from yeast for both nuclear and cytoplasmic polyadenylate-binding proteins: domain structure and expression. Cell 45(6):827–835
Sachs AB, Davis RW, Kornberg RD (1987) A single domain of yeast poly(A)-binding protein is necessary and sufficient for RNA binding and cell viability. Mol Cell Biol 7(9):3268–3276
Schisa JA, Pitt JN, Priess JR (2001) Analysis of RNA associated with P granules in germ cells of *C. elegans* adults. Development 128(8):1287–1298
Schmid M, Kuchler B, Eckmann CR (2009) Two conserved regulatory cytoplasmic poly(A) polymerases, GLD-4 and GLD-2, regulate meiotic progression in *C. elegans*. Genes Dev 23(7):824–836. doi:10.1101/gad.494009
Schumacher B, Hanazawa M, Lee MH, Nayak S, Volkmann K, Hofmann ER, Hengartner M, Schedl T, Gartner A (2005) Translational repression of *C. elegans* p53 by GLD-1 regulates DNA damage-induced apoptosis. Cell 120(3):357–368. doi:10.1016/j.cell.2004.12.009
Segal SP, Graves LE, Verheyden J, Goodwin EB (2001) RNA-regulated TRA-1 nuclear export controls sexual fate. Dev Cell 1(4):539–551
Seydoux G, Fire A (1994) Soma-germline asymmetry in the distributions of embryonic RNAs in *Caenorhabditis elegans*. Development 120(10):2823–2834
She X, Xu X, Fedotov A, Kelly WG, Maine EM (2009) Regulation of heterochromatin assembly on unpaired chromosomes during *Caenorhabditis elegans* meiosis by components of a small RNA-mediated pathway. PLoS Genet 5(8):e1000624. doi:10.1371/journal.pgen.1000624
Sheth U, Pitt J, Dennis S, Priess JR (2010) Perinuclear P granules are the principal sites of mRNA export in adult *C. elegans* germ cells. Development 137(8):1305–1314. doi:10.1242/dev.044255
Shimotohno K, Kodama Y, Hashimoto J, Miura KI (1977) Importance of 5′-terminal blocking structure to stabilize mRNA in eukaryotic protein synthesis. Proc Natl Acad Sci USA 74(7):2734–2738
Simmer F, Moorman C, van der Linden AM, Kuijk E, van den Berghe PV, Kamath RS, Fraser AG, Ahringer J, Plasterk RH (2003) Genome-wide RNAi of *C. elegans* using the hypersensitive rrf-3 strain reveals novel gene functions. PLoS Biol 1(1):E12. doi:10.1371/journal.pbio.0000012
Song A, Labella S, Korneeva NL, Keiper BD, Aamodt EJ, Zetka M, Rhoads RE (2010) A *C. elegans* eIF4E-family member upregulates translation at elevated temperatures of mRNAs encoding MSH-5 and other meiotic crossover proteins. J Cell Sci 123(Pt 13):2228–2237. doi:10.1242/jcs.063107
Sonnichsen B, Koski LB, Walsh A, Marschall P, Neumann B, Brehm M, Alleaume AM, Artelt J, Bettencourt P, Cassin E, Hewitson M, Holz C, Khan M, Lazik S, Martin C, Nitzsche B, Ruer M, Stamford J, Winzi M, Heinkel R, Roder M, Finell J, Hantsch H, Jones SJ, Jones M, Piano F, Gunsalus KC, Oegema K, Gonczy P, Coulson A, Hyman AA, Echeverri CJ (2005) Full-genome RNAi profiling of early embryogenesis in *Caenorhabditis elegans*. Nature 434(7032):462–469. doi:10.1038/nature03353
Squirrell JM, Eggers ZT, Luedke N, Saari B, Grimson A, Lyons GE, Anderson P, White JG (2006) CAR-1, a protein that localizes with the mRNA decapping component DCAP-1, is required for

cytokinesis and ER organization in *Caenorhabditis elegans* embryos. Mol Biol Cell 17(1):336–344. doi:10.1091/mbc.E05-09-0874

Srayko M, Buster DW, Bazirgan OA, McNally FJ, Mains PE (2000) MEI-1/MEI-2 katanin-like microtubule severing activity is required for *Caenorhabditis elegans* meiosis. Genes Dev 14(9):1072–1084

Starck J (1977) Autoradiographic study of RNA-synthesis in Caenorhabditis-elegans (Bergerac Variety) oogenesis. Biol Cellulaire 30(2):181

Starck J, Gibert MA, Brun J, Bosch C (1983) Ribosomal-RNA synthesis and processing during oogenesis of the free living nematode Caenorhabditis-elegans. Comp Biochem Phys B 75(4):575–580

Stebbins-Boaz B, Hake LE, Richter JD (1996) CPEB controls the cytoplasmic polyadenylation of cyclin, Cdk2 and c-mos mRNAs and is necessary for oocyte maturation in Xenopus. EMBO J 15(10):2582–2592

Stebbins-Boaz B, Cao Q, de Moor CH, Mendez R, Richter JD (1999) Maskin is a CPEB-associated factor that transiently interacts with elF-4E. Mol Cell 4(6):1017–1027

Strome S (2005) Specification of the germ line. WormBook:1–10. doi:10.1895/wormbook.1.9.1

Stumpf CR, Kimble J, Wickens M (2008) A *Caenorhabditis elegans* PUF protein family with distinct RNA binding specificity. RNA 14(8):1550–1557. doi:10.1261/rna.1095908

Subramaniam K, Seydoux G (2003) Dedifferentiation of primary spermatocytes into germ cell tumors in *C. elegans* lacking the pumilio-like protein PUF-8. Curr Biol 13(2):134–139

Suh N, Jedamzik B, Eckmann CR, Wickens M, Kimble J (2006) The GLD-2 poly(A) polymerase activates gld-1 mRNA in the *Caenorhabditis elegans* germ line. Proc Natl Acad Sci USA 103(41):15108–15112. doi:10.1073/pnas.0607050103

Suh N, Crittenden SL, Goldstrohm A, Hook B, Thompson B, Wickens M, Kimble J (2009) FBF and its dual control of gld-1 expression in the *Caenorhabditis elegans* germline. Genetics 181(4):1249–1260. doi:10.1534/genetics.108.099440

Tabara H, Yigit E, Siomi H, Mello CC (2002) The dsRNA binding protein RDE-4 interacts with RDE-1, DCR-1, and a DExH-box helicase to direct RNAi in *C. elegans*. Cell 109(7):861–871

Tarun SZ Jr, Sachs AB (1995) A common function for mRNA 5′ and 3′ ends in translation initiation in yeast. Genes Dev 9(23):2997–3007

Thompson SR, Goodwin EB, Wickens M (2000) Rapid deadenylation and Poly(A)-dependent translational repression mediated by the *Caenorhabditis elegans* tra-2 3′ untranslated region in Xenopus embryos. Mol Cell Biol 20(6):2129–2137

Thompson BE, Bernstein DS, Bachorik JL, Petcherski AG, Wickens M, Kimble J (2005) Dose-dependent control of proliferation and sperm specification by FOG-1/CPEB. Development 132(15):3471–3481. doi:10.1242/dev.01921

Tursun B, Patel T, Kratsios P, Hobert O (2011) Direct conversion of *C. elegans* germ cells into specific neuron types. Science 331(6015):304–308. doi:10.1126/science.1199082

Valverde R, Edwards L, Regan L (2008) Structure and function of KH domains. FEBS J 275(11): 2712–2726. doi:10.1111/j.1742-4658.2008.06411.x

Vernet C, Artzt K (1997) STAR, a gene family involved in signal transduction and activation of RNA. Trends Genet 13(12):479–484

Wang JT, Seydoux S (2012) Germ cell specification. Advances in Experimental Medicine and Biology 757:17– 39. (Chap. 2, this volume) Springer, New York

Wang L, Eckmann CR, Kadyk LC, Wickens M, Kimble J (2002) A regulatory cytoplasmic poly(A) polymerase in *Caenorhabditis elegans*. Nature 419(6904):312–316. doi:10.1038/nature01039

Wang X, Zhao Y, Wong K, Ehlers P, Kohara Y, Jones SJ, Marra MA, Holt RA, Moerman DG, Hansen D (2009a) Identification of genes expressed in the hermaphrodite germ line of *C. elegans* using SAGE. BMC Genomics 10:213. doi:10.1186/1471-2164-10-213

Wang Y, Opperman L, Wickens M, Hall TM (2009b) Structural basis for specific recognition of multiple mRNA targets by a PUF regulatory protein. Proc Natl Acad Sci USA 106(48):20186–20191. doi:10.1073/pnas.0812076106

Wells SE, Hillner PE, Vale RD, Sachs AB (1998) Circularization of mRNA by eukaryotic translation initiation factors. Mol Cell 2(1):135–140

Wickens M, Bernstein DS, Kimble J, Parker R (2002) A PUF family portrait: 3′UTR regulation as a way of life. Trends Genet 18(3):150–157

Wightman B, Ha I, Ruvkun G (1993) Posttranscriptional regulation of the heterochronic gene lin-14 by lin-4 mediates temporal pattern formation in *C. elegans*. Cell 75(5):855–862

Wilkie GS, Dickson KS, Gray NK (2003) Regulation of mRNA translation by 5′- and 3′-UTR-binding factors. Trends Biochem Sci 28(4):182–188

Wright JE, Gaidatzis D, Senften M, Farley BM, Westhof E, Ryder SP, Ciosk R (2011) A quantitative RNA code for mRNA target selection by the germline fate determinant GLD-1. EMBO J 30(3):533–545. doi:10.1038/emboj.2010.334

Yu X, Vought VE, Conradt B, Maine EM (2006) Eukaryotic translation initiation factor 5B activity regulates larval growth rate and germline development in *Caenorhabditis elegans*. Genesis 44(9):412–418. doi:10.1002/dvg.20232

Zanetti S, Puoti A (2012) Sex determination in the *C. elegans* germline. Advances in Experimental Medicine and Biology 757:41–69. (Chap. 3, this volume) Springer, New York

Zanin E, Pacquelet A, Scheckel C, Ciosk R, Gotta M (2010) LARP-1 promotes oogenesis by repressing fem-3 in the *C. elegans* germline. J Cell Sci 123(Pt 16):2717–2724. doi:10.1242/jcs.066761

Zhang B, Gallegos M, Puoti A, Durkin E, Fields S, Kimble J, Wickens MP (1997) A conserved RNA-binding protein that regulates sexual fates in the *C. elegans* hermaphrodite germ line. Nature 390(6659):477–484. doi:10.1038/37297

Chapter 9
Germ Cell Apoptosis and DNA Damage Responses

Aymeric Bailly and Anton Gartner

Abstract In the past 12 years, since the first description of *C. elegans* germ cell apoptosis, this area of research rapidly expanded. It became evident that multiple genetic pathways lead to the apoptotic demise of germ cells. We are only beginning to understand how these pathways that all require the CED-9/Bcl-2, Apaf-1/CED-4 and CED-3 caspase core apoptosis components are regulated. Physiological apoptosis, which likely accounts for the elimination of more than 50% of all germ cells, even in unperturbed conditions, is likely to be required to maintain tissue homeostasis. The best-studied pathways lead to DNA damage-induced germ cell apoptosis in response to a variety of genotoxic stimuli. This apoptosis appears to be regulated similar to DNA damage-induced apoptosis in the mouse germ line and converges on p53 family transcription factors. DNA damage response pathways not only lead to apoptosis induction, but also directly affect DNA repair, and a transient cell cycle arrest of mitotic germ cells. Finally, distinct pathways activate germ cell apoptosis in response to defects in meiotic recombination and meiotic chromosome pairing.

Keywords Apoptosis • DNA repair • DNA damage response • Meiosis checkpoints • Recombination

A. Bailly
Centre de Recherche de Biochimie Macromoleculaire, CNRS UMR 5237,
1919 Route de Mende, 34293 Montpellier Cedex 5, France

A. Gartner (✉)
Wellcome Trust Centre for Gene Regulation and Expression, The University of Dundee,
Dow Street, Dundee DD1 5EH, UK
e-mail: a.gartner@dundee.ac.uk

T. Schedl (ed.), *Germ Cell Development in C. elegans*, Advances in Experimental Medicine and Biology 757, DOI 10.1007/978-1-4614-4015-4_9,

9.1 Introduction

Apoptosis is a highly conserved mechanism to eliminate superfluous or damaged cells in multicellular eukaryotes. Apoptosis is remarkably prevalent in germ cells. It is estimated that out of ~7,000,000 germ cells occurring in the ovaries of 5 months old human embryos only ~300,000 remain at an age of 7 years, and less than a thousand oocytes are left in the years before ovarian senescence; for review, see Tilly (2001) and Morita and Tilly (1999). It thus appears that apoptosis is especially important during germ cell development. This is likely to be the case because germ cells and somatic cells are subjected to distinct selective pressures. Somatic cells are optimized to contribute to the fitness for one generation, whereas germ cells are optimized both for the maintenance of eternal proliferative potential and for maintaining pluripotency to allow for differentiation after fertilization. Surveillance mechanisms are thus likely to be extremely important in the germline to ensure quality control. Germ cell apoptosis might also be required to maintain germline tissue homeostasis. This is especially challenging given the enormous size of mature oocytes compared to other germ cells. The dichotomy between somatic cells and germ cells is most extreme in the worm life cycle. Somatic cells are entirely post-mitotic and invariantly derive from the fertilized zygote by a very limited number of cell cycles, rather than being continuously replenished by adult somatic stem cells as in most other animals. Thus, while somatic tissues may tolerate high levels of DNA damage or other environmental insults, germline-specific checkpoints might have evolved to guard germ cells from acquiring deleterious mutations that could be passed from one generation to the next.

We summarize our knowledge of germ cell apoptosis in the *C. elegans* germline. Since the first report on worm germ cell apoptosis in 1999 it became clear that multiple pathways are required (Gumienny et al. 1999). A basal level of germ cell apoptosis, termed physiological germ cell apoptosis, occurs in the absence of external stimuli and appears to be linked to maintaining tissue homeostasis (Gartner et al. 2000). Furthermore, germ cells are culled by genetically distinct checkpoints that monitor genome integrity, meiotic recombination, and meiotic chromosome synapsis (Gartner et al. 2000; Bhalla and Dernburg 2005). At present we know most about DNA damage response pathways leading to apoptosis. Finally, stresses such as starvation, heat shock, bacterial infection, or a high glucose diet also trigger excessive germ cell apoptosis (Salinas et al. 2006; Angelo and Van Gilst 2009; Aballay and Ausubel 2001; Choi 2011).

9.2 The Core Apoptosis Pathway in *C. elegans*

Seminal studies on *C. elegans* apoptosis occurring during the invariant development of the worm were important for our general understanding of apoptosis induction. EGL-1, CED-4, and CED-3 are conserved proteins required for the vast majority of the 131 cell deaths that occur during the invariant development; for review, see

Conradt and Xue (2005) and Conradt (2009). The sole worm Bcl2 family member, CED-9, is required to protect healthy cells form undergoing apoptosis. *ced-9* loss-of-function mutants lead to the induction of apoptosis in cells that are normally destined to survive, while a *ced-9* gain-of-function mutant was shown to be defective for apoptosis induction. In the worm, apoptosis in somatic tissues is thought to typically involve the transcriptional up-regulation of the BH3-only domain protein EGL-1 in cells destined to die. Combinations of transcriptional regulators have been found to be required for the transcriptional induction of *egl-1* in specific cells destined to die. EGL-1 antagonizes CED-9, which in turn antagonizes the activity of the pro-apoptotic CED-4, the Apaf1 homolog. According to the current model, mitochondrial CED-9 directly binds pro-apoptotic CED-4 to inhibit apoptosis induction. Upon transcriptional induction, EGL-1 binding to CED-9 leads to the disruption of the CED-9/CED-4 complex and to the release of CED-4 from mitochondria. This event leads to the oligomerization of CED-4 and to the ultimate activation of apoptosis induction through CED-3 caspase activation. However, more recent evidence suggests that there might not be a direct interaction of CED-9 and CED-4 in cells destined to survive, implying a more complex mode of regulation of CED-9 and CED-4 (Pourkarimi et al. 2012). In the worm, in contrast to germ cell apoptosis, somatic apoptosis appears to be part of a developmental process and neither occurs in adult animals nor in response to environmental stimuli.

9.3 *C. elegans* Germ Cell Apoptosis

When *C. elegans* germ cell apoptosis was first described in 1999 (Gumienny et al. 1999), it was observed that a steady-state level of ~1–5 apoptotic corpses occur, the number of which increases with the age of the worm. This apoptosis, termed physiological germ cell death apoptosis, occurs in the absence of any stress and was estimated to eliminate approximately 50% of all germ cells. Indeed this number, initially measured by the number of corpses observed at a given time in relation to the number of oocytes that are laid, might be even higher. More recent estimates of germ cell proliferation rates indicate that approximately 20 germ cells are produced every hour, while only ~3 oocytes are laid (Fox et al. 2011). Apoptosis only occurs in the female and not in the male germline and is thus only present in hermaphrodites (Gumienny et al. 1999; Gartner et al. 2000; Jaramillo-Lambert et al. 2010). Mutational analysis indicated that a female germline is required for apoptosis, while the gender of somatic tissues does not affect germ cell apoptosis competency. Within the germline, apoptosis is restricted to the late stage of pachytene cells, residing just distal of the loop of the germline, close to entering the diplotene stage of meiosis. During late-stage pachytene, meiotic cells have completed meiotic recombination and the majority of meiosis-induced DNA double-strand breaks are repaired (see Lui and Colaiácovo 2012, Chap. 6). With the exception of late stage, nearly mature oocytes, germ "cells" are not fully enclosed by a plasma membrane, leaving an opening to a shared cytoplasmic space at the centre of the germline called the rachis. The first morphologically visible step of germ cell apoptosis is the complete cellularization of the affected cell (Gumienny et al. 1999). Such "cellularization" is likely to be needed to restrict full caspase activation to single apoptotic cells, thus preventing the spreading of apoptosis to neighbouring cells. However, this mechanism might not be perfect, as apoptotic germ cell corpses tend to occur in clusters. While the molecular mechanism of apoptotic corpse engulfment is the same as for somatic corpses, only sheath cells but not germ cells are able to engulf apoptotic cells. At present we do not know why apoptosis is restricted to late pachytene stage germ cells.

Due to the invariant nature of the somatic development of the worm, the extent of developmental apoptosis can be precisely measured (Schwartz 2007). As the germline is a dynamic tissue, the interpretation of apoptosis steady-state levels is more complicated. Factors such as the rates of apoptotic corpse-engulfment and germ cell proliferation, the precise age of the worm, and even the temperature have to be considered when reporting the number of germ cell corpses, ideally done as time course experiments. Also the possibility that bacterial contamination might trigger germ cell apoptosis has to be considered (Aballay and Ausubel 2001; Anton Gartner, unpublished observation). Methods for detecting germ cell corpses and for labelling them with GFP markers or specific dyes have been described (Gartner et al. 2004, 2008). Nevertheless, it is generally not possible to discriminate between small differences in the frequency of germ cell apoptosis.

9.4 Physiological Germ Cell Apoptosis

Germ cell apoptosis generally requires CED-3 and CED-4 and excessive apoptosis occurs in the absence of CED-9 (Fig. 9.1), but there are important genetic differences between physiological germ cell apoptosis and somatic cell death. Physiological cell death does not require EGL-1, and a *ced-9* gain-of-function mutant (*n1950*), which blocks somatic apoptosis, does not have any effect on the germline (Gumienny et al. 1999). Therefore, physiological germ cell death involves the induction of the core apoptosis pathway by an unknown mechanism. Indeed, it is not even clear whether physiological germ cell apoptosis is triggered by single or by multiple distinct genetic pathways. Several transcriptional regulators are known to generally affect the level of germ cell apoptosis by regulating the expression of core apoptosis proteins in the germline (Fig. 9.1). This regulation affects physiological apoptosis, and DNA damage-induced apoptosis, and thus likely affects all forms of germ cell death. *C. elegans* Pax 2/5/8 transcription factors, mutations of which cause excessive apoptosis, are required for efficient *ced-9* transcription in the germline and in somatic tissues (Park et al. 2006). Conversely, the *C. elegans* RB retinoblastoma susceptibility protein ortholog, LIN-35, mutants of which confer reduced germ cell apoptosis, is required for normal levels of germ cell apoptosis, likely by regulating the level of *ced-9* transcription (Schertel and Conradt 2007). In addition, worm subunits of the E2F transcription factor promote physiological germ cell apoptosis by increasing *ced-3* and *ced-4* transcription in germ cells (Schertel and Conradt 2007).

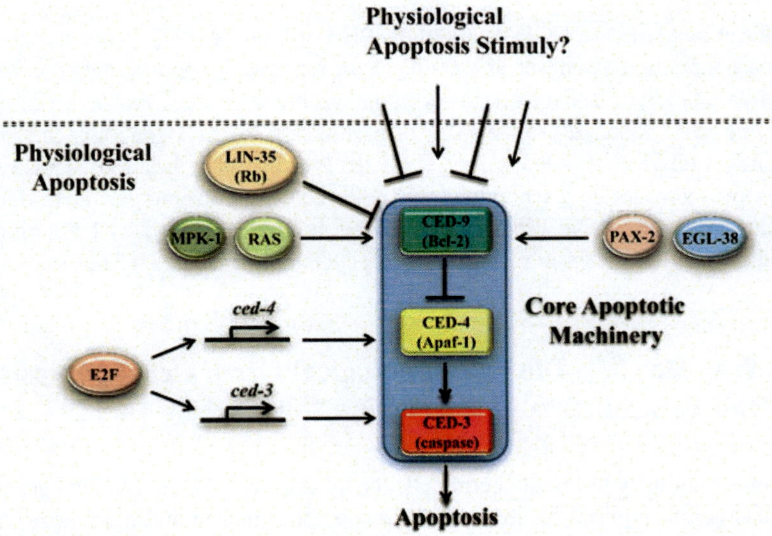

Fig. 9.1 Physiological germ cell apoptosis

Finally, there is evidence that MAP kinase signalling might be required for physiological germ cell death (Gumienny et al. 1999). These studies are complicated by the fact that MAP kinase signalling is required for the transition to the proximal pachytene stage where apoptosis occurs (Lee et al. 2007). Thus, the absence of physiological germ cell apoptosis in MAP kinase mutants might be related to a developmental defect rather than the lack of apoptosis induction. However, excessive apoptosis occurs when mutants of the MAP kinase pathway are combined with a *ced-9* loss-of-function mutant (Gumienny et al. 1999). Therefore, MAP kinase signalling defective germlines show apoptosis competency and the lack of apoptosis in MAP kinase signalling defective mutants might be a genuine lack of physiological germ cell death. Consistent with this view, excessive apoptosis occurs when MAP kinase signalling is elevated in mutants of the *lip-1* MAP kinase phosphatase (at high temperature) or the *gla-3* RNA binding protein which can affect MAP kinase signalling (Kritikou et al. 2006). However, there is also emerging evidence that MAP kinase signalling might be needed to repress germ cell apoptosis. Germ cell nuclei accumulate in the temperature sensitive *mpk-1(ga111)* mutant, 24 h after shifting to the restrictive temperature of 25°C, and RNAi inactivation of putative MAP kinase substrates increases germ cell apoptosis observed in *mpk-1(ga111)* mutants but not in wild-type worms (Arur et al. 2011). Given that MAP kinase signalling is regulating multiple aspects of germline differentiation, it still remains to be established whether this pathway directly affects physiological germ cell apoptosis, or directs germ cells to stages that are more or less susceptible to pro-apoptotic signals (also see below for ionizing irradiation-induced apoptosis).

Inactivation of multiple proteins increases the level of germ cell apoptosis and it was argued that various RNA binding proteins, such as the RGG-box protein CAR-1, the DEAD box helicase CGH-1, and the CPEB ortholog CPB-3, prevent excessive apoptosis induction (Boag et al. 2005). However, inactivation of these and other genes by mutation or RNAi leads to additional germline defects, such as an extended pachytene zone, reduced progeny, full or partial sterility, or a generally misshapen germline (Audhya et al. 2005; Green et al. 2011; Squirrell et al. 2006). Thus excessive apoptosis might be a secondary consequence of pleiotropic germ cell defects. In many cases, excessive germ cell apoptosis might be caused by triggering the DNA damage checkpoint in response to cell cycle perturbation (see below).

9.5 Why Do Cells Die by Physiological Germ Cell Apoptosis and Evolutionary Conservation

Why do the majority of female germ cells die by apoptosis? Physiological germ cell apoptosis is not required for general oogenesis, but the size and number of viable oocytes can be reduced when apoptosis is blocked (Gumienny et al. 1999). Smaller oocytes and a reduction in oocyte viability are especially prominent in old apoptosis-defective germlines (Andux and Ellis 2008). Using female animals generated by mutants blocking sperm generation, reduced oocyte size and increased lethality has been correlated. Such females can be mated at different stages to generate fertilized embryos, and it could be shown that oocyte quality declines with the age of the mother. It will be interesting to determine whether oocyte survival is increased in long-lived worm mutants. It is unlikely that under normal conditions a large proportion of germ cells are culled because they are genetically compromised or because the meiotic recombination or synapsis checkpoints are engaged as apoptosis is only slightly or not reduced in mutants affected by those checkpoints (see below) (Andux and Ellis 2008). However, a slightly increased incidence of male progeny, indicative of meiotic problems, is observed in old animals (Luo et al. 2010). At present the most attractive model proposed in the initial *C. elegans* germ cell apoptosis paper is that apoptotic germ cells might act as nurse cells to provide nutrients and cytoplasmic material for a limited number of oocytes. The finding that the quality of oocytes and size is decreased in old apoptosis-defective mutants supports this model. This occurs even though egg laying rates in apoptosis-defective mutants are not increased, a finding that, albeit not discussed in the literature, indicates that the rate of germ cell proliferation is reduced to compensate for the lack of apoptosis. Similarly oocyte quality declines in mutants where corpse engulfment is largely defective, consistent with a model that recycling apoptotic germ cells is important for the generation of fully grown, healthy oocytes (Andux and Ellis 2008). The use of oocytes as nurse cells appears to be evolutionarily conserved. In the fruit fly, egg chambers contain 16 interconnected cells (for review, see Buszczak and Cooley 2000). All but one of these cells serve as nurse cells and nutrients are transported to the single surviving oocyte while the remaining cells succumb to apoptosis. In mammals, developing oocytes are also interconnected and occur in cysts (Pepling 2006; Pepling and Spradling 1998, 2001). At around birth two thirds of oocytes die by apoptosis and surviving oocytes form primordial follicles. Nevertheless, at least in *C. elegans,* physiological germ cell apoptosis might not solely be explained by dying germ cells acting as nurse cells. Consistent with the nurse cell hypothesis it has been observed that cytoplasmic streaming occurs in rachis, commencing from the mid-pachytene area to the last not yet fully cellularized oocyte (Wolke et al. 2007). However, the transport of nutrients also occurs from healthy pachytene stage cells. Thus, germ cells do not necessarily have to die to act as nurse cells, and death might have merely evolved to cull the majority of those ~20 germ cells that are being produced every hour in the proliferative zone of the germline to the mere ~3 oocytes that are eventually laid. It is unclear why this system is so wasteful, given that the same result could be obtained by merely reducing germ cell proliferation. The extent of apoptosis might correlate with the number of nurse cells that are needed to produce high-quality oocytes at a rate of ~3 per hour; under optimal growth conditions to produce as many progeny as fast as possible. Alternatively, apoptosis might be part of the mechanism that allows worms to quickly adapt to maintain germ cell homeostasis in response to environmental challenges. The observed increase of germ cell apoptosis upon starvation makes sense to focus limited resources to ensure the survival of some oocytes (which are post-pachytene and thus cannot undergo apoptosis) at the expense of culling pachytene cells (Salinas et al. 2006; Angelo and Van Gilst 2009). Since mitotic germ stem cells do not have the potential to die by apoptosis, down-regulation of germ cell proliferation would still waste the majority of later germ cells if limiting resources were equally distributed amongst developing oocytes. Indeed when L4 stage worms are starved, they appear to be able to cease germ cell net proliferation and preserve the viability of some of their fertilized eggs (Angelo and Van Gilst 2009). Apoptosis-defective worms kept under the same conditions have a larger germline, but the chance of embryo survival is reduced. All in all, it will be interesting to study the mechanisms that lead to germline homeostasis under normal and stressful conditions.

9.6 DNA Damage Checkpoint Signalling

The fact that germ cell apoptosis occurs, together with the notion that the germline is the only proliferative tissue in the worm, encouraged searching for DNA damage-induced apoptosis in the *C. elegans* germline (Gartner et al. 2000). Genome maintenance and the correct duplication of the genetic information are constantly challenged by the exposure to genotoxic agents, such as environmental toxins, or UV and ionizing irradiation. Equally, mistakes in DNA replication or mis-incorporation of nucleotides, oxidative stress or reactive metabolites lead to base modification and DNA lesions that have to be repaired. In the context of the germline, transposon activity, which results in transient DNA double-strand breakage, equally has to be dealt with. In response to DNA damage, cells activate DNA damage response pathways that will coordinate various DNA repair pathways best suited to repair specific lesions (for review, see Jackson and Bartek

2009). At the same time DNA damage response pathways lead to a transient cell cycle arrest, in order to allow for DNA repair before mutations might be fixed. Finally, when the level of DNA damage is excessive, cells might choose to permanently cease proliferation by entering the senescence program. Alternatively, cells might undergo apoptosis. This is thought to eliminate cells whose continued existence might become detrimental for the survival of the entire organism. Within a year of the initial observation of *C. elegans* germ cells apoptosis it was shown that germ cells can be induced to undergo apoptosis in response to genotoxic stress, and upon failure to undergo meiotic recombination (see below) (Gartner et al. 2000). Apoptosis is maximally induced by genotoxic agents that cause DNA double-strand breaks. Increased apoptosis is also observed with agents that primarily act by conferring base modifications such as EMS or ENU, or upon UV treatment that leads to the formation of tyrosine and thymidine dimers (Gartner et al. 2000; Derry et al. 2007; Stergiou et al. 2007). Similar to physiological cell death, DNA damage-induced apoptosis is restricted to late pachytene stage cells and requires the core apoptotic machinery formed by CED-3 and CED-4 proteins (Gartner et al. 2000). However, this type of inducible cell death also requires the BH3-only protein EGL-1 and the second worm BH3-only protein CED-13 contributes to DNA damage-induced germ cell apoptosis (Schumacher et al. 2005a). Like developmental apoptosis but in contrast to physiological germ cell death, DNA damage-induced apoptosis is blocked by the *ced-9 n1950* gain-of-function mutant (Gartner et al. 2000). At the same time a second DNA damage response, the transient cell cycle arrest of mitotically proliferating germ cells, was described (Gartner et al. 2000). This transient cell cycle arrest, which under typical conditions last from 10 to 20 h, leads to a distinct phenotype under DIC optics. Germ cells arrest cell proliferation, but continue to grow, thus leading to a mitotic germline where all cells are enlarged. At least in response to ionizing irradiation, cell cycle arrest largely occurs in the G2 cell cycle stage. Finally, the most important assay to score for DNA damage response and DNA repair pathway activity is to assess the relative survival of embryos, as well as the reduced numbers of oocytes laid that are generated from germlines treated with genotoxic agents (Gartner et al. 2000, 2004; Bailly et al. 2010).

9.7 Upstream DNA Damage Checkpoint Signalling in the *C. elegans* Germline

Various mutants differentially affect DNA damage responses. In core apoptosis pathway mutants, DNA damage-induced apoptosis but not any other damage responses is affected (Gartner et al. 2000). Other mutants, the most prominent example of which is the worm p53-like protein CEP-1 solely affects DNA damage-induced apoptosis (see below) (Derry et al. 2001; Schumacher et al. 2001). A third class of mutants are generally defective in both DNA damage-induced cell cycle arrest and apoptosis, and in most cases are also repair defective, as measured by reduced progeny survival (Gartner et al. 2000). DNA repair mutants tend to show excessive apoptosis, likely because persistent DNA damage leads to constitutive checkpoint activation. In the worm, DNA damage response and repair pathways dealing with DNA double-strand breaks (DSBs) are best characterized. DSBs that are induced, for example by ionizing radiation (IR), are the most serious type of DNA lesion. However, DSBs are also generated during natural cellular programmes, such as in meiosis by the conserved topoisomerase II-like enzyme SPO-11 (Keeney et al. 1997). DSB repair in the worm is extensively covered by a recent review and also in a chapter of this book (Lui and Colaiácovo 2012, Chap. 6; Lemmens and Tijsterman 2011). Two major pathways repair DSBs; non-homologous end joining (NHEJ) and homologous recombination (HR). NHEJ is an error-prone DNA repair mechanism that is mostly restricted to the G1 cell cycle stage. It involves the ligation of the broken DNA ends and is mostly used during somatic development of the worm. HR predominates in the germline. In somatic cells and mitotically dividing germ cells, the sister chromatid is taken as a template. During meiosis, HR is a more tricky undertaking. Meiotic cells must differentiate between the homologous chromosome as a template to facilitate homologous meiotic recombination, as opposed to using the sister chromatid to facilitate DNA double-strand break repair (Couteau and Zetka 2011; Bickel et al. 2010; Adamo et al. 2008).

The two related protein kinases ATM and ATR are critically important regulators of DNA damage responses in cells. ATM is "directly" activated by double-strand breaks (for review, see Jackson and Bartek 2009). ATR-activation occurs when double-strand breaks are recessed to generate single-stranded DNA (ssDNA). ATR recruitment to ssDNA requires protein-A and ATRIP. In *C. elegans* DNA damage responses are mediated largely by ATL-1, the worm ATR homolog, while no ATRIP-like molecules have been described so far. In addition, the related ATM kinase has only a minor role in DNA damage signalling. CtIP, a protein related to yeast Sae2, acts in conjunction with the MRN (MRE11, RAD50, NBS1) complex to resect the double-strand breaks which yields 3′ overhangs that can initiate homologous recombination. This processing event is crucial for the initiation of both repair and DNA damage signalling. It is not known whether the *C. elegans* CtIP, which is required for the repair of SPO-11 generated DNA double-strand breaks, has such a role in the worm (Penkner et al. 2007). The MRE-11 nuclease is required for ATL-1 loading (Garcia-Muse and Boulton 2005) but DNA damage checkpoint response defects have not been reported in the *mre-11* worm mutant (Chin and Villeneuve 2001). A DNA damage-specific clamp loader, comprised of Rad17 in a complex with the four smallest RFC (Replication Factor C) subunits, recruits a PCNA (proliferating cell nuclear antigen)-like complex referred to as "9-1-1" complex to the dsDNA–ssDNA transition at resected DNA ends. The heterotrimeric 9-1-1 complex, composed of RAD9, RAD1 and HUS1, is phosphorylated by ATR and is needed for full ATR-activation in yeast and vertebrate systems. In *C. elegans* the corresponding mutants are defective in triggering DNA damage checkpoint responses upon IR treatment (Gartner et al. 2000; Boulton et al. 2002; Hofmann et al. 2002; Boerckel et al. 2007). At the same time these genes are involved in telomere replication, possibly by being required for recruiting telomerase. Interestingly, it was recently shown that sister chromatid cohesion is not only required for efficient repair of meiotic DSBs but that cohesin is also required for triggering DNA damage-induced germ cell apoptosis (Lightfoot et al. 2011). This function is likely to act very upstream in the DNA damage response cascade that leads to apoptosis induction, as HUS-1 is not appropriately localized to chromosomes upon treatment with ionizing irradiation (Lightfoot et al. 2011). ATM- and ATR-activation lead to the phosphorylation of downstream targets including Chk1 and Chk2, which initiates a secondary wave of phosphorylation events. In the worm *chk-1,* like *atl-1* is required for all checkpoint responses and for the maintenance of genome stability in unchallenged germlines (Garcia-Muse and Boulton 2005). In contrast, *chk-2* does not affect IR-dependent checkpoint responses, but is rather required for meiotic chromosome pairing (MacQueen and Villeneuve 2001). Interestingly, however *chk-2* is required for DNA damage-induced apoptosis in response to UV treatment (Stergiou et al. 2007).

C. elegans genetics was instrumental in defining new conserved genes generally affecting checkpoint responses or specifically being required for checkpoint-dependent apoptosis. *clk-2/rad-5* was initially found in *C. elegans* and shown to be involved in both the S-phase checkpoint (which senses stalled replication forks) and the IR-induced apoptosis as well as cell cycle arrest response (Ahmed et al. 2001). Two alleles affecting the same locus were separately isolated. *clk-2* alleles are temperature sensitive but display a dramatic defect in response to IR-induced damage as well as replication stress at the "permissive" temperature. At the restrictive temperature, *clk-2* mutations are sterile and embryonic lethal (Ahmed et al. 2001). In the embryo, a checkpoint-dependent delay in cell cycle timing occurs, indicating that *clk-2* might be required for the integrity of DNA replication (Moser et al. 2009). Genome instability also occurs in *clk-2* mutant germlines, but strong alleles lead to a late G2-like cell cycle arrest. Subsequent studies on mammalian and yeast CLK-2 orthologs, referred to as TEL2 in yeasts, established that CLK-2/TEL2 is required for the full activation of all PIKKs-type protein kinases, a class of protein kinases that includes ATM/*atm-1* and ATL-1, thus explaining checkpoint-signalling phenotypes (Hayashi et al. 2007; Kanoh and Yanagida 2007; Hurov et al. 2010; Takai et al. 2010; Kaizuka et al. 2010). The current model suggests that CLK-2/TEL2 might have a chaperon-like function required for the full activation of those kinases (Horejsi et al. 2010). It remains to be seen whether CLK-2/TEL2 has further functions, as the above model is not fully consistent with the *C. elegans* genetics data: During embryogenesis, depleting ATL-1 and CLK-2 reveals that they have opposing functions, in the germline the cell cycle arrest phenotype conferred by *atl-1* inactivation leads to excessive genome instability while this is not the case for the cell cycle arrest phenotype observed in *clk-2* null mutants (Moser et al. 2009).

DNA damage signalling is generally triggered during the early steps of DNA double-strand break repair. Signalling is mediated by ATM directly bound to DNA double-strand breaks and/or by ATR, which often enforces the initial signal. The full activation of ATR requires the initial nucleolytic processing of DNA double-strand breaks, leading to extended single-stranded DNA stretches, which ultimately act as the landing platform for ATR. Little is known about how the persistence of late-stage recombination events imparts on DNA damage checkpoint signalling and

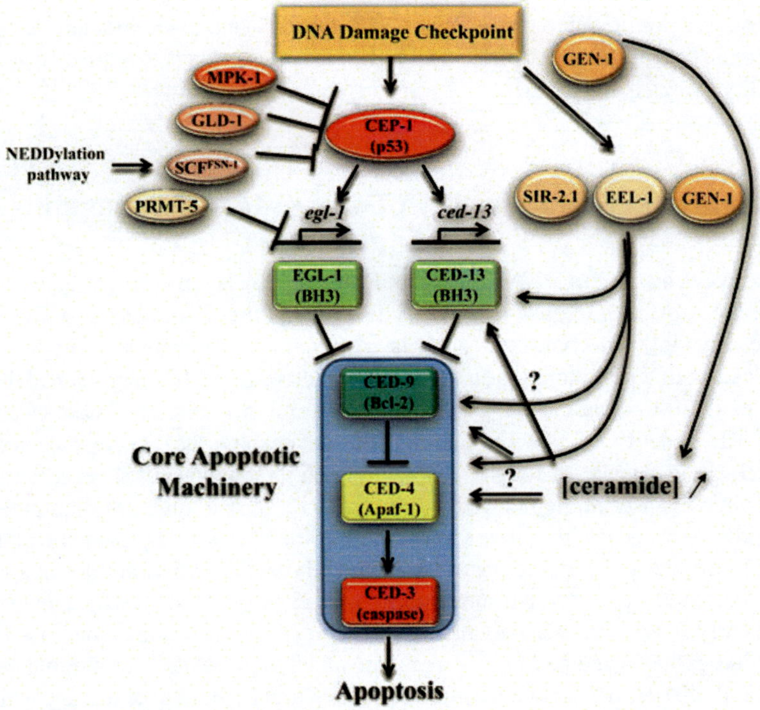

Fig. 9.2 DNA damage-induced germ cell apoptosis

apoptosis induction. An important indication that such signalling occurs came from the analysis of *C. elegans gen-1*, which likely acts as a Holliday Junction resolvase (Bailly et al. 2010). Holliday Junctions are four-way DNA structures that arise during recombinational repair between homologous chromosomes and which can be resolved by symmetrical cleavage (Ip et al. 2008; West 2009). *gen-1* was initially isolated based on an unbiased genetic screen for mutants required for DNA damage signalling and DNA double-strand break repair (Fig. 9.2). The repair defect associated with *gen-1* mutation can be explained by a defect in resolving Holliday Junction DNA double-strand breaks intermediates. However, the DNA damage signalling defect conferred by *gen-1* was surprising as the initial nucleolytic processing events thought to trigger checkpoint induction appeared to be normal in the mutant. Not only this, cytological and biochemical evidence indicates that the canonical ATM/ATR-dependent checkpoint pathways are activated in *gen-1* mutants. Thus GEN-1, besides its function in DNA repair, appears to act in a noncanonical DNA damage response signalling cascade. This pathway possibly senses the completion or the absence of Holliday Junction processing. It will be interesting to further dissect how GEN-1 affects DNA damage signalling. The biochemical activity of the mammalian GEN1 resolvase has been extensively characterized (Rass et al. 2010). At present no checkpoint function of mammalian GEN1 or yeast YEN1 was reported, and its depletion was not linked to repair defect. This is likely due to redundancy, as GEN1 depletion in cells defective for Blooms syndrome helicase leads to increased genome instability (Wechsler et al. 2011).

9.8 Pathways Leading to DNA Damage-Induced Apoptosis

The pathways leading to DNA damage-induced apoptosis are surprisingly complex and are the subject of intense research. The idea that the *C. elegans* germline recapitulates apoptotic mechanisms found in mammalian cells was reinforced by the identification of a p53 homolog in the worm genome, *cep-1* (*C. elegans p*53-like *1*) (Derry et al. 2001; Schumacher et al. 2001; for recent reviews, see Rutkowski et al. 2010). The identity of *cep-1* remained elusive despite the *C. elegans* genome sequence being known for several years, based on its low level of sequence conservation, which is essentially restricted to its DNA-binding domain. In mammals, the p53 family of transcription factors includes p63 and p73 and phylogenetic analysis that included many invertebrate species indicated that the radiation into those three groups occurred during vertebrate evolution (Rutkowski et al. 2010). However, it was recently shown that placozoans, which are very simple animals, and ticks have Mdm2-like proteins and p53 family members much more related to mammalian p53 (Lane et al. 2010a, b). Mammalian p53 acts as a transcriptional activator, which is essential for genome stability and the elimination of damaged cells by apoptosis. Human p53 or components of the p53 pathway are mutated in the majority of human cancers (for a recent review series, see Lane and Levine 2010). p63 and p73 have roles in stem cell proliferation, epithelial and neuronal differentiation and also have roles in the DNA damage response. p63 possibly encodes for the more ancient form of the p53 family and might be thus most related to *cep-1*. Besides structural considerations this notion is further supported by the more recent finding that p63 acts analogously to CEP-1 to specifically affect DNA damage-induced apoptosis in the female mammalian germline (Suh et al. 2006; Ou et al. 2007). The *C. elegans* p53-like protein CEP-1 turned out to be a direct transcriptional activator of *egl-1* and a second BH3-only protein called CED-13 (Schumacher et al. 2005a) (Fig. 9.2). *egl-1* and *ced-13* are the strongest *cep-1*-dependent genes induced by IR (Greiss et al. 2008a). EGL-1 induction plays a predominant role in DNA damage-induced apoptosis, while CED-13 appears to be important for UV-induced apoptosis (Schumacher et al. 2005a; Stergiou et al. 2007). Conceptually, the worm CEP-1 pathway acts analogously to the mammalian p53-dependent apoptosis induction of the BH3 proteins Puma and Noxa, which in turn leads to apoptosis induction analogous to EGL-1 by Bcl-2/CED-9 inactivation (Jeffers et al. 2003; Villunger et al. 2003; Shibue et al. 2003; Oda et al. 2000). The *cep-1* null mutant worms do not have any overt developmental or germline-specific phenotype but are fully defective in DNA damage-induced apoptosis. *cep-1* mutations do not affect DNA damage-induced cell cycle arrest and only show a slight reduction in progeny survival upon ionizing irradiation (Derry et al. 2001; Schumacher et al. 2001). Thus CEP-1 does not appear to have a prominent DNA repair function in response to ionizing irradiation. Interestingly, CEP-1 seems to be required for both UV-induced cell cycle arrest and apoptosis (Derry et al. 2007; Stergiou et al. 2007). Like mammalian p53, CEP-1 is phosphorylated in response to ionizing irradiation but the importance of this has not been identified (Schumacher et al. 2005b; Gao et al. 2008). The identification of CEP-1 and the tremendous interest in the mammalian p53 family prompted searches for CEP-1 regulatory mechanisms. Below, some of these mechanisms are described but other reviews are more comprehensive and include all components identified (Gartner et al. 2008; Rutkowski et al. 2010). Studying CEP-1 regulation in an intact developmental system allows the addressing not only of how CEP-1 function is activated or inactivated, but also how apoptosis induction might be restricted to a certain germ cell compartment. Importantly, this developmental system also allowed the defining of a cell non-autonomous mechanism by which radiation-induced apoptosis is regulated (see below).

9.9 CEP-1 Regulation

During normal development, or in the absence of DNA damage, the activity of human p53 has to be kept at bay to prevent inappropriate apoptosis induction. The best-characterized negative regulator of p53 is MDM2, an E3 ubiquitin ligase required for p53 degradation. Mdm-2 knockout mice die by excessive apoptosis, but Mdm2, p53 double knockout mice survive as excessive apoptosis is blocked (de Oca et al. 1995; Jones et al. 1995). There does not appear to be an Mdm2 homolog in worms, but several mechanisms restrict CEP-1 activation. A screen for mutations hypersensitized for CEP-1-dependent ionizing irradiation- induced apoptosis (Fig. 9.2) revealed a hypermorphic mutation of *gld-1*, a gene required for many facets of germline development (Schumacher et al. 2005b; Francis et al. 1995a, b; Jones et al. 1996). GLD-1 acts primarily as a translational repressor by binding to the 3'UTR of its targeted genes, hundreds of which are known (Lee and Schedl 2001; Wright et al. 2011; Carmel et al. 2010). The *gld-1(op236)* mutant at the permissive temperature, unlike all other *gld-1* mutants that have been isolated previously, does not affect overall germline development or germ cell identity, and germ cell apoptosis in the absence of ionizing irradiation is nearly normal. Biochemical analysis indicated that *cep-1* mRNA is a target of GLD-1 and that *cep-1* binding to GLD-1(*op236*) is compromised. In contrast, other tested GLD-1 targets are still bound to this mutant protein. GLD-1 expression is highest in early and mid-pachytene, but starts to decline as cells pass through the late pachytene stage where apoptosis normally occurs. Conversely, CEP-1 protein is absent in early and mid-pachytene but gradually increases in late pachytene. In *op236* mutants, CEP-1 expression is higher and extends more distally. This does not increase *cep-1*-dependent apoptosis under unchallenged conditions but sensitizes germ cells towards DNA damage-induced apoptosis. It is likely that this mechanism helps to ensure that DNA damage-induced apoptosis does not occur earlier in mid-pachytene.

In this stage, meiotic recombination intermediates occurring naturally and in the presence of CEP-1 could trigger apoptosis induction. The same unbiased genetic screen uncovered another CEP-1 regulatory module. Increased activation of the *C. elegans* MPK ortholog MPK-1, resulting from either loss of the *lip-1* phosphatase or activation of *let-60* Ras, results in enhanced *cep-1*-dependent, DNA damage-induced apoptosis (Rutkowski et al. 2011) (Fig. 9.2). Conversely MPK-1 is required for DNA damage-induced germ cell apoptosis. MPK-1 appears to affect CEP-1 in several ways. MPK-1 signalling regulates the apoptotic competency of germ cells by restricting CEP-1 protein expression to cells in late pachytene. MPK-1 signalling regulates CEP-1 expression in part by regulating the levels of GLD-1, a translational repressor of CEP-1, but also via a GLD-1-independent mechanism. In addition, MPK-1 is phosphorylated and activated upon ionizing radiation in late pachytene germ cells, and MPK-1-dependent CEP-1 activation may directly lead to the activation of CEP-1 as MPK-1 and CEP-1 interact in a yeast two-hybrid assay. Thus, GLD-1 and MPK-1 signalling might act primarily by setting the responsiveness to DNA damage-induced apoptosis, and by restricting apoptosis induction to late-stage pachytene cells. However, MAP signalling is also activated in response to DNA damage and seems to play an additional, more direct, role in CEP-1 regulation. In *gld-1(op236)* mutants, as well as in mutant situations that lead to excessive MPK-1 signalling, ionizing irradiation-induced apoptosis is enhanced while no obvious DNA repair defects occur. Thus, it is likely that these factors directly impinge on CEP-1 regulation and expression rather than generally causing excessive apoptosis triggered by the failure to repair DNA lesions.

Other regulatory mechanisms have been shown to impact on CEP-1 transcriptional regulation. The *p*rotein a*r*ginine *m*ethyl*t*ransferase *5* (PRMT-5) acts as a negative regulator of the CEP-1 function by binding and methylating the conserved transcriptional co-factor CBP-1 (Yang et al. 2009) (Fig. 9.2). Mammalian p300/CBP-1 acts as transcriptional co-factor for p53. Worms carrying a deletion in the *prmt-5* gene display an enhanced apoptotic response upon treatment with IR that fully depends on *cep-1*. In addition, PRMT-5 physically interacts with CEP-1 and CBP-1 in an in vitro pull-down assay as well as in an expression system in mammalian cells, and importantly can directly methylate CBP-1. These studies suggest that CEP-1 function as a transcriptional factor requires the co-factor CBP-1, which may be negatively regulated by the methyltransferase activity of PMRT-5. Given the absence of MDM2 it appears that other ubiquitin E3 ligases might regulate CEP-1 turnover. Skp1/cullin/F-box (SCF) ubiquitin ligases are activated by neddylation of the cullin subunit. The cullin 3 SCF^{FSN-1} complex is required to dampen the apoptosis response, and likely acts in the same pathway as neddylation (Gao et al. 2008). In addition, the level of CEP-1 is significantly increased in *fsn-1* mutants, suggesting that the E3 ubiquitin ligase complex SCF^{FSN-1} negatively regulates CEP-1 activity in *C. elegans* germline, possibly by the degradation of CEP-1 (Fig. 9.2). Intriguingly, the human FSN-1 homolog FBXO45 promotes the degradation of p73 in tissue culture-based experiments (Peschiaroli et al. 2009) (Fig. 9.2). Nevertheless, at present it is not clear whether the genetic interaction of SCF^{FSN-1} and *cep-1* is direct. We know little about CEP-1 post-translational modifications. It will be important to identify the modifications and to determine their importance in knock-in experiments, which should be possible using transposon excision-based gene replacement technologies (Frokjaer-Jensen et al. 2008).

9.10 Pathways Parallel to CEP-1

Uncovering the CEP-1 pathway provided a simple model for DNA damage-induced apoptosis: DNA damage, or recombination intermediates would be sensed and signalling would lead to CEP-1 transcriptional activation. This in turn would lead to the transcriptional induction of EGL-1 and CED-13, which in turn would trigger the core apoptosis pathway by the direct binding to CED-9. In theory such a model is sufficient to explain apoptosis induction in the germline. It therefore came as a surprise that mutations exist where DNA damage-induced apoptosis is largely abrogated, while CEP-1-dependent transcription still occurs (Fig. 9.2). Studying such factors is a tricky undertaking as care must be taken not to mistake reduced germ cell proliferation, or a reduced number of apoptosis competent cells with a genuine defect in apoptosis induction; however, such controls are possible.

The first report of such a situation was based on the analysis of SIR-2.1, a histone deacetylase which is homologous to human SIRT1 (Greiss et al. 2008b) (Fig. 9.2). Worms deleted for the *sir-2.1* locus display a complete defect in apoptosis response upon IR treatment. While germ cell proliferation in *sir-2.1* mutants is slightly reduced, apoptosis competency is not affected. For instance, the basal level of physiological germ cell death is not affected; more apoptosis occurs in mutants that are thought to increase physiological germ cell apoptosis, and most importantly the depletion of *ced-9* leads to excessive apoptosis induction to a degree comparable to wild type. Thus, in *sir-2.1* mutants, DNA damage-induced apoptosis seems to be genuinely affected, and the reduction in apoptosis induction is almost as strong as in *cep-1* mutants, where *egl-1* transcription is not induced. In summary, these data suggest that SIR-2.1 acts in parallel or downstream to CEP-1. A likely target of SIR-2.1 might be the EGL-1 protein, which is only required for DNA damage-induced germ cell apoptosis. Equally CED-9 could be affected. The *ced-9 n1950* gain-of-function allele does not affect physiological germ cell death, while DNA damage-induced apoptosis is compromised; thus, there are likely differential modes of CED-9 regulation. Given that CED-4 and CED-3 are required for all forms of germ cell apoptosis, at the first glance they appear as less likely targets unless their regulation is much more complicated than anticipated. We do not know how SIR-2.1 acts mechanistically, but there are intriguing cytological correlations, possibly hinting at CED-4. SIR-2.1 translocates from the nucleus to the cytoplasm during early apoptosis. This translocation occurs very early in apoptosis where the nuclear envelope still appears to be intact. In addition such translocation, which correlates with CED-4 accumulation at the nuclear periphery, also occurs in apoptosis execution defective *ced-3* germ cells. These findings indicate that such translocation might have a functional role rather than being a mere consequence of apoptosis induction. During translocation, SIR-2.1 appears to transiently co-localize with CED-4 at the nuclear periphery. In a separate study, the HECT-domain E3 ligase EEL-1, which is homologous to human Huwe1/ARF-BP1/Mule, was shown to genetically behave like SIR-2.1, as does KRI-1 the ortholog of human *KRIT1/CCM1,* a gene frequently mutated in the neurovascular disease cerebral cavernous malformation (for cell non-autonomous apoptosis regulation, see below) (Ross et al. 2011; Ito et al. 2010) (Fig. 9.2). We do not know whether these genes act in the same genetic pathway or what their molecular function is. Intriguingly, the human EEL-1 homolog Mule was found to regulate levels of the mammalian anti-apoptotic Mcl1, a Bcl2-family protein. Unfortunately, we do not know whether EEL-1 directly binds to CED-9 or EGL-1.

In the damage response field, ATR and ATM are considered to be the most upstream signalling factors that affect all DNA damage responses including apoptosis. This assumption is likely to be an oversimplification, especially when it comes to apoptosis induction (Deng et al. 2008). Ceramide is a central molecule in sphingolipid metabolism and critical for plasma membrane integrity. The role of ceramides in apoptosis induction has been controversial in mammalian cells (Kolesnick and Hannun 1999; Perry and Hannun 1998; Hofmann and Dixit 1998, 1999). The cellular level of ceramide increases after treatment with diverse apoptotic stimuli such as IR, UV light or TNF. Blocking ceramide synthesis reduced apoptosis in various human cell lines. Some reports suggest that ceramides can form channels in mitochondrial outer membranes and promote the release of pro-apoptotic factors, an obligate step in mammalian apoptosis execution. In worms, mutations blocking ceramide biogenesis completely abrogate IR-induced apoptosis and this defect appears to be rescued by micro-injecting long-chain ceramides into the worm gonad, and evidence was provided that ceramide might accumulate upon apoptosis induction (Deng et al. 2008) (Fig. 9.2). Importantly, blocking ceramide does not affect *egl-1* and *ced-13* as is the case for the aforementioned factors and, like SIR-2.1, was shown not to affect developmental and physiological germ cell apoptosis.

It will be interesting to see whether these and other factors affecting DNA damage-induced apoptosis induction independent of CEP-1 act in a single pathway. This is probably unlikely. However, even "linear" DNA damage response pathways might be more complicated. For instance, as mentioned above, GEN-1 is also required for all DNA damage checkpoint responses including apoptosis, but *egl-1* induction is not compromised (Bailly et al. 2010). A further question that arises is why multiple pathways are needed to allow for full apoptosis induction? It might be possible that this is a fail-safe mechanism to ensure that germ cells are not inappropriately eliminated. Such a mechanism makes sense, as DNA lesions occur, even in the absence of exogenous DNA damaging agents. Furthermore, programmed DNA breaks and repair intermediates occur during meiosis, as discussed below. Why nature would be so careful about not losing cells by apoptosis, even when under normal conditions the majority of cells are eliminated by apoptosis remains a mystery.

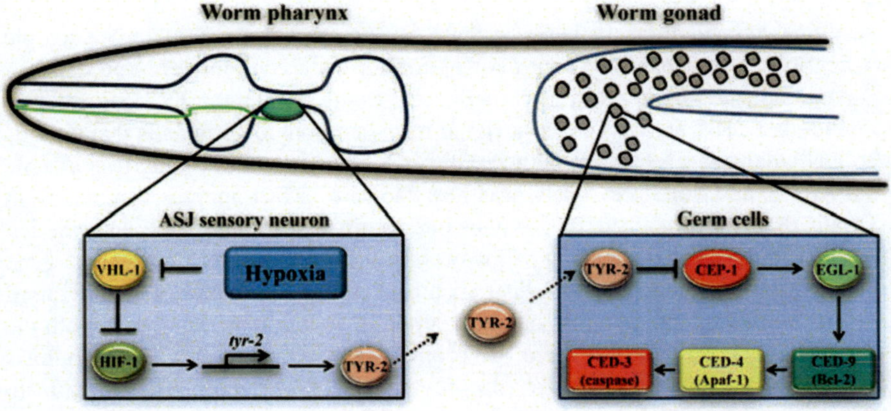

Fig. 9.3 Cell non-autonomous regulation of germ cell apoptosis

9.11 Apoptosis Induction by Cell Non-Autonomous Mechanism

Although the worm germline provides a powerful model system to investigate stress-induced apoptosis, most of the studies have focused on cell-autonomous response to cellular stress, assuming that pro-apoptotic stimuli and their effects occur in the same cell. This is not necessarily always the case and the worm provides a model where stress responses can be analyzed in a developmental context. The Hengartner lab has recently reported a cell non-autonomous way of regulating germ cells apoptosis showing that hypoxia perceived in neuronal cells blocks cell death induction in the germline (Sendoel et al. 2010). The cellular response to hypoxia is highly conserved and one of the critical factors is the transcriptional activator HIF-1. Under normoxic conditions, HIF-1 is prolyl-hydroxylated and this modification is required for its binding to VHL, the substrate-binding subunit of an E3 ligase that mediates HIF-1 degradation (Fig. 9.3). Upon exposure to low oxygen pressure, HIF-1 is dehydroxylated and becomes resistant to VHL-mediated degradation. HIF-1 stabilization induces the expression of various hypoxia-specific genes. The authors show that hypoxia is a robust repressor of DNA damage-induced apoptosis. Thus, worms grown under low oxygen conditions do not trigger germ cell apoptosis after treatment with IR. Low oxygen leads to HIF-1 stabilization, and HIF-1 stabilization also occurs in *vhl-1* mutants even under normoxic conditions (Fig. 9.3). A *vhl-1* mutant mimics HIF-1 activation normally occurring in hypoxic conditions. Scanning through HIF target genes, the authors found that a mutation of the *tyr-2* tyrosinase restores IR-dependent apoptosis (Fig. 9.3). The surprise came when the authors looked for the localization of TYR-2. *tyr-2* is expressed in epidermis/hypodermis and uterine muscle cells under normoxic conditions but hypoxia induces *tyr-2* expression only in the amphid sensory neurons, located in the head of the worm. Laser-based ablation of these two neurons completely restores the ability of germ cells to respond to IR under low oxygen condition, reinforcing the idea that the signal somehow transits between these two compartments. Finally, the ectopic expression of *tyr-2* under the control of a germline-specific promoter impedes DNA damage-induced apoptosis in germ cells, suggesting that either TYR-2 itself or a product of TYR-2 activity carries a signal from neuronal to germ cells that inhibits the DNA damage checkpoint pathway under low oxygen conditions. It is of note that KRI-1, described previously, was also shown to affect apoptosis by acting in somatic tissues (Ito et al. 2010). It will be interesting to see where it acts and whether it relates to the hypoxia-sensing pathway. Also the worm retinoblastoma-like gene *lin-35,* discussed in the physiological apoptosis section and which is also required for *egl-1* independent apoptosis induction upon IR function, is required in both the somatic gonad and the germline to promote constitutive germ cell apoptosis (Schertel and Conradt 2007). We can only speculate why worms would want to block apoptosis under hypoxic conditions. As an organism found in soil and rotten fruits, it is not unlikely that they encounter a low oxygen environment.

9.12 Meiotic Recombination and Chromosome Synapsis Checkpoints

During meiosis (see Lui and Colaiácovo 2012, Chap. 6), gametes are produced by two sequential cell divisions that first separate homologous chromosomes (meiosis I) and then sister chromatids (meiosis II). During prophase of meiosis I, homologous chromosomes align (Fig. 9.4, left panels) and undergo meiotic recombination. Recombination initiation is triggered by SPO-11-generated DNA double-strand breaks and is followed by the formation of the synaptonemal complex, a proteinacious structure which tightly links paired homologs and blocks excessive cross-over recombination (Fig. 9.4, right panels). In the worm this cross-over interference is "perfect" and generally only a single cross-over recombination event occurs on each chromosome. During meiotic prophase the integrity of both meiotic recombination and meiotic chromosome synapsis is monitored by checkpoints, and such checkpoint activation leads to a cell cycle arrest phenotype in yeast and apoptosis in animals. In most organisms meiotic recombination and chromosome pairing are genetically linked. In contrast, in the fly and the worm, homologous chromosome pairing occurs in the absence of meiotic recombination, thus allowing these two events to be separately studied.

9.13 Meiotic Recombination Checkpoint and Template Preference

The *C. elegans* recombination checkpoint was uncovered in worms by inactivating genes required for homologous recombination such as *rad-51,* which leads to a high level of apoptosis (Gartner et al. 2000). This cell death is induced by unprocessed

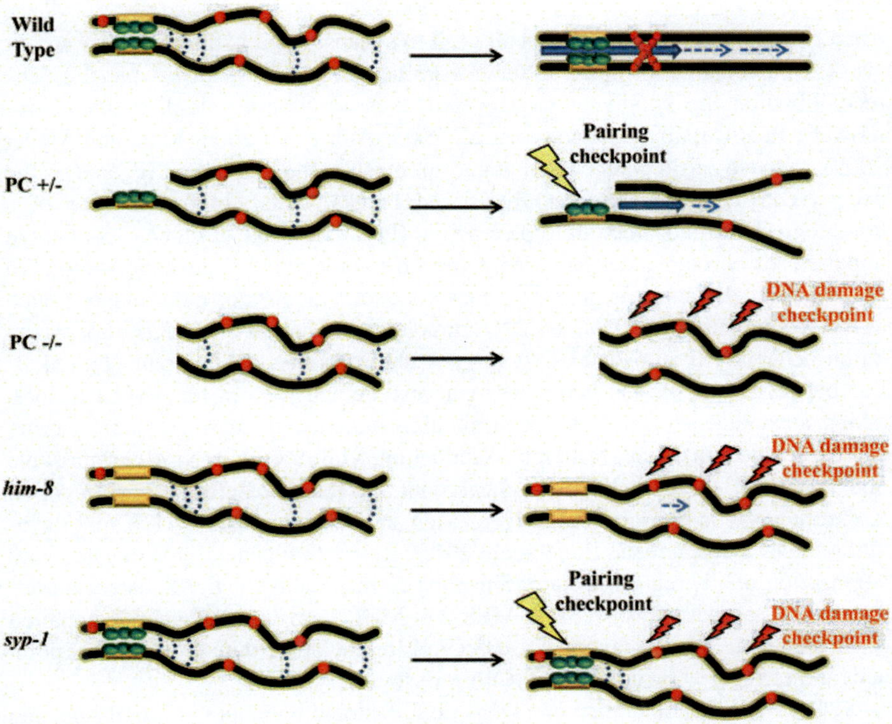

Fig. 9.4 Meiotic recombination and pairing checkpoints. Diagram adapted from Meier and Gartner (2006). Chromosomes are indicated in *black*. The *left panel* show early transition zone stage chromosomes. The corresponding *right panels* show corresponding mid-pachytene chromosomes. Pairing centres are indicated in *yellow*, HIM-8 is indicated in *green*. The localization of PCH-2 was not reported. Presynaptic, synapsis-independent pairing is indicated by *dotted blue lines*. Synapsis and its directionality is indicated by *blue arrows*. The *dotted blue arrow* in the three lower panels indicates no or residual synapsis. DNA double-strand breaks marked by RAD-51 foci are depicted in *red*. *Yellow* and *red arrows* indicate the triggering of the synapsis and recombination checkpoint, respectively

recombination intermediates as excessive apoptosis induction is blocked in *spo-11* mutations. Since then, mutations in multiple genes required for homologous recombination have been shown to trigger this checkpoint. At present, the meiotic recombination and the DNA double-strand break-induced checkpoint pathway cannot be genetically separated. In both cases checkpoint genes like *mrt-2* and *clk-2*, and importantly also *cep-1* p53-like are required for apoptosis induction (Gartner et al. 2000). The recombination checkpoint is also activated when individual meiotic chromosomes completely fail to pair. This occurs when both pairing centres (PCs), which are *cis*-elements needed to initiate chromosome pairing and to stabilize synapsis, are deleted or when the proteins that bind to those specific PCs and mediate pairing, like HIM-8 for the X-chromosome PC, are missing (Bhalla and Dernburg 2005) (Fig. 9.4, third and forth panel). The recombination checkpoint is also triggered when chromosomes globally fail to synapse, as it is the case for *syp-1* or *syp-2* synaptonemal complex mutants (see below, Fig. 9.4, lowest panels). The recombination checkpoint can be triggered, even when only a single meiotic chromosome fails to pair (Colaiacovo et al. 2003; Alpi et al. 2003; Mets and Meyer 2009). In those pairing defective mutants apoptosis is thought to be triggered by the excessive SPO-11-generated double-strand breaks, which in the absence of a homologous chromosome fail to be repaired (Fig. 9.4). Male germ cells contain an unpaired sex chromosome (XO). Why does this not lead to apoptosis induction? In males germ cell apoptosis generally does not occur, albeit molecular markers such as the accumulation of ATR-1 on DNA double-strand breaks or CHK-1 phosphorylation occur when autosomes (but not sex chromosomes) fail to pair (Jaramillo-Lambert et al. 2010). XO germ cells can also be analyzed in female germ lines, taking advantage of *fem-3* sex determination mutants. In such a context pairing defects of autosomes leads to apoptosis induction, while the unpaired X-chromosome does not lead to recombination checkpoint triggered apoptosis. The unpaired X-chromosome is transcriptionally repressed and bears the histone H3K9m2 methylation mark (Checchi and Engebrecht 2011), a signature generally associated with a transcripitionally repressed state. When the corresponding methyltransferase Met-2, or two other methyltransferases, MES-2 or MET-1 are depleted a single unpaired X-chromosome triggers apoptosis. H3K9m2 methylation thus blocks checkpoint activation triggered by unrepaired X-chromosomes.

SPO-11 induces more than one DNA double-strand break per chromosome, and these excessive breaks are repaired by gene conversion events initially by taking the homologous chromosome as a template (see below). In late pachytene, both the homologous chromosome and the sister chromatid can be used for repair. The former mode of HR is required for meiotic cross-over recombination and gene conversion. Both *C. elegans brc-1,* the worm homolog of the mammalian breast and ovarian cancer susceptibility gene BRCA1, and the structural maintenance of chromosomes (SMC) proteins SMC-5 and SMC-6 were shown to be specifically required for meiotic sister-chromatid recombination (Bickel et al. 2010; Adamo et al. 2008). In the corresponding mutants, no overt defect in meiotic cross-over recombination can be observed and chiasmata occur as in wild type. In contrast, homolog-independent repair defects caused by the failure to use the sister chromatid as template can be detected in *smc-5/6* and *brc-1* mutants. A chromosome fragmentation phenotype arises in those mutants when inter-homolog repair is blocked.

There appears to be a template switch occurring during late pachytene. In wild-type *spo-11*-dependent RAD-51 foci are generated in the transition zone and are generally repaired in the early and mid-pachytene stages. In pairing defective mutants, foci stay, but appear to be repaired in the very late pachytene stage (independent of non-homologous end-joining), likely by sister chromatid templated repair (Adamo et al. 2008; Colaiacovo et al. 2003; Alpi et al. 2003). It is intriguing that germ cell apoptosis induction also affects these very late-stage pachytene cells. It will be interesting to further address if this switch in HR repair templates is linked to apoptosis induction. In summary, it is possible that DNA double-strand breaks that fail to be repaired by cross-over recombination or homolog templated gene conversation, can be repaired from the sister chromatid. If this fails, apoptosis appears to be the last resort to prevent passing on a DNA lesion to the next generation. Interestingly, the failure to cull affected pachytene cells upon recombination checkpoint activation, which can be conferred by *cep-1* or by *ced-4* apoptosis-defective mutants, leads to an increased rate of meiotic chromosome missegregation (Bhalla and Dernburg 2005; Jaramillo-Lambert et al. 2010; Jaramillo-Lambert and Engebrecht 2010). Thus, it appears possible that chromosomes that fail to properly repair excessive SPO-11-generated double-strand breaks are preferentially culled. It appears that repair from the sister chromatid is the preferred mode of repair for exogenously inflicted DNA double-strand breakage (Couteau and Zetka 2011). A recent report showed that synapsed late-stage pachytene chromosomes transiently desynapse upon treatment with ionizing irradiation. This separation, which correlates with the deacetylation on H2A Lysine 5 in the affected chromosomal domains, is thought to ensure repair by the sister-chromatid (Couteau and Zetka 2011).

9.14 The Meiotic Pairing Checkpoint

The landmark study describing the *C. elegans* synapsis checkpoint used the X-chromosome pairing centre, which is deleted by the *meDf2* deficiency (Bhalla and Dernburg 2005) (Fig. 9.4, second panel from top). X-chromosome pairing is perturbed but not eliminated in strains hemizygous for the X-chromosome PC in early meiotic prophase. This initial pairing defect reduces the frequency of synapsis to approximately 45%. Worms, hemizygous for *meDf2,* show the same excessive level of germ cell apoptosis as *meDf2* X-chromosome PC homozygous mutants that trigger the recombination checkpoint. However, the excessive apoptosis in the PC hemizygous mutant is not *spo-11* or *cep-1* dependent, and apoptosis induction rather requires the *C. elegans* homolog of Pch2p, a yeast protein previously implicated in meiotic checkpoint control (San-Segundo and Roeder 1999, 2000). As it is the case for germ cells that activate the recombination checkpoint, germ cells in pairing defective mutants are preferentially eliminated: In *meDf2/+* hermaphrodites, 60% of meiotic nuclei exhibit unsynapsed X-chromosomes, but only ~6% of their self-progeny are males. The low percentage of males, which is indicative of sex chromosome missegregation, can only be explained if cells carrying unsynapsed chromosomes are preferentially culled by apoptosis. Consistent with this notion, more males occur in *ced-4* and *pch-2* mutants. The authors argue that a PC not engaged in chromosome pairing is required to trigger the synapsis checkpoint as the synapsis checkpoint is not engaged in PC homozygous *him-8* mutants. Conversely, if DNA double-strand break intermediates indeed trigger the DNA damage checkpoint, why is the DNA damage checkpoint not activated in PC hemizygotes as opposed to PC or *him-8* homozygote mutants? The argument is that the failure of DNA damage checkpoint activation could be due to the repression of the DNA damage checkpoint once the pairing checkpoint is activated. Alternatively, the reduced double-strand breaks in

PC hemizygous mutants might not be sufficient to trigger the DNA recombination checkpoint. Consistent with this hypothesis, the meiotic DNA damage and pairing checkpoint can be activated simultaneously in *syp-1* and *syp-2* mutants (Fig. 9.4, lowest panels). In these mutants, synapsis independent chromosome pairing occurs but synaptonemal complex formation is globally abrogated on all chromosomes. In *C. elegans*, *pch-2* has no reported roles in regulating unperturbed apoptosis. In budding yeast and mice, where *PCH2* was reported to affect the pachytene checkpoint that monitors both recombination of chromosome pairing, *PCH2* has recently shown to also affect multiple functions occurring during unperturbed meiosis in yeast and mice (Roig et al. 2010; Zanders et al. 2011; Zanders and Alani 2009; Joshi et al. 2009). Mutant phenotypes include alternations in the rate and timing of meiotic recombination, alteration in the level of cross-over interference, in the level of meiotic chromosome condensation, and in the extend of synapsis. It will be interesting to learn how PCH-2 triggers germ cell apoptosis and to elucidate the interplay between pairing and recombination checkpoints.

9.15 Concluding Remarks

Male germ cells fully activate the recombination checkpoint cascade, and CEP-1-dependent *egl-1* transcription is induced (Jaramillo-Lambert et al. 2010). While all apoptosis proteins are expressed in the male, the CED-3 caspase is not induced. Thus it appears that in males checkpoint induced repair might be very efficient and possibly compensates for failure to cull damaged cells. Why would apoptosis induction occur in the female and not in the male germline? Why is nature so wasteful to trigger such extensive levels of germ cell death? Why is germ cell apoptosis regulation so intricate? Why can only a subset of germ cell undergo apoptosis even though cell death proteins appear to be expressed throughout the germline? We do not have answers to those questions. Furthermore, we are only beginning to understand how apoptosis, DNA damage, recombination and pairing checkpoints are controlled. It is difficult to address these questions in the context of germline development in mammals. The worm system will continue to provide important insights. It will be interesting to directly assess mutagenesis rates in male and female germlines, and to establish to what extend they depend on the ability to induce apoptosis.

Acknowledgments We apologize for not being able to cite all papers relating to germ cell apoptosis and *C. elegans* germline genome stability. Comprehensive reviews on *C. elegans* DNA double-strand break repair apoptosis and CEP-1 have recently been published and are cited throughout the text. Aymeric Bailly was funded by the Association pour la Recherche contre le Cancer (ARC), INSERM-Avenir from Dimitris Xirodimas, CNRS, University of Montpellier 1. Research in the Gartner lab was funded by a CR-UK Career Development Award (C11852/A4500) and by a Wellcome Trust Senior Research Fellowship (0909444/Z/09/Z) and by a Wellcome Trust Project grant (081923/Z/07/Z). We are grateful to Ulrike Gartner and Alexander Holmes for proofreading and Dimitris Xirodimas for supporting A. Bailly.

References

Aballay A, Ausubel FM (2001) Programmed cell death mediated by ced-3 and ced-4 protects *Caenorhabditis elegans* from Salmonella typhimurium-mediated killing. Proc Natl Acad Sci USA 98(5):2735–2739. doi:10.1073/pnas.041613098041613098 [pii]

Adamo A, Montemauri P, Silva N, Ward JD, Boulton SJ, La Volpe A (2008) BRC-1 acts in the inter-sister pathway of meiotic double-strand break repair. EMBO Rep 9(3):287–292. doi:7401167 [pii]10.1038/sj.embor.7401167

Ahmed S, Alpi A, Hengartner MO, Gartner A (2001) *C. elegans* RAD-5/CLK-2 defines a new DNA damage checkpoint protein. Curr Biol 11(24):1934–1944

Alpi A, Pasierbek P, Gartner A, Loidl J (2003) Genetic and cytological characterization of the recombination protein RAD-51 in *Caenorhabditis elegans*. Chromosoma 112(1):6–16

Andux S, Ellis RE (2008) Apoptosis maintains oocyte quality in aging *Caenorhabditis elegans* females. PLoS Genet 4(12):e1000295. doi:10.1371/journal.pgen.1000295

Angelo G, Van Gilst MR (2009) Starvation protects germline stem cells and extends reproductive longevity in *C. elegans*. Science 326(5955):954–958. doi:1178343 [pii]10.1126/science.1178343

Arur S, Ohmachi M, Berkseth M, Nayak S, Hansen D, Zarkower D, Schedl T (2011) MPK-1 ERK controls membrane organization in *C. elegans* oogenesis via a sex-determination module. Dev Cell 20(5):677–688. doi:S1534-5807(11)00160-2 [pii]10.1016/j.devcel.2011.04.009

Audhya A, Hyndman F, McLeod IX, Maddox AS, 3rd Yates JR, Desai A, Oegema K (2005) A complex containing the Sm protein CAR-1 and the RNA helicase CGH-1 is required for embryonic cytokinesis in Caenorhabditis elegans. J Cell Biol 171(2):267–279. doi:jcb.200506124 [pii]10.1083/jcb.200506124

Bailly AP, Freeman A, Hall J, Declais AC, Alpi A, Lilley DM, Ahmed S, Gartner A (2010) The *Caenorhabditis elegans* homolog of Gen1/Yen1 resolvases links DNA damage signaling to DNA double-strand break repair. PLoS Genet 6(7):e1001025. doi:10.1371/journal.pgen.1001025

Bhalla N, Dernburg AF (2005) A conserved checkpoint monitors meiotic chromosome synapsis in *Caenorhabditis elegans*. Science 310(5754):1683–1686. doi:310/5754/1683 [pii]10.1126/science.1117468

Bickel JS, Chen L, Hayward J, Yeap SL, Alkers AE, Chan RC (2010) Structural maintenance of chromosomes (SMC) proteins promote homolog-independent recombination repair in meiosis crucial for germ cell genomic stability. PLoS Genet 6(7):e1001028. doi:10.1371/journal.pgen.1001028

Boag PR, Nakamura A, Blackwell TK (2005) A conserved RNA-protein complex component involved in physiological germline apoptosis regulation in *C. elegans*. Development 132(22):4975–4986. doi:dev.02060 [pii]10.1242/dev.02060

Boerckel J, Walker D, Ahmed S (2007) The *Caenorhabditis elegans* Rad17 homolog HPR-17 is required for telomere replication. Genetics 176(1):703–709. doi:genetics.106.070201 [pii]10.1534/genetics.106.070201

Boulton SJ, Gartner A, Reboul J, Vaglio P, Dyson N, Hill DE, Vidal M (2002) Combined functional genomic maps of the *C. elegans* DNA damage response. Science 295(5552):127–131

Buszczak M, Cooley L (2000) Eggs to die for: cell death during Drosophila oogenesis. Cell Death Differ 7(11):1071–1074. doi:10.1038/sj.cdd.4400755

Carmel AB, Wu J, Lehmann-Blount KA, Williamson JR (2010) High-affinity consensus binding of target RNAs by the STAR/GSG proteins GLD-1, STAR-2 and Quaking. BMC Mol Biol 11:48. doi:1471-2199-11-48 [pii]10.1186/1471-2199-11-48

Checchi PM, Engebrecht J (2011) *Caenorhabditis elegans* histone methyltransferase MET-2 shields the male X chromosome from checkpoint machinery and mediates meiotic sex chromosome inactivation. PLoS Genet 7(9):e1002267. doi:10.1371/journal.pgen.1002267PGENETICS-D-11-00601 [pii]

Chin GM, Villeneuve AM (2001) *C. elegans* mre-11 is required for meiotic recombination and DNA repair but is dispensable for the meiotic G(2) DNA damage checkpoint. Genes Dev 15(5):522–534

Choi SS (2011) High glucose diets shorten lifespan of *Caenorhabditis elegans* via ectopic apoptosis induction. Nutr Res Pract 5(3):214–218. doi:10.4162/nrp. 2011.5.3.214

Colaiacovo MP, MacQueen AJ, Martinez-Perez E, McDonald K, Adamo A, La Volpe A, Villeneuve AM (2003) Synaptonemal complex assembly in *C. elegans* is dispensable for loading strand-exchange proteins but critical for proper completion of recombination. Dev Cell 5(3):463–474. doi:S1534580703002326 [pii]

Conradt B (2009) Genetic control of programmed cell death during animal development. Annu Rev Genet 43:493–523. doi:10.1146/annurev.genet.42.110807.091533

Conradt B, Xue D (2005) Programmed cell death. WormBook:1–13. doi:10.1895/wormbook.1.32.1

Couteau F, Zetka M (2011) DNA damage during meiosis induces chromatin remodeling and synaptonemal complex disassembly. Dev Cell 20(3):353–363. doi:S1534-5807(11)00043-8 [pii]10.1016/j.devcel.2011.01.015

de Oca M, Luna R, Wagner DS, Lozano G (1995) Rescue of early embryonic lethality in mdm2-deficient mice by deletion of p53. Nature 378(6553):203–206. doi:10.1038/378203a0

Deng X, Yin X, Allan R, Lu DD, Maurer CW, Haimovitz-Friedman A, Fuks Z, Shaham S, Kolesnick R (2008) Ceramide biogenesis is required for radiation-induced apoptosis in the germ line of *C. elegans*. Science 322(5898):110–115. doi:322/5898/110 [pii]10.1126/science.1158111

Derry WB, Putzke AP, Rothman JH (2001) *Caenorhabditis elegans* p53: role in apoptosis, meiosis, and stress resistance. Science 294(5542):591–595

Derry WB, Bierings R, van Iersel M, Satkunendran T, Reinke V, Rothman JH (2007) Regulation of developmental rate and germ cell proliferation in *Caenorhabditis elegans* by the p53 gene network. Cell Death Differ 14(4):662–670. doi:4402075 [pii]10.1038/sj.cdd.4402075

Fox PM, Vought VE, Hanazawa M, Lee MH, Maine EM, Schedl T (2011) Cyclin E and CDK-2 regulate proliferative cell fate and cell cycle progression in the *C. elegans* germline. Development 138(11):2223–2234. doi:138/11/2223 [pii]10.1242/dev.059535

Francis R, Maine E, Schedl T (1995a) Analysis of the multiple roles of gld-1 in germline development: interactions with the sex determination cascade and the glp-1 signaling pathway. Genetics 139(2):607–630

Francis R, Barton MK, Kimble J, Schedl T (1995b) gld-1, a tumor suppressor gene required for oocyte development in *Caenorhabditis elegans*. Genetics 139(2):579–606

Frokjaer-Jensen C, Davis MW, Hopkins CE, Newman BJ, Thummel JM, Olesen SP, Grunnet M, Jorgensen EM (2008) Single-copy insertion of transgenes in *Caenorhabditis elegans*. Nat Genet 40(11):1375–1383. doi:ng.248 [pii]10.1038/ng.248

Gao MX, Liao EH, Yu B, Wang Y, Zhen M, Derry WB (2008) The SCF FSN-1 ubiquitin ligase controls germline apoptosis through CEP-1/p53 in *C. elegans*. Cell Death Differ 15(6):1054–1062. doi:cdd200830 [pii]10.1038/cdd.2008.30

Garcia-Muse T, Boulton SJ (2005) Distinct modes of ATR activation after replication stress and DNA double-strand breaks in *Caenorhabditis elegans*. EMBO J 24(24):4345–4355

Gartner A, Milstein S, Ahmed S, Hodgkin J, Hengartner MO (2000) A conserved checkpoint pathway mediates DNA damage-induced apoptosis and cell cycle arrest in *C. elegans*. Mol Cell 5(3):435–443

Gartner A, MacQueen AJ, Villeneuve AM (2004) Methods for analyzing checkpoint responses in *Caenorhabditis elegans*. Methods Mol Biol 280:257–274

Gartner A, Boag PR, Blackwell TK (2008) Germline survival and apoptosis. WormBook:1–20

Green RA, Kao HL, Audhya A, Arur S, Mayers JR, Fridolfsson HN, Schulman M, Schloissnig S, Niessen S, Laband K, Wang S, Starr DA, Hyman AA, Schedl T, Desai A, Piano F, Gunsalus KC, Oegema K (2011) A high-resolution *C. elegans* essential gene network based on phenotypic profiling of a complex tissue. Cell 145(3):470–482. doi:S0092-8674(11)00371-0 [pii]10.1016/j.cell.2011.03.037

Greiss S, Schumacher B, Grandien K, Rothblatt J, Gartner A (2008a) Transcriptional profiling in *C. elegans* suggests DNA damage dependent apoptosis as an ancient function of the p53 family. BMC Genomics 9:334

Greiss S, Hall J, Ahmed S, Gartner A (2008b) *C. elegans* SIR-2.1 translocation is linked to a proapoptotic pathway parallel to cep-1/p53 during DNA damage-induced apoptosis. Genes Dev 22(20):2831–2842

Gumienny TL, Lambie E, Hartwieg E, Horvitz HR, Hengartner MO (1999) Genetic control of programmed cell death in the *Caenorhabditis elegans* hermaphrodite germline. Development 126(5):1011–1022

Hayashi T, Hatanaka M, Nagao K, Nakaseko Y, Kanoh J, Kokubu A, Ebe M, Yanagida M (2007) Rapamycin sensitivity of the Schizosaccharomyces pombe tor2 mutant and organization of two highly phosphorylated TOR complexes by specific and common subunits. Genes Cells 12(12):1357–1370. doi:GTC1141 [pii]10.1111/j.1365-2443.2007.01141.x

Hofmann K, Dixit VM (1998) Ceramide in apoptosis—does it really matter? Trends Biochem Sci 23(10):374–377. doi:S0968-0004(98)01289-4 [pii]

Hofmann K, Dixit VM (1999) Reply to kolesnick and hannun, and perry and hannun. Trends Biochem Sci 24(6):227. doi:S0968-0004(99)01410-3 [pii]

Hofmann ER, Milstein S, Boulton SJ, Ye M, Hofmann JJ, Stergiou L, Gartner A, Vidal M, Hengartner MO (2002) *Caenorhabditis elegans* HUS-1 is a DNA damage checkpoint protein required for genome stability and EGL-1-mediated apoptosis. Curr Biol 12(22):1908–1918

Horejsi Z, Takai H, Adelman CA, Collis SJ, Flynn H, Maslen S, Skehel JM, de Lange T, Boulton SJ (2010) CK2 phospho-dependent binding of R2TP complex to TEL2 is essential for mTOR and SMG1 stability. Mol Cell 39(6):839–850. doi:S1097-2765(10)00673-8 [pii]10.1016/j.molcel.2010.08.037

Hurov KE, Cotta-Ramusino C, Elledge SJ (2010) A genetic screen identifies the Triple T complex required for DNA damage signaling and ATM and ATR stability. Genes Dev 24(17):1939–1950. doi:24/17/1939 [pii]10.1101/gad.1934210

Ip SC, Rass U, Blanco MG, Flynn HR, Skehel JM, West SC (2008) Identification of Holliday junction resolvases from humans and yeast. Nature 456(7220):357–361. doi:nature07470 [pii]10.1038/nature07470

Ito S, Greiss S, Gartner A, Derry WB (2010) Cell-nonautonomous regulation of *C. elegans* germ cell death by kri-1. Curr Biol 20(4):333–338. doi:S0960-9822(09)02156-3 [pii]10.1016/j.cub.2009.12.032

Jackson SP, Bartek J (2009) The DNA-damage response in human biology and disease. Nature 461(7267):1071–1078. doi:nature08467 [pii]10.1038/nature08467

Jaramillo-Lambert A, Engebrecht J (2010) A single unpaired and transcriptionally silenced X chromosome locally precludes checkpoint signaling in the *Caenorhabditis elegans* germ line. Genetics 184(3):613–628. doi:genetics.109.110338 [pii]10.1534/genetics.109.110338

Jaramillo-Lambert A, Harigaya Y, Vitt J, Villeneuve A, Engebrecht J (2010) Meiotic errors activate checkpoints that improve gamete quality without triggering apoptosis in male germ cells. Curr Biol 20(23):2078–2089. doi:S0960-9822(10)01270-4 [pii]10.1016/j.cub.2010.10.008

Jeffers JR, Parganas E, Lee Y, Yang C, Wang J, Brennan J, MacLean KH, Han J, Chittenden T, Ihle JN, McKinnon PJ, Cleveland JL, Zambetti GP (2003) Puma is an essential mediator of p53-dependent and -independent apoptotic pathways. Cancer Cell 4(4):321–328. doi:S1535610803002447 [pii]

Jones SN, Roe AE, Donehower LA, Bradley A (1995) Rescue of embryonic lethality in Mdm2-deficient mice by absence of p53. Nature 378(6553):206–208. doi:10.1038/378206a0

Jones AR, Francis R, Schedl T (1996) GLD-1, a cytoplasmic protein essential for oocyte differentiation, shows stage- and sex-specific expression during *Caenorhabditis elegans* germline development. Dev Biol 180(1):165–183. doi:S0012-1606(96)90293-6 [pii]10.1006/dbio.1996.0293

Joshi N, Barot A, Jamison C, Borner GV (2009) Pch2 links chromosome axis remodeling at future crossover sites and crossover distribution during yeast meiosis. PLoS Genet 5(7):e1000557. doi:10.1371/journal.pgen.1000557

Kaizuka T, Hara T, Oshiro N, Kikkawa U, Yonezawa K, Takehana K, Iemura S, Natsume T, Mizushima N (2010) Tti1 and Tel2 are critical factors in mammalian target of rapamycin complex assembly. J Biol Chem 285(26):20109–20116. doi:M110.121699 [pii]10.1074/jbc.M110.121699

Kanoh J, Yanagida M (2007) Tel2: a common partner of PIK-related kinases and a link between DNA checkpoint and nutritional response? Genes Cells 12(12):1301–1304. doi:GTC1142 [pii]10.1111/j.1365-2443.2007.01142.x

Keeney S, Giroux CN, Kleckner N (1997) Meiosis-specific DNA double-strand breaks are catalyzed by Spo11, a member of a widely conserved protein family. Cell 88(3):375–384

Kolesnick R, Hannun YA (1999) Ceramide and apoptosis. Trends Biochem Sci 24(6):224–225. doi:S0968-0004(99)01408-5 [pii] (author reply 227)

Kritikou EA, Milstein S, Vidalain PO, Lettre G, Bogan E, Doukoumetzidis K, Gray P, Chappell TG, Vidal M, Hengartner MO (2006) *C. elegans* GLA-3 is a novel component of the MAP kinase MPK-1 signaling pathway required for germ cell survival. Genes Dev 20(16):2279–2292. doi:20/16/2279 [pii]10.1101/gad.384506

Lane D, Levine A (2010) p53 research: the past thirty years and the next thirty years. Cold Spring Harb Perspect Biol 2(12):a000893. doi:cshperspect.a000893 [pii]10.1101/cshperspect.a000893

Lane DP, Cheok CF, Brown CJ, Madhumalar A, Ghadessy FJ, Verma C (2010a) The Mdm2 and p53 genes are conserved in the Arachnids. Cell Cycle 9(4):748–754. doi:10616 [pii]

Lane DP, Cheok CF, Brown C, Madhumalar A, Ghadessy FJ, Verma C (2010b) Mdm2 and p53 are highly conserved from placozoans to man. Cell Cycle 9(3):540–547. doi:10516 [pii]

Lee MH, Schedl T (2001) Identification of in vivo mRNA targets of GLD-1, a maxi-KH motif containing protein required for *C. elegans* germ cell development. Genes Dev 15(18):2408–2420. doi:10.1101/gad.915901

Lee MH, Ohmachi M, Arur S, Nayak S, Francis R, Church D, Lambie E, Schedl T (2007) Multiple functions and dynamic activation of MPK-1 extracellular signal-regulated kinase signaling in *Caenorhabditis elegans* germline development. Genetics 177(4):2039–2062. doi:177/4/2039 [pii]10.1534/genetics.107.081356

Lemmens BB, Tijsterman M (2011) DNA double-strand break repair in *Caenorhabditis elegans*. Chromosoma 120(1):1–21. doi:10.1007/s00412-010-0296-3

Lightfoot J, Testori S, Barroso C, Martinez-Perez E (2011) Loading of meiotic cohesin by SCC-2 is required for early processing of DSBs and for the DNA damage checkpoint. Curr Biol. doi:S0960-9822(11)00776-7 [pii]10.1016/j.cub.2011.07.007

Lui DY, Colaiácovo MP (2012) Meiotic development in *C. elegans*. Advances in Experimental Medicine and Biology 757:133–170. (Chap. 6, this volume) Springer, New York

Luo S, Kleemann GA, Ashraf JM, Shaw WM, Murphy CT (2010) TGF-beta and insulin signaling regulate reproductive aging via oocyte and germline quality maintenance. Cell 143(2):299–312. doi:S0092-8674(10)01061-5 [pii]10.1016/j.cell.2010.09.013

MacQueen AJ, Villeneuve AM (2001) Nuclear reorganization and homologous chromosome pairing during meiotic prophase require *C. elegans* chk-2. Genes Dev 15(13):1674–1687. doi:10.1101/gad.902601

Meier B, Gartner A (2006) Meiosis: checking chromosomes pair up properly. Curr Biol 16(7):R249–R251. doi:S0960-9822(06)01250-4 [pii]10.1016/j.cub.2006.03.002

Mets DG, Meyer BJ (2009) Condensins regulate meiotic DNA break distribution, thus crossover frequency, by controlling chromosome structure. Cell 139(1):73–86. doi:S0092-8674(09)00915-5 [pii]10.1016/j.cell.2009.07.035

Morita Y, Tilly JL (1999) Oocyte apoptosis: like sand through an hourglass. Dev Biol 213(1):1–17. doi:10.1006/dbio.1999.9344S0012-1606(99)99344-2 [pii]

Moser SC, von Elsner S, Bussing I, Alpi A, Schnabel R, Gartner A (2009) Functional dissection of *Caenorhabditis elegans* CLK-2/TEL2 cell cycle defects during embryogenesis and germline development. PLoS Genet 5(4):e1000451. doi:10.1371/journal.pgen.1000451

Oda E, Ohki R, Murasawa H, Nemoto J, Shibue T, Yamashita T, Tokino T, Taniguchi T, Tanaka N (2000) Noxa, a BH3-only member of the Bcl-2 family and candidate mediator of p53-induced apoptosis. Science 288(5468):1053–1058. doi:8508 [pii]

Ou HD, Lohr F, Vogel V, Mantele W, Dotsch V (2007) Structural evolution of C-terminal domains in the p53 family. EMBO J 26(14):3463–3473. doi:7601764 [pii]10.1038/sj.emboj.7601764

Park D, Jia H, Rajakumar V, Chamberlin HM (2006) Pax2/5/8 proteins promote cell survival in *C. elegans*. Development 133(21):4193–4202. doi:dev.02614 [pii]10.1242/dev.02614

Penkner A, Portik-Dobos Z, Tang L, Schnabel R, Novatchkova M, Jantsch V, Loidl J (2007) A conserved function for a *Caenorhabditis elegans* Com1/Sae2/CtIP protein homolog in meiotic recombination. EMBO J 26(24):5071–5082. doi:7601916 [pii]10.1038/sj.emboj.7601916

Pepling ME (2006) From primordial germ cell to primordial follicle: mammalian female germ cell development. Genesis 44(12):622–632. doi:10.1002/dvg.20258

Pepling ME, Spradling AC (1998) Female mouse germ cells form synchronously dividing cysts. Development 125(17):3323–3328

Pepling ME, Spradling AC (2001) Mouse ovarian germ cell cysts undergo programmed breakdown to form primordial follicles. Dev Biol 234(2):339–351. doi:10.1006/dbio.2001.0269S0012-1606(01)90269-6 [pii]

Perry DK, Hannun YA (1998) The role of ceramide in cell signaling. Biochim Biophys Acta 1436(1–2):233–243. doi:S0005-2760(98)00145-3 [pii]

Peschiaroli A, Scialpi F, Bernassola F, Pagano M, Melino G (2009) The F-box protein FBXO45 promotes the proteasome-dependent degradation of p73. Oncogene 28(35):3157–3166. doi:onc2009177 [pii]10.1038/onc.2009.177

Pourkarimi E, Greiss S, Gartner A (2012) Evidence that CED-9/Bcl2 and CED-4/Apaf-1 localization is not consistent with the current model for *C. elegans* apoptosis induction. Cell Death Differ 19(3):406–415. doi:cdd2011104 [pii]10.1038/cdd.2011.104

Rass U, Compton SA, Matos J, Singleton MR, Ip SC, Blanco MG, Griffith JD, West SC (2010) Mechanism of Holliday junction resolution by the human GEN1 protein. Genes Dev 24(14):1559–1569. doi:24/14/1559 [pii]10.1101/gad.585310

Roig I, Dowdle JA, Toth A, de Rooij DG, Jasin M, Keeney S (2010) Mouse TRIP13/PCH2 is required for recombination and normal higher-order chromosome structure during meiosis. PLoS Genet 6(8). doi:10.1371/journal.pgen.1001062

Ross AJ, Li M, Yu B, Gao MX, Derry WB (2011) The EEL-1 ubiquitin ligase promotes DNA damage-induced germ cell apoptosis in *C. elegans*. Cell Death Differ 18(7):1140–1149. doi:cdd2010180 [pii]10.1038/cdd.2010.180

Rutkowski R, Hofmann K, Gartner A (2010) Phylogeny and function of the invertebrate p53 superfamily. Cold Spring Harb Perspect Biol 2(7):a001131. doi:cshperspect.a001131 [pii]10.1101/cshperspect.a001131

Rutkowski R, Dickinson R, Stewart G, Craig A, Schimpl M, Keyse SM, Gartner A (2011) Regulation of *Caenorhabditis elegans* p53/CEP-1-dependent germ cell apoptosis by Ras/MAPK signaling. PLoS Genet 7(8):e1002238

Salinas LS, Maldonado E, Navarro RE (2006) Stress-induced germ cell apoptosis by a p53 independent pathway in *Caenorhabditis elegans*. Cell Death Differ 13(12):2129–2139. doi:4401976 [pii]10.1038/sj.cdd.4401976

San-Segundo PA, Roeder GS (1999) Pch2 links chromatin silencing to meiotic checkpoint control. Cell 97(3):313–324. doi:S0092-8674(00)80741-2 [pii]

San-Segundo PA, Roeder GS (2000) Role for the silencing protein Dot1 in meiotic checkpoint control. Mol Biol Cell 11(10):3601–3615

Schertel C, Conradt B (2007) *C. elegans* orthologs of components of the RB tumor suppressor complex have distinct pro-apoptotic functions. Development 134(20):3691–3701. doi:dev.004606 [pii]10.1242/dev.004606

Schumacher B, Hofmann K, Boulton S, Gartner A (2001) The *C. elegans* homolog of the p53 tumor suppressor is required for DNA damage-induced apoptosis. Curr Biol 11(21):1722–1727

Schumacher B, Schertel C, Wittenburg N, Tuck S, Mitani S, Gartner A, Conradt B, Shaham S (2005a) *C. elegans* ced-13 can promote apoptosis and is induced in response to DNA damage. Cell Death Differ 12(2):153–161

Schumacher B, Hanazawa M, Lee MH, Nayak S, Volkmann K, Hofmann ER, Hengartner M, Schedl T, Gartner A (2005b) Translational repression of *C. elegans* p53 by GLD-1 regulates DNA damage-induced apoptosis. Cell 120(3):357–368

Schwartz HT (2007) A protocol describing pharynx counts and a review of other assays of apoptotic cell death in the nematode worm *Caenorhabditis elegans*. Nat Protoc 2(3): 705–714. doi:nprot.2007.93 [pii]10.1038/nprot.2007.93

Sendoel A, Kohler I, Fellmann C, Lowe SW, Hengartner MO (2010) HIF-1 antagonizes p53-mediated apoptosis through a secreted neuronal tyrosinase. Nature 465(7298):577–583. doi:nature09141 [pii]10.1038/nature09141

Shibue T, Takeda K, Oda E, Tanaka H, Murasawa H, Takaoka A, Morishita Y, Akira S, Taniguchi T, Tanaka N (2003) Integral role of Noxa in p53-mediated apoptotic response. Genes Dev 17(18):2233–2238. doi:10.1101/gad.11036031103603 [pii]

Squirrell JM, Eggers ZT, Luedke N, Saari B, Grimson A, Lyons GE, Anderson P, White JG (2006) CAR-1, a protein that localizes with the mRNA decapping component DCAP-1, is required for cytokinesis and ER organization in *Caenorhabditis elegans* embryos. Mol Biol Cell 17(1):336–344. doi:E05-09-0874 [pii]10.1091/mbc.E05-09-0874

Stergiou L, Doukoumetzidis K, Sendoel A, Hengartner MO (2007) The nucleotide excision repair pathway is required for UV-C-induced apoptosis in *Caenorhabditis elegans*. Cell Death Differ 14(6):1129–1138. doi:4402115 [pii]10.1038/sj.cdd.4402115

Suh EK, Yang A, Kettenbach A, Bamberger C, Michaelis AH, Zhu Z, Elvin JA, Bronson RT, Crum CP, McKeon F (2006) p63 protects the female germ line during meiotic arrest. Nature 444(7119):624–628. doi:nature05337 [pii]10.1038/nature05337

Takai H, Xie Y, de Lange T, Pavletich NP (2010) Tel2 structure and function in the Hsp90-dependent maturation of mTOR and ATR complexes. Genes Dev 24(18):2019–2030. doi:gad.1956410 [pii]10.1101/gad.1956410

Tilly JL (2001) Commuting the death sentence: how oocytes strive to survive. Nat Rev Mol Cell Biol 2(11):838–848. doi:10.1038/3509908635099086 [pii]

Villunger A, Michalak EM, Coultas L, Mullauer F, Bock G, Ausserlechner MJ, Adams JM, Strasser A (2003) p53- and drug-induced apoptotic responses mediated by BH3-only proteins puma and noxa. Science 302(5647):1036–1038. doi:10.1126/science.10900721090072 [pii]

Wechsler T, Newman S, West SC (2011) Aberrant chromosome morphology in human cells defective for Holliday junction resolution. Nature 471(7340):642–646. doi:nature09790 [pii]10.1038/nature09790

West SC (2009) The search for a human Holliday junction resolvase. Biochem Soc Trans 37(Pt 3):519–526. doi:BST0370519 [pii]10.1042/BST0370519

Wolke U, Jezuit EA, Priess JR (2007) Actin-dependent cytoplasmic streaming in *C. elegans* oogenesis. Development 134(12):2227–2236. doi:dev.004952 [pii]10.1242/dev.004952

Wright JE, Gaidatzis D, Senften M, Farley BM, Westhof E, Ryder SP, Ciosk R (2011) A quantitative RNA code for mRNA target selection by the germline fate determinant GLD-1. EMBO J 30(3):533–545. doi:emboj2010334 [pii]10.1038/emboj.2010.334

Yang M, Sun J, Sun X, Shen Q, Gao Z, Yang C (2009) *Caenorhabditis elegans* protein arginine methyltransferase PRMT-5 negatively regulates DNA damage-induced apoptosis. PLoS Genet 5(6):e1000514. doi:10.1371/journal.pgen.1000514

Zanders S, Alani E (2009) The pch2Delta mutation in baker's yeast alters meiotic crossover levels and confers a defect in crossover interference. PLoS Genet 5(7):e1000571. doi:10.1371/journal.pgen.1000571

Zanders S, Brown MS, Chen C, Alani E (2011) Pch2 modulates chromatid partner choice during meiotic double-strand break repair in *Saccharomyces cerevisiae*. Genetics 188(3):511–521. doi:genetics.111.129031 [pii]10.1534/genetics.111.129031

Chapter 10
Control of Oocyte Growth and Meiotic Maturation in *Caenorhabditis elegans*

Seongseop Kim, Caroline Spike, and David Greenstein

Abstract In sexually reproducing animals, oocytes arrest at diplotene or diakinesis and resume meiosis (meiotic maturation) in response to hormones. Chromosome segregation errors in female meiosis I are the leading cause of human birth defects, and age-related changes in the hormonal environment of the ovary are a suggested cause. *Caenorhabditis elegans* is emerging as a genetic paradigm for studying hormonal control of meiotic maturation. The meiotic maturation processes in *C. elegans* and mammals share a number of biological and molecular similarities. Major sperm protein (MSP) and luteinizing hormone (LH), though unrelated in sequence, both trigger meiotic resumption using somatic $G\alpha_s$-adenylate cyclase pathways and soma–germline gap-junctional communication. At a molecular level, the oocyte responses apparently involve the control of conserved protein kinase pathways and post-transcriptional gene regulation in the oocyte. At a cellular level, the responses include cortical cytoskeletal rearrangement, nuclear envelope breakdown, assembly of the acentriolar meiotic spindle, chromosome segregation, and likely changes important for fertilization and the oocyte-to-embryo transition. This chapter focuses on signaling mechanisms required for oocyte growth and meiotic maturation in *C. elegans* and discusses how these mechanisms coordinate the completion of meiosis and the oocyte-to-embryo transition.

Keywords Oogenesis • Meiosis • Meiotic maturation signaling • Major sperm protein • G-protein signaling • Meiotic resumption • Meiotic chromosome segregation • Meiotic spindle

---

S. Kim • C. Spike • D. Greenstein (✉)
Department of Genetics, Cell Biology, and Development, University of Minnesota, Minneapolis, MN 55455, USA
e-mail: green959@umn.edu

T. Schedl (ed.), *Germ Cell Development in C. elegans*, Advances in Experimental Medicine and Biology 757, DOI 10.1007/978-1-4614-4015-4_10,

10.1 Overview

10.1.1 Meiosis and the Meiotic Maturation Divisions

Cells of the germ line form gametes and establish an unbroken chain between generations. The physical links in this chain, forged by the union of gametes at fertilization, are dependent on the faithful execution of meiosis. Meiosis ensures the formation of euploid embryos by halving the number of chromosomes contributed by each gamete (see Chap. 6, Lui and Colaiácovo 2012). Despite this universal requirement in the sexual reproduction of eukaryotic organisms, meiosis is regulated differently in oocytes and spermatocytes. Whereas spermatocytes proceed through the meiotic divisions uninterrupted (see Chap. 7, Chu and Shakes 2012), oocytes almost invariably arrest once, and sometimes twice following premeiotic DNA replication and meiotic recombination, depending on the species. This unique characteristic of oocyte meiosis, as well as its close temporal association with fertilization, was recognized early by developmental biologists, who coined the term "meiotic maturation" for the nuclear and cytoplasmic changes occurring in oocytes just before zygote formation (Wilson 1925; Masui and Clarke 1979). Oocyte meiotic maturation is defined by the transition between diakinesis and metaphase of meiosis I and is accompanied by nuclear envelope breakdown, rearrangement of the cortical cytoskeleton, and meiotic spindle assembly (Fig. 10.1). The timing of the meiotic divisions with respect to fertilization varies among species (Fig. 10.1). Despite these differences in timing, molecular underpinnings of oocyte meiotic maturation are conserved among different animals. The discovery of Maturation

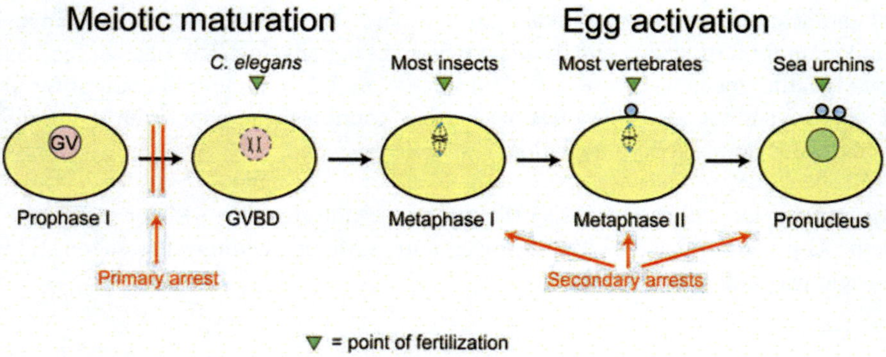

Fig. 10.1 Oocyte meiotic maturation and egg activation. The oocytes of most animals arrest in meiotic prophase I (primary arrest), and resume meiosis (meiotic maturation) in response to hormonal stimulation (starfish, 1-methyladenine; *Xenopus*, progesterone; *C. elegans*, MSP; humans, LH). Meiotic maturation is accompanied by germinal vesicle breakdown (GVBD), cortical cytoskeletal reorganization, and meiotic spindle assembly (*blue circles* are polar bodies). The point of fertilization is species dependent—in *C. elegans*, fertilization happens when the mature oocyte enters the spermatheca at ovulation. Oocytes of many species undergo secondary arrests before fertilization, which triggers egg activation, as indicated

promoting factor (MPF; Masui and Markert 1971; Masui 2001) in studies of meiotic maturation in amphibian oocytes provides a prime example. Genetic and biochemical analysis of the cell cycle, together with MPF purification, demonstrated that cyclin-dependent protein kinases are universal regulators of meiotic and mitotic cell cycle progression in eukaryotes (Morgan 2007).

This review, like its antecedent chapters, focuses on the regulation of key developmental events in the germ line, here oocyte growth and meiotic maturation, in the nematode *Caenorhabditis elegans*. As an experimental system, *C. elegans* is a relative newcomer to this area of reproductive and developmental biology, yet the worm offers a number of advantages, including the ability to observe the events of oocyte growth and meiotic maturation in intact living animals and the potential to perturb normal development by mutational analysis and RNA interference. Intercellular signaling between gametes, and between the soma and germ line, regulates oocyte meiotic maturation and is a focal point of current research in this field.

10.1.2 Maturation-Promoting Factor

In *C. elegans*, as for all examined species, MPF is a master regulator of cell cycle progression during oocyte meiotic maturation (Boxem et al. 1999; Burrows et al. 2006). Because studies in vertebrate systems established the foundation for the meiotic maturation field, these studies are briefly reviewed to provide context for discussion of the *C. elegans* system. Classic studies of amphibian oocyte meiotic maturation by Yoshio Masui first led to the discovery of MPF (Masui and Markert 1971; Tunquist and Maller 2003). MPF consists of the Cdk1 catalytic subunit and the cyclin B regulatory subunit (Dunphy et al. 1988; Gautier et al. 1988; Lohka et al. 1988; Gautier et al. 1990). The cyclin B/Cdk1 protein kinase is inactive in immature oocytes due to inhibitory CDK phosphorylations at Thr14 and Tyr15 catalyzed by the Wee1 or Myt1 kinases (Fig. 10.2; Kornbluth et al. 1994; Mueller et al. 1995). In *Xenopus*, these inhibitory phosphorylations are removed by the conserved Cdc25 phosphatase following stimulation with progesterone (Kumagai and Dunphy 1991), which leads to nuclear envelope breakdown. An initial signal for MPF activation is amplified by a positive feedback loop in which the active CDK promotes the inactivation of its inhibitors, Wee1 and Myt1 (Walter et al. 2000; Peter et al. 2002), and stimulates its activator, Cdc25 (Kumagai and Dunphy 1996). The Greatwall kinase is a new component of the pathway for MPF activation (Yu et al. 2006; Zhao et al. 2008). Itself an MPF substrate, Greatwall inhibits protein phosphatase 2A (PP2A) activity, which in turn is an inhibitor of Cdc25. Greatwall exerts its effects by phosphorylating the PP2A inhibitor α-endosulfine (Castilho et al. 2009; Mochida et al. 2010), which is required for meiotic maturation in *Drosophila* (Von Stetina et al. 2008). Since Greatwall is not found in *C. elegans*, but other pathway components are conserved (Fig. 10.2) another kinase might contribute to MPF activation loop in this organism. In fact, the *C. elegans*, polo-like kinase PLK-1 is required for timely germinal vesicle breakdown (GVBD) and the completion of meiosis (Chase et al. 2000).

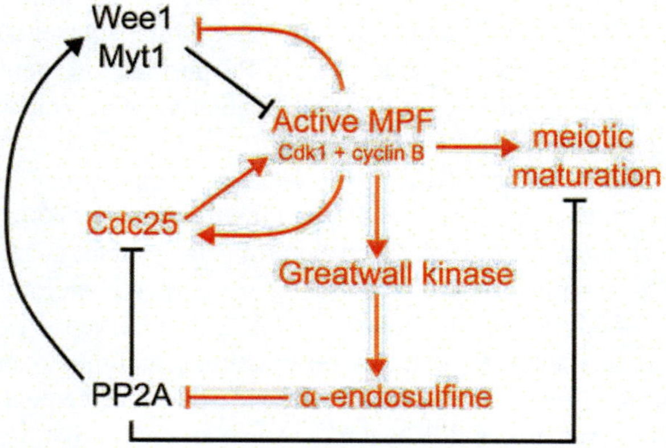

Fig. 10.2 Pathways regulating MPF activity. Activation loops resulting in amplified MPF activation and meiotic maturation are in *red*. Regulators inhibiting MPF activation, and inhibited by the MPF-dependent feed-forward activation loops, are in *black*. The Greatwall kinase is not conserved in *C. elegans*, but other components are conserved: the PP2A catalytic subunit is LET-92; the PP2A structural subunit is PAA-1; there are several PP2A regulatory subunits (PPTR-1, PPTR-2, RSA-1, SUR-6, C06G1.5, F47B8.3, T22D1.5); and α-endosulfine is K10C3.2

Active MPF phosphorylates substrates that function in key cellular processes of meiotic maturation including nuclear envelope breakdown, chromosome condensation, and spindle assembly. Subsequently, M-phase exit and anaphase chromosome segregation require the function of a multi-subunit E3 ubiquitin ligase called the anaphase promoting complex or cyclosome (APC/C; Peters 2002), which promotes cyclin B degradation and MPF inactivation.

10.1.3 Translational Regulation and Meiotic Maturation

The regulation of translation is a critical aspect of meiotic maturation (Mendez and Richter 2001). The oocytes of most animals possess translationally repressed or "masked" mRNAs that are translated upon meiotic resumption or following fertilization. Progesterone-induced oocyte meiotic maturation in frogs requires new protein synthesis but not new transcription (Smith and Ecker 1969). Translation of several key regulatory proteins promotes meiotic progression, including the MPF subunit cyclin B (Hochegger et al. 2001; Haccard and Jessus 2006), the novel cyclin Ringo/Speedy (Ferby et al. 1999; Lenormand et al. 1999), and Mos, which functions as a mitogen-activated protein kinase kinase kinase (MAPKKK; Sagata et al. 1988; Dupre et al. 2002; Haccard and Jessus 2006). In *Xenopus*, mitogen-activated protein kinase (MAPK) activates Cdc25 to promote meiotic maturation (Wang et al. 2007), and there is evidence for positive feedback, ensuring an all-or-none response (Liang et al. 2007). Members of the Aurora A family of serine/threonine protein kinases also play critical roles during oocyte maturation. The Eg2 aurora family kinase is phosphorylated and activated soon after progesterone stimulation in

Xenopus oocytes (Andresson and Ruderman 1998). The Eg2 kinase phosphorylates the cytoplasmic polyadenylation element binding protein (CPEB) on Ser174 to promote polyadenylation and translation of *mos* mRNA (Mendez et al. 2000). Biochemical studies in the *Xenopus* system suggest that functionally redundant translational regulatory pathways control MPF activation in response to the meiotic maturation hormone (Haccard and Jessus 2006). The ability to conduct genetic analysis in the *C. elegans* system might prove helpful in teasing apart functional redundancies in meiotic maturation regulatory pathways.

While the regulation of meiotic maturation in the *Xenopus* system is still being actively investigated, the current view is that cytoplasmic polyadenylation of mRNAs promotes translation by promoting a conformation in which the initiation factors eIF4E and eIF4G are productively engaged (Richter 2007). In mice, inactivation of CPEB using oocyte-specific RNAi reduced fertility and led to a variety of defects including premature meiotic maturation, parthenogenesis, and defective folliculogenesis (Racki and Richter 2006). It is clear that other translational control factors, besides CPEB, are critical for oocyte meiosis. For example, *Xenopus* Pumilio-2 is required for Ringo/Speedy translation in response to progesterone, which in turn results in the activation of CPEB and *mos* mRNA translation (Padmanabhan and Richter 2006).

In addition to the control of translation by the regulation of cytoplasmic polyadenylation, studies of mouse oocytes suggest the involvement of small RNA pathways in the completion of meiosis. An analysis of fertility defects of Dicer-deficient and Argonaute 2-deficient oocytes was reported in the mouse (Murchison et al. 2007; Tang et al. 2007; Kaneda et al. 2009). These studies suggest that translational regulation by microRNAs in the germ line is important for normal oocyte meiotic maturation and zygotic development. While there has been considerable progress in elucidating the biochemistry of kinase cascades and translational control, it is less clear how hormonal signaling and soma–germline interactions tie into these intracellular processes to regulate meiotic maturation. The following section reviews studies that establish *C. elegans* as a genetic model system for studying the regulation of oocyte meiotic maturation by hormonal signaling.

10.2 Meiotic Maturation in *C. elegans*

10.2.1 Timing of Events

Oocyte meiotic maturation can be viewed by time-lapse videomicroscopy of living animals (Ward and Carrel 1979). The timing of landmark events during meiotic maturation (McCarter et al. 1999) and the ultrastructure of the proximal gonad (Hall et al. 1999; Hall and Altun 2008) have been described. The nuclear envelope of the most proximal oocyte breaks down ~5 min prior to ovulation as it enters meiotic M-phase from prophase (Fig. 10.3; McCarter et al. 1999). During maturation, the

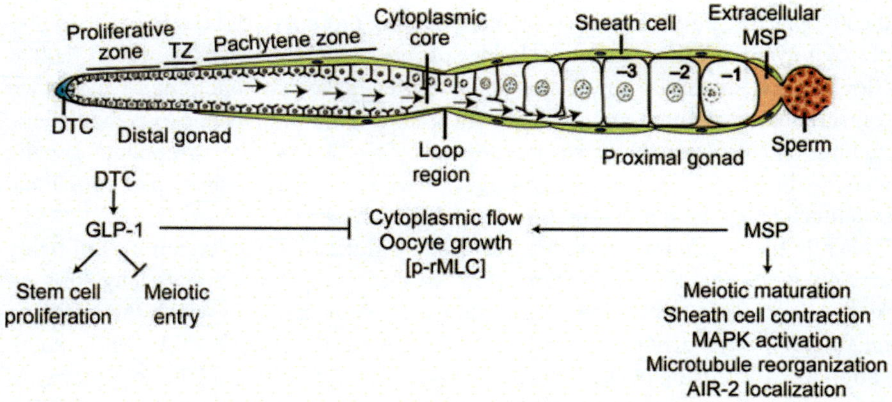

Fig. 10.3 A model for the control of germline proliferation, oocyte growth, and meiotic maturation by GLP-1/Notch and MSP signaling. One of the two adult hermaphrodite gonad arms is depicted (indicated are: *DTC* distal tip cell; *TZ* transition zone; and proximal oocytes, –1 to –3). Arrows indicate cytoplasmic flow for oocyte growth. p-rMLC is shown in *brackets* to indicate that MSP signaling is sufficient to promote rMLC phosphorylation in the germline, whereas the *glp-1* pathway appears dispensable

oocyte also undergoes a structural change termed cortical rearrangement (McCarter et al. 1999). These changes within the oocyte coincide with a reproducible sequence of somatic motor events mediated by the contractile proximal sheath cells and the distal spermatheca resulting in ovulation. As the nuclear envelope breaks down, microtubules gain access to the highly condensed bivalents and the acentriolar meiotic spindle begins to assemble (Yang et al. 2003). Fertilization appears to occur rapidly upon oocyte entry into the spermatheca (Ward and Carrel 1979; Samuel et al. 2001). The fertilized embryo enters the uterus approximately 4 min after ovulation and the meiotic divisions are completed there within approximately 30 min of nuclear envelope breakdown, with the MI division taking approximately 20 min and the MII division taking approximately 10 min (McCarter et al. 1999).

10.2.2 *C. elegans Sperm Use Major Sperm Proteins to Promote Meiotic Maturation*

In a landmark study, McCarter et al. (1999) showed that a sperm-associated signal promotes oocyte meiotic maturation and contraction of the follicle-like gonadal sheath cells, prior to and independent of fertilization. In females (genetically altered XX animals that produce no sperm), oocytes mature and are ovulated at low rates (<0.1 maturations per gonad arm per hour). Mating of spermless females to wild-type males, or fertilization incompetent sperm-defective (*spe*) mutant males, restores a normal rate of oocyte maturation (~2.5 maturations per gonad arm per hour; McCarter et al. 1999). Oocytes also appear to exhibit an increase in metabolic activity in the presence of sperm (Yang et al. 2010). A gonad arm in a female produces

~15 diakinesis arrested oocytes. Thus, sperm signaling is not required for germline development up to the diakinesis stage. However, in the absence of sperm, the assembly-line production of oocytes halts. The continued presence of sperm is required for continued oogenesis. As explained below, in addition to promoting meiotic maturation, sperm stimulate the cytoplasmic flows that drive oocyte growth.

In nematodes, the ancestral mating system appears to be male–female, whereas hermaphroditism and parthenogenesis are derived alternative reproductive strategies (Cho et al. 2004; Kiontke et al. 2004; Kiontke and Fitch 2005; Woodruff et al. 2010). The regulation of meiotic maturation by sperm in many nematodes might function to preserve germline resources, for if oocytes undergo meiotic maturation in the absence of sperm, they rapidly lose competence for fertilization, become endomitotic, and are expelled from the animal through the vulva. The stimulation of meiotic resumption by sperm provides an efficient means for uniting developmentally fertilization-competent sperm and oocyte.

The sperm signal for oocyte meiotic maturation was identified biochemically (Miller et al. 2001). The surprising result was that the major sperm proteins (MSPs), central cytoskeletal elements required for the actin-independent motility of nematode spermatozoa (Bottino et al. 2002), have a dual role in *C. elegans* reproduction, functioning as hormones for oocyte meiotic maturation and gonadal sheath cell contraction (Miller et al. 2001). An in vivo bioassay was used in which sperm-conditioned medium or sperm lysates were injected into the uterus of unmated *fog-2* female animals, and oocyte meiotic maturation and gonadal sheath cell contraction were monitored by time-lapse videomicroscopy. Using meiotic maturation as a functional readout, the bioactive factors were purified to homogeneity with reversed-phase high-performance liquid chromatography and shown to contain only the MSPs by mass spectrometry. Injection of anti-MSP antibodies into the uterus of hermaphrodites results in a reduction in ovulation rates (Miller et al. 2001), consistent with the hypothesis that MSPs function as endogenous signals. A multigene family of 28 genes encodes MSP proteins that share approximately 97–100% identity (Burke and Ward 1983; Klass et al. 1984; Ward et al. 1988). Several MSP isoforms, including MSP-38, MSP-77, and MSP-142, have been expressed in bacteria, either with or without N-terminal 6His-Tags, and shown to induce meiotic maturation and sheath cell contraction when injected into the uterus at a range of concentrations (25–100 nM) (Miller et al. 2001; Govindan et al. 2009). The actual concentration range for signaling is likely considerably lower because the initial volume of injected MSP solution (~50 pl) subsequently diffuses within the uterus, spermatheca, and gonad arm.

The identification of a signaling role for MSP raised the question of how spermatozoa, devoid of ribosomes, endoplasmic reticulum (ER), and Golgi, release a cytoskeletal protein lacking a signal sequence. Non-motile spermatids and motile spermatozoa appear to bud MSP-containing vesicles to signal sheath cells and oocytes (Kosinski et al. 2005). MSP vesicles appear to be labile structures, which provide a potential basis for release of MSP in an extracellular form that exhibits a graded distribution in the gonad.

MSP is a bipartite signal for meiotic maturation and sheath cell contraction because a synthetic peptide consisting of evolutionarily conserved C-terminal 21 amino acids stimulates sheath cell contraction at normal levels, but not meiotic maturation (Miller et al. 2001; Yin et al. 2004). By contrast, an MSP deletion derivative lacking this C-terminal region can promote meiotic maturation, but is less efficient in its ability to promote sheath cell contraction (Miller et al. 2001). This result suggests that MSP possesses two separable signaling functions and therefore likely activates distinct signal transduction pathways for meiotic maturation and sheath contraction (Miller et al. 2001). Consistent with this prediction and as described below, meiotic maturation requires $G\alpha_s$-adenylate cyclase-protein kinase A (PKA) signaling in the gonadal sheath cells, whereas sheath cell contraction requires EGL-30/$G\alpha_q$ signaling (Govindan et al. 2009). Divergent MSP-related proteins, called VAPs (for VAMP-associated protein) share only ~20% amino acid identity with nematode MSP yet fold into similar seven-stranded immunoglobulin β sandwich-like structures (Kaiser et al. 2005). Human, *Drosophila* and *C. elegans* VAPs are able to signal meiotic maturation and sheath contraction using the *C. elegans* bioassay (Tsuda et al. 2008). A phylogenetic analysis of MSP domain-containing proteins indicates that the VAP clade is widely distributed among plants, animals, and protists, but that three additional clades are largely nematode specific (Tarr and Scott 2005). Recent results in *Drosophila, C. elegans*, and mammals suggest that some VAP family members can function as extracellular signals (Tsuda et al. 2008).

10.2.3 Molecular Readouts of MSP Signaling

MSP promotes meiotic maturation and sheath cell contraction and activates signaling pathways that are important for both processes. These readouts include activation of MAPK in the germ line, reorganization of oocyte microtubules, localization of the AIR-2 Aurora B kinase to chromatin, reorganization of ribonucleoprotein complexes in oocytes, and stimulation of actomyosin-based motility. Each of these readouts of MSP signaling is now discussed.

10.2.3.1 Activation of MPK-1 MAPK in the Germ Line

In *C. elegans*, the MPK-1 MAPK pathway functions in multiple developmental processes in the germ line, including the specification of the male germline fate in males and hermaphrodites, progression through pachytene, the negative regulation of physiological apoptosis in the germ line, and the control of oocyte growth and meiotic maturation (Lee et al. 2007; Arur et al. 2009). MPK-1 appears to have multiple targets that function in diverse cellular processes in the germ line, including morphogenesis and cellular organization of the gonad, oocyte growth control, and oocyte organization and differentiation (Lee et al. 2007; Arur et al. 2009). Proximal oocytes exhibit MAPK activation in the presence of sperm, as detected using antibodies that specifically recognize the diphosphorylated-activated form of MPK-1 MAPK (dpMPK-1; Miller et al. 2001; Page et al. 2001; Lee et al. 2007).

Injection of MSP into the uterus of unmated females is sufficient to generate dpMPK-1 in oocytes within 40 min of injection (Miller et al. 2001; Jud et al. 2008). MAPK activation plays critical roles in regulating cytoplasmic and nuclear events of meiotic maturation in invertebrates and vertebrates (Ferrell 1999a, b; Haccard and Jessus 2006; Liang et al. 2007), presumably through phosphorylation of specific targets. While the exact functions of MAPK in meiotic maturation are not fully defined at a mechanistic level, the identification of MAPK substrates will provide multiple handles on this problem (Arur et al. 2009, 2011).

10.2.3.2 Microtubule Reorganization Prior to Fertilization

The meiotic spindles of most animal oocytes are distinctive from mitotic spindles of somatic cells in terms of their mechanism of assembly, their function, and their modes of regulation. The female meiotic spindles of many species are both acentriolar and anastral (Albertson and Thomson 1993). Instead of relying on centrosomes for assembly of a bipolar spindle, the meiotic chromatin functions to nucleate microtubules, which self-organize through incompletely understood mechanisms that involve sorting by microtubule motors and microtubule dynamics (Heald et al. 1996; Matthies et al. 1996; Walczak et al. 1998; Skold et al. 2005). Assembly of a bipolar meiotic spindle through chromatin-dependent mechanisms requires that the cytoplasmic microtubules gain access to the nuclear environment, which occurs upon GVBD during the meiotic maturation process. MSP causes GVBD, which as suggested by studies in other systems (Kalab et al. 2011), might expose spindle assembly proteins to a RanGTP gradient, thereby driving spindle assembly.

Interestingly, MSP might promote meiotic spindle assembly through effects on microtubule dynamics that occur prior to GVBD. MSP is sufficient to trigger the reorganization of cytoplasmic microtubules in the oocyte prior to GVBD and fertilization (Harris et al. 2006). When MSP is absent, as in females or older hermaphrodites, microtubules are enriched at the proximal and distal cortices of oocytes. In mated females or younger hermaphrodites, microtubules are dispersed evenly in a net-like fashion throughout the cytoplasm of proximal oocytes. A quantitative assay for oocyte microtubule reorganization was used to show that purified MSP is sufficient to direct cytoskeletal remodeling in the oocyte. The presence of MSP affects the localization and density of growing plus ends, as well as their directionality of movement.

10.2.3.3 Localization of the AIR-2 Aurora B Kinase to Oocyte Chromatin

The establishment, maintenance, and stepwise loss of sister chromatid cohesion are essential for faithful meiotic chromosome segregation. The AIR-2 Aurora B kinase provides a potential link between the sister chromatid cohesion maintenance apparatus and MSP signaling. Chromatin localization of the AIR-2 to the meiotic chromosomes of proximal oocytes depends on the presence of sperm (Schumacher et al.

1998) and MSP is sufficient to promote AIR-2 chromatin localization (Govindan et al. 2009). During anaphase I, sister chromatid cohesion at the short arm of the bivalent, which mediates interhomolog association, is selectively removed to allow separation of homologs. By contrast, sister chromatid cohesion at the long arm of the bivalent must be maintained until anaphase II. *C. elegans* has holocentric chromosomes and microtubules appear to attach to both lateral and poleward ends of meiotic chromosomes (Albertson and Thomson 1993; Howe et al. 2001; Wignall and Villeneuve 2009). The structure and function of holocentric *C. elegans* chromosomes depends on both conserved and novel factors (Maddox et al. 2004; Zetka 2009). A novel protein, LAB-1 is required to protect sister chromatid cohesion during meiosis I (de Carvalho et al. 2008). LAB-1 binds to the long arm of the bivalent and excludes the AIR-2 Aurora B kinase. AIR-2 localizes to the interface between homologs where it phosphorylates the REC-8 meiosis-specific cohesin kleisin subunit promoting its cleavage by separase at anaphase I (Kaitna et al. 2000; Rogers et al. 2002). *C. elegans* has three meiosis-specific kleisin paralogs, *rec-8*, *coh-3*, and *coh-4*, which are required for sister chromatid cohesion during meiosis yet carry out specific functions (Severson et al. 2009). Among these, REC-8 is uniquely required for maintaining sister chromatid cohesion after meiosis I (Severson et al. 2009). Thus, when sperm are present for fertilization, oocytes prepare in advance to initiate chromosome segregation. MSP signaling might be integrated with oocyte intrinsic mechanisms to maintain a high fidelity of chromosome segregation.

10.2.3.4 Reorganization of Ribonucleoprotein Particles

Cortically-localized aggregates of ribonucleoprotein particles (RNPs) form in the oocytes of unmated females or in older adult hermaphrodites that have depleted their self sperm (Schisa et al. 2001; Jud et al. 2007, 2008; Noble et al. 2008; Patterson et al. 2011). The large RNPs that accumulate in arrested oocytes might function to translationally repress and preserve mRNAs that are needed for meiotic maturation and early embryo development. MSP was shown to be sufficient to promote the dissolution of large RNP foci in oocytes (Jud et al. 2008).

10.2.4 The Gonadal Sheath Cells Regulate Oocyte Meiotic Maturation

The gonadal sheath cells function as the major determinant of all described germline responses to the MSP hormone (Miller et al. 2003; Govindan et al. 2006, 2009; Harris et al. 2006; Jud et al. 2008; Nadarajan et al. 2009). A role for the gonadal sheath cells in regulating meiotic maturation was suggested from an analysis of the POU-homeobox gene *ceh-18*, which is expressed in the gonadal sheath cells and required for their proper differentiation and function (Greenstein et al.

1994; McCarter et al. 1997; Rose et al. 1997). Oocytes in *ceh-18* mutant females exhibit a defect in meiotic arrest and undergo MAPK activation, meiotic maturation, and ovulation despite the absence of sperm (Greenstein et al. 1994; Miller et al. 2003; Govindan et al. 2006; Suzuki and Han 2006).

10.2.4.1 Antagonistic Gα_s and G$\alpha_{o/i}$ Pathways Regulate Meiotic Maturation

Besides their essential mechanical role in the process of ovulation (see below), the gonadal sheath cells regulate meiotic maturation through both inhibitory and stimulatory pathways. In the absence of sperm, the sheath cells are critical for inhibiting meiotic maturation, thereby preventing maturation, ovulation, and ultimately wastage of metabolically costly oocytes. An RNAi screen in a *fog-2* female background was used to define negative regulatory pathways that inhibit meiotic maturation in the absence of sperm (Govindan et al. 2006). This screen identified *goa-1*, which encodes G$\alpha_{o/i}$, as a strong negative regulator of meiotic maturation. *goa-1* is expressed in the germ line and somatic gonad, but the use of somatic gonad RNAi-deficient strains indicates that *goa-1* functions in the somatic gonad to inhibit meiotic maturation in the absence of sperm (Govindan et al. 2006).

The gonadal sheath cells also have an essential function in promoting meiotic maturation when sperm are present. This activation function is mediated by components of the Gα_s-adenylate cyclase-protein kinase A pathway (Govindan et al. 2006, 2009). RNAi to the stimulatory Gα_s protein, encoded by *gsa-1*, prevents oocytes from undergoing meiotic maturation despite the presence of sperm (Govindan et al. 2006). *gsa-1* is an essential gene; *gsa-1* null mutants die as L1 stage larvae (Korswagen et al. 1997). Genetic mosaic analysis using a *gsa-1* null mutant demonstrated that *gsa-1* is required in somatic cells of the gonad for meiotic maturation (Govindan et al. 2009). Importantly, *gsa-1* was shown to be dispensable in germ cells for meiotic maturation, and there was no evidence for maternal contribution of *gsa-1* to early development. *acy-4*, which encodes one of four *C. elegans* adenylate cyclases, is required for meiotic maturation (Fig. 10.4). Genetic mosaic analysis established that *acy-4* functions in the gonadal sheath cells to promote meiotic maturation (Govindan et al. 2009). Likewise, genetic mosaic analysis revealed that *kin-1*, which encodes the catalytic subunit of cAMP-dependent PKA, is required in the gonadal sheath cells for meiotic maturation (S. Kim, J.A. Govindan, and D. Greenstein, unpublished results). As expected, *acy-4* acts by regulating PKA activity because a mutation in *kin-2*, which encodes the cAMP-binding regulatory subunit of PKA, suppresses the sterility of *acy-4* null mutants (Govindan et al. 2009).

In female animals, PKA activity must be kept off to prevent meiotic maturation. Constitutively-activated *gsa-1* alleles, *acy-4* over-expression, or inactivation of *kin-2* all derepress meiotic maturation in female backgrounds (Govindan et al. 2006, 2009). In the regulation of meiotic maturation, GOA-1/G$\alpha_{o/i}$ appears to function as an inhibitory G protein because mutations in *gsa-1* and *acy-4* are epistatic to *goa-1* (Govindan et al. 2009). That *goa-1* can regulate meiotic maturation in the

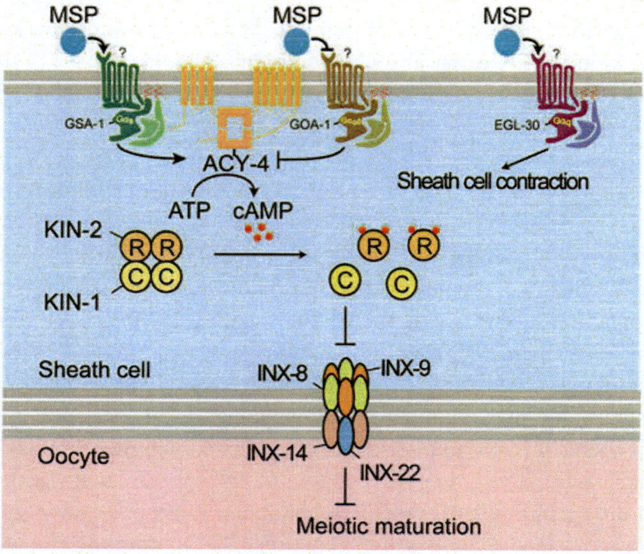

Fig. 10.4 A model for the regulation of meiotic maturation by G protein signaling. Somatic Gα_s-adenylate cyclase-protein kinase A signaling is required for oocyte meiotic maturation. EGL-30/Gα_q promotes sheath cell contraction. G$\alpha_{o/i}$ and sheath–oocyte gap junctions function as inhibitors of meiotic maturation. *Gray lines* are the lipid bilayers of the sheath cell and oocyte

hermaphrodite background is established from experiments in which expression of a constitutively-activated G$\alpha_{o/i}$ protein under control of its endogenous promoter in a multi-copy array significantly reduced meiotic maturation rates (Govindan et al. 2006). Thus, the antagonistic interaction between G$\alpha_{o/i}$ and Gα_s might serve to couple meiotic maturation rates to sperm availability. *gsa-1* and *acy-4* are not required for sheath cell contractions; however, EGL-30/Gaq was shown to be necessary for sheath contractions (Govindan et al. 2009). Thus, multiple sheath cell G protein pathways appear to be required for germline meiotic maturation responses to sperm (Fig. 10.4).

Two models were considered to explain the requirement for Gα_s-adenylate cyclase-PKA signaling in the gonadal sheath cells for MSP responses in the germ line (Govindan et al. 2009). Gα_s-adenylate cyclase-PKA signaling might affect the competence of oocytes to respond to MSP. In this scenario, the Gα_s pathway would not directly sense the MSP gradient, but would enable oocytes or sheath cells to respond via other receptor pathways. This model is difficult to reconcile with multiple lines of experimental evidence. Most importantly, activation of Gα_s-adenylate cyclase-PKA signaling in the sheath cells by multiple means is sufficient to drive meiotic maturation in the absence of MSP (Govindan et al. 2009). The possibility that Gα_s-adenylate cyclase signaling has an earlier developmental role in the gonadal sheath cell lineages was excluded by the finding that phosphodiesterase inhibitors suppress the sterility of *acy-4* mutant adults. The possibility that MSP is unavailable to bind MSP receptors or that the receptors themselves are not expressed was excluded by MSP binding and localization studies. Thus, the model currently favored is one in which unidentified MSP receptors on gonadal sheath cells are G protein-coupled receptors (GPCRs). In this scenario, $G\alpha_s$-coupled receptors would trigger meiotic maturation, $G\alpha_{o/i}$-coupled receptors would inhibit meiotic maturation in the absence of MSP, and $G\alpha_q$-coupled receptors would promote sheath cell contraction (Fig. 10.4). The identification of the sheath cell MSP receptors will represent a critical test of this model.

10.2.4.2 Gap-Junctional Communication and the Control of Meiotic Maturation

The conclusion that the gonadal sheath cells function as the major initial MSP sensors raises the question of how this information is communicated to the germ line. Transmission electron microscopy and freeze-fracture analysis revealed that sheath cells form gap junctions with oocytes (Hall et al. 1999). Gap junctions were observed at regions of extensive contact between sheath cell somata and oocytes. In addition, gap junctions were seen at sites where finger-like sheath cell processes extended between oocytes. Proximal sheath cells are also connected to one another via gap junctions (Hall et al. 1999).

inx-14 and *inx-22,* which encode innexin/pannexin gap junction proteins, negatively regulate meiotic maturation, oocyte MAPK activation, oocyte microtubule reorganization, and the localization of AIR-2 to chromatin in the absence of MSP (Govindan et al. 2006, 2009; Harris et al. 2006; Whitten and Miller 2007). *inx-14* also regulates the assembly of RNP granules in arrested oocytes (J. Schisa, unpublished results). INX-14 and INX-22 are expressed in the germ line and co-localize at plaque-like structures at the interface between oocytes and sheath cells (Govindan et al. 2009), consistent with the possibility that they are components of sheath–oocyte gap junctions. Recently, sheath cell components of sheath–oocyte gap junctions have been identified as *inx-8* and *inx-9* (T. Starich and D. Greenstein, unpublished results). *inx-8* and *inx-9* promoter fusion constructs showed expression in the gonadal sheath cells (Starich et al. 2001; Altun et al. 2009), and this observation has been confirmed using specific antibodies and rescuing GFP protein fusions (T. Starich and D. Greenstein, unpublished results). INX-8 and INX-9 expression is also observed in the somatic gonad progenitors Z1 and Z4, the DTCs, and cells of the somatic gonadal primordium. *inx-8* and *inx-9* are found in the same operon and share approximately 87% identity at the protein sequence level. Deletion of either *inx-8* or *inx-9* produces no apparent mutant phenotype; however, inactivation of both genes causes sterility in both hermaphrodites and males. In the double mutant, few Z2/Z3 germ cell descendents are observed, which appears to reflect a defect in germ cell proliferation (T. Starich and D. Greenstein, unpublished results). A similar sterile phenotype is also observed in *inx-14* deletion alleles in both sexes (Govindan et al. 2009).

The observation that a germline innexin, *inx-14*, and the somatic gonadal innexins, *inx-8/inx-9*, share a common sterile phenotype suggests that soma–germline gap junctions may play a more global role in the gonad beyond mediating communication between oocytes and gonadal sheath cells. Thus, there appears to be an early role for soma–germline gap junctions that are needed for germ cell proliferation and a later role that functions in the regulation of meiotic maturation. Apparently, *inx-22* is not required for the early germ cell proliferation or survival role as a deletion allele is fertile (Whitten and Miller 2007). However, *inx-22* is found in an operon with *inx-21*, which is also expressed in the germ line as detected using specific antibodies, and *inx-21(RNAi)* in an *inx-22* mutant, but not a wild-type background, results in a sterile phenotype (T. Starich and D. Greenstein, unpublished results). It will be important to address whether specific small molecules move through these gap junctions to control meiotic maturation and germ cell proliferation.

Genetic epistasis analysis was used to examine the relationship between *inx-22*, which is a negative regulator of meiotic maturation and *gsa-1* and *acy-4*, which are positive regulators. Genetic mosaic analysis was used to reduce *gsa-1*(+) function in the somatic gonad in an *inx-22* background (Govindan et al. 2009). The experimental observation was that an *inx-22* mutation suppresses the sterility defect caused by loss of *gsa-1*(+) function in the somatic gonad, suggesting that gap-junction proteins function downstream of $G\alpha_s$ signaling. Neither *inx-22* nor *inx-14* and *inx-22* depletion suppresses *acy-4* null sterility. Possibly, unidentified targets of $G\alpha_s$-ACY-4 signaling might therefore regulate meiotic maturation in parallel with gap junction proteins. Alternatively, it might not be possible to eliminate gap-junctional communication between oocytes and sheath cells with available genetic tools on account of the germline proliferation function of the gonadal innexins.

Remarkably, the meiotic maturation process in *C. elegans* and mammals share a number of molecular and biological similarities (see Govindan et al. 2009 for an in-depth discussion and complete references). For example, MSP and LH, though unrelated in sequence, both trigger meiotic resumption using somatic $G\alpha_s$-adenylate cyclase-PKA pathways and soma-to-germline gap-junctional communication. Mural granulosa cells on the periphery of the follicle express the LH receptor, which is a GPCR. Cumulus granulosa cells form gap junctions with the oocyte using specialized extensions, called transzonal projections, which penetrate the zona pellucida and reach the oocyte cell surface. In both systems, interfering with the function of soma-to-germline gap junctions permits meiotic maturation in the absence of the maturation hormone. At a molecular level, the oocyte responses apparently involve the control of conserved protein kinase pathways and post-transcriptional gene regulation in the oocyte. At a cellular level, the responses include nuclear envelope breakdown, cortical cytoskeletal rearrangement, assembly of the acentriolar meiotic spindle, chromosome segregation, and likely changes important for fertilization and the oocyte-to-embryo transition. A major difference between the systems is that $G\alpha_s$-adenylate cyclase-PKA signaling also has a function within the oocyte to maintain meiotic arrest in vertebrates and mammals. In *C. elegans*, $G\alpha_s$-ACY-4 signaling functions exclusively in the gonadal sheath cells to promote meiotic maturation, as established by genetic mosaic analysis.

10.2.4.3 Regulation of Meiotic Maturation by VAB-1 MSP/Eph Receptor Signaling

Whereas the sheath cell MSP receptors, proposed to be GPCRs (Govindan et al. 2009, see below), have as of yet eluded detection, prior work identified the VAB-1 Eph receptor as an oocyte MSP receptor (Miller et al. 2003; Corrigan et al. 2005; Govindan et al. 2006; Cheng et al. 2008). Adult hermaphrodite *vab-1* null mutant animals are fertile (George et al. 1998), exhibit normal rates of meiotic maturation, and respond to MSP (Miller et al. 2003). However, unmated *vab-1* null mutant females modestly derepress meiotic maturation (Miller et al. 2003; Corrigan et al. 2005; Govindan et al. 2006; Cheng et al. 2008). The regulated endocytic trafficking of the VAB-1 MSP/Eph receptor appears to be a factor in the regulation of oocyte meiotic maturation. In the absence of the MSP ligand the VAB-1 Eph receptor inhibits meiotic maturation while either in or in transit to the endocytic recycling compartment (ERC; Cheng et al. 2008). The localization of VAB-1::GFP in oocytes to the RAB-11-positive ERC was shown to be antagonized by MSP signaling. Interestingly, $G\alpha_s$-adenylate cyclase-PKA signaling in the gonadal sheath cells was shown to be required for the trafficking of VAB-1::GFP to the oocyte plasma membrane from the ERC when MSP is present. Thus, the VAB-1 MSP/Eph receptor appears to play a non-essential modulatory role, in contrast to the $G\alpha_s$-adenylate cyclase-PKA pathway, which is required for meiotic maturation. The VAB-1 receptor pathways might contribute to the robustness of the response to sperm.

10.2.5 Control of Meiotic Maturation and the Regulation of Translation

The TIS11-type CCCH zinc finger domain-containing proteins OMA-1 and OMA-2, hereafter referred to as OMA proteins, are redundantly required for oocyte meiotic maturation and ovulation (Detwiler et al. 2001; Shimada et al. 2002). In *oma-1; oma-2* double mutants, MAPK activation is not sustained, nuclear envelope breakdown does not occur properly, and AIR-2 fails to localize to oocyte chromatin. Although the mechanism by which the OMA proteins promote meiotic maturation remains to be determined, they function upstream of the conserved cell cycle regulators WEE-1.3 and CDK-1 (Detwiler et al. 2001). *wee-1.3(RNAi)* in *oma-1; oma-2* double mutants can drive oocytes into M-phase, however fertilization does not occur (Detwiler et al. 2001; Burrows et al. 2006). OMA proteins have been shown to repress the translation of *nos-2* and *zif-1* in oocytes (Jadhav et al. 2008; Guven-Ozkan et al. 2010). OMA proteins bind to the 3′-UTRs of *nos-2* and *zif-1* (Jadhav et al. 2008; Guven-Ozkan et al. 2010). Repression of *zif-1* in oocytes also requires the eIF4E-binding protein SPN-2 (Guven-Ozkan et al. 2010). Neither *nos-2* nor *zif-1* is required for meiotic maturation, yet their regulation might indicate a general function for OMA proteins in regulating translation in oocytes.

OMA proteins are multifunctional—they interact with TAF-4, a subunit of TFIID, to repress RNA polymerase II-mediated transcription in the zygote and the germline blastomere P1 (Guven-Ozkan et al. 2008). This transcriptional repression function of OMA proteins is likely not relevant for the regulation of meiotic maturation because this activity only manifests upon phosphorylation by the dual-specificity tyrosine-phosphorylation-regulated protein kinase MBK-2. MBK-2 only becomes active in oocytes upon meiotic maturation (Stitzel et al. 2006; Cheng et al. 2009). In fact, phosphorylation of OMA-1 by MBK-2 was shown to displace SPN-2 from the *zif-1* 3-UTR, thereby alleviating translational repression (Guven-Ozkan et al. 2010). In *C. elegans*, as in many species, fully grown oocytes appear to be transcriptionally inactive (Starck 1977; Gibert et al. 1984; Schisa et al. 2001; Walker et al. 2007). Thus, translational control by OMA proteins might play an important role in meiotic maturation. Because the *C. elegans* germ line develops as a syncytium, it is difficult to assess directly whether meiotic maturation requires translation in oocytes. That translational regulation is critical for oogenesis is clear. For example, the translational regulators GLD-1 and GLD-2, which function downstream in the GLP-1 signaling pathway in the stem cell vs. meiotic development decision (see Chap. 4, Hansen and Schedl 2012; Chap. 8, Nousch and Eckmann 2012) are each individually required for the formation of normal oocytes (Francis et al. 1995; Kadyk and Kimble 1998). Further, IFE-1, one of the three eIF4E isoforms, promotes the accumulation of *oma-1* mRNA on polysomes in developing oocytes (Henderson et al. 2009). Further analysis of the OMA proteins will likely continue to provide important insights into the role of translational control during meiotic maturation.

10.2.6 Regulation of Ovulation

C. elegans ovulation is an attractive physiological model for how intercellular signaling influences the behavior of smooth muscle. In worms, myoepithelial cells form a smooth-muscle structure known as the gonadal sheath, and their coordinated function is needed for ovulation. The ability to observe the function of the gonadal sheath cells and their dynamic interaction with cells of an epithelial tube, the spermatheca, provides an ideal format for analyzing the interplay between cell structure and intercellular communication. Nonetheless, the study of ovulation is complicated due to the involvement of many genes and multiple tissues (Iwasaki et al. 1996; Kostic et al. 2003; Aono et al. 2004; Gissendanner et al. 2008; Pilipiuk et al. 2009).

During ovulation, the proximal gonadal sheath cells contract rapidly, the distal constriction of the spermatheca dilates, and sheath cells pull the distal spermatheca over the mature oocyte. The maturing oocyte signals its own ovulation in two ways: it modulates sheath contractions, which includes an increase in contraction rate and intensity during ovulation, and it induces spermathecal dilation during ovulation (Iwasaki et al. 1996; McCarter et al. 1999). Mutations that lead to defective ovulation cause an endomitotic oocyte (Emo) phenotype (Iwasaki et al. 1996). When oocytes are retained in the gonad arm due to defective ovulation, they undergo multiple rounds of nuclear envelope breakdown (M-phase entry) and S-phase, and become highly polyploid.

10.2.6.1 EGF Receptor and IP_3 Signaling for Ovulation

The signal from the maturing oocyte that controls dilation of the distal spermatheca is thought to be LIN-3/EGF, which triggers LET-23/EGF receptor signaling in the distal spermatheca and possibly the sheath cells (Clandinin et al. 1998; McCarter 1998). LIN-3/LET-23 signaling in the gonadal sheath cells also promotes ovulatory contractions (Yin et al. 2004). The *let-23* pathway required for spermathecal dilation during ovulation is *let-60/ras*-independent and involves a downstream IP_3-mediated pathway (Clandinin et al. 1998). Mutations in two genes, *lfe-1/itr-1* and *lfe-2*, respectively gain-of-function and loss-of-function, were isolated in a genetic screen for suppressors of *let-23* sterility. *lfe-1/itr-1* and *lfe-2* encode an inositol (1, 4, 5) triphosphate receptor and an inositol (1, 4, 5) triphosphate-3-kinase, respectively (Clandinin et al. 1998). These results suggest that spermathecal dilation is likely to be dependent on calcium release regulated by IP_3. Consistent with this possibility, spermathecal dilation requires the function of a Ca^{2+} release-activated Ca^{2+} channel expressed in sheath and spermathecal cells (Yan et al. 2006; Lorin-Nebel et al. 2007). Further, a mutant allele of *ipp-5*, which encodes an inositol 5-phosphatase, predicted to lower IP_3 levels, exhibits an unusual ovulation phenotype in which the spermatheca overextends, thereby ovulating two oocytes per cycle (Bui and Sternberg 2002).

IP_3 signaling also plays an important role in sheath cells. A reduction-of-function mutation in *itr-1* disrupts both basal sheath cell contractions in response to a synthetic MSP C-terminal peptide and ovulatory contractions (Yin et al. 2004). Phospholipase C (PLC)-mediated hydrolysis of the membrane lipid phosphatidylinositol 4,5-bisphosphate (PIP_2) generates IP_3. Both *plc-1* and *plc-3* are required for ovulation (Kariya et al. 2004; Yin et al. 2004; Vazquez-Manrique et al. 2008), as is phosphatidylinositol-4-phosphate 5′ kinase, encoded by *ppk-1*, which is needed for the synthesis of PIP_2 (Xu et al. 2007). GFP reporter constructs indicate that PLC-3 and PPK-1 are expressed in sheath and spermathecal cells (Yin et al. 2004; Xu et al. 2007), whereas PLC-1 is expressed only in the spermatheca (Kariya et al. 2004; Yin et al. 2004). PLC-3 promotes both the basal and ovulatory sheath cell contractions (Yin et al. 2004). By contrast, PLC-1 is not required for basal or ovulatory sheath cell contractions, consistent with a role in the spermatheca. Indeed, mutations in *plc-1* or *plc-1(RNAi)* cause spermathecal entry and exit defects (Kariya et al. 2004; Yin et al. 2004). PLC-1 expression in the spermatheca requires the FOS-1/JUN-1 heterodimeric transcriptional activator (Hiatt et al. 2009). RNAi of *fos-1* or *jun-1* disrupts ovulation and this defect is rescued by expression of PLC-1 in the spermatheca (Hiatt et al. 2009).

The precise connection between LET-23 activation, likely in sheath and spermathecal cells, and IP_3 generation remains to be determined. Signaling effectors likely functioning upstream of PLC-3 and PLC-1 activation include the Rho/Rac-family guanine nucleotide exchange factor VAV-1 (Norman et al. 2005), RHO-1 GTPase (McMullan and Nurrish 2011), and the ARK-1 tyrosine kinase (Hopper et al. 2000). How the sheath and spermathecal cells coordinate their behaviors during ovulation is unclear. Not only must mature oocytes enter the spermatheca at ovulation, but fertilized embryos must also exit the spermatheca in a timely fashion. Specific disruptions in IP_3 signaling and the actin cytoskeleton reduce fertility by interfering with exit of the fertilized embryo from the spermatheca, thereby disrupting the reproductive assembly line (Kariya et al. 2004; Kovacevic and Cram 2010).

10.2.6.2 Yolk Lipoprotein Metabolism and Ovulation

Several intersecting lines of evidence provide an indication that lipid signaling might play a role in promoting ovulation. Growing oocytes take up yolk lipoprotein particles produced by the intestine (Kimble and Sharrock 1983) by a process of receptor-mediated endocytosis (Grant and Hirsh 1999). Intestinal cells secrete yolk lipoprotein particles into the pseudocoelom which then pass through the gonadal basal lamina and pores in the gonadal sheath cells to gain access to the oocyte surface (Hall et al. 1999). *rme-2* encodes a member of the LDL receptor superfamily and is the oocyte yolk receptor (Grant and Hirsh 1999). *rme-2* is required for yolk uptake and transport of cholesterol into oocytes (Grant and Hirsh 1999; Matyash et al. 2001). Interestingly, *rme-2* mutants display ovulation defects (Grant and Hirsh 1999), though the basis for these defects has been mysterious. The heterodimeric E2F EFL-1/DPL-1 transcription factor is required for the expression of *rme-2* in the germ line (Chi and Reinke 2006) and expression of *rme-2* using the germline-specific *pie-1* promoter partially rescues the ovulation defects of strong loss-of-function *dpl-1* mutants (Chi and Reinke 2009). Oocytes utilize yolk lipoprotein particles to generate polyunsaturated fatty acid derivatives, including F-series prostaglandins, to generate signals that promote sperm guidance to the spermatheca (Kubagawa et al. 2006; Edmonds et al. 2010). Possibly, lipid signals dependent on *rme-2* might also function to promote ovulation. Such a possibility might explain the finding that the EGRH-1 transcription factor is required in the intestine for normal ovulation (Clary and Okkema 2010). *egrh-1* mutants might affect the quantity or quality of yolk lipoprotein particles received by oocytes, thereby affecting ovulation. Alternatively, the disruption of a major pathway for endocytosis and lipid transport in *rme-2* mutants might have deleterious side effects on membrane trafficking in oocytes. Since *rme-2* mutant oocytes are small and misshapen, the effects could even be more indirect. Further work will be needed to explain fully the role of *rme-2* in ovulation.

10.2.6.3 Smooth Muscle Structure and Function and Ovulation

The myoepithelial gonadal sheath cells provide a valuable system for studying the cell biology and function of a smooth muscle cell type (Strome 1986; McCarter et al. 1997; Rose et al. 1997; Hall et al. 1999; Ono et al. 2007). Mutations that perturb actomyosin contractility in sheath cells can cause an Emo phenotype and infertility (Myers et al. 1996; Wissmann et al. 1999; Ono and Ono 2004; Ono et al. 2008). However, some mutations that cause hypercontractility of the sheath cells also result in infertility. For instance, worms double mutant for *tni-1* and *unc-27*, two troponin I isoforms expressed in the sheath cells, display defects in oocyte production, meiotic maturation, and ovulation (Obinata et al. 2010). In the future, studies of smooth muscle physiology in the context of ovulation will no doubt benefit from continued improvements in methods for measuring calcium concentrations and for recording ion channel activity in the gonad (Rutledge et al. 2001; Samuel et al. 2001).

10.3 Control of Oocyte Growth and Coordination with Meiotic Maturation

Fertility depends on germline stem cell proliferation, meiosis and gametogenesis, yet how these key transitions are coordinated is unclear. Recent data suggest that the continued presence of sperm maintains the adult hermaphrodite gonad in an active reproductive mode. The MSP hormone promotes the production and growth of oocytes and works in concert with the GLP-1/Notch pathway to regulate an optimal allocation of germline stem cells into oocytes. This section focuses on the cellular processes of pachytene progression, oocyte growth, and their regulation by MSP signaling.

10.3.1 Pachytene Progression and Oogenesis

Once germ cells have entered meiosis, progression through the pachytene stage in both sexes requires the germline function of genes of the MAP kinase signaling pathway, including *let-60/ras, ksr-2, lin-45/raf, mek-2/mapkk,* and *mpk-1/mapk* (Church et al. 1995; Ohmachi et al. 2002; Lee et al. 2007). An increase in dpMPK-1 levels is observed midway through the pachytene stage, consistent with the conclusion that activated MPK-1 is needed for germ cells to progress from an early (or distal) to a late (or proximal) pachytene stage (Lee et al. 2007). The continued presence of sperm in the gonad is needed for MPK-1 activation in this region of the gonad. Young adult *fog-2(oz40)* females (8 h after the L4 to adult molt) exhibit dpMPK-1 in the proximal pachytene region and a single distal oocyte at the −6 position (Lee et al. 2007). This distal activation of MPK-1 occurs independently of the presence of sperm. By contrast, older *fog-2(oz40)* females (20 h after the L4 to adult molt) display undetectable dpMPK-1 levels in the germ line (Lee et al. 2007). This result fits well with the observation that the flux of germ cells through pachytene is significantly reduced in unmated adult females (Jaramillo-Lambert et al. 2007). *acy-4* mutant hermaphrodites produce fewer oocytes than the wild type and exhibit dpMPK-1 in the pachytene region in the early adult stage but not later (Govindan et al. 2009), suggesting a role for the sheath cells in maintaining MAPK activation in the pachytene region. Cell ablation studies are also consistent with the possibility that cells in the sheath/spermathecal lineages promote progression through pachytene (McCarter et al. 1997). One model is that the distal sheath cells (pairs 1 and 2) might produce a secondary signal to trigger MPK-1 activation in the pachytene region, thereby promoting meiotic prophase progression. In this model, the production of this hypothesized secondary signal would be dependent on the continued presence of MSP in the gonad, as sensed by the proximal sheath cells via the $G\alpha_s$-adenylate cyclase pathway.

10.3.2 Oocyte Growth

Prior to cellularization at late stages of gametogenesis, *C. elegans* germ cells maintain a connection to the cytoplasmic core of the gonad (rachis), the contents of which are shared with developing oocytes (Hirsh et al. 1976; Wolke et al. 2007). In adult hermaphrodites, germ cells that exit pachytene either differentiate as oocytes or undergo apoptosis (Gumienny et al. 1999). Female meiotic germ cells destined for apoptosis might function as nurse cells (Gumienny et al. 1999; Jaramillo-Lambert et al. 2007) by contributing mRNA, protein, and cellular organelles to growing oocytes, which are transcriptionally quiescent themselves (see Chap. 9 on germ cell apoptosis and DNA damage responses, Bailly and Gartner 2012). Oocytes in the loop region grow primarily by receiving actomyosin-dependent flow from the core cytoplasm (Wolke et al. 2007), and yolk uptake in the most proximal oocytes also contributes to their growth (Grant and Hirsh 1999; Wolke et al. 2007).

The pathways that control incomplete cytokinesis in the distal germ line, and that promote cellularization of developing oocytes in the proximal germ line, are incompletely understood. Several genes with important functions in cytokinesis are required for oocyte cellularization, including *mlc-4*, *mel-11,* and *cyk-1*, which encode the regulatory light chain of non-muscle myosin, a myosin phosphatase regulatory subunit and an actin regulator, respectively (Swan et al. 1998; Shelton et al. 1999; Piekny and Mains 2002). The anillin ANI-2 is required for developing oocytes to maintain their connection to the core cytoplasm (Maddox et al. 2005). Thus, *C. elegans* oogenesis appears to provide a useful experimental system for addressing the cell biological mechanisms by which interconnected cysts form and breakdown, as occurs in mammalian female germ cell development (Pepling and Spradling 1998). It will be essential to determine the basic cell biological principles underlying incomplete cytokinesis and oocyte cellularization before the field can achieve a relatively complete understanding of the regulatory mechanisms.

Nonetheless, several studies have begun to address how intercellular signaling regulates oocyte growth and cellularization. The PTP-2 protein tyrosine phosphatase functions in the germ line as a negative regulator of oocyte growth; *ptp-2* mutant oocytes grow abnormally large in the presence of sperm (Gutch et al. 1998; Yang et al. 2010). PTP-2 is required for the MPK-1 MAPK activation in the germ line, which might explain the growth defect because oocytes also grow abnormally large when a temperature-sensitive *mpk-1* mutant is upshifted (Lee et al. 2007). The small oocyte phenotype observed in the constitutively-activated *let-60/ras*(*ga89*) allele depends on the presence of sperm in the gonad (Lee et al. 2007). Two identified MPK-1 substrates function to restrict oocyte growth, whereas eight identified substrates promote growth (Arur et al. 2009). Analysis of the growth-promoting substrates is complicated by the effects on dpMPK-1 levels.

10.3.2.1 Actomyosin-Dependent Cytoplasmic Streaming

Oocytes in the loop region of the gonad grow by receiving flow from the cytoplasmic core (Wolke et al. 2007). This flow was shown to be dependent on the actomyosin cytoskeleton, but independent of microtubules. Cytoplasmic streaming requires the continued presence of sperm in the gonad but does not depend on meiotic maturation (Wolke et al. 2007). In young adult females observed right after the L4 to adult molt, flows were observed and were therefore independent of sperm. By contrast, adult females observed on the second day of adulthood did not exhibit flows, however mating restored the flows. Thus, initial growth of oocytes in the young adult stage is independent of sperm, but sperm needs to be continually available for additional oocytes to form. This result provides an explanation for the original observation that sperm promote oocyte production (Ward and Carrel 1979).

A series of elegant oil injection studies suggested that the oocytes generate the forces driving cytoplasmic streaming (Fig. 10.5). Neither germ cell apoptosis nor sheath cell contraction is required for the cytoplasmic flow. While the mechanism of force generation is unclear, several testable models were proposed (Fig. 10.5; Wolke et al. 2007).

10.3.2.2 MSP Signaling, the Sheath Cells, and the Control of Cytoplasmic Streaming

The MSP hormone is sufficient to promote the sustained actomyosin-dependent cytoplasmic streaming that drives oocyte growth (Govindan et al. 2009). Injection of MSP into unmated females that do not exhibit flows (e.g., at 24 h post L4) caused the flows to resume. Efficient oocyte production and cytoplasmic streaming require $G\alpha_s$-adenylate cyclase signaling in the gonadal sheath cells. Thus the gonadal sheath cells coordinate oocyte growth and meiotic maturation with sperm availability. Phosphorylation of the regulatory myosin light chain (rMLC) increases the ATPase activity of nonmuscle myosin and is required for myosin motor function, which is

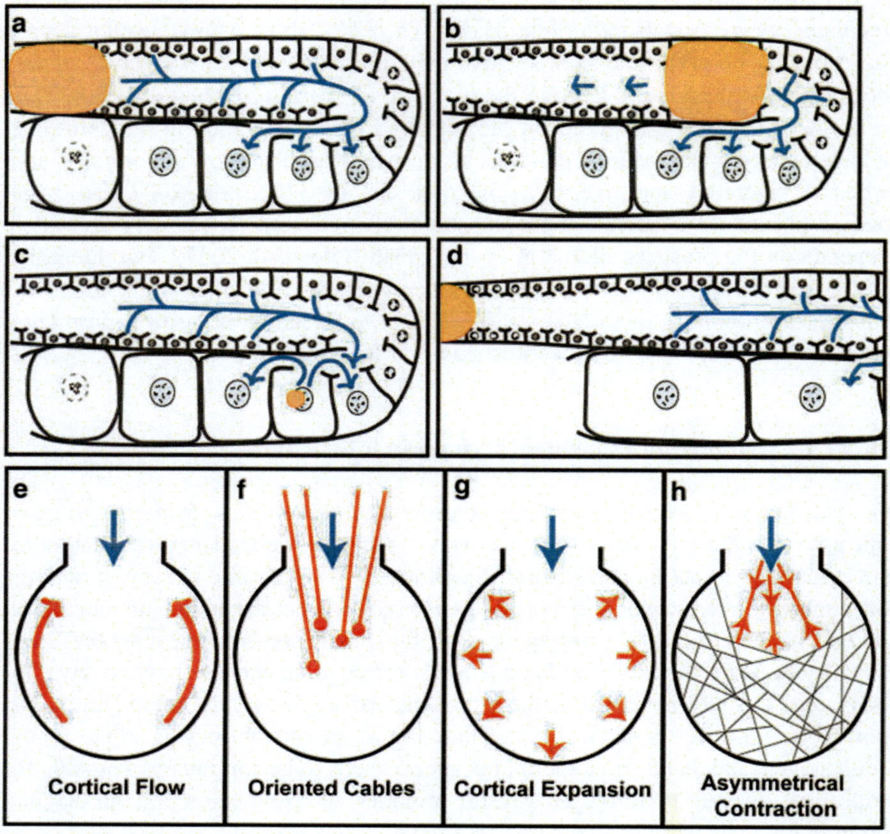

Fig. 10.5 Cytoplasmic streaming drives oocyte growth. Oil injection studies informative for probing the mechanism of oocyte growth (**a**–**d**). Injection in the pachytene region (**a**) or just prior the loop (**b**) does not interfere with flow within the proximal gonad arm; damaging an oocyte by oil injection disrupts flow to that oocyte and flows are rerouted to neighboring oocytes (**c**). These results suggest that oocytes generate the force for flow (Wolke et al. 2007). (**d**) Distal oil injection causes a large oocyte phenotype and suppresses the small oocyte phenotype of *glp-1*(*gf*) mutants, suggesting that DTC signaling via GLP-1 in the distal arm regulates MSP-dependent oocyte growth (Nadarajan et al. 2009). (**e**–**h**) Models proposed by Wolke et al. (2007) for the generation of gonadal flow by the oocyte actomyosin cytoskeleton. Simplified oocytes receiving flow (*blue arrows*) from the cytoplasmic core of the *C. elegans* germ line are diagrammed. Cortical flow in the opposite direction draws material into the oocyte (**e**), similar to a model of cytoplasmic streaming proposed for early *C. elegans* embryos. Oriented actin cables bring material into oocytes (**f**). Expansion of the oocyte cortex draws material into the oocyte (**g**). A stable actomyosin network in the oocyte (*black lines*) is coupled to dynamic actomyosin network (*red lines and arrows*) to bring material into oocytes

needed for the actomyosin-dependent cytoplasmic streaming that drives oocyte growth (Fig. 10.3; Wolke et al. 2007). Phosphorylated rMLC (p-rMLC) was detected in the germ line and gonadal sheath cells of hermaphrodites and mated females, but not 2-day-old unmated adult females (Nadarajan et al. 2009). MSP injection was shown to induce p-rMLC formation rapidly (within 15 min) throughout the germ line and in the gonadal sheath cells (Nadarajan et al. 2009). Sustained gonadal flows and p-rMLC formation also require the function of Gα_s-adenylate cyclase signaling in the gonadal sheath cells. This activity of MSP appears to be important for promoting oocyte growth (see below; Nadarajan et al. 2009). Thus, MSP appears to promote oocyte meiotic maturation in part though coordination of several of its component processes. Genetic evidence suggests that MSP and Gα_s-adenylate cyclase signaling regulate oocyte growth and meiotic maturation in part by antagonizing gap-junctional communication between sheath cells and oocytes.

10.3.2.3 GLP-1/Notch Signaling and the Control of Sperm-Dependent Oocyte Growth

Surprisingly, a genetic screen for mutations that cause oocytes to grow abnormally large in the presence of sperm recovered reduction-of-function alleles of *glp-1/Notch* (Nadarajan et al. 2009). GLP-1/Notch signaling restricts the growth of oocytes to the proper size in response to the MSP signal. Germline GLP-1 activity negatively regulates MSP-dependent cytoplasmic streaming, whereas it promotes oocyte cellularization. Several observations suggest that GLP-1 functions in the distal germ line in response to DTC signaling to regulate oocyte growth. Laser ablation of the DTC and co-depletion of the DTC-expressed GLP-1 ligands LAG-2 and APX-1 cause a large oocyte phenotype. Oil injection studies, similar to those first performed by Wolke et al. (2007), showed that oocytes grow abnormally large when the proliferative zone is occluded by oil injection (Nadarajan et al. 2009). Distal oil injection not only suppressed the small oocyte phenotype of the *glp-1(ar202)* gain-of-function mutant, but also phenocopied a reduction in *glp-1* function. These results suggest that GLP-1/Notch functions in the proliferative zone to regulate MSP-dependent oocyte growth. Consistent with this possibility, normal oocyte growth was shown to require the function of several mediators of GLP-1 signaling, including the LAG-1 transcription factor and the FBF-1/2 RNA-binding proteins.

Several lines of evidence suggested that the role of *glp-1* in oocyte growth is separable from the proliferation versus meiotic entry decision (Nadarajan et al. 2009). One notable finding was that blocking apoptosis strongly suppressed the *glp-1* oocyte growth defect without affecting the premature meiotic entry defect. It is not clear whether the *ced-3* and *ced-4* requirement for the *glp-1* large oocyte phenotype involves their apoptotic function, or perhaps a cell death-independent function. The results of Nadarajan et al. (2009) suggest a model in which two major signaling centers in the adult hermaphrodite gonad, distal GLP-1 signaling and proximal MSP signaling, work in opposition to regulate the differentiation of germ cells into functional oocytes (Fig. 10.3). In this model, the adult hermaphrodite gonad is rapidly switched into a reproductive mode by the MSP hormone. MSP signaling provides the impetus for oocyte growth, differentiation, and the completion of meiosis, and GLP-1 signaling both provides the raw material for gametogenesis and modulates additional processes that restrict oocyte growth, including cytoplasmic streaming and oocyte cellularization.

10.4 Meiotic Maturation and Regulation of the Oocyte-to-Embryo Transition

In many animals, fertilization triggers egg activation (the process whereby the oocyte completes the meiotic divisions), the blocks to polyspermy, and activation of the embryonic program (Runft et al. 2002; Horner and Wolfner 2008). Fertilization is extensively discussed in Chap. 11 (Marcello et al. 2012). Recent findings provide views into how signaling mechanisms initiated during meiotic maturation and extending through fertilization control the completion of meiosis and prepare the zygote for embryonic patterning. In *C. elegans*, fertilization (Goldstein and Hird 1996) and the downstream PAR proteins (Kemphues et al. 1988) are required for the establishment of embryonic polarity. Polarity establishment in the *C. elegans* embryo occurs via a cue associated with the sperm centrosome complex and microtubules (St Johnston and Ahringer 2010; Nance and Zallen 2011). Chapter 12 (Robertson and Lin 2012) discusses the function of the OMA proteins in the oocyte-to-embryo transition and the role of MBK-2 in controlling the degradation of maternal proteins after completion of meiosis. Here, we focus on the role of meiotic maturation and fertilization in promoting the completion of meiosis.

10.4.1 Function and Regulation of the Oocyte Meiotic Spindle

In many animal oocytes, the microtubule arrays of the meiotic spindle form independently of a centriole-containing centrosome. The short barrel-shaped meiotic spindles of *C. elegans* oocytes form by microtubule nucleation around meiotic chromatin and are both acentriolar and anastral (Fig. 10.6; Albertson and Thomson 1993; Howe et al. 2001; Yang et al. 2003). Time-lapse videomicroscopic observations of meiotic chromosomes and spindles indicate that meiosis I spindle assembly initiates prior to ovulation when the oocyte is in the gonad arm (Yang et al. 2003). Both meiotic divisions are then completed in the uterus following fertilization (Albertson and Thomson 1993; McCarter et al. 1999; Yang et al. 2003).

The *C. elegans* oocyte is emerging as a powerful model for studying the assembly, regulation, and function of the meiotic spindle (Muller-Reichert et al. 2010; Fabritius et al. 2011). One salient feature of this system is that meiotic spindle assembly is a microtubule-driven process, and F-actin is not required for translocation of the spindle to the cortex (Yang et al. 2003). This contrasts with the situation in mouse oocytes in which cortical movement of the meiotic spindle is actin based (Na and Zernicka-Goetz 2006; Dumont et al. 2007; Li et al. 2008; Schuh and Ellenberg 2008). This property appears to provide some experimental flexibility for the separate dissection of meiotic spindle assembly and cytokinesis. At the same time, it cautions against generalizing results to all instances of acentrosomal spindle assembly.

During oogenesis, the maternal centrioles are eliminated, disappearing in the diplotene stage (Zhou et al. 2009; Mikeladze-Dvali et al. 2012), and embryogenesis thus depends on the sperm-supplied centriole pair (Wolf et al. 1978; O'Connell et al. 2001). An early study using co-suppression methodology suggested that centrosome

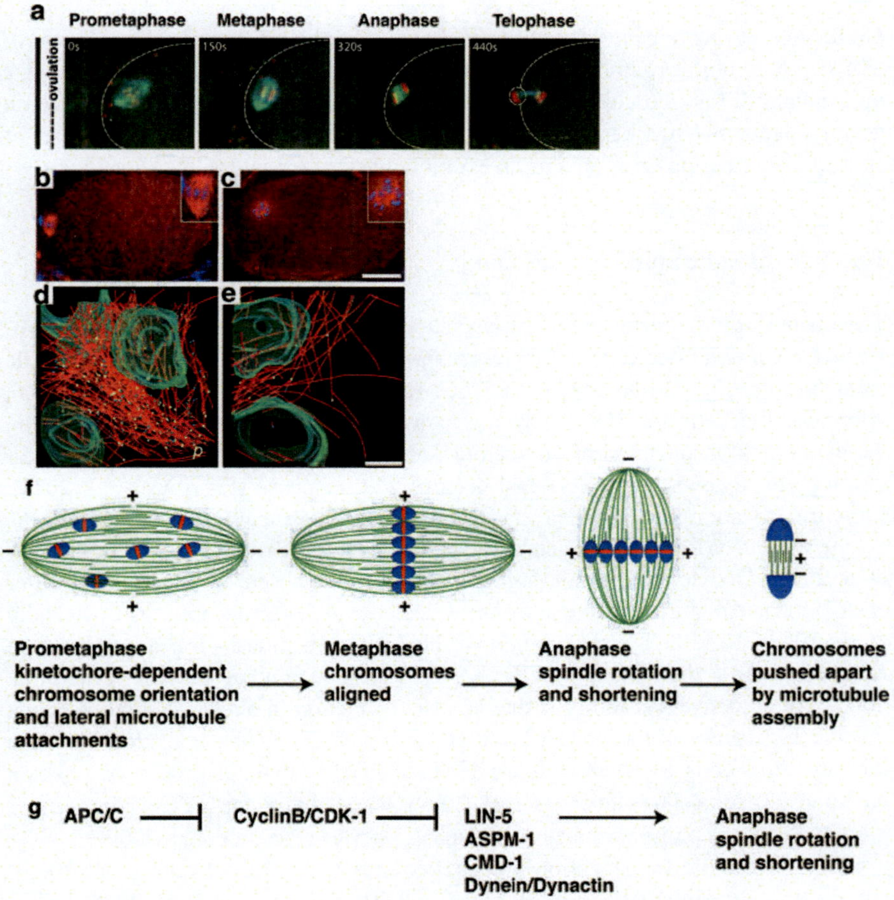

Fig. 10.6 Assembly and function of the oocyte meiotic spindle. (**a**) Time-lapse series of MI visualized with GFP::β-tubulin and mCherry::Histone fusions (unpublished data generously provided by Valerie Osterberg, Sara Christensen, and Bruce Bowerman). (**b, c**) The meiotic spindle in wild-type (**b**) and *mei-1(null)* (**c**) embryos, detected using antibodies to tubulin (*red*) and DAPI (*blue*) to stain DNA. (**d, e**) Three-dimensional reconstruction of portions of the wild-type (**d**) and *mei-1(null)* (**e**) meiotic spindle assembled from tomographic data sets. Microtubules are shown in *red*, with their pole-proximal ends marked by *white spheres* and their pole-distal ends marked with *blue spheres*; chromatin is in *green*. Note, in the wild type, many microtubules terminated before reaching the pole (p) or the chromatin. By contrast, the *mei-1(null)* mutant spindle contains a disorganized array of longer microtubules. Panels (**b–e**) are from Srayko et al. (2006) and are used with permission. (**f**) Models for chromosome alignment and segregation. Chromosomes orient and align using kinetochore-dependent (Dumont et al. 2010) and lateral attachments (Wignall and Villeneuve 2009). The cup-like kinetochores are shown in *blue* and the mid-region enriched for components of the chromosomal passenger complex is shown in *red*. (**g**) A pathway for the control of anaphase spindle rotation and shortening (Ellefson and McNally 2011)

elimination during oogenesis requires the function of the *cki-2* cyclin-dependent protein kinase inhibitor (Kim and Roy 2006). However, subsequent analysis of a cki-2 null mutant did not confirm an essential role for this gene in centriole elimination during oogenesis (Buck et al. 2009; Mikeladze-Dvali et al. 2012). The mechanism of centriole elimination thus remains an exciting mystery for future investigations.

10.4.1.1 Meiotic Spindle Positioning

The meiotic spindle forms in close association with the cortex to facilitate the extrusion of small polar bodies thereby preserving oocyte cytoplasm. The meiotic spindle assembles parallel to the cortex and then rotates to be perpendicular (Fig. 10.6; Albertson and Thomson 1993). Two mechanisms appear to be particularly important for the cortical localization of the meiotic spindle, the first occurring before nuclear envelope breakdown and the second occurring at anaphase (Fabritius et al. 2011). Early translocation involves the distal (away from the spermatheca) migration of the oocyte nucleus prior to nuclear envelope breakdown (McCarter et al. 1999; McNally et al. 2010). Oocyte nuclear migration can occur in the absence of sperm (McCarter et al. 1999) and requires the function of *mpk-1* (Lee et al. 2007) and kinesin-1 (McNally et al. 2010). Kinesin-1 activity requires the UNC-116 kinesin-1 heavy chain, the KLC-1 kinesin-1 light chain, and the KCA-1 kinesin-1 binding protein (Yang et al. 2005). An attractive possibility is that kinesin-1 function in nuclear migration might involve the function of the ZYG-12 KASH domain protein (Zhou et al. 2009) because kinesin-1 functions together with the UNC-83 KASH domain protein to position nuclei in somatic cells (Meyerzon et al. 2009). Interestingly, the ZYG-12 interacting partner, SUN-1 is phosphorylated in diakinesis oocytes (Penkner et al. 2009).

The late spindle translocation process requires the APC/C (Yang et al. 2005), cytoplasmic dynein (Ellefson and McNally 2009; van der Voet et al. 2009), and a protein complex containing the NUMA-related LIN-5, abnormal spindle-like, microcephaly-associated ASPM-1, and calmodulin CMD-1 (van der Voet et al. 2009). Late translocation involves a 90° rotation of the meiotic spindle (Albertson and Thomson 1993; Yang et al. 2005; Ellefson and McNally 2009; van der Voet et al. 2009). Kinesin-1 has also been proposed to anchor the meiotic spindle to the cortex until completion of MII (Yang et al. 2005; McNally et al. 2010). Recent data indicate that cyclin B/Cdk1 inhibits meiotic spindle rotation and anaphase spindle shortening. By promoting cyclin B degradation, the APC/C couples spindle rotation and chromosome segregation (Fig. 10.6; Ellefson and McNally 2011).

10.4.1.2 MEI-1 and Meiotic Spindle Assembly

The assembly of a bipolar female meiotic spindle requires the function of two interacting genes, *mei-1* and *mei-2* (Mains et al. 1990; Clandinin and Mains 1993; Clark-Maguire and Mains 1994a, b; Srayko et al. 2000). *mei-1* and *mei-2*, respectively, encode the p60 and p80 subunits of katanin (Clark-Maguire and Mains 1994a;

Srayko et al. 2000), a dimeric microtubule severing AAA-ATPase first purified from sea urchin eggs (McNally and Vale 1993). Katanin function also controls meiotic spindle length and spindle shortening (McNally et al. 2006). Although *mei-1* and *mei-2* are required for the normal segregation of chromosomes during female meiosis, they are dispensable during male meiosis. Electron microscopy (EM) with tomography and 3-D modeling were used to compare microtubule organization in the oocyte meiotic spindle in the wild type and a *mei-1* null mutant (Fig. 10.6; Srayko et al. 2006). In the wild type, microtubule ends were distributed throughout the spindle, not just exclusively at the poles. Thus, the meiotic spindle appears to assemble as a network of interdigitating microtubules. In the *mei-1* mutant, fewer microtubules surrounded the chromatin. These microtubules appeared disorganized and did not sort into a bipolar spindle. The chromatin-associated microtubules in the *mei-1* mutant were also longer than in the wild type. These findings suggest that katanin functions to increase microtubule number and density in the vicinity of meiotic chromatin. Consistent with this model, meiotic spindles in a reduction-of-function *mei-2* mutant are significantly longer than in the wild type (McNally et al. 2006). In the EM tomography study, structural evidence for microtubule severing was observed, leading to a model in which katanin promotes bipolar meiotic spindle assembly by increasing the local density of microtubule polymers (Srayko et al. 2006). An exciting recent finding is that the viable and fertile MEI-1(A338S) mutant, which is defective in microtubule severing when expressed in *Xenopus* cells, is proficient in bipolar meiotic spindle assembly, spindle rotation, and post-rotation spindle shortening in the worm (McNally and McNally 2011). MEI-1(A338S) meiotic spindles, however, were markedly longer than the wild type (McNally and McNally 2011). A model was proposed in which katanin contributes to the assembly of the bipolar meiotic spindle through a combination of microtubule binding and bundling activities, in combination with microtubule severing (McNally and McNally 2011).

10.4.1.3 Chromosome Alignment and Segregation on the Meiotic Spindle

In *C. elegans* oocytes, kinetochore proteins localize to cup-like structures, which encase the meiotic chromosomes, except for the mid-region, which is enriched for components of the chromosomal passenger complex, including AIR-2 (Fig. 10.6; Howe et al. 2001; Rogers et al. 2002; Monen et al. 2005; Dumont et al. 2010). Ultrastructural analyses of *C. elegans* oocyte meiotic spindles have not detected exclusive end-on attachments between kinetochores and microtubules, raising questions about the basis for chromosomal attachment, alignment, and segregation (Howe et al. 2001; Srayko et al. 2006). Interestingly, the chromokinesin, KLP-19, which generates a polar ejection force promoting chromosome congression on the metaphase plate during mitosis, localizes to the mid-bivalent region during meiosis (Powers et al. 2004; Wignall and Villeneuve 2009; Dumont et al. 2010). Time-lapse imaging of KLP-19-depeleted embryos revealed a spatial dispersion of meiotic chromosomes at anaphase I and an instability in the late anaphase spindle (Dumont et al. 2010). KLP-19 first becomes concentrated on meiotic chromosomes in the most proximal oocytes, but the sperm dependence of this localization has not been investigated. The BimC-related kinesin BMK-1 localizes to meiotic spindles in an AIR-2-dependent fashion. A *bmk-1* deletion allele and *bmk-1(RNAi)* do not disrupt meiosis and mitosis (Bishop et al. 2005; Saunders et al. 2007), a finding that contrasts with the essential requirement for BimC kinesins for maintenance of bipolar spindle structure and spindle pole separation in other systems (Kashina et al. 1997). By contrast, the Klp2-related kinesin KLP-18 is required for meiotic spindle bipolarity (Segbert et al. 2003; McNally et al. 2006; Wignall and Villeneuve 2009).

Recently it was observed that lateral microtubule spindles ensheath meiotic chromosomes and appear to promote their biorientation on the meiotic spindle (Wignall and Villeneuve 2009). This new model envisages a major role for lateral microtubule bundles and localized microtubule motors, such as KLP-19, in promoting chromosome congression and biorientation (Wignall and Villeneuve 2009). There is strong evidence that meiotic kinetochore proteins do play a role in orienting meiotic chromosomes, however (Dumont et al. 2010). The early function of kinetochores in aligning meiotic chromosomes on the metaphase plate (Dumont et al. 2010) might facilitate lateral attachments by microtubule bundles (Wignall and Villeneuve 2009). Interestingly, meiotic chromosomes separate at anaphase in the absence of kinetochore function; microtubules forming between the meiotic chromosomes appear to push them apart (Dumont et al. 2010). Current models for meiotic spindle assembly and function appear to rely on the meiotic spindle having microtubules of a defined polarity (Fig. 10.6), which has not been determined. In *Drosophila*, the female meiosis I spindle contains microtubules of mixed polarity (Liang et al. 2009). It will be crucial to address the polarity of microtubules in the oocyte before, during, and after meiotic maturation.

10.4.2 Fertilization and the Completion of Meiosis

In *C. elegans*, fertilization is required for the proper completion of the two successive meiotic divisions (Ward and Carrel 1979; McNally and McNally 2005). When the oocyte undergoes meiotic maturation but is not fertilized, as happens in fertilization-defective mutants, meiotic spindle assembly and anaphase I occurs normally and on schedule; however, the spindle midzone persists, the first polar body does not form, and the meiosis II spindle fails to assemble (McNally and McNally 2005). Instead, these unfertilized oocytes undergo multiple cycles of nuclear reformation, S-phase, and nuclear envelope breakdown, becoming highly polyploidy (Ward and Carrel 1979). No cell division is apparently possible in the absence of the sperm-contributed centrioles. The degradation of cyclin B that accompanies M-phase exit in many systems (Murray 2004) is incomplete if fertilization is blocked (McNally and McNally 2005). Fertilization might therefore be required for activation of the ZYG-11/CUL-2 E3 ubiquitin ligase, which is required for progression through meiosis II and the degradation of cyclins B1 and B3 after their functions have been completed (Liu et al. 2004; Sonneville and Gonczy 2004; Deyter et al. 2010).

As in many animals (Horner and Wolfner 2008), an increase in cytosolic calcium levels accompanies fertilization in *C. elegans* (Samuel et al. 2001). A calcium signal at fertilization triggers APC/C and exit from meiosis II in *Xenopus* (Tunquist and Maller 2003; Liu and Maller 2005; Rauh et al. 2005). In *C. elegans*, the APC/C and separase are required for the completion of meiosis I (Furuta et al. 2000; Golden et al. 2000; Siomos et al. 2001; Davis et al. 2002; Shakes et al. 2011). Because the APC/C is required for meiosis I, it has been difficult to determine whether it also functions in meiosis II or polarity establishment. An analysis of conditional alleles at semi-permissive temperatures (Shakes et al. 2003) showed that the embryonic polarity defects of hypomorphic APC/C alleles (Rappleye et al. 2002) are likely due to effects on cell cycle progression during meiosis I (Shakes et al. 2003).

Activation of the APC/C is independent of fertilization in *C. elegans* because oocytes that undergo meiotic maturation and ovulation but not fertilization complete the metaphase-to-anaphase transition in meiosis I (McNally and McNally 2005). The APC/C and separase appear to have distinct roles in promoting cortical granule exocytosis required for formation of the eggshell (Sato et al. 2006; Bembenek et al. 2007). Yet, formation of the eggshell requires fertilization. Eggshell formation requires components of a cortical complex including three protein tyrosine phosphatase-like proteins (EGG-3, EGG-4, and EGG-5), and chitin synthetase (CHS-1) (Johnston et al. 2006; Maruyama et al. 2007; Parry et al. 2009). EGG-3, EGG-4, EGG-5, and CHS-1 are also required for the completion of meiosis, polar body formation, and the block to polyspermy (Johnston et al. 2006, 2010; Maruyama et al. 2007; Parry et al. 2009).

SPE-11 is a key sperm-derived factor needed for the completion of meiosis and eggshell formation (Hill et al. 1989; Browning and Strome 1996; McNally and McNally 2005). Fertilization with sperm from homozygous *spe-11* mutant males causes paternal-effect lethality due to a failure to produce polar bodies and defective cytokinesis in meiosis I and II (McNally and McNally 2005). The mutant embryos also exhibit defects in embryonic polarity (Hill et al. 1989). *spe-11* mutants can assemble meiosis I and II spindles; however, the anaphase I chromosomes appear to collapse. Because *spe-11* mutants assemble a meiosis II spindle, but unfertilized oocytes do not, the sperm must make unidentified contributions to meiosis II spindle assembly (McNally and McNally 2005). *spe-11* mutants were also observed to exhibit an incompletely penetrant polyspermy phenotype (Johnston et al. 2010), perhaps owing to the defect in eggshell synthesis. SPE-11 does not appear to exhibit conservation at the primary amino acid sequence level and its biochemical activity is unknown, though it exhibits a perinuclear localization pattern in sperm (Browning and Strome 1996). Because treatments that interfere with actin polymerization, such as Latrunculin A application or profilin (RNAi), phenocopy aspects of *spe-11* mutants (Yang et al. 2003), SPE-11 might regulate actin dynamics. Meiotic cytokinesis appears to involve a novel structural mechanism involving the ANI-2 anillin to ensure the formation of small polar bodies (Dorn et al. 2010). Future studies of SPE-11 and actin regulators may provide insights into the general cell biology of meiotic cytokinesis.

10.5 Conclusions and Future Directions

10.5.1 Control of Oogenesis in the Wild Type

A decade ago, only five paragraphs were devoted to the control of oocyte growth and meiotic maturation in a review of *C. elegans* gonadogenesis (Hubbard and Greenstein 2000). It is clear that there has been immense progress in this field since then. As the field endeavors to obtain a complete understanding of oogenesis in the wild type, it is clear that studies utilizing forward and reverse genetics and genomics will continue to play a prominent role. A central question concerns the connection between the sheath cell signaling pathway (Fig. 10.4) and activation of protein kinase pathways in the germ line that are critical for meiotic maturation (e.g., the MPK-1 and CDK-1 pathways). Whether translational regulation in the germ line is a key target of meiotic maturation signaling is as of yet unclear. A major goal for the field is to connect phenotypes with cell biological and biochemical mechanisms. Yet, there are vast areas of the cell biology of oogenesis of which we have only limited knowledge. Models for how oocytes grow and cellularize are quite incomplete at cell biological and biochemical levels. Likewise, it is unclear how oocyte meiotic maturation is spatially restricted to the most proximal oocyte. A challenge for the field will be to achieve integration between the component parts of oogenesis. It is now clear that there is a connection between the presence of sperm in the gonad, oocyte growth, and downstream responses of germline stem cells to DTC signaling. It will be crucial to elucidate the dependencies between germline events and to understand the mechanisms and checkpoints by which they are controlled. Forward genetic screens and RNAi approaches (e.g., Green et al. 2011) will continue to provide entry points for unraveling the mechanisms of oogenesis.

10.5.2 Oogenesis and Reproductive Aging

The control of oogenesis in *C. elegans* and mammals share a number of biological and mechanistic similarities. Yet, there are also fundamental differences. A central issue is the extent to which female reproductive aging—the decline in fertility as a function of time—can be modeled in the worm. Chromosome missegregation in female meiosis I is the leading cause of miscarriage and congenital birth defects in humans, such as Down syndrome (Hassold and Hunt 2001). This "maternal-age effect" is a major barrier to human reproduction. In human females, meiotic recombination occurs exclusively in the embryo, and there is no generally accepted evidence for the existence of actual stem cells in the adult mammalian ovary (Zuckerman 1951; Peters et al. 1962; Telfer et al. 2005; Bristol-Gould et al. 2006; Eggan et al. 2006). This absence of stem cells in the adult human ovary is a major difference between the mammalian and *C. elegans* systems (Hansen and Schedl 2012, Chap. 4); this difference is important for considering the origins of reproductive aging. Additionally, mammalian oocytes arrest in diplotene for years, but these oocytes are transcriptionally active; the importance of transcription in the oocyte is illustrated by mutant phenotypes obtained from generating conditional gene knock-outs using the oocyte-specific Zp3 driver (Sun et al. 2008). By contrast, in *C. elegans* oocytes are transcriptionally inactive and arrest for days. In mammals, arrested oocytes are recruited each cycle, which then undergo considerable growth and development—folliculogenesis and imprinting occur in the time window before meiotic maturation. In *C. elegans* females, arrested oocytes are poised to mature as soon as the sperm signal appears, though mating stimulates oocyte production and growth, which is appropriate for the worm's assembly-line reproductive strategy. There may be key differences in how aging impinges on these biological processes.

Reproductive aging is a new frontier in the *C. elegans* germline field. In considering how aging might impact the fidelity of the oocyte meiotic maturation divisions in the worm, it is necessary to address the complication that defects could occur in earlier events, extending back to the behavior of the germline stem cells themselves. In female mammals by contrast, chiasmata established in utero link duplicated homologs by virtue of sister chromatid cohesion. Sister chromatid cohesion must be maintained through the entirety of reproductive life. Premature loss of sister-chromatid cohesion is a likely cause of chromosome segregation errors during meiosis I in human females (Wolstenholme and Angell 2000), and this view is supported by an analysis of mouse models (Revenkova et al. 2004, 2010; Hodges et al. 2005; Chiang et al. 2010; Lister et al. 2010).

Several pioneering studies have begun to address reproductive aging in the *C. elegans* female germ line. For example, Hughes et al. (2007) found that dietary restriction and reduced insulin/IGF-1 signaling delayed reproductive aging. Interestingly, the age-related decline in the reproductive system was observed to be independent of progeny production. In these studies, feminization of the germ line was not observed to delay reproductive aging, once sperm were supplied by mating at later times. It is important to point out, however, that the germ line and somatic gonad actively function to inhibit meiotic maturation when sperm are absent. The energetics of this inhibition in comparison with fertility is unknown, but one would imagine it might be less costly, especially since oocyte production stops in the absence of sperm.

Several observations suggest that nutrition impacts reproductive aging and oocyte quality. Andux and Ellis (2008) reported that oocyte quality declines as worms age, and that this decline affects both prophase-arrested oocytes and newly forming oocytes in older animals as well. An especially interesting finding was that germline apoptosis functions to preserve oocyte quality in aging worms. One might imagine that germline apoptosis might preserve oocyte quality by culling defective oocytes. However, mutations that specifically block DNA-damage-induced apoptosis do not lower oocyte quality. Therefore, a model was presented in which apoptosis in the germ line optimizes the allocation of resources between developing oocytes.

In addition to insulin/IGF-1 signaling, the TGFβ Sma/Mab signaling was observed to affect reproductive aging and oocyte quality (Luo et al. 2009, 2010). Mutations in the TGFβ Sma/Mab were observed to delay reproductive aging and preserve oocyte quality. Since the TGFβ Sma/Mab pathway affects somatic cell growth, one of several possibilities is that this pathway might affect the relative allocation of nutritional resources between the germ line and soma. Heretofore, most genes affecting reproductive aging have been identified using a candidate gene approach. A recent advance is the use of forward genetics in combination with quantitative screening strategies to identify mutations affecting reproductive aging (Hughes et al. 2011).

An exciting recent finding was the observation that starvation of L4-stage hermaphrodites can bring about a state of "adult reproductive diapause" in which the entire germ line, except the germline stem cells, undergoes resorption in a process involving apoptosis (Angelo and Van Gilst 2009). The state of adult reproductive diapause results from a starvation response of oogenic germ cells (Seidel and Kimble 2011). Repopulation of the germ line and fertility was observed after refeeding (Angelo and Van Gilst 2009; Seidel and Kimble 2011). This work provides a new experimental model for investigating the impacts of aging and nutritive status on reproduction.

C. elegans is also emerging as a model for studying the reproductive impact of environmental toxicants. In mice, there is evidence that endocrine disrupters and environmental toxicants, such as bisphenol, can perturb the fidelity of meiotic chromosome segregation and that the final stages of oocyte growth and meiotic maturation are particularly sensitive (Hunt et al. 2003). However, bisphenol's effects are not specific for late events of oogenesis because fetal exposure also disrupts synaptonemal complex assembly (Susiarjo et al. 2007). In *C. elegans*, bisphenol perturbs early events in meiosis, including double strand break repair, as well as the kinetics of remodeling the oocyte bivalents (Allard and Colaiacovo 2010). These recent findings highlight the potential of the *C. elegans* system for addressing the origin of aneuploidies and miscarriage in humans. For this potential to be realized, a comprehensive understanding of the normal mechanisms controlling oocyte growth and meiotic maturation and their integration into the overall germline program is essential.

As this new field develops, it is clear that standardized assays and biomarkers for reproductive aging and oocyte quality would facilitate the comparison of data between labs. Normal germline development and function in young animals is complex and involves soma–germline interactions and coordination between multiple cellular processes. The study of reproductive aging is expected to be at least as complicated. Normal *C. elegans* development is remarkable for its reproducibility. By contrast, extensive phenotypic and functional variance in reproduction is observed in populations of aging worms. In the future, we anticipate exciting progress in the study of reproductive aging through the interaction of *C. elegans* researchers studying germline development and aging.

Acknowledgments We are grateful to Valerie Osterberg, Sara Christensen, and Bruce Bowerman for providing Fig. 10.6a. Thanks to Swathi Arur, Bruce Bowerman, Andy Golden, Frank McNally, Michael Miller, Martin Srayko, and Jennifer Schisa for providing comments on the manuscript. Thanks to Bruce Bowerman, Amaranath Govindan, Tim Schedl, Jennifer Schisa, and Todd Starich for communication of unpublished results. This work was supported by NIH grants GM65115 and GM57173 to D.G.

References

Albertson DG, Thomson JN (1993) Segregation of holocentric chromosomes at meiosis in the nematode, *Caenorhabditis elegans*. Chromosome Res 1(1):15–26

Allard P, Colaiacovo MP (2010) Bisphenol A impairs the double-strand break repair machinery in the germline and causes chromosome abnormalities. Proc Natl Acad Sci USA 107(47):20405–20410

Altun ZF, Chen B, Wang ZW, Hall DH (2009) High resolution map of *Caenorhabditis elegans* gap junction proteins. Dev Dyn 238(8):1936–1950

Andresson T, Ruderman JV (1998) The kinase Eg2 is a component of the *Xenopus* oocyte progesterone-activated signaling pathway. EMBO J 17(19):5627–5637

Andux S, Ellis RE (2008) Apoptosis maintains oocyte quality in aging *Caenorhabditis elegans* females. PLoS Genet 4(12):e1000295

Angelo G, Van Gilst MR (2009) Starvation protects germline stem cells and extends reproductive longevity in *C. elegans*. Science 326(5955):954–958

Aono S, Legouis R, Hoose WA, Kemphues KJ (2004) PAR-3 is required for epithelial cell polarity in the distal spermatheca of *C. elegans*. Development 131(12):2865–2874

Arur S, Ohmachi M, Nayak S, Hayes M, Miranda A, Hay A, Golden A, Schedl T (2009) Multiple ERK substrates execute single biological processes in *Caenorhabditis elegans* germ-line development. Proc Natl Acad Sci USA 106(12):4776–4781

Arur S, Ohmachi M, Berkseth M, Nayak S, Hansen D, Zarkower D, Schedl T (2011) MPK-1 ERK controls membrane organization in *C. elegans* oogenesis via a sex-determination module. Dev Cell 20(5):677–688

Bailly A, Gartner A (2012) Germ cell apoptosis and DNA damage responses. Advances in Experimental Medicine and Biology 757:249–276. (Chap. 9, this volume) Springer, New York

Bembenek JN, Richie CT, Squirrell JM, Campbell JM, Eliceiri KW, Poteryaev D, Spang A, Golden A, White JG (2007) Cortical granule exocytosis in *C. elegans* is regulated by cell cycle components including separase. Development 134(21):3837–3848

Bishop JD, Han Z, Schumacher JM (2005) The *Caenorhabditis elegans* Aurora B kinase AIR-2 phosphorylates and is required for the localization of a BimC kinesin to meiotic and mitotic spindles. Mol Biol Cell 16(2):742–756

Bottino D, Mogilner A, Roberts T, Stewart M, Oster G (2002) How nematode sperm crawl. J Cell Sci 115(Pt 2):367–384

Boxem M, Srinivasan DG, van den Heuvel S (1999) The *Caenorhabditis elegans* gene *ncc-1* encodes a *cdc2*-related kinase required for M phase in meiotic and mitotic cell divisions, but not for S phase. Development 126(10):2227–2239

Bristol-Gould SK, Kreeger PK, Selkirk CG, Kilen SM, Mayo KE, Shea LD, Woodruff TK (2006) Fate of the initial follicle pool: empirical and mathematical evidence supporting its sufficiency for adult fertility. Dev Biol 298(1):149–154

Browning H, Strome S (1996) A sperm-supplied factor required for embryogenesis in *C. elegans*. Development 122(1):391–404

Buck SH, Chiu D, Saito RM (2009) The cyclin-dependent kinase inhibitors, *cki-1* and *cki-2*, act in overlapping but distinct pathways to control cell cycle quiescence during *C. elegans* development. Cell Cycle 8(16):2613–2620

Bui YK, Sternberg PW (2002) *Caenorhabditis elegans* inositol 5-phosphatase homolog negatively regulates inositol 1,4,5-triphosphate signaling in ovulation. Mol Biol Cell 13(5):1641–1651

Burke DJ, Ward S (1983) Identification of a large multigene family encoding the major sperm protein of *Caenorhabditis elegans*. J Mol Biol 171(1):1–29

Burrows AE, Sceurman BK, Kosinski ME, Richie CT, Sadler PL, Schumacher JM, Golden A (2006) The *C. elegans* Myt1 ortholog is required for the proper timing of oocyte maturation. Development 133(4):697–709

Castilho PV, Williams BC, Mochida S, Zhao Y, Goldberg ML (2009) The M phase kinase Greatwall (Gwl) promotes inactivation of PP2A/B55delta, a phosphatase directed against CDK phosphosites. Mol Biol Cell 20(22):4777–4789

Chase D, Serafinas C, Ashcroft N, Kosinski M, Longo D, Ferris DK, Golden A (2000) The polo-like kinase PLK-1 is required for nuclear envelope breakdown and the completion of meiosis in *Caenorhabditis elegans*. Genesis 26(1):26–41

Cheng H, Govindan JA, Greenstein D (2008) Regulated trafficking of the MSP/Eph receptor during oocyte meiotic maturation in *C. elegans*. Curr Biol 18(10):705–714

Cheng KC, Klancer R, Singson A, Seydoux G (2009) Regulation of MBK-2/DYRK by CDK-1 and the pseudophosphatases EGG-4 and EGG-5 during the oocyte-to-embryo transition. Cell 139(3):560–572

Chi W, Reinke V (2006) Promotion of oogenesis and embryogenesis in the *C. elegans* gonad by EFL-1/DPL-1 (E2F) does not require LIN-35 (pRB). Development 133(16):3147–3157

Chi W, Reinke V (2009) DPL-1 (DP) acts in the germ line to coordinate ovulation and fertilization in *C. elegans*. Mech Dev 126(5–6):406–416

Chiang T, Duncan FE, Schindler K, Schultz RM, Lampson MA (2010) Evidence that weakened centromere cohesion is a leading cause of age-related aneuploidy in oocytes. Curr Biol 20(17):1522–1528

Cho S, Jin SW, Cohen A, Ellis RE (2004) A phylogeny of *Caenorhabditis* reveals frequent loss of introns during nematode evolution. Genome Res 14(7):1207–1220

Chu DS, Shakes DC (2012) Spermatogenesis. Advances in Experimental Medicine and Biology 757:171–203. (Chap. 7, this volume) Springer, New York

Church DL, Guan KL, Lambie EJ (1995) Three genes of the MAP kinase cascade, *mek-2*, *mpk-1/sur-1* and *let-60* ras, are required for meiotic cell cycle progression in *Caenorhabditis elegans*. Development 121(8):2525–2535

Clandinin TR, Mains PE (1993) Genetic studies of *mei-1* gene activity during the transition from meiosis to mitosis in *Caenorhabditis elegans*. Genetics 134(1):199–210

Clandinin TR, DeModena JA, Sternberg PW (1998) Inositol trisphosphate mediates a RAS-independent response to LET-23 receptor tyrosine kinase activation in *C. elegans*. Cell 92(4):523–533

Clark-Maguire S, Mains PE (1994a) *mei-1*, a gene required for meiotic spindle formation in *Caenorhabditis elegans*, is a member of a family of ATPases. Genetics 136(2):533–546

Clark-Maguire S, Mains PE (1994b) Localization of the *mei-1* gene product of *Caenorhaditis elegans*, a meiotic-specific spindle component. J Cell Biol 126(1):199–209

Clary LM, Okkema PG (2010) The EGR family gene *egrh-1* functions non-autonomously in the control of oocyte meiotic maturation and ovulation in *C. elegans*. Development 137(18):3129–3137

Corrigan C, Subramanian R, Miller MA (2005) Eph and NMDA receptors control Ca2+/calmodulin-dependent protein kinase II activation during *C. elegans* oocyte meiotic maturation. Development 132(23):5225–5237

Davis ES, Wille L, Chestnut BA, Sadler PL, Shakes DC, Golden A (2002) Multiple subunits of the *Caenorhabditis elegans* anaphase-promoting complex are required for chromosome segregation during meiosis I. Genetics 160(2):805–813

de Carvalho CE, Zaaijer S, Smolikov S, Gu Y, Schumacher JM, Colaiacovo MP (2008) LAB-1 antagonizes the Aurora B kinase in *C. elegans*. Genes Dev 22(20):2869–2885

Detwiler MR, Reuben M, Li X, Rogers E, Lin R (2001) Two zinc finger proteins, OMA-1 and OMA-2, are redundantly required for oocyte maturation in *C. elegans*. Dev Cell 1(2):187–199

Deyter GM, Furuta T, Kurasawa Y, Schumacher JM (2010) *Caenorhabditis elegans* cyclin B3 is required for multiple mitotic processes including alleviation of a spindle checkpoint-dependent block in anaphase chromosome segregation. PLoS Genet 6(11):e1001218

Dorn JF, Zhang L, Paradis V, Edoh-Bedi D, Jusu S, Maddox PS, Maddox AS (2010) Actomyosin tube formation in polar body cytokinesis requires Anillin in *C. elegans*. Curr Biol 20(22):2046–2051

Dumont J, Million K, Sunderland K, Rassinier P, Lim H, Leader B, Verlhac MH (2007) Formin-2 is required for spindle migration and for the late steps of cytokinesis in mouse oocytes. Dev Biol 301(1):254–265

Dumont J, Oegema K, Desai A (2010) A kinetochore-independent mechanism drives anaphase chromosome separation during acentrosomal meiosis. Nat Cell Biol 12(9):894–901

Dunphy WG, Brizuela L, Beach D, Newport J (1988) The *Xenopus* cdc2 protein is a component of MPF, a cytoplasmic regulator of mitosis. Cell 54(3):423–431

Dupre A, Jessus C, Ozon R, Haccard O (2002) Mos is not required for the initiation of meiotic maturation in *Xenopus* oocytes. EMBO J 21(15):4026–4036

Edmonds JW, Prasain JK, Dorand D, Yang Y, Hoang HD, Vibbert J, Kubagawa HM, Miller MA (2010) Insulin/FOXO signaling regulates ovarian prostaglandins critical for reproduction. Dev Cell 19(6):858–871

Eggan K, Jurga S, Gosden R, Min IM, Wagers AJ (2006) Ovulated oocytes in adult mice derive from non-circulating germ cells. Nature 441(7097):1109–1114

Ellefson ML, McNally FJ (2009) Kinesin-1 and cytoplasmic dynein act sequentially to move the meiotic spindle to the oocyte cortex in *Caenorhabditis elegans*. Mol Biol Cell 20(11):2722–2730

Ellefson ML, McNally FJ (2011) CDK-1 inhibits meiotic spindle shortening and dynein-dependent spindle rotation in *C. elegans*. J Cell Biol 193(7):1229–1244

Fabritius AS, Ellefson ML, McNally FJ (2011) Nuclear and spindle positioning during oocyte meiosis. Curr Opin Cell Biol 23(1):78–84

Ferby I, Blazquez M, Palmer A, Eritja R, Nebreda AR (1999) A novel p34(cdc2)-binding and activating protein that is necessary and sufficient to trigger G(2)/M progression in *Xenopus* oocytes. Genes Dev 13(16):2177–2189

Ferrell JE Jr (1999a) *Xenopus* oocyte maturation: new lessons from a good egg. Bioessays 21(10):833–842

Ferrell JE Jr (1999b) Building a cellular switch: more lessons from a good egg. Bioessays 21(10):866–870

Francis R, Maine E, Schedl T (1995) Analysis of the multiple roles of *gld-1* in germline development: interactions with the sex determination cascade and the *glp-1* signaling pathway. Genetics 139(2):607–630

Furuta T, Tuck S, Kirchner J, Koch B, Auty R, Kitagawa R, Rose AM, Greenstein D (2000) EMB-30: an APC4 homologue required for metaphase-to-anaphase transitions during meiosis and mitosis in *Caenorhabditis elegans*. Mol Biol Cell 11(4):1401–1419

Gautier J, Norbury C, Lohka M, Nurse P, Maller J (1988) Purified maturation-promoting factor contains the product of a *Xenopus* homolog of the fission yeast cell cycle control gene cdc2+. Cell 54(3):433–439

Gautier J, Minshull J, Lohka M, Glotzer M, Hunt T, Maller JL (1990) Cyclin is a component of maturation-promoting factor from *Xenopus*. Cell 60(3):487–494

George SE, Simokat K, Hardin J, Chisholm AD (1998) The VAB-1 Eph receptor tyrosine kinase functions in neural and epithelial morphogenesis in *C. elegans*. Cell 92(5):633–643

Gibert MA, Starck J, Beguet B (1984) Role of the gonad cytoplasmic core during oogenesis of the nematode *Caenorhabditis elegans*. Biol Cell 50(1):77–85

Gissendanner CR, Kelley K, Nguyen TQ, Hoener MC, Sluder AE, Maina CV (2008) The *Caenorhabditis elegans* NR4A nuclear receptor is required for spermatheca morphogenesis. Dev Biol 313(2):767–786

Golden A, Sadler PL, Wallenfang MR, Schumacher JM, Hamill DR, Bates G, Bowerman B, Seydoux G, Shakes DC (2000) Metaphase to anaphase (mat) transition-defective mutants in *Caenorhabditis elegans*. J Cell Biol 151(7):1469–1482

Goldstein B, Hird SN (1996) Specification of the anteroposterior axis in *Caenorhabditis elegans*. Development 122(5):1467–1474

Govindan JA, Cheng H, Harris JE, Greenstein D (2006) Galphao/i and Galphas signaling function in parallel with the MSP/Eph receptor to control meiotic diapause in *C. elegans*. Curr Biol 16(13):1257–1268

Govindan JA, Nadarajan S, Kim S, Starich TA, Greenstein D (2009) Somatic cAMP signaling regulates MSP-dependent oocyte growth and meiotic maturation in *C. elegans*. Development 136(13):2211–2221

Grant B, Hirsh D (1999) Receptor-mediated endocytosis in the *Caenorhabditis elegans* oocyte. Mol Biol Cell 10(12):4311–4326

Green RA, Kao HL, Audhya A, Arur S, Mayers JR, Fridolfsson HN, Schulman M, Schloissnig S, Niessen S, Laband K, Wang S, Starr DA, Hyman AA, Schedl T, Desai A, Piano F, Gunsalus KC, Oegema K (2011) A high-resolution *C. elegans* essential gene network based on phenotypic profiling of a complex tissue. Cell 145(3):470–482

Greenstein D, Hird S, Plasterk RH, Andachi Y, Kohara Y, Wang B, Finney M, Ruvkun G (1994) Targeted mutations in the *Caenorhabditis elegans* POU homeo box gene *ceh-18* cause defects in oocyte cell cycle arrest, gonad migration, and epidermal differentiation. Genes Dev 8(16):1935–1948

Gumienny TL, Lambie E, Hartwieg E, Horvitz HR, Hengartner MO (1999) Genetic control of programmed cell death in the *Caenorhabditis elegans* hermaphrodite germline. Development 126(5):1011–1022

Gutch MJ, Flint AJ, Keller J, Tonks NK, Hengartner MO (1998) The *Caenorhabditis elegans* SH2 domain-containing protein tyrosine phosphatase PTP-2 participates in signal transduction during oogenesis and vulval development. Genes Dev 12(4):571–585

Guven-Ozkan T, Nishi Y, Robertson SM, Lin R (2008) Global transcriptional repression in *C. elegans* germline precursors by regulated sequestration of TAF-4. Cell 135(1):149–160

Guven-Ozkan T, Robertson SM, Nishi Y, Lin R (2010) *zif-1* translational repression defines a second, mutually exclusive OMA function in germline transcriptional repression. Development 137(20):3373–3382

Haccard O, Jessus C (2006) Redundant pathways for Cdc2 activation in *Xenopus* oocyte: either cyclin B or Mos synthesis. EMBO Rep 7(3):321–325

Hall DH, Altun Z (2008) *C. elegans* atlas. Cold Spring Harbor Laboratory Press, Cold Spring Harbor

Hall DH, Winfrey VP, Blaeuer G, Hoffman LH, Furuta T, Rose KL, Hobert O, Greenstein D (1999) Ultrastructural features of the adult hermaphrodite gonad of *Caenorhabditis elegans*: relations between the germ line and soma. Dev Biol 212(1):101–123

Hansen D, Schedl T (2012) Stem cell proliferation versus meiotic fate decision in *C. elegans*. Advances in Experimental Medicine and Biology 757:71–99. (Chap. 4, this volume) Springer, New York

Harris JE, Govindan JA, Yamamoto I, Schwartz J, Kaverina I, Greenstein D (2006) Major sperm protein signaling promotes oocyte microtubule reorganization prior to fertilization in *Caenorhabditis elegans*. Dev Biol 299(1):105–121

Hassold T, Hunt P (2001) To err (meiotically) is human: the genesis of human aneuploidy. Nat Rev 2(4):280–291

Heald R, Tournebize R, Blank T, Sandaltzopoulos R, Becker P, Hyman A, Karsenti E (1996) Self-organization of microtubules into bipolar spindles around artificial chromosomes in *Xenopus* egg extracts. Nature 382(6590):420–425

Henderson MA, Cronland E, Dunkelbarger S, Contreras V, Strome S, Keiper BD (2009) A germline-specific isoform of eIF4E (IFE-1) is required for efficient translation of stored mRNAs and maturation of both oocytes and sperm. J Cell Sci 122(Pt 10):1529–1539

Hiatt SM, Duren HM, Shyu YJ, Ellis RE, Hisamoto N, Matsumoto K, Kariya K, Kerppola TK, Hu CD (2009) *Caenorhabditis elegans* FOS-1 and JUN-1 regulate *plc-1* expression in the spermatheca to control ovulation. Mol Biol Cell 20(17):3888–3895

Hill DP, Shakes DC, Ward S, Strome S (1989) A sperm-supplied product essential for initiation of normal embryogenesis in *Caenorhabditis elegans* is encoded by the paternal-effect embryonic-lethal gene, *spe-11*. Dev Biol 136(1):154–166

Hirsh D, Oppenheim D, Klass M (1976) Development of the reproductive system of *Caenorhabditis elegans*. Dev Biol 49(1):200–219

Hochegger H, Klotzbucher A, Kirk J, Howell M, le Guellec K, Fletcher K, Duncan T, Sohail M, Hunt T (2001) New B-type cyclin synthesis is required between meiosis I and II during *Xenopus* oocyte maturation. Development 128(19):3795–3807

Hodges CA, Revenkova E, Jessberger R, Hassold TJ, Hunt PA (2005) SMC1beta-deficient female mice provide evidence that cohesins are a missing link in age-related nondisjunction. Nat Genet 37(12):1351–1355

Hopper NA, Lee J, Sternberg PW (2000) ARK-1 inhibits EGFR signaling in *C. elegans*. Mol Cell 6(1):65–75

Horner VL, Wolfner MF (2008) Transitioning from egg to embryo: triggers and mechanisms of egg activation. Dev Dyn 237(3):527–544

Howe M, McDonald KL, Albertson DG, Meyer BJ (2001) HIM-10 is required for kinetochore structure and function on *Caenorhabditis elegans* holocentric chromosomes. J Cell Biol 153(6):1227–1238

Hubbard EJ, Greenstein D (2000) The *Caenorhabditis elegans* gonad: a test tube for cell and developmental biology. Dev Dyn 218(1):2–22

Hughes SE, Evason K, Xiong C, Kornfeld K (2007) Genetic and pharmacological factors that influence reproductive aging in nematodes. PLoS Genet 3(2):e25

Hughes SE, Huang C, Kornfeld K (2011) Identification of mutations that delay somatic or reproductive aging of *Caenorhabditis elegans*. Genetics 189(1):341–356

Hunt PA, Koehler KE, Susiarjo M, Hodges CA, Ilagan A, Voigt RC, Thomas S, Thomas BF, Hassold TJ (2003) Bisphenol a exposure causes meiotic aneuploidy in the female mouse. Curr Biol 13(7):546–553

Iwasaki K, McCarter J, Francis R, Schedl T (1996) *emo-1*, a *Caenorhabditis elegans* Sec61p gamma homologue, is required for oocyte development and ovulation. J Cell Biol 134(3):699–714

Jadhav S, Rana M, Subramaniam K (2008) Multiple maternal proteins coordinate to restrict the translation of *C. elegans nanos-2* to primordial germ cells. Development 135(10):1803–1812

Jaramillo-Lambert A, Ellefson M, Villeneuve AM, Engebrecht J (2007) Differential timing of S phases, X chromosome replication, and meiotic prophase in the *C. elegans* germ line. Dev Biol 308(1):206–221

Johnston WL, Krizus A, Dennis JW (2006) The eggshell is required for meiotic fidelity, polar-body extrusion and polarization of the *C. elegans* embryo. BMC Biol 4:35

Johnston WL, Krizus A, Dennis JW (2010) Eggshell chitin and chitin-interacting proteins prevent polyspermy in *C. elegans*. Curr Biol 20(21):1932–1937

Jud M, Razelun J, Bickel J, Czerwinski M, Schisa JA (2007) Conservation of large foci formation in arrested oocytes of *Caenorhabditis* nematodes. Dev Genes Evol 217(3):221–226

Jud MC, Czerwinski MJ, Wood MP, Young RA, Gallo CM, Bickel JS, Petty EL, Mason JM, Little BA, Padilla PA, Schisa JA (2008) Large P body-like RNPs form in *C. elegans* oocytes in response to arrested ovulation, heat shock, osmotic stress, and anoxia and are regulated by the major sperm protein pathway. Dev Biol 318(1):38–51

Kadyk LC, Kimble J (1998) Genetic regulation of entry into meiosis in *Caenorhabditis elegans*. Development 125(10):1803–1813

Kaiser SE, Brickner JH, Reilein AR, Fenn TD, Walter P, Brunger AT (2005) Structural basis of FFAT motif-mediated ER targeting. Structure 13(7):1035–1045

Kaitna S, Mendoza M, Jantsch-Plunger V, Glotzer M (2000) Incenp and an aurora-like kinase form a complex essential for chromosome segregation and efficient completion of cytokinesis. Curr Biol 10(19):1172–1181

Kalab P, Solc P, Motlik J (2011) The role of RanGTP gradient in vertebrate oocyte maturation. Results Probl Cell Differ 53:235–267

Kaneda M, Tang F, O'Carroll D, Lao K, Surani MA (2009) Essential role for Argonaute2 protein in mouse oogenesis. Epigenetics Chromatin 2(1):9

Kariya K, Bui YK, Gao X, Sternberg PW, Kataoka T (2004) Phospholipase Cepsilon regulates ovulation in *Caenorhabditis elegans*. Dev Biol 274(1):201–210

Kashina AS, Rogers GC, Scholey JM (1997) The bimC family of kinesins: essential bipolar mitotic motors driving centrosome separation. Biochim Biophys Acta 1357(3):257–271

Kemphues KJ, Priess JR, Morton DG, Cheng NS (1988) Identification of genes required for cytoplasmic localization in early *C. elegans* embryos. Cell 52(3):311–320

Kim DY, Roy R (2006) Cell cycle regulators control centrosome elimination during oogenesis in *Caenorhabditis elegans*. J Cell Biol 174(6):751–757

Kimble J, Sharrock WJ (1983) Tissue-specific synthesis of yolk proteins in *Caenorhabditis elegans*. Dev Biol 96(1):189–196

Kiontke K, Fitch DH (2005) The phylogenetic relationships of *Caenorhabditis* and other rhabditids. WormBook:1–11

Kiontke K, Gavin NP, Raynes Y, Roehrig C, Piano F, Fitch DH (2004) *Caenorhabditis* phylogeny predicts convergence of hermaphroditism and extensive intron loss. Proc Natl Acad Sci USA 101(24):9003–9008

Klass MR, Kinsley S, Lopez LC (1984) Isolation and characterization of a sperm-specific gene family in the nematode *Caenorhabditis elegans*. Mol Cell Biol 4(3):529–537

Kornbluth S, Sebastian B, Hunter T, Newport J (1994) Membrane localization of the kinase which phosphorylates p34cdc2 on threonine 14. Mol Biol Cell 5(3):273–282

Korswagen HC, Park JH, Ohshima Y, Plasterk RH (1997) An activating mutation in a *Caenorhabditis elegans* Gs protein induces neural degeneration. Genes Dev 11(12):1493–1503

Kosinski M, McDonald K, Schwartz J, Yamamoto I, Greenstein D (2005) *C. elegans* sperm bud vesicles to deliver a meiotic maturation signal to distant oocytes. Development 132(15):3357–3369

Kostic I, Li S, Roy R (2003) *cki-1* links cell division and cell fate acquisition in the *C. elegans* somatic gonad. Dev Biol 263(2):242–252

Kovacevic I, Cram EJ (2010) FLN-1/filamin is required for maintenance of actin and exit of fertilized oocytes from the spermatheca in *C. elegans*. Dev Biol 347(2):247–257

Kubagawa HM, Watts JL, Corrigan C, Edmonds JW, Sztul E, Browse J, Miller MA (2006) Oocyte signals derived from polyunsaturated fatty acids control sperm recruitment in vivo. Nat Cell Biol 8(10):1143–1148

Kumagai A, Dunphy WG (1991) The cdc25 protein controls tyrosine dephosphorylation of the cdc2 protein in a cell-free system. Cell 64(5):903–914

Kumagai A, Dunphy WG (1996) Purification and molecular cloning of Plx1, a Cdc25-regulatory kinase from *Xenopus* egg extracts. Science 273(5280):1377–1380

Lee MH, Ohmachi M, Arur S, Nayak S, Francis R, Church D, Lambie E, Schedl T (2007) Multiple functions and dynamic activation of MPK-1 extracellular signal-regulated kinase signaling in *Caenorhabditis elegans* germline development. Genetics 177(4):2039–2062

Lenormand JL, Dellinger RW, Knudsen KE, Subramani S, Donoghue DJ (1999) Speedy: a novel cell cycle regulator of the G2/M transition. EMBO J 18(7):1869–1877

Li H, Guo F, Rubinstein B, Li R (2008) Actin-driven chromosomal motility leads to symmetry breaking in mammalian meiotic oocytes. Nat Cell Biol 10(11):1301–1308

Liang CG, Su YQ, Fan HY, Schatten H, Sun QY (2007) Mechanisms regulating oocyte meiotic resumption: roles of mitogen-activated protein kinase. Mol Endocrinol 21(9):2037–2055

Liang ZY, Hallen MA, Endow SA (2009) Mature *Drosophila* meiosis I spindles comprise microtubules of mixed polarity. Curr Biol 19(2):163–168

Lister LM, Kouznetsova A, Hyslop LA, Kalleas D, Pace SL, Barel JC, Nathan A, Floros V, Adelfalk C, Watanabe Y, Jessberger R, Kirkwood TB, Hoog C, Herbert M (2010) Age-related meiotic segregation errors in mammalian oocytes are preceded by depletion of cohesin and Sgo2. Curr Biol 20(17):1511–1521

Liu J, Maller JL (2005) Calcium elevation at fertilization coordinates phosphorylation of XErp1/Emi2 by Plx1 and CaMK II to release metaphase arrest by cytostatic factor. Curr Biol 15(16):1458–1468

Liu J, Vasudevan S, Kipreos ET (2004) CUL-2 and ZYG-11 promote meiotic anaphase II and the proper placement of the anterior-posterior axis in *C. elegans*. Development 131(15):3513–3525

Lohka MJ, Hayes MK, Maller JL (1988) Purification of maturation-promoting factor, an intracellular regulator of early mitotic events. Proc Natl Acad Sci USA 85(9):3009–3013

Lorin-Nebel C, Xing J, Yan X, Strange K (2007) CRAC channel activity in *C. elegans* is mediated by Orai1 and STIM1 homologues and is essential for ovulation and fertility. J Physiol 580 (Pt 1):67–85

Lui DY, Colaiácovo MP (2012) Meiotic development in *C. elegans*. Advances in Experimental Medicine and Biology 757:133–170. (Chap. 6, this volume) Springer, New York

Luo S, Shaw WM, Ashraf J, Murphy CT (2009) TGF-beta Sma/Mab signaling mutations uncouple reproductive aging from somatic aging. PLoS Genet 5(12):e1000789

Luo S, Kleemann GA, Ashraf JM, Shaw WM, Murphy CT (2010) TGF-beta and insulin signaling regulate reproductive aging via oocyte and germline quality maintenance. Cell 143(2):299–312

Maddox PS, Oegema K, Desai A, Cheeseman IM (2004) "Holo"er than thou: chromosome segregation and kinetochore function in *C. elegans*. Chromosome Res 12(6):641–653

Maddox AS, Habermann B, Desai A, Oegema K (2005) Distinct roles for two *C. elegans* anillins in the gonad and early embryo. Development 132(12):2837–2848

Mains PE, Kemphues KJ, Sprunger SA, Sulston IA, Wood WB (1990) Mutations affecting the meiotic and mitotic divisions of the early *Caenorhabditis elegans* embryo. Genetics 126(3):593–605

Marcello MR, Singaravelu G, Singson A (2012) Fertilization. Advances in Experimental Medicine and Biology 757:321–350. (Chap. 11, this volume) Springer, New York

Maruyama R, Velarde NV, Klancer R, Gordon S, Kadandale P, Parry JM, Hang JS, Rubin J, Stewart-Michaelis A, Schweinsberg P, Grant BD, Piano F, Sugimoto A, Singson A (2007) EGG-3 regulates cell-surface and cortex rearrangements during egg activation in *Caenorhabditis elegans*. Curr Biol 17(18):1555–1560

Masui Y (2001) From oocyte maturation to the in vitro cell cycle: the history of discoveries of maturation-promoting factor (MPF) and cytostatic factor (CSF). Differentiation 69(1):1–17

Masui Y, Clarke HJ (1979) Oocyte maturation. Int Rev Cytol 57:185–282

Masui Y, Markert CL (1971) Cytoplasmic control of nuclear behavior during meiotic maturation of frog oocytes. J Exp Zool 177(2):129–145

Matthies HJ, McDonald HB, Goldstein LS, Theurkauf WE (1996) Anastral meiotic spindle morphogenesis: role of the non-claret disjunctional kinesin-like protein. J Cell Biol 134(2):455–464

Matyash V, Geier C, Henske A, Mukherjee S, Hirsh D, Thiele C, Grant B, Maxfield FR, Kurzchalia TV (2001) Distribution and transport of cholesterol in *Caenorhabditis elegans*. Mol Biol Cell 12(6):1725–1736

McCarter J (1998) The regulation of oocyte maturation and ovulation in *Caenorhabditis elegans*. PhD Thesis, Washington University

McCarter J, Bartlett B, Dang T, Schedl T (1997) Soma-germ cell interactions in *Caenorhabditis elegans*: multiple events of hermaphrodite germline development require the somatic sheath and spermathecal lineages. Dev Biol 181(2):121–143

McCarter J, Bartlett B, Dang T, Schedl T (1999) On the control of oocyte meiotic maturation and ovulation in *Caenorhabditis elegans*. Dev Biol 205(1):111–128

McMullan R, Nurrish SJ (2011) The RHO-1 RhoGTPase modulates fertility and multiple behaviors in adult *C. elegans*. PLoS One 6(2):e17265

McNally KL, McNally FJ (2005) Fertilization initiates the transition from anaphase I to metaphase II during female meiosis in *C. elegans*. Dev Biol 282(1):218–230

McNally KP, McNally FJ (2011) The spindle assembly function of *Caenorhabditis elegans* katanin does not require microtubule-severing activity. Mol Biol Cell 22(9):1550–1560

McNally FJ, Vale RD (1993) Identification of katanin, an ATPase that severs and disassembles stable microtubules. Cell 75(3):419–429

McNally K, Audhya A, Oegema K, McNally FJ (2006) Katanin controls mitotic and meiotic spindle length. J Cell Biol 175(6):881–891

McNally KL, Martin JL, Ellefson M, McNally FJ (2010) Kinesin-dependent transport results in polarized migration of the nucleus in oocytes and inward movement of yolk granules in meiotic embryos. Dev Biol 339(1):126–140

Mendez R, Richter JD (2001) Translational control by CPEB: a means to the end. Nat Rev Mol Cell Biol 2(7):521–529

Mendez R, Hake LE, Andresson T, Littlepage LE, Ruderman JV, Richter JD (2000) Phosphorylation of CPE binding factor by Eg2 regulates translation of *c-mos* mRNA. Nature 404(6775):302–307

Meyerzon M, Fridolfsson HN, Ly N, McNally FJ, Starr DA (2009) UNC-83 is a nuclear-specific cargo adaptor for kinesin-1-mediated nuclear migration. Development 136(16):2725–2733

Mikeladze-Dvali T, von Tobel L, Strnad P, Knott G, Leonhardt H, Schermelleh L, Gönczy P (2012) Analysis of centriole elimination during *C. elegans* oogenesis. Development 139(9): 1670–1679

Miller MA, Nguyen VQ, Lee MH, Kosinski M, Schedl T, Caprioli RM, Greenstein D (2001) A sperm cytoskeletal protein that signals oocyte meiotic maturation and ovulation. Science 291(5511):2144–2147

Miller MA, Ruest PJ, Kosinski M, Hanks SK, Greenstein D (2003) An Eph receptor sperm-sensing control mechanism for oocyte meiotic maturation in *Caenorhabditis elegans*. Genes Dev 17(2):187–200

Mochida S, Maslen SL, Skehel M, Hunt T (2010) Greatwall phosphorylates an inhibitor of protein phosphatase 2A that is essential for mitosis. Science 330(6011):1670–1673

Monen J, Maddox PS, Hyndman F, Oegema K, Desai A (2005) Differential role of CENP-A in the segregation of holocentric *C. elegans* chromosomes during meiosis and mitosis. Nat Cell Biol 7(12):1248–1255

Morgan DO (2007) The cell cycle: principles of control. New Science Press, London

Mueller PR, Coleman TR, Dunphy WG (1995) Cell cycle regulation of a *Xenopus* Wee1-like kinase. Mol Biol Cell 6(1):119–134

Muller-Reichert T, Greenan G, O'Toole E, Srayko M (2010) The elegans of spindle assembly. Cell Mol Life Sci 67(13):2195–2213

Murchison EP, Stein P, Xuan Z, Pan H, Zhang MQ, Schultz RM, Hannon GJ (2007) Critical roles for Dicer in the female germline. Genes Dev 21(6):682–693

Murray AW (2004) Recycling the cell cycle: cyclins revisited. Cell 116(2):221–234

Myers CD, Goh PY, Allen TS, Bucher EA, Bogaert T (1996) Developmental genetic analysis of troponin T mutations in striated and nonstriated muscle cells of *Caenorhabditis elegans*. J Cell Biol 132(6):1061–1077

Na J, Zernicka-Goetz M (2006) Asymmetric positioning and organization of the meiotic spindle of mouse oocytes requires CDC42 function. Curr Biol 16(12):1249–1254

Nadarajan S, Govindan JA, McGovern M, Hubbard EJ, Greenstein D (2009) MSP and GLP-1/Notch signaling coordinately regulate actomyosin-dependent cytoplasmic streaming and oocyte growth in *C. elegans*. Development 136(13):2223–2234

Nance J, Zallen JA (2011) Elaborating polarity: PAR proteins and the cytoskeleton. Development 138(5):799–809

Noble SL, Allen BL, Goh LK, Nordick K, Evans TC (2008) Maternal mRNAs are regulated by diverse P body-related mRNP granules during early *Caenorhabditis elegans* development. J Cell Biol 182(3):559–572

Norman KR, Fazzio RT, Mellem JE, Espelt MV, Strange K, Beckerle MC, Maricq AV (2005) The Rho/Rac-family guanine nucleotide exchange factor VAV-1 regulates rhythmic behaviors in *C. elegans*. Cell 123(1):119–132

Nousch M, Eckmann CR (2012) Translational control in the *C. elegans* germ line. Advances in Experimental Medicine and Biology 757:205–247. (Chap. 8, this volume) Springer, New York

Obinata T, Ono K, Ono S (2010) Troponin I controls ovulatory contraction of non-striated actomyosin networks in the *C. elegans* somatic gonad. J Cell Sci 123(Pt 9):1557–1566

O'Connell KF, Caron C, Kopish KR, Hurd DD, Kemphues KJ, Li Y, White JG (2001) The *C. elegans zyg-1* gene encodes a regulator of centrosome duplication with distinct maternal and paternal roles in the embryo. Cell 105(4):547–558

Ohmachi M, Rocheleau CE, Church D, Lambie E, Schedl T, Sundaram MV (2002) *C. elegans ksr-1* and *ksr-2* have both unique and redundant functions and are required for MPK-1 ERK phosphorylation. Curr Biol 12(5):427–433

Ono K, Ono S (2004) Tropomyosin and troponin are required for ovarian contraction in the *Caenorhabditis elegans* reproductive system. Mol Biol Cell 15(6):2782–2793

Ono K, Yu R, Ono S (2007) Structural components of the nonstriated contractile apparatuses in the *Caenorhabditis elegans* gonadal myoepithelial sheath and their essential roles for ovulation. Dev Dyn 236(4):1093–1105

Ono K, Yamashiro S, Ono S (2008) Essential role of ADF/cofilin for assembly of contractile actin networks in the *C elegans* somatic gonad. J Cell Sci 121(Pt 16):2662–2670

Padmanabhan K, Richter JD (2006) Regulated Pumilio-2 binding controls RINGO/Spy mRNA translation and CPEB activation. Genes Dev 20(2):199–209

Page BD, Guedes S, Waring D, Priess JR (2001) The *C. elegans* E2F- and DP-related proteins are required for embryonic asymmetry and negatively regulate Ras/MAPK signaling. Mol Cell 7(3):451–460

Parry JM, Velarde NV, Lefkovith AJ, Zegarek MH, Hang JS, Ohm J, Klancer R, Maruyama R, Druzhinina MK, Grant BD, Piano F, Singson A (2009) EGG-4 and EGG-5 link events of the oocyte-to-embryo transition with meiotic progression in *C. elegans*. Curr Biol 19(20):1752–1757

Patterson JR, Wood MP, Schisa JA (2011) Assembly of RNP granules in stressed and aging oocytes requires nucleoporins and is coordinated with nuclear membrane blebbing. Dev Biol 353(2):173–185

Penkner AM, Fridkin A, Gloggnitzer J, Baudrimont A, Machacek T, Woglar A, Csaszar E, Pasierbek P, Ammerer G, Gruenbaum Y, Jantsch V (2009) Meiotic chromosome homology search involves modifications of the nuclear envelope protein Matefin/SUN-1. Cell 139(5):920–933

Pepling ME, Spradling AC (1998) Female mouse germ cells form synchronously dividing cysts. Development 125(17):3323–3328

Peter M, Labbe JC, Doree M, Mandart E (2002) A new role for Mos in *Xenopus* oocyte maturation: targeting Myt1 independently of MAPK. Development 129(9):2129–2139

Peters JM (2002) The anaphase-promoting complex: proteolysis in mitosis and beyond. Mol Cell 9(5):931–943

Peters H, Levy E, Crone M (1962) Deoxyribonucleic acid synthesis in oocytes of mouse embryos. Nature 195:915–916

Piekny AJ, Mains PE (2002) Rho-binding kinase (LET-502) and myosin phosphatase (MEL-11) regulate cytokinesis in the early *Caenorhabditis elegans* embryo. J Cell Sci 115(Pt 11):2271–2282

Pilipiuk J, Lefebvre C, Wiesenfahrt T, Legouis R, Bossinger O (2009) Increased IP3/Ca2+ signaling compensates depletion of LET-413/DLG-1 in *C. elegans* epithelial junction assembly. Dev Biol 327(1):34–47

Powers J, Rose DJ, Saunders A, Dunkelbarger S, Strome S, Saxton WM (2004) Loss of KLP-19 polar ejection force causes misorientation and missegregation of holocentric chromosomes. J Cell Biol 166(7):991–1001

Racki WJ, Richter JD (2006) CPEB controls oocyte growth and follicle development in the mouse. Development 133(22):4527–4537

Rappleye CA, Tagawa A, Lyczak R, Bowerman B, Aroian RV (2002) The anaphase-promoting complex and separin are required for embryonic anterior-posterior axis formation. Dev Cell 2(2):195–206

Rauh NR, Schmidt A, Bormann J, Nigg EA, Mayer TU (2005) Calcium triggers exit from meiosis II by targeting the APC/C inhibitor XErp1 for degradation. Nature 437(7061):1048–1052

Revenkova E, Eijpe M, Heyting C, Hodges CA, Hunt PA, Liebe B, Scherthan H, Jessberger R (2004) Cohesin SMC1 beta is required for meiotic chromosome dynamics, sister chromatid cohesion and DNA recombination. Nat Cell Biol 6(6):555–562

Revenkova E, Herrmann K, Adelfalk C, Jessberger R (2010) Oocyte cohesin expression restricted to predictyate stages provides full fertility and prevents aneuploidy. Curr Biol 20(17):1529–1533

Richter JD (2007) CPEB: a life in translation. Trends Biochem Sci 32(6):279–285

Robertson S, Lin R (2012) The oocyte-to-embryo transition. Advances in Experimental Medicine and Biology, 757:351–372. (Chap. 12, this volume) Springer, New York

Rogers E, Bishop JD, Waddle JA, Schumacher JM, Lin R (2002) The aurora kinase AIR-2 functions in the release of chromosome cohesion in *Caenorhabditis elegans* meiosis. J Cell Biol 157(2):219–229

Rose KL, Winfrey VP, Hoffman LH, Hall DH, Furuta T, Greenstein D (1997) The POU gene *ceh-18* promotes gonadal sheath cell differentiation and function required for meiotic maturation and ovulation in *Caenorhabditis elegans*. Dev Biol 192(1):59–77

Runft LL, Jaffe LA, Mehlmann LM (2002) Egg activation at fertilization: where it all begins. Dev Biol 245(2):237–254

Rutledge E, Bianchi L, Christensen M, Boehmer C, Morrison R, Broslat A, Beld AM, George AL, Greenstein D, Strange K (2001) CLH-3, a ClC-2 anion channel ortholog activated during meiotic maturation in *C. elegans* oocytes. Curr Biol 11(3):161–170

Sagata N, Oskarsson M, Copeland T, Brumbaugh J, Vande Woude GF (1988) Function of *c-mos* proto-oncogene product in meiotic maturation in *Xenopus* oocytes. Nature 335(6190):519–525

Samuel AD, Murthy VN, Hengartner MO (2001) Calcium dynamics during fertilization in *C. elegans*. BMC Dev Biol 1:8

Sato K, Sato M, Audhya A, Oegema K, Schweinsberg P, Grant BD (2006) Dynamic regulation of caveolin-1 trafficking in the germ line and embryo of *Caenorhabditis elegans*. Mol Biol Cell 17(7):3085–3094

Saunders AM, Powers J, Strome S, Saxton WM (2007) Kinesin-5 acts as a brake in anaphase spindle elongation. Curr Biol 17(12):R453–R454

Schisa JA, Pitt JN, Priess JR (2001) Analysis of RNA associated with P granules in germ cells of *C. elegans* adults. Development 128(8):1287–1298

Schuh M, Ellenberg J (2008) A new model for asymmetric spindle positioning in mouse oocytes. Curr Biol 18(24):1986–1992

Schumacher JM, Golden A, Donovan PJ (1998) AIR-2: An Aurora/Ipl1-related protein kinase associated with chromosomes and midbody microtubules is required for polar body extrusion and cytokinesis in *Caenorhabditis elegans* embryos. J Cell Biol 143(6):1635–1646

Segbert C, Barkus R, Powers J, Strome S, Saxton WM, Bossinger O (2003) KLP-18, a Klp2 kinesin, is required for assembly of acentrosomal meiotic spindles in *Caenorhabditis elegans*. Mol Biol Cell 14(11):4458–4469

Seidel HS, Kimble J (2011) The oogenic germline starvation response in *C. elegans*. PLoS One 6(12):e28074

Severson AF, Ling L, van Zuylen V, Meyer BJ (2009) The axial element protein HTP-3 promotes cohesin loading and meiotic axis assembly in *C. elegans* to implement the meiotic program of chromosome segregation. Genes Dev 23(15):1763–1778

Shakes DC, Sadler PL, Schumacher JM, Abdolrasulnia M, Golden A (2003) Developmental defects observed in hypomorphic anaphase-promoting complex mutants are linked to cell cycle abnormalities. Development 130(8):1605–1620

Shakes DC, Allen AK, Albert KM, Golden A (2011) *emb-1* encodes the APC16 subunit of the *Caenorhabditis elegans* anaphase-promoting complex. Genetics 189(2):549–560

Shelton CA, Carter JC, Ellis GC, Bowerman B (1999) The nonmuscle myosin regulatory light chain gene *mlc-4* is required for cytokinesis, anterior-posterior polarity, and body morphology during *Caenorhabditis elegans* embryogenesis. J Cell Biol 146(2):439–451

Shimada M, Kawahara H, Doi H (2002) Novel family of CCCH-type zinc-finger proteins, MOE-1, -2 and –3, participates in *C. elegans* oocyte maturation. Genes Cells 7(9):933–947

Siomos MF, Badrinath A, Pasierbek P, Livingstone D, White J, Glotzer M, Nasmyth K (2001) Separase is required for chromosome segregation during meiosis I in *Caenorhabditis elegans*. Curr Biol 11(23):1825–1835

Skold HN, Komma DJ, Endow SA (2005) Assembly pathway of the anastral *Drosophila* oocyte meiosis I spindle. J Cell Sci 118(Pt 8):1745–1755

Smith LD, Ecker RE (1969) Role of the oocyte nucleus in physiological maturation in *Rana pipiens*. Dev Biol 19(3):281–309

Sonneville R, Gonczy P (2004) *zyg-11* and *cul-2* regulate progression through meiosis II and polarity establishment in *C. elegans*. Development 131(15):3527–3543

Srayko M, Buster DW, Bazirgan OA, McNally FJ, Mains PE (2000) MEI-1/MEI-2 katanin-like microtubule severing activity is required for *Caenorhabditis elegans* meiosis. Genes Dev 14(9):1072–1084

Srayko M, O'Toole ET, Hyman AA, Muller-Reichert T (2006) Katanin disrupts the microtubule lattice and increases polymer number in *C. elegans* meiosis. Curr Biol 16(19):1944–1949

St Johnston D, Ahringer J (2010) Cell polarity in eggs and epithelia: parallels and diversity. Cell 141(5):757–774

Starck J (1977) Radioautographic study of RNA synthesis in *Caenorhabditis elegans* (Bergerac Variety) oogenesis. Biol Cell 30:181–182

Starich T, Sheehan M, Jadrich J, Shaw J (2001) Innexins in *C. elegans*. Cell Comm Adhesion 8(4–6):311–314

Stitzel ML, Pellettieri J, Seydoux G (2006) The *C. elegans* DYRK kinase MBK-2 marks oocyte proteins for degradation in response to meiotic maturation. Curr Biol 16(1):56–62

Strome S (1986) Fluorescence visualization of the distribution of microfilaments in gonads and early embryos of the nematode *Caenorhabditis elegans*. J Cell Biol 103(6 Pt 1):2241–2252

Sun QY, Liu K, Kikuchi K (2008) Oocyte-specific knockout: a novel in vivo approach for studying gene functions during folliculogenesis, oocyte maturation, fertilization, and embryogenesis. Biol Reprod 79(6):1014–1020

Susiarjo M, Hassold TJ, Freeman E, Hunt PA (2007) Bisphenol A exposure in utero disrupts early oogenesis in the mouse. PLoS Genet 3(1):e5

Suzuki Y, Han M (2006) Genetic redundancy masks diverse functions of the tumor suppressor gene PTEN during *C. elegans* development. Genes Dev 20(4):423–428

Swan KA, Severson AF, Carter JC, Martin PR, Schnabel H, Schnabel R, Bowerman B (1998) *cyk-1*: a *C. elegans* FH gene required for a late step in embryonic cytokinesis. J Cell Sci 111(Pt 14):2017–2027

Tang F, Kaneda M, O'Carroll D, Hajkova P, Barton SC, Sun YA, Lee C, Tarakhovsky A, Lao K, Surani MA (2007) Maternal microRNAs are essential for mouse zygotic development. Genes Dev 21(6):644–648

Tarr DE, Scott AL (2005) MSP domain proteins. Trends Parasitol 21(5):224–231

Telfer EE, Gosden RG, Byskov AG, Spears N, Albertini D, Andersen CY, Anderson R, Braw-Tal R, Clarke H, Gougeon A, McLaughlin E, McLaren A, McNatty K, Schatten G, Silber S, Tsafriri A (2005) On regenerating the ovary and generating controversy. Cell 122(6):821–822

Tsuda H, Han SM, Yang Y, Tong C, Lin YQ, Mohan K, Haueter C, Zoghbi A, Harati Y, Kwan J, Miller MA, Bellen HJ (2008) The amyotrophic lateral sclerosis 8 protein VAPB is cleaved, secreted, and acts as a ligand for Eph receptors. Cell 133(6):963–977

Tunquist BJ, Maller JL (2003) Under arrest: cytostatic factor (CSF)-mediated metaphase arrest in vertebrate eggs. Genes Dev 17(6):683–710

van der Voet M, Berends CW, Perreault A, Nguyen-Ngoc T, Gonczy P, Vidal M, Boxem M, van den Heuvel S (2009) NuMA-related LIN-5, ASPM-1, calmodulin and dynein promote meiotic spindle rotation independently of cortical LIN-5/GPR/Galpha. Nat Cell Biol 11(3):269–277

Vazquez-Manrique RP, Nagy AI, Legg JC, Bales OA, Ly S, Baylis HA (2008) Phospholipase C-epsilon regulates epidermal morphogenesis in *Caenorhabditis elegans*. PLoS Genet 4(3):e1000043

Von Stetina JR, Tranguch S, Dey SK, Lee LA, Cha B, Drummond-Barbosa D (2008) alpha-Endosulfine is a conserved protein required for oocyte meiotic maturation in *Drosophila*. Development 135(22):3697–3706

Walczak CE, Vernos I, Mitchison TJ, Karsenti E, Heald R (1998) A model for the proposed roles of different microtubule-based motor proteins in establishing spindle bipolarity. Curr Biol 8(16):903–913

Walker AK, Boag PR, Blackwell TK (2007) Transcription reactivation steps stimulated by oocyte maturation in *C. elegans*. Dev Biol 304(1):382–393

Walter SA, Guadagno SN, Ferrell JE Jr (2000) Activation of Wee1 by p42 MAPK in vitro and in cycling *Xenopus* egg extracts. Mol Biol Cell 11(3):887–896

Wang R, He G, Nelman-Gonzalez M, Ashorn CL, Gallick GE, Stukenberg PT, Kirschner MW, Kuang J (2007) Regulation of Cdc25C by ERK-MAP kinases during the G2/M transition. Cell 128(6):1119–1132

Ward S, Carrel JS (1979) Fertilization and sperm competition in the nematode *Caenorhabditis elegans*. Dev Biol 73(2):304–321

Ward S, Burke DJ, Sulston JE, Coulson AR, Albertson DG, Ammons D, Klass M, Hogan E (1988) Genomic organization of major sperm protein genes and pseudogenes in the nematode *Caenorhabditis elegans*. J Mol Biol 199(1):1–13

Whitten SJ, Miller MA (2007) The role of gap junctions in *Caenorhabditis elegans* oocyte maturation and fertilization. Dev Biol 301(2):432–446

Wignall SM, Villeneuve AM (2009) Lateral microtubule bundles promote chromosome alignment during acentrosomal oocyte meiosis. Nat Cell Biol 11(7):839–844

Wilson EB (1925) The cell in development and heredity. Macmillan, New York

Wissmann A, Ingles J, Mains PE (1999) The *Caenorhabditis elegans mel-11* myosin phosphatase regulatory subunit affects tissue contraction in the somatic gonad and the embryonic epidermis and genetically interacts with the Rac signaling pathway. Dev Biol 209(1):111–127

Wolf N, Hirsh D, McIntosh JR (1978) Spermatogenesis in males of the free-living nematode, *Caenorhabditis elegans*. J Ultrastruct Res 63(2):155–169

Wolke U, Jezuit EA, Priess JR (2007) Actin-dependent cytoplasmic streaming in *C. elegans* oogenesis. Development 134(12):2227–2236

Wolstenholme J, Angell RR (2000) Maternal age and trisomy–a unifying mechanism of formation. Chromosoma 109(7):435–438

Woodruff GC, Eke O, Baird SE, Felix MA, Haag ES (2010) Insights into species divergence and the evolution of hermaphroditism from fertile interspecies hybrids of *Caenorhabditis* nematodes. Genetics 186(3):997–1012

Xu X, Guo H, Wycuff DL, Lee M (2007) Role of phosphatidylinositol-4-phosphate 5′ kinase (*ppk-1*) in ovulation of *Caenorhabditis elegans*. Exp Cell Res 313(11):2465–2475

Yan X, Xing J, Lorin-Nebel C, Estevez AY, Nehrke K, Lamitina T, Strange K (2006) Function of a STIM1 homologue in *C. elegans*: evidence that store-operated Ca2+ entry is not essential for oscillatory Ca2+ signaling and ER Ca2+ homeostasis. J Gen Physiol 128(4):443–459

Yang HY, McNally K, McNally FJ (2003) MEI-1/katanin is required for translocation of the meiosis I spindle to the oocyte cortex in *C elegans*. Dev Biol 260(1):245–259

Yang HY, Mains PE, McNally FJ (2005) Kinesin-1 mediates translocation of the meiotic spindle to the oocyte cortex through KCA-1, a novel cargo adapter. J Cell Biol 169(3):447–457

Yang Y, Han SM, Miller MA (2010) MSP hormonal control of the oocyte MAP kinase cascade and reactive oxygen species signaling. Dev Biol 342(1):96–107

Yin X, Gower NJ, Baylis HA, Strange K (2004) Inositol 1,4,5-trisphosphate signaling regulates rhythmic contractile activity of myoepithelial sheath cells in *Caenorhabditis elegans*. Mol Biol Cell 15(8):3938–3949

Yu J, Zhao Y, Li Z, Galas S, Goldberg ML (2006) Greatwall kinase participates in the Cdc2 autoregulatory loop in *Xenopus* egg extracts. Mol Cell 22(1):83–91

Zetka M (2009) Homologue pairing, recombination and segregation in *Caenorhabditis elegans*. Genome Dyn 5:43–55

Zhao Y, Haccard O, Wang R, Yu J, Kuang J, Jessus C, Goldberg ML (2008) Roles of Greatwall kinase in the regulation of cdc25 phosphatase. Mol Biol Cell 19(4):1317–1327

Zhou K, Rolls MM, Hall DH, Malone CJ, Hanna-Rose W (2009) A ZYG-12-dynein interaction at the nuclear envelope defines cytoskeletal architecture in the *C. elegans* gonad. J Cell Biol 186(2):229–241

Zuckerman S (1951) The number of oocytes in the mature ovary. Recent Prog Horm Res 6:63–109

Chapter 11
Fertilization

Matthew R. Marcello, Gunasekaran Singaravelu, and Andrew Singson

Abstract Fertilization—the fusion of gametes to produce a new organism—is the culmination of a multitude of intricately regulated cellular processes. In *Caenorhabditis elegans*, fertilization is highly efficient. Sperm become fertilization competent after undergoing a maturation process during which they become motile, and the plasma membrane protein composition is reorganized in preparation for interaction with the oocyte. The highly specialized gametes begin their interactions by signaling to one another to ensure that fertilization occurs when they meet. The oocyte releases prostaglandin signals to help guide the sperm to the site of fertilization, and sperm secrete a protein called major sperm protein (MSP) to trigger oocyte maturation and ovulation. Upon meeting one another in the spermatheca, the sperm and oocyte fuse in a specific and tightly regulated process. Recent studies are providing new insights into the molecular basis of this fusion process. After fertilization, the oocyte must quickly transition from the relative quiescence of oogenesis to a phase of rapid development during the cleavage divisions of early embryogenesis. In addition, the fertilized oocyte must prevent other sperm from fusing with it as

M.R. Marcello (✉)
Waksman Institute, 190 Frelinghuysen Road, Piscataway, NJ 08854, USA

Department of Genetics, Rutgers University, 145 Bevier Road,
Piscataway, NJ 08854, USA

Department of Molecular Genetics, Microbiology and Immunology,
UMDNJ-Robert Wood Johnson Medical School, 675 Hoes Lane,
Piscataway, NJ 08854, USA
e-mail: mmarcello@waksman.rutgers.edu

G. Singaravelu • A. Singson
Waksman Institute, 190 Frelinghuysen Road, Piscataway, NJ 08854, USA

Department of Genetics, Rutgers University, 145 Bevier Road,
Piscataway, NJ 08854, USA
e-mail: guna@waksman.rutgers.edu; singson@waksman.rutgers.edu

T. Schedl (ed.), *Germ Cell Development in C. elegans*, Advances in Experimental Medicine and Biology 757, DOI 10.1007/978-1-4614-4015-4_11,

well as produce an eggshell for protection during external development. This chapter will review the nature and regulation of the various cellular processes of fertilization, including the development of fertilization competence, gamete signaling, sperm–oocyte fusion, the oocyte to embryo transition, and production of an eggshell to protect the developing embryo.

Keywords Fertilization • Oocyte-to-embryo transition • Egg activation • Spermiogenesis • Sperm activation • Eggshell • Polyspermy • Sperm migration • PUFA-derived signals • Major sperm protein (MSP)

11.1 Overview of Fertilization in *Caenorhabditis elegans*

Fertilization is the central molecular process of sexual reproduction. Despite this, the cellular and molecular mechanisms that govern sperm–egg fusion are not well understood. *C. elegans* serves as an excellent model system for dissecting the molecular and genetic processes of fertilization, as well as those that prepare the gametes for fertilization and ensure proper embryogenesis. Importantly, fertilization in *C. elegans* shares many characteristics with fertilization in other species, including mammals. In both *C. elegans* and mammalian fertilization, sperm activation occurs within the reproductive tract, fertilization occurs internally, and egg activation includes a polyspermy block, cortical granule exocytosis, meiotic resumption, and the activation and degradation of selected maternal mRNAs and proteins (Singson et al. 2008; Horner and Wolfner 2008; Marcello and Singson 2010; Stitzel and Seydoux 2007).

C. elegans has two sexes: hermaphrodite and male (Nigon and Dougherty 1949; Madl and Herman 1979). Hermaphrodite *C. elegans* are essentially females that produce and store sperm early in their life cycle before producing oocytes (Ward and Carrel 1979; Hirsh et al. 1976). Hermaphrodites produce approximately 100–175 immature sperm, or spermatids, per gonad arm during the final larval stage (L4) and then switch strictly to oogenesis for the remainder of their life (see Chap. 3, Zanetti and Puoti 2012; Ward and Carrel 1979; Hirsh et al. 1976).

Male-derived sperm are introduced to the hermaphrodite reproductive tract through the uterus during copulation (L'Hernault 2006; Stanfield and Villeneuve 2006). Immature sperm, or spermatids, are stored in the seminal vesicle of males and undergo a final maturation process, called sperm activation or spermiogenesis, as they are ejaculated into the hermaphrodite (Hirsh et al. 1976; Klass et al. 1976; Kimble and Hirsh 1979; Shakes and Ward 1989). During sperm activation, the spermatids mature to a motile, amoeboid sperm capable of fertilizing the oocyte (see Chap. 7, Chu and Shakes 2012). Both the sperm and oocyte send signals to increase the efficiency of fertilization. Oocytes release polyunsaturated fatty acid-derived signals to attract sperm to the oocytes and sperm then migrate to the spermatheca, the site of fertilization (Fig. 11.1) (Kubagawa et al. 2006). *C. elegans* sperm do not contain any actin and instead employ another cytoskeletal protein, major sperm

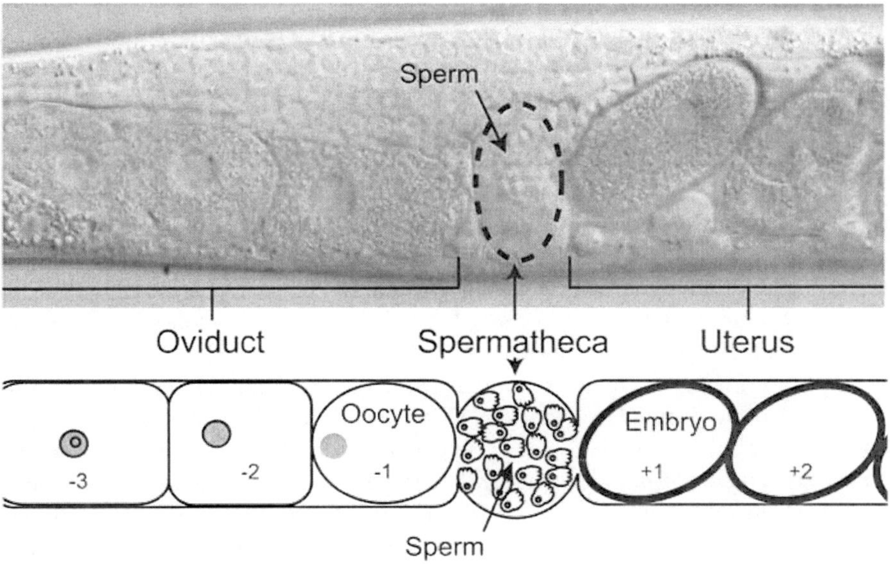

Fig. 11.1 The *C. elegans* reproductive tract. The *top image* shows a light micrograph of the *C. elegans* reproductive tract with focus on the spermatheca. The *bottom image* is a schematic of the same region. The proximal oocyte (−1) is ovulated from the oviduct into the spermatheca. Sperm are stored in the spermatheca as they await the ovulating oocyte. After fertilization, the embryo is pushed into the uterus (+1)

protein (MSP), to mediate motility (Roberts and Ward 1982). In addition to serving as the cytoskeletal protein that is responsible for sperm locomotion, MSP is also released from the sperm and stimulates meiotic maturation and ovulation of the oocyte into the spermatheca (Miller et al. 2001).

Fertilization triggers a number of rapid changes in the oocyte. The first detectable change in the oocyte after fertilization is an increase in intracellular calcium (Samuel et al. 2001). The oocyte then forms an actin-rich cap over the site of fertilization, completes meiosis, extrudes an eggshell, and initiates embryonic development (Parry et al. 2009; Maruyama et al. 2007). If the oocyte-to-embryo transition occurs normally, the embryo will divide in the uterus until approximately the 30-cell stage and then will be laid out of the vulva (egg-laying) (Hirsh et al. 1976).

Hermaphrodites are able to self-fertilize when males are not present in the population (L'Hernault 2006; Ward and Carrel 1979; Hirsh et al. 1976). The hermaphrodite-derived spermatids that are produced during L4 are stored in the proximal gonad and are also pushed into the spermatheca when the first oocytes are ovulated (McCarter et al. 1999; Singson 2001). In contrast to sperm activation of male-derived sperm, hermaphrodite-derived spermatids undergo sperm activation when they are deposited into the spermatheca (L'Hernault 2006). Fertilization in *C. elegans* is very efficient and almost all of the hermaphrodite-derived sperm fertilize the egg (Singson 2001).

If both male- and hermaphrodite-derived sperm are present in the spermatheca, the male-derived sperm are competitively superior to hermaphrodite-derived sperm and will preferentially fertilize the oocytes (L'Hernault 2006; Ward and Carrel 1979). It is not completely understood how male-derived sperm outcompete hermaphrodite-derived sperm (LaMunyon and Ward 1998). One hypothesis is that male-derived sperm are larger than hermaphrodite-derived sperm because of the differences in the physiological environment of male gonad; this size difference may aid in gaining access to the oocytes (Baldi et al. 2011; LaMunyon and Ward 1998). Interestingly, males preferentially mate with older hermaphrodites that have depleted their own sperm populations, and it is hypothesized that hermaphrodites produce a mating cue that attracts males when no activated sperm are present (Morsci et al. 2011).

The experimental techniques available to study the multitude of cellular processes that occur during fertilization in *C. elegans* make it an attractive model system (see Geldziler et al. 2011 for an extensive review of techniques). The most widely used method for understanding the mechanism of fertilization in *C. elegans* is genetic analysis (Geldziler et al. 2011). Forward genetic screens for fertility mutants in *C. elegans* are very powerful because it is relatively easy to isolate mutations in sperm-specific genes, and mutants can be isolated that have defects only in fertilization and/or egg activation, not spermatogenesis, spermiogenesis, or oogenesis (L'Hernault et al. 1988; Singson et al. 1998). RNAi has been helpful in identifying genes in the oocyte that are necessary for fertilization (Geldziler et al. 2004). However, many questions still remain about molecular basis of *C. elegans* fertilization. This chapter will outline the current status of the field of fertilization *C. elegans* and provide some perspective on the future directions of the field.

11.2 Preparing Sperm for Interaction with the Oocyte

11.2.1 Sperm Activation

C. elegans sperm must undergo a final maturation process, called sperm activation (spermiogenesis), to be fertilization competent. Sperm activation is a rapid post-meiotic differentiation that converts round immotile spermatids into mature amoeboid sperm (see Chap. 7, Chu and Shakes 2012; Shakes and Ward 1989) (Fig. 11.2). Hermaphrodite-derived spermatids undergo sperm activation when they are pushed into the spermatheca during the first ovulations and male-derived spermatids undergo sperm activation when they mix with the seminal fluid during ejaculation (L'Hernault 2006; Stanfield and Villeneuve 2006). Regardless of the origin of the sperm, two events take place that are critical to render the sperm fertilization competent: (1) membranous organelles (MO), which are specialized secretory vesicles derived from the Golgi, must fuse with the plasma membrane to release their glycoprotein contents and add critical transmembrane proteins to the sperm plasma membrane, and (2) sperm must gain the ability to crawl by extending a pseudopod from one side of the cell body (Fig. 11.2) (Shakes and Ward 1989; Roberts and

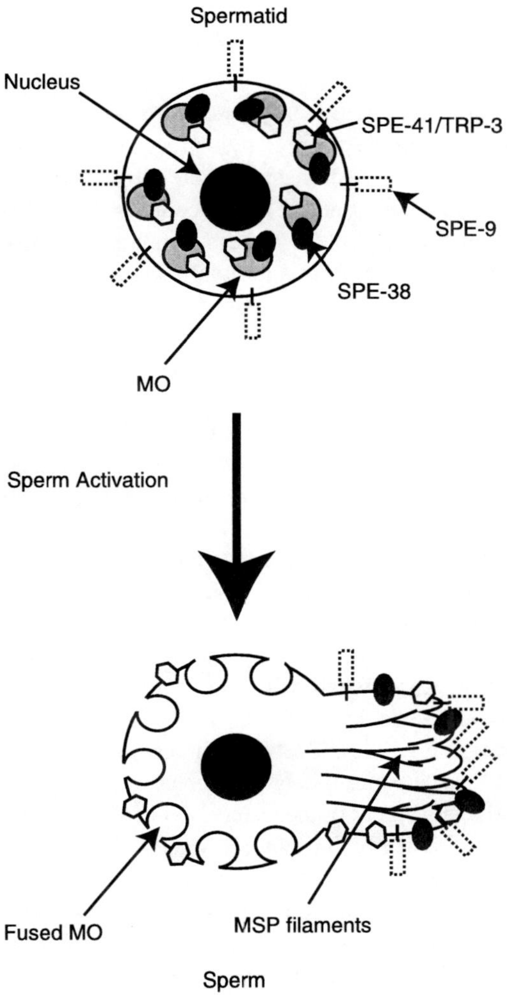

Fig. 11.2 Sperm activation in *C. elegans*. In order for sperm to be fertilization competent, they must undergo a maturation process called sperm activation (spermiogenesis). During sperm activation, membranous organelles (MO) in the immature spermatid fuse with the plasma membrane, resulting in membrane reorganization. The sperm membrane protein SPE-9 is always present in the plasma membrane but relocalizes exclusively to the pseudopod after sperm activation. SPE-38 and SPE-41/TRP-3 are both present in MOs in spermatids and change localization after MO fusion. After MO fusion, SPE-38 is present only on the pseudopod of the sperm while SPE-41/TRP-3 is present throughout the plasma membrane. In addition to MO fusion, the spermatid also extrudes a pseudopod on one side of the cell body during sperm activation. The pseudopod allows for sperm movement, which is necessary for locating the oocyte

Streitmatter 1984; Roberts and Ward 1982). Sperm activation (spermiogenesis) is covered extensively in Chap. 7 by D.S. Chu and D.C. Shakes. This chapter will focus on the mechanisms of sperm locomotion that are necessary for sperm to fertilize the ovulated oocyte in the spermatheca.

11.2.2 MSP Polymerization and Sperm Motility

Sperm motility is crucial for sperm to reach and fertilize the oocyte. The movement of the amoeboid sperm is made possible by a protrusion at its leading edge, adhesion to the substrate, and retraction near the cell body (Shimabukuro et al. 2011). Sperm from *C. elegans* and other nematodes accomplish this without actin or motor proteins (Roberts and Stewart 2000). Instead, nematodes employ an MSP-based motility system (King et al. 1994). In addition to signaling oocyte meiotic maturation and ovulation, MSP is also the structural component responsible for sperm movement (King et al. 1994; Smith 2006). MSP accounts for more than 40% of the cytosolic protein in nematode sperm (Smith 2006).

To better understand the mechanism of the MSP-based motility of nematode sperm, many studies were performed using the parasitic nematode *Ascaris suum* (Italiano et al. 2001). *A. suum* sperm are large and easy to isolate in large numbers, which is advantageous for imaging and allows for more convenient protein purification and biochemical analyses (L'Hernault and Roberts 1995). Moreover, the development of a cell-free filament assembly system from *A. suum* lysate has allowed for the reconstitution of purified motility factors *in vitro* (Italiano et al. 1996).

The three-dimensional structure of *C. elegans* MSP is very similar to that of *A. suum* MSP (Baker et al. 2002). X-ray crystallography and NMR have shown that MSP folds into a seven-stranded beta sandwich that resembles an immunoglobulin-like fold (Bullock et al. 1996a, b; Haaf et al. 1996). MSP monomers self-assemble to form dimers (Haaf et al. 1996). The MSP dimers are then polymerized to form helical subfilaments that are ultimately assembled into larger fibers (Haaf et al. 1996). In *A. suum,* the movement of the pseudopod is ATP dependent and is accomplished by the continual assembly of bundles of MSP filaments at the leading edge and concurrent disassembly of the filaments near the cell body (Miao et al. 2003).

Assembly of the MSP fibers at the leading edge is mediated by a crucial membrane protein and multiple phosphorylation events. A critical protein in MSP nucleation and assembly is MSP polymerization organizing protein (MPOP), a 48 kDa protein present on the inner leaflet of the plasma membrane (LeClaire et al. 2003). MPOP is selectively phosphorylated by an unknown tyrosine kinase at the leading edge, after which phospho-MPOP recruits MSP polymerization-activation kinase (MPAK) to the leading edge of the pseudopod, where MPAK then phosphorylates MFP2 (MSP fiber protein 2) (Yi et al. 2007). The phosphorylated form of MFP2 is present on the MSP fiber complexes in sperm and in the reconstituted MSP fibers *in vitro* (Buttery et al. 2003). MFP2 is thought to increase the rate of fiber formation by cross-linking MSP filaments or recruiting other proteins required for filament formation (Buttery et al. 2003; Grant et al. 2005). Together, these proteins regulate MSP assembly at the leading edge and are necessary for the pseudopod protrusions.

The disassembly of MSP filaments is as important as assembly in regulating pseudopod movement and is regulated by a putative protein phosphatase 2A homolog known as PP2A (Yi et al. 2009). PP2A is localized near the cell body, which is the site of MSP depolymerization, and when PP2A is activated, it promotes the retraction of the MSP fiber (Yi et al. 2009). One of the substrates of PP2A is MFP3 (MSP fiber protein 3), which in its phosphorylated form binds with MSP filaments and stabilizes it (Yi et al. 2009). The dephosphorylation of MFP3 by PP2A causes MFP3 to detach from the MSP bundles, which causes the destabilization and depolymerization of MSP fibers (Yi et al. 2009). MFP1 (MSP fiber protein 1) also negatively regulates the elongation of MSP fibers in vitro but its relationship with PP2A and MFP3 is unknown (Buttery et al. 2003). The depolymerization of the MSP fibers is crucial for the forward movement of the cell body and amoeboid movement of sperm. Disassembly of MSP fibers leads to a gradual decrease in the optical density of filaments. Depolymerizing the MSP network produces sufficient force to propel the pseudopod (Shimabukuro et al. 2011).

An increasing number of studies have focused on understanding MSP dynamics in *C. elegans*. Batchelder et al. found that membrane tension plays an important role in motility, and they postulate that increasing membrane tension in the direction of movement streamlines MSP polymerization and optimizes motility (Batchelder et al. 2011). Phosphorylation of MSP has been observed in *C. elegans* and MSP dynamics seem to be the same, but the regulatory proteins between *C. elegans* and *A. suum* are not conserved (Fraire-Zamora et al. 2011). In *C. elegans*, homologs to proteins that regulate MSP assembly (MPAK, MPOP, or MFP2) have not been identified. Instead, SPE-6, a casein kinase I homolog, may play a role in MSP assembly (see Chap. 7, Chu and Shakes 2012; Varkey et al. 1993). MSP disassembly in *C. elegans* involves phosphatases, but not PP2A as in *A. suum*. As a potential alternative to PP2A, *C. elegans* employs two nearly identical sperm-specific PP1 phosphatases, GSP-3 and GSP-4 (GSP-3/4), for motility (Wu et al. 2011). GSP-3/4 are hypothesized to regulate the spatial disassembly of MSP and sperm from *C. elegans* lacking GSP-3/4 are immotile and have defects in pseudopod development (Wu et al. 2011). PP1 phosphatases are also necessary for sperm development and fertility in mice (Varmuza et al. 1999; Oppedisano et al. 2002; Chakrabarti et al. 2007; Huang and Vijayaraghavan 2004; Soler et al. 2009; Wu et al. 2011).

MSP plays diverse roles during fertilization and is necessary for generating the amoeboid movement of nematode sperm that is necessary for the sperm to migrate through the reproductive tract. Understanding MSP regulation will provide insight into how protein polymerization can generate and force and membrane protrusions (Smith 2006).

11.2.3 The Role of pH in Regulating Sperm Activation and MSP Polymerization

A change in intracellular pH can trigger sperm activation. Addition of weak bases, such as triethanolamine, is sufficient to activate *C. elegans* sperm in vitro (Ward et al. 1983). Similarly, the addition of vas deferens extract can activate *A. suum* sperm in vitro (Abbas and Cain 1979). Within 15 s of the addition of vas deferens extract, the intracellular pH rises from 6.25 to 6.50 in *A. suum* sperm, suggesting that sperm activation is accompanied by an increase in intracellular pH (King et al. 1994). Understanding the mechanism by which pH triggers sperm activation might provide insights into how ion channel regulation can affect vesicle fusion events and morphology changes.

Alterations in intracellular pH also have effects on the polymerization of MSP. Treatment of *A. suum* sperm with a weak acid promotes the disassembly of MSP fibers and removal of the weak acid promotes the re-assembly of MSP fibers (King et al. 1994). Additionally, in activated *A. suum* sperm, a pH gradient is formed between the leading edge and the base of the pseudopod: the intracellular pH at the leading edge and base of the pseudopod are 6.35 and 6.17, respectively (King et al. 1994). In *A. suum* sperm, higher pH at the leading edge stimulates the phosphorylation of MPOP (see Sect. 11.2.2), which in turn assists in the nucleation of MSP fibers at the leading edge (Buttery et al. 2003).

11.3 Sperm and Oocyte Communication

11.3.1 Regulation of Sperm Guidance by Oocyte Polyunsaturated Fatty Acid-Derived Signals

For *C. elegans*, guiding sperm to and subsequently maintaining sperm within the spermatheca is crucial for maximal reproduction. Regardless of whether sperm originate from a hermaphrodite or male, sperm must be guided back into the spermatheca after they are pushed into the uterus by newly fertilized embryos on their way to the uterus (Ward and Carrel 1979).

In *C. elegans*, oocytes attract sperm to the spermatheca by releasing polyunsaturated fatty acid (PUFA)-derived prostaglandin signals produced by oocytes (Edmonds et al. 2010; Kubagawa et al. 2006). PUFAs are synthesized from dietary precursors by lipid anabolic enzymes. These enzymes, the so-called *fat* genes, include *fat-2*, a Δ12-desaturase that generates 18- and 20-carbon PUFAs (Watts and Browse 2002). In *C. elegans,* PUFAs produced within the intestine are initially incorporated into the yolk, which is then taken up by oocytes via receptor-mediated endocytosis using the low-density lipoprotein (LDL) receptor RME-2 (receptor mediated endocytosis-2) (Kubagawa et al. 2006; Grant and Hirsh 1999). Once the PUFAs enter the oocytes, they are converted into prostaglandins and eventually secreted into the reproductive tract (Edmonds et al. 2010). Notably, a number of signaling pathways in mammals are also regulated by prostaglandins, including many processes in mammalian reproduction such as ovulation and fertilization (Han et al. 2010).

Other *Caenorhabditis* species may also use prostaglandins for directional migration of sperm into the spermatheca (Singaravelu and Singson 2011). When males from one *Caenorhabditis* species were allowed to mate with hermaphrodites of another closely related species, the ejaculated sperm successfully reached the spermatheca, implying that directed sperm migration within the *Caenorhabditis* genus employs an evolutionarily conserved signal (Hill and L'Hernault 2001; Singaravelu and Singson 2011).

11.3.2 Secretion of MSP by Sperm to Induce Oocyte Maturation and Gonadal Sheath Cell Contraction

In *C. elegans* sperm, the MSP provides the signal to induce oocyte meiotic maturation and ovulation (see Chap. 10, Kim et al. 2012). MSP has two separate signaling functions (Miller et al. 2001). The 20 C-terminal amino acids of MSP promote sheath cell contraction that leads to ovulation and MSP lacking these 20 amino acids promote oocyte maturation (Miller et al. 2001). Using sperm-derived signals to regulate ovulation rates increases the probability that fertilization will take place, as oocytes are less likely to enter an empty spermatheca (McCarter et al. 1999; Miller et al. 2001). MSP is secreted from sperm through an unconventional vesicle budding process (Kosinski et al. 2005). The MSP signal persists as long as sperm are present (Miller et al. 2001). The signal for the first ovulation in a hermaphrodite originates from the spermatids, which have MSP and can signal for oocyte maturation and sheath contraction (McCarter et al. 1999).

11.4 Sperm–Oocyte Fusion

11.4.1 spe and egg Genes

In order for sperm and oocytes to fuse, the gametes must be able to recognize one another and have the fusion machinery localized and assembled properly to mediate fertilization. Precise recognition is mediated by specific interactions between proteins present on the surfaces of the sperm and oocyte. Mutations disrupting the fusion machinery on either gamete will impair fertilization but development is expected to be otherwise normal. Genes that are essential for sperm development, differentiation, or function during fusion are classified as *spe* genes (also previously referred to as *fer* genes) (also see Chap. 7, Chu and Shakes 2012; Singson 2001; Parry and Singson 2011). *egg* genes are defined as genes that are essential for the oocyte's ability to promote sperm fusion or respond to fertilization but are not necessary for proper oogenesis (Singson 2001; Parry and Singson 2011). The number of genes known to be necessary for fertilization is low in all animal species and every gene that is identified and cloned will add a significant amount of information to our knowledge of how fertilization is mediated.

11.4.2 Sperm Genes Necessary for Fertilization

spe-9 was the first sperm gene necessary for fertilization to have been cloned (Singson et al. 1998). Sperm from animals with *spe-9* mutations undergo normal spermatogenesis and sperm activation and have normal morphology and motility, but are unable to fertilize oocytes (Singson et al. 1998). Subsequently, all sperm genes that are fertilization-defective but undergo normal sperm activation are members of the *spe-9* class of mutants. Currently, the *spe-9* class of mutants consists of seven mutants (*spe-9, spe-13, spe-36, spe-38, spe-41/trp-3, spe-42,* and *fer-14*) (Table 11.1) (Singson et al. 1998; Kroft et al. 2005; Chatterjee et al. 2005; Xu and Sternberg 2003; L'Hernault et al. 1988; Putiri et al. 2004; L'Hernault 2006; Nishimura and L'Hernault 2010). Four of the genes responsible for the *spe-9* class phenotype have been cloned (*spe-9, spe-38, spe-41/trp-3*, and *spe-42*), and we will focus on what is known about the function of these four genes.

spe-9: SPE-9 is a single-pass transmembrane protein with a large extracellular domain and a short cytoplasmic tail (Singson et al. 1998). The extracellular domain of SPE-9 has ten epidermal growth factor (EGF) repeats that are thought to mediate protein–protein interactions (Balzar et al. 2001; Singson et al. 1998). SPE-9 localization is dynamic: in spermatids, SPE-9 is localized uniformly over the entire plasma membrane, but after sperm activation, it is predominantly localized to the pseudopod (Fig. 11.2) (Zannoni et al. 2003).

spe-38: SPE-38 is a nematode specific, novel four-pass transmembrane protein that resembles many small tetraspanins which are associated with cell–cell interaction (Chatterjee et al. 2005). In spermatids, SPE-38 localizes to membranous organelles (MOs) (Chatterjee et al. 2005). In mature sperm, SPE-38 relocalizes to the pseudopod, similar to SPE-9 (Fig. 11.2) (Chatterjee et al. 2005).

spe-41/trp-3: SPE-41/TRP-3 is a transient receptor potential channel and is conserved from *C. elegans* to humans (Xu and Sternberg 2003). TRP-3 homologs have been implicated in a variety of pathologies, including cancer, heart disease, kidney disease, and pain (Dietrich et al. 2010; Kiselyov et al. 2007; Bodding 2007; Eder and Molkentin 2011; Chung et al. 2011). Like SPE-38, SPE-41/TRP-3 localizes to the MO in spermatids; however, SPE-41/TRP-3 localizes uniformly all over the plasma membrane in mature sperm and does not show pseudopod-only localization as do SPE-38 and SPE-9 (Fig. 11.2) (Xu and Sternberg 2003). In *Drosophila*, TRP channels have independent roles as a scaffold and calcium channel (Wang et al. 2005). Future areas of interest include determining whether the calcium channel activity of SPE-41/TRP-3 is separable from its role in fertility and whether SPE-41/TRP-3 plays a role in mediating the calcium increase observed during fertilization.

spe-42: SPE-42 is a novel six-pass transmembrane protein whose homologs are found in many metazoans, such as flies, mouse, and human (Kroft et al. 2005). SPE-42 contains a DC-STAMP (dendritic cell-specific transmembrane protein) domain, which is required for cell fusion of osteoclasts in mammals (Mensah et al. 2010; Miyamoto 2006). It has been hypothesized that the DC-STAMP domain of SPE-42

Table 11.1 Genes necessary for fertilization and the oocyte-to-embryo transition

Gene	Encoded protein	Function
Fertilization		
spe-9	EGF repeat (10)-containing protein with a short intracellular domain	Fertilization (in sperm)
spe-13	Unknown	Fertilization (in sperm)
spe-36	Unknown	Fertilization (in sperm)
spe-38	Novel four-pass transmembrane protein	Fertilization (in sperm)
spe-41/trp-3	Calcium-permeable cation channels that are members of the TRPC (transient receptor potential canonical) subfamily of TRP channels	Fertilization; calcium influx (in sperm)
spe-42	Novel sperm-specific seven-pass transmembrane protein; two functional domains: DC-STAMP (dendritic cell-specific transmembrane protein) and C4C4-type RING finger	Fertilization (in sperm)
fer-14	Unknown	Fertilization (in sperm)
egg-1/2	LDL-receptor repeat-containing protein	Fertilization (in oocytes)
Oocyte-to-embryo transition		
mbk-2	DYRK (dual-specificity Yak1-related) kinase	Early embryo development; Substrates include MEI-1, OMA-1, OMA-2, MEX-5, and MEX-6
egg-3	Member of the protein tyrosine phosphatase-like (PTPL) family	Polar body emission; eggshell extrusion; MBK-2 regulation
egg-4/5	Member of the protein tyrosine phosphatase-like (PTPL) family	Polar body emission; eggshell extrusion; MBK-2 regulation; block to polyspermy
chs-1	Chitin synthase	Chitin polymerization; MBK-2, EGG-3, and EGG-4/5 localization
spe-11	Novel protein	Polar body emission; eggshell extrusion
cyk-4	Rho GAP (Rho guanosine triphosphatase (GTPase) activating protein)	Establishes anteroposterior axis along with small GTPase RhoA (Ras homolog gene family, member A) and ECT-2 (a RhoA guanine nucleotide-exchange factor)
sep-1	Separase cysteine protease	CG exocytosis; eggshell formation
rab-11.1	Small GTPase homologous to the Rab GTPases	CG exocytosis; eggshell formation
syn-4	Syntaxin-related t-SNARE	CG exocytosis; eggshell formation
gna-2	Glucosamine 6-phosphate *N*-acetyltransferase	Synthesis of UDP-*N*-acetylglucosamine (UDP-GlcNAc), *N*-acetylgalactosamine (UDP-GalNAc), and chitin; eggshell formation

could be used for membrane fusion events, including sperm–egg fusion; however, the mechanism by which DC-STAMP domains mediate fusion is unclear, and currently there is no evidence to support this hypothesis (Wilson et al. 2011; Kroft et al. 2005). The C-terminal tail of SPE-42 forms a RING finger domain that has eight conserved cysteine residues that are thought to coordinate Zn^{2+} ions; a mutation in any of these cysteine residues abolishes SPE-42 function (Wilson et al. 2011).

At present, not much is known about how the *spe-9* class of genes interact on the protein level. Recent work has shown that SPE-38 and SPE-41/TRP-3 interact both genetically and physically (Singaravelu et al. 2012). In *spe-38* mutant spermatozoa, SPE-41/TRP-3 is trapped in the MO, suggesting that SPE-38 is required for the trafficking of SPE-41/TRP-3 following sperm activation. In contrast, the localization of SPE-38 is unperturbed in *spe-41/trp-3* mutant sperm, suggesting that SPE-38 regulates SPE-41/TRP-3 trafficking and not vice versa. Split ubiquitin yeast-two hybrid results show that SPE-38 and SPE-41/TRP-3 can interact with each other. Further studies could elucidate whether SPE-38 modulates channel activity of SPE-41/TRP-3.

The SPE-9 class proteins may potentially function in sperm to mediate either recognizing, adhering, or fusing with the oocyte. All SPE-9 class proteins are membrane proteins and thus have the potential to be a sperm ligand. SPE-9 is the most likely candidate to serve as a sperm ligand because of the presence of multiple EGF repeats and the selective localization of SPE-9 to the pseudopod (Singson et al. 1998). Proteins that localize to the pseudopod are the best candidates to mediate fusion with the oocyte, as the pseudopod is thought to make initial contact with the oocyte (Zannoni et al. 2003). Delineating whether these SPE-9 class proteins function to organize the sperm membrane appropriately or as an adhesion molecule or fusogen will provide insight into the molecular basis of fertilization.

11.4.3 Oocyte Genes Necessary for Fertilization

Currently, only two genes on the oocyte are known to be necessary for fertilization: *egg-1* and *egg-2* (*egg-1/2*) (Kadandale et al. 2005). EGG-1 and EGG-2 are single-pass transmembrane proteins that contain low-density lipoprotein (LDL) receptor repeats on the plasma membrane of the oocyte (Kadandale et al. 2005). EGG-1 and EGG-2 are 67% identical and are likely to be functionally redundant (Kadandale et al. 2005). In *C. elegans*, the hermaphrodites lacking function *egg-1* and *egg-2*, either by mutation or RNAi, lay unfertilized oocytes and have severe reductions in progeny production (Kadandale et al. 2005; Lee and Schedl 2001; Maeda et al. 2001; Johnston et al. 2010). Analysis of *egg-1(tm1071)* indicates that oocytes lacking *egg-1* are ovulated and make contact with sperm, but a majority of the oocytes are not fertilized (Kadandale et al. 2005; Johnston et al. 2010). Proteins related to LDL receptors bind with diverse ligands such as lipoproteins, viruses, and signaling molecules, and thus could mediate fertilization (Nykjaer and Willnow 2002). The identification of additional transmembrane proteins in the oocyte is paramount to

Table 11.2 Species-specific fertilization of the *elegans* group

	Fem./herm.			
Male	*C. elegans*	*C. briggsae*	*C. remanei*	*C. brenneri*
C. elegans	X	–	+	+
C. briggsae	–	X	+	+
C. remanei	–	–	X	+
C. brenneri	Not tested	Not tested	–	X

our understanding of *C. elegans* fertilization and could provide insight into how fertilization is mediated in all species.

11.4.4 Species-Specific Fertilization in Caenorhabditis

Fertilization in the *Caenorhabditis* species is specific; however, a selected number *Caenorhabditis* species are able to fertilize the oocytes of other species (Table 11.2) (Baird et al. 1992; Sudhaus and Kiontke 2007; Baird and Yen 2000; Hill and L'Hernault 2001; Kiontke and Fitch 2005). The *elegans* group of *Caenorhabditis* contains four species: *C. elegans, C. briggsae, C. remanei,* and *C. brenneri* (Kiontke and Fitch 2005). *C. elegans* is actually a sister species of a clade (a species and all of its descendents) made up of *C. briggsae, C. remanei,* and *C. brenneri* (Kiontke and Fitch 2005). Like *C. elegans, C. briggsae* is a hermaphrodite–male species while *C. remanei* and *C. brenneri* are female–male species (Kiontke and Fitch 2005). Interestingly, *C. brenneri* oocytes can be fertilized by sperm from any of the three other members of the *elegans* group, indicating that *C. brenneri* oocytes possess receptor(s) that can recognize ligand(s) on *C. elegans*, *C. briggsae*, and *C. remanei* sperm and mediate fusion (Baird and Yen 2000). On the other hand, *C. elegans* and *C. briggsae* oocytes can only be fertilized by sperm from their own species (Baird et al. 1992). Despite the fact that fertilization can occur between species of the *elegans* group, these fertilization events do not result in fertile offspring, indicating that there is post-zygotic incompatibility (Baird et al. 1992; Sudhaus and Kiontke 2007; Baird and Yen 2000; Sudhaus et al. 2007). Studying the species specificity of fertilization in the *elegans* group may provide clues as to which proteins mediate sperm–egg fusion as well as insight into the evolutionary origins of the species.

11.5 Egg Activation

The transition from an oocyte to an embryo is comprised of two steps: meiotic maturation and egg activation (Stitzel and Seydoux 2007). During meiotic maturation, the oocyte exits meiotic prophase, initiates the meiotic division phase, and becomes fertilization competent (see Chap. 10, Kim et al. 2012; Stitzel and Seydoux 2007;

Horner and Wolfner 2008; Yamamoto et al. 2006). Egg activation is the transition of a fertilized egg to a developing embryo. Egg activation entails a number of dynamic cellular processes: completion of meiosis, maternal protein degradation, cytoskeletal rearrangements, activation of embryogenesis, and eggshell formation (Stitzel and Seydoux 2007; Horner and Wolfner 2008).

11.5.1 Regulation of Meiotic Resumption

In *C. elegans* it is critical that meiotic resumption and cell cycle progression are tightly regulated because of the continual production of oocytes and rapid fertilization (McCarter et al. 1999). In many animals, oocytes arrest in late meiotic prophase to temporally control utilization and often have a second arrest during the meiotic divisions; in most vertebrates, oocytes arrest in prophase I until they are cyclically recruited and then arrest again in metaphase II as they wait for a fertilization-competent sperm (Masui and Clarke 1979; Eppig 1996; McCarter et al. 1999). However, in *C. elegans*, oocytes are produced continually in the presence of sperm, so there is no need for oocytes to arrest in meiotic prophase (although they do arrest in late meiotic prophase in the absence of sperm) (McCarter et al. 1999). Additionally, since sperm signal both maturation and ovulation, there is no need to arrest the oocyte cell cycle in order to wait for the arrival of competent sperm (McCarter et al. 1999; Miller et al. 2001). In fact, in the absence of sperm, oocytes progress to anaphase I (McNally and McNally 2005). As a result, meiotic resumption and completion as well as egg activation events associated with fertilization occur concurrently, making tight regulatory control critical for proper embryogenesis.

In *C. elegans*, fertilization occurs as the oocyte enters the spermatheca, after germinal vesicle breakdown and while the meiosis I spindle is assembling (McNally and McNally 2005) (Fig. 11.3a). Fertilization is accompanied by an increase in free cytosolic calcium levels (Samuel et al. 2001). The increase in cytoplasmic calcium levels at fertilization occurs in other animals; however, the functional consequence of this change in calcium dynamics in *C. elegans* needs to be more fully explored (Samuel et al. 2001; Horner and Wolfner 2008). In *Xenopus*, a calcium signal triggers the exit from meiosis II by activating the anaphase-promoting complex APC/C (Tunquist et al. 2002; Liu and Maller 2005; Rauh et al. 2005). Like *Xenopus*, the

Fig. 11.3 (continued) The MBK2-, EGG-3, EGG-4/5, CHS-1 complex is still in the cortex in metaphase I but disassembles as the embryo transitions to meiotic metaphase II (**c**). During the transition to metaphase II, EGG-3, EGG-4/5, and SPE-11 are all necessary for polar body extrusion and eggshell production. Eggshell production is also dependent upon CHS-1 to polymerize chitin and upon the release of chondroitin proteoglycans to the lipid layer via CG exocytosis during this transition. At meiotic metaphase II, MBK-2, EGG-3, and CHS-1 are localized to the cytoplasmic puncta. As the embryo enters mitosis (**d**), the complex dissociates and MBK-2 is free to phosphorylate its substrates and promote embryo development. The eggshell is impermeable by mitotic entry and consists of three layers: the outer vitelline layer, the middle chitin layer, and the inner lipid-rich layer

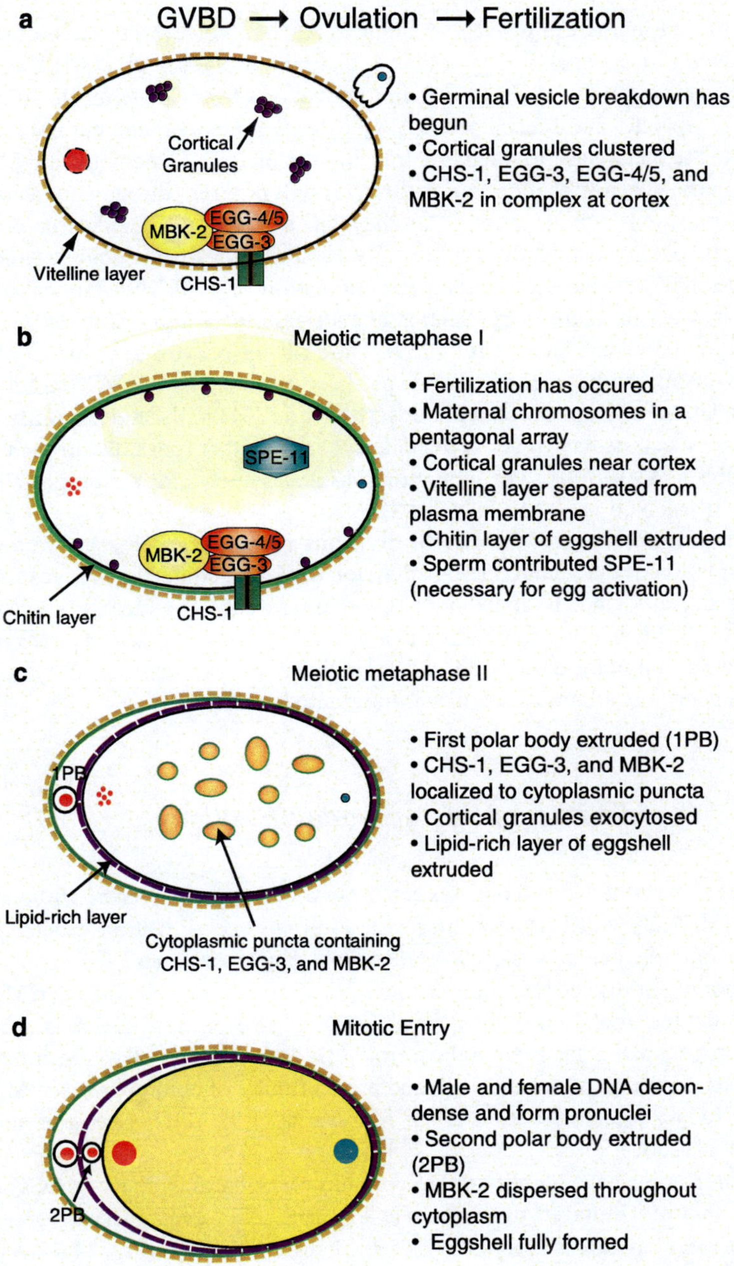

Fig. 11.3 Oocyte-to-embryo transition in *C. elegans*. During germinal vesicle breakdown (GVBD) and ovulation as well as immediately before fertilization (**a**), MBK-2, EGG-3, and EGG-4/5 are localized at the cortex through interactions with CHS-1 and cortical granules (CGs) are clustered. SPE-11 is supplied by the sperm during fertilization and is necessary for polar body extrusion and eggshell production. In meiotic metaphase I (**b**), CGs redistribute to the cortex to prepare for exo cytosis.

completion of meiosis in *C. elegans* requires APC/C; however, the role of calcium is unknown (Furuta et al. 2000; Golden et al. 2000; Siomos et al. 2001; Davis et al. 2002; Shakes et al. 2011). In *C. elegans*, activation of APC/C is fertilization independent (McNally and McNally 2005). Oocytes that are ovulated but not fertilized complete the metaphase-to-anaphase transition in meiosis I (McNally and McNally 2005). In the absence of fertilization, oocytes will progress through anaphase I but cannot transition to metaphase II, as they fail to form polar bodies or assemble meiosis II spindles (McNally and McNally 2005). Unfertilized oocytes, which lack centrioles/MTOC from sperm, undergo continuous cycles of DNA replication without cytokinesis, termed endoreplication or endo-mitotic reduplication (McNally and McNally 2005; Ward and Carrel 1979). Unfertilized oocytes are able to progress through anaphase I because cyclin B is partially degraded by APC/C, but full cyclin B degradation will not occur without fertilization (McNally and McNally 2005). Fertilization acts as the trigger to degrade cyclin B further (possibly through activation of the ZYG-11/CUL-2 E3 ubiqutin ligase) and allows for the progression into meiosis II (McNally and McNally 2005).

The rapid succession of the meiotic divisions and the requirement of fertilization for completion of the second meiotic division represent unique challenges for regulation, especially since both divisions rely on many common factors (i.e., cyclin B and APC/C). Identifying the factors that account for and regulate the two meiotic divisions is important as will be understanding how the increase in cytoplasmic calcium is originated, transmitted, and interpreted.

11.5.2 Coordination of Egg Activation Events

As noted above, unfertilized oocytes can progress through anaphase I without fertilization (McNally and McNally 2005). In addition, egg activation events such as maternal protein degradation, cytoskeletal rearrangements, and the activation of early embryogenesis cues can also occur without fertilization (McNally and McNally 2005; Pellettieri et al. 2003; Stitzel et al. 2006). However, fertilization is necessary for the proper spatial and temporal control of these events as well as the completion of meiosis, secretion of the eggshell, and proper timing of embryogenesis (McNally and McNally 2005; Parry et al. 2009; Maruyama et al. 2007; Cheng et al. 2009; Stitzel et al. 2007).

Recent studies have identified three pseudo-phosphatases, EGG-3, EGG-4, and EGG-5, that are required to coordinate egg activation events (Maruyama et al. 2007; Stitzel et al. 2007; Parry et al. 2009; Cheng et al. 2009) (Fig. 11.3). As pseudo-phosphatases, EGG-3, EGG-4, and EGG-5 contain inactive protein tyrosine phosphatase (PTP) domains that allow them to bind to phosphotyrosine residues without promoting their hydrolysis (Maruyama et al. 2007; Stitzel et al. 2007; Parry et al. 2009; Cheng et al. 2009). *egg-4* and *egg-5* are functionally redundant genes with 99.2% amino acid identity; for this reason, this pair of genes and their products is referred to as *egg-4/5* and EGG-4/5 (Parry et al. 2009). In the absence of EGG-3 and EGG-4/5, fertilized oocytes fail to extrude polar bodies, form a proper eggshell, or yield embryos that develop properly (Maruyama et al. 2007; Parry et al. 2009).

Exactly how EGG-3 and EGG-4/5 coordinate meiotic completion, fertilization, and embryogenesis is not well understood. However, it is clear that one of the main functions of EGG-3 and EGG-4/5 is to regulate MBK-2 (minibrain kinase-2), a kinase responsible for embryonic development (Maruyama et al. 2007; Stitzel et al. 2007; Parry et al. 2009; Cheng et al. 2009; Pellettieri et al. 2003). MBK-2 is a kinase that modifies many oocyte proteins, which are essential for multiple steps of embryonic development (Pellettieri et al. 2003). Known MBK-2 substrates include MEI-1, OMA-1, OMA-2, MEX-5, and MEX-6 (see Chap. 12, Robertson and Lin 2012; Guven-Ozkan et al. 2008; Nishi et al. 2008; Pang et al. 2004; Pellettieri et al. 2003; Quintin et al. 2003). MEI-1, a homolog of the microtubule severing protein katanin p60, is necessary for meiotic spindle function, and it is signaled for degradation through phosphorylation by MBK-2 (Bowerman and Kurz 2006; Lu and Mains 2007; Stitzel et al. 2006). OMA-1 and OMA-2 prevent precocious embryogenesis by sequestering the general transcription factor component TAF-4 (TATA-binding protein associated factor 4) in the cytoplasm, thus repressing transcription in early embryonic blastomeres (Detwiler et al. 2001; Guven-Ozkan et al. 2008). Phosphorylation by MBK-2 activates OMA-1 and OMA-2 and facilitates OMA-1/2 interaction with TAF-4 (Detwiler et al. 2001; Guven-Ozkan et al. 2008). MEX-5 and MEX-6 regulate embryonic polarity (Schubert et al. 2000). Phosophorylation by MBK-2 is a prerequisite for MEX-5 and MEX-6 activation by the polo kinases PLK-1 and PLK-2 (Schubert et al. 2000; Nishi et al. 2008). The variety of MBK-2 substrates necessitates tight control of MBK-2 activity by EGG-3 and EGG-4/5 in order to prevent aberrant spatial and temporal activation of many critical factors in embryogenesis.

MBK-2 activity is controlled by EGG-3 and EGG-4/5 in two ways. First, MBK-2 is physically sequestered to the oocyte cortex in a complex with EGG-3 and EGG-4/5 that is localized to the plasma membrane through an interaction between the integral membrane protein CHS-1 and EGG-3 (Harris et al. 2000; Maruyama et al. 2007; Veronico et al. 2001; Zhang et al. 2005; Singson et al. 2008) (Fig. 11.3b). Secondly, EGG-4/5 uses its PTP domain to bind the phosphotyrosine in the activation loop of MBK-2, thereby inhibiting MBK-2 by blocking access to substrates and reducing catalytic activity (Cheng et al. 2009).

The full release of MBK-2 inhibition is also accomplished sequentially. First, MBK-2 is phosphorylated by the MPF component CDK-1 (Cheng et al. 2009). Second, at meiotic anaphase I, EGG-3 is targeted for and degraded by the ubiquitin ligase containing APC/C. These events lead to the dissociation of the MBK-2–EGG-3–EGG-4/5 complex, thereby allowing MBK-2 to be released and to act on its substrates (Stitzel et al. 2007) (Fig. 11.3d). The regulation of MBK-2 by the EGG complex is dependent upon cell cycle progression but not fertilization. However, embryogenesis will ultimately fail because, without fertilization, the oocyte DNA will undergo endoreplication because of a number of reasons, including the inability to form polar bodies, assemble meiosis II spindles, or degrade cyclin B further (Maruyama et al. 2007; Stitzel et al. 2007; Cheng et al. 2009; Parry et al. 2009).

Unlike other factors in this cascade, EGG-3 and EGG-4/5 regulate fertilization-independent events, including MBK-2 activity, as well as fertilization-dependent events, including actin cap formation, polar body extrusion, and secretion of the eggshell (Maruyama et al. 2007; Stitzel et al. 2007; Cheng et al. 2009; Parry et al. 2009; McNally and McNally 2005). Future directions include the identification of additional regulatory elements necessary for the coordination of egg activation events to fertilization, as well as signal transduction pathways that promote meiotic completion and eggshell exocytosis.

11.5.3 The Role of Sperm in Egg Activation and Embryogenesis

Fertilization provides multiple sperm components that are necessary to coordinate meiotic progression and egg activation and to ensure proper embryogenesis. Each haploid sperm has a single pair of centrioles and, after fertilization, these centrioles duplicate and ultimately generate the two active centrosomes that establish the two poles of the embryo's first mitotic spindle (Albertson 1984; Dammermann et al. 2008; Pelletier et al. 2006; Albertson and Thomson 1993). *C. elegans* sperm also supply the centrosome that specifies the anterior–posterior axis (O'Connell et al. 2000; Cowan and Hyman 2004; Hamill et al. 2002; Wallenfang and Seydoux 2000). In addition to providing organelles, sperm provide proteins necessary for egg activation (SPE-11) and embryogenesis (CYK-4) (Browning and Strome 1996; Jenkins et al. 2006).

spe-11 is predicted to encode a sperm-specific novel cytoplasmic protein that exhibits a perinuclear localization pattern (Browning and Strome 1996). Sperm produced by *spe-11* mutants are capable of fertilizing oocytes but are unable to activate them, and oocytes fertilized by *spe-11* mutants have defects in meiosis (spindle orientation and cytokinesis) and embryogenesis, do not secrete an eggshell, and do not produce polar bodies (Hill et al. 1989; McNally and McNally 2005). *spe-11* paternal-effect lethal embryos can assemble the meiosis I spindle; however, the anaphase I chromosomes do not segregate and instead collapse back together and individualize (McNally and McNally 2005). Despite the defect in chromosome segregation at anaphase I, *spe-11* embryos are able to form a meiosis II spindle and undergo normal anaphase II separation, ultimately producing two female pronuclei (McNally and McNally 2005). The assembly of a meiosis II spindle in *spe-11* embryos indicates a role for sperm contributions to meiosis spindle assembly and cytokinesis, as unfertilized oocytes do not assemble a meiosis II spindle (McNally and McNally 2005). Using transgenes to ectopically supply *spe-11* gene function through maternal expression in oocytes, rescues *spe-11* function suggesting it acts when it is incorporated into the embryo after fertilization (Browning and Strome 1996).

In *C. elegans*, the sperm-enriched *cyk-4* gene product, a Rho guanosine triposphatase (GTPase)-activating protein (GAP), is necessary for establishing anterior–posterior polarity (Portereiko et al. 2004; Portereiko and Mango 2001; Jenkins et al. 2006). CYK-4 is enriched in the MO of sperm and is detected in the posterior cortex of the one-cell embryo after fertilization (Jenkins et al. 2006). CYK-4, along with the small GTPase RhoA (Ras homolog gene family, member A) and ECT-2, a RhoA guanine nucleotide-exchange factor, modulate myosin light-chain activity to create an actomyosin gradient (Jenkins et al. 2006). The actomyosin gradient is necessary to establish the anterior domain of the embryo and properly localizes the polarity proteins PAR-3 and PAR-6 (Jenkins et al. 2006). Additional sperm proteins that regulate and coordinate egg activation are yet to be identified, but it is clear that sperm entry plays an important role in these processes (McNally and McNally 2005).

11.5.4 Additional Sperm Contributions

Paternal mitochondria: When sperm–egg fusion occurs, paternal mitochondria in the sperm are incorporated into the oocyte cytoplasm. To ensure that the maternal mitochondrial DNA (mtDNA) is inherited in *C. elegans*, paternal mitochondria are eliminated by autophagy in the oocyte (Sato and Sato 2011; Al Rawi et al. 2011). Autophagy allows for the degradation of cytoplasmic proteins and organelles by sequestering them to autophagosomes and eventually lysosomes where they are broken down (Nakatogawa et al. 2009; Sato and Sato 2011; Mizushima 2007; Xie and Klionsky 2007; Al Rawi et al. 2011). Two recent reports found that autophagosomes engulf paternal mitochondria and that the mitochondria are degraded during early embryogenesis (Sato and Sato 2011; Al Rawi et al. 2011). Both reports also found that *lgg-1*, a gene necessary for autophagosome formation, was necessary for clearance of paternal mitochondria (Sato and Sato 2011; Al Rawi et al. 2011). LGG-1 and LGG-2 are orthologs of *S. cerevisae* Atg8 and mammalian LC3 (Nakatogawa et al. 2009; Mizushima 2007; Xie and Klionsky 2007; Alberti et al. 2010). *lgg-1* null mutant homozygotes were able to produce fertilized eggs; however, 36% of the eggs were unable to hatch and 59% of the embryos that did hatch died at the L1 larval stage (Sato and Sato 2011). These findings provide evidence that paternal mitochondria are actively degraded and that degradation, not dilution of paternal mtDNA by an excess of maternal mtDNA, is the primary mode of ensuring maternal mtDNA inheritance (Sato and Sato 2011; Al Rawi et al. 2011). This paradigm may also hold true in mammals, as mouse autophagy has been shown to be necessary for preimplantation development (Tsukamoto et al. 2008; Sato and Sato 2011). Moreover, the paternal mitochondria in mouse sperm, which are located in the midpiece of the flagellum, stain positively with an anti-LC3 antibody after fertilization indicating that autophagy plays a role in the degradation of paternal mitochondria (Al Rawi et al. 2011).

peel-1/zeel-1 element: Sperm also deliver another other component that is actually toxic if the proper antidote is not present in the embryo (Seidel et al. 2011; Seidel et al. 2008). The *C. elegans* gene *peel-1* (*paternal effect epistatic embryonic lethal-1*) encodes a four-pass transmembrane protein that localizes to MOs in spermatids and is delivered to the embryo during fertilization (Seidel et al. 2011). PEEL-1 is a potent toxin that disrupts the development of muscle and epidermal tissue during late embryogenesis (Seidel et al. 2011). To neutralize the toxicity of

PEEL-1 the embryo transiently expresses *zeel-1* (*zygotic epistatic embryonic lethal-1*) (Seidel et al. 2011). ZEEL-1 is six-pass transmembrane protein, the transmembrane domain of which is necessary for its function as an antidote to PEEL-1 (Seidel et al. 2011). The cause of PEEL-1 toxicity and the mechanism by which ZEEL-1 mitigates PEEL-1 toxicity are unknown (Seidel et al. 2011). Seidel et al. hypothesize that ZEEL-1 could promote the degradation of PEEL-1 or prevent PEEL-1 from binding to its target (Seidel et al. 2011).

Despite its deleterious effects on zygotic development, *peel-1* maintains its place in the genome through an inseparable association with its antidote *zeel-1* (Seidel et al. 2008, 2011). *peel-1* and *zeel-1* are adjacent genes and cannot be separated by homologous recombination; they are referred to as the *peel-1/zeel-1* element (Seidel et al. 2008). The *peel-1/zeel-1* element persists because, in *C. elegans*, homozygosity, not heterozygosity, is most common as a result of a low number of males in the population (Seidel et al. 2011). The evolutionary origins of the *peel-1/zeel-1* element are unknown (Seidel et al. 2008, 2011). The *peel-1/zeel-1* element is intriguing from the perspective of fertilization and embryogenesis because it may provide insight into how sperm-supplied factors are interpreted by and influence zygotic transcription and embryonic development.

11.6 The Eggshell

11.6.1 Eggshell Structure

The chitin eggshell is deposited and assembled after fertilization and is essential for *C. elegans* embryogenesis (Rappleye et al. 1999; Bembenek et al. 2007; McNally and McNally 2005). Structural support provided by the eggshell allows for the completion of meiosis, polar body extrusion, and establishment of embryo polarity, and it also forms an osmotic barrier that protects the developing embryo from osmotic changes (Rappleye et al. 1999; Siomos et al. 2001; Kaitna et al. 2000; McNally and McNally 2005). The *C. elegans* eggshell is made of three layers: the outer vitelline layer, the middle chitin layer, and the inner lipid-rich layer (Rappleye et al. 1999). At the time of fertilization, the outer vitelline layer that surrounds the developing oocyte begins to separate from the plasma membrane as the chitin layer is formed (Rappleye et al. 1999; Bembenek et al. 2007). The middle chitin layer is formed during metaphase I and provides the eggshell with its mechanical strength (Rappleye et al. 1999; Maruyama et al. 2007). Chitin is formed through the polymerization of UDP-*N*-acteylglucosamine (UDP-GlcNAc), which is catalyzed by the oocyte membrane protein CHS-1, and the chitin-binding protein CBD-1 is necessary for the formation of a continuous chitin layer (Rappleye et al. 1999; Zhang et al. 2005; Maruyama et al. 2007; Johnston et al. 2010). The lipid-rich inner layer of the eggshell begins to form during anaphase I and provides osmotic and mechanical strength (Rappleye et al. 1999; Bembenek et al. 2007; Sato et al. 2008). The eggshell performs diverse roles for the *C. elegans* embryo and, without proper eggshell deposition, the embryo is not able to divide properly and is susceptible to many mechanical and osmotic stresses (Rappleye et al. 1999; McNally and McNally 2005).

11.6.2 Cortical Granule Exocytosis

Cortical granules (CGs) are Golgi-derived secretory vesicles that are stored at the cortex of oocytes (Horner and Wolfner 2008). In *C. elegans*, CGs are necessary for proper eggshell formation (Bembenek et al. 2007; Sato et al. 2008). CG exocytosis occurs during anaphase I in response to APC/C activation—not fertilization—and requires a number of cell cycle components, including the *C. elegans* ortholog separase-1 SEP-1 as well as the small GTPase RAB-11.1 and the target-SNARE SYN-4 (Bembenek et al. 2007; Sato et al. 2008). CGs contain chondroitin proteoglycans and, upon exocytosis, these chondroitin proteoglycans are released to the extracellular space surrounding the embryo (Hwang and Horvitz 2002; Bembenek et al. 2007; Sato et al. 2008).

The role of chondroitin proteoglycans in eggshell formation is unclear. Two chondroitin proteoglycans, CPG-1 and CPG-2, are known to bind chitin, thus suggesting a role for CPG-1 and CPG-2 in the formation or maintenance of the chitin layer (Bembenek et al. 2007; Olson et al. 2006). However, chondroitin proteoglycans are present in CGs and are not exocytosed until after the chitin layer has formed during anaphase I, implicating a role for chondroitin proteoglycans in the formation of the lipid-rich layer (Bembenek et al. 2007; Sato et al. 2008). Determining the precise localization and role for chondroitin proteoglycans in eggshell formation will aid in understanding how the eggshell provides osmotic and mechanical strength.

11.6.3 Membrane Block to Polyspermy

C. elegans have a robust block to polyspermy (Ward and Carrel 1979). In addition to the ovulated oocyte, the spermatheca also contains multiple sperm in an extremely tight space (Ward and Carrel 1979). Despite contact with multiple sperm, *C. elegans* oocytes are only fertilized by a single sperm (Ward and Carrel 1979). Once fertilization occurs, the oocyte prevents additional sperm from entering (Ward and Carrel 1979). Recently, embryos with multiple fused sperm have been described; however, these polyspermic events occur in a very small percentage of the embryos examined (Parry et al. 2009; Johnston et al. 2010). In *C. elegans*, polyspermy has been observed after the depletion of *chs-1, gna-2*, or *egg-4/5* (Parry et al. 2009; Johnston et al. 2010). As discussed earlier, CHS-1 is necessary for the polymerization of the chitin component UDP-*N*-acteylglucosamine (Zhang et al. 2005). *gna-2* encodes a GLD-regulated glucosamine-6-P N acetyltransferase that supplies UDP-*N*-acetyl glucosamine for chitin biosynthesis (Johnston et al. 2006). The role of *chs-1* and *gna-2* in chitin synthesis indicates a potential role for chitin in the block to polyspermy.

In other organisms, cortical granule exocytosis is known to contribute to polyspermy blocks (Wessel et al. 2001). The connection between CG exocytosis and the block to polyspermy in *C. elegans* is unclear, but evidence from *egg-4/5* embryos indicates that CG exocytosis does not act in the block to polyspermy (Parry et al. 2009). *egg-4/5* embryos seem to have normal CG exocytosis but polyspermic embryos can still be detected (Parry et al. 2009). However, since the majority of *egg-4/5* embryos are not polyspermic, there may be other contributions to the block to polyspermy or more subtle details about CG exocytosis that are not yet understood (Parry et al. 2009).

11.7 Conclusion

Fertilization in *C. elegans* involves a vast number of cell processes that must be executed with high fidelity so as to ensure the successful propagation of the species. First, *C. elegans* must produce haploid gametes with extraordinary differences in morphology and function to ensure they are able to locate one another. When the gametes finally do meet, they will only fuse if they are from the correct species despite the fact sperm from other species look similar and can enter the reproductive tract and migrate to the site of fertilization. Finally, the oocyte must only allow one sperm to enter, sort out which components of the sperm are necessary for embryogenesis, and then initiate embryonic development. Many questions remain about how fertilization is regulated in *C. elegans*. What are the signal transduction pathways that allow sperm to respond to prostaglandin signals from the oocyte? How is the developmental program regulated during the oocyte-to-embryo transition? How does the eggshell form? How is the rapid block to polyspermy triggered? Finally, the central question remains—how is sperm–egg fusion mediated?

The genetic and molecular basis of fertilization is not well understood. The study of fertilization in *C. elegans* provides an excellent opportunity to understand the molecular functions of the genes that are necessary for fertilization and to discover previously unpredicted genes necessary for this fundamental process of sexual reproduction. The study of mammalian fertilization has been difficult, partially because the ability to obtain and successfully manipulate gametes is challenging. *C. elegans* is an emerging model system to study fertilization. Many parallels have already been identified between cellular processes in *C. elegans* and other species. The molecular and genetic techniques available for use in *C. elegans* allow for relatively easy gene discovery. It is possible that many of the genes that are necessary for fertilization in *C. elegans* will also be important in mammalian systems because of the conserved features of sexual reproduction.

Acknowledgments We thank the anonymous reviewers and Drs. Tim Schedl, Diane Shakes, and David Greenstein for their helpful comments and suggestions. We also thank the members of the Singson laboratory, Roopa Kalyanaraman Marcello, and Dr. Sara K. Olson for critical comments and editorial assistance. This work was supported by a grant from the National Institutes of Health to AS [R01HD054681-06]. MRM is a UMDNJ-RWJMS IRACDA Postdoctoral Fellow supported by NIH Grant GM093854.

References

Abbas M, Cain GD (1979) *In vitro* activation and behavior of the ameboid sperm of *Ascaris suum* (Nematoda). Cell Tissue Res 200(2):273–284

Al Rawi S, Louvet-Vallee S, Djeddi A, Sachse M, Culetto E, Hajjar C, Boyd L, Legouis R, Galy V (2011) Postfertilization autophagy of sperm organelles prevents paternal mitochondrial DNA transmission. Science 334(6059):1144–1147. doi:science.1211878 [pii] 10.1126/science.1211878

Alberti A, Michelet X, Djeddi A, Legouis R (2010) The autophagosomal protein LGG-2 acts synergistically with LGG-1 in dauer formation and longevity in *C. elegans*. Autophagy 6(5). doi:12252 [pii]

Albertson DG (1984) Formation of the first cleavage spindle in nematode embryos. Dev Biol 101(1):61–72. doi:0012-1606(84)90117-9 [pii]

Albertson DG, Thomson JN (1993) Segregation of holocentric chromosomes at meiosis in the nematode, *Caenorhabditis elegans*. Chromosome Res 1(1):15–26

Baird SE, Yen WC (2000) Reproductive isolation in *Caenorhabditis*: terminal phenotypes of hybrid embryos. Evol Dev 2(1):9–15

Baird SE, Sutherlin ME, Emmons SW (1992) Reproductive isolation in Rhabditidae (Nematoda: Secernentea); mechanisms that isolate six species of three genera. Evolution 46(3):585–594

Baker AM, Roberts TM, Stewart M (2002) 2.6 A resolution crystal structure of helices of the motile major sperm protein (MSP) of *Caenorhabditis elegans*. J Mol Biol 319(2):491–499. doi:10.1016/S0022-2836(02)00294-2 S0022-2836(02)00294-2 [pii]

Baldi C, Viviano J, Ellis RE (2011) A bias caused by ectopic development produces sexually dimorphic sperm in nematodes. Curr Biol 21(16):1416–1420. doi:S0960-9822(11)00835-9 [pii] 10.1016/j.cub.2011.07.034

Balzar M, Briaire-de Bruijn IH, Rees-Bakker HA, Prins FA, Helfrich W, de Leij L, Riethmuller G, Alberti S, Warnaar SO, Fleuren GJ, Litvinov SV (2001) Epidermal growth factor-like repeats mediate lateral and reciprocal interactions of Ep-CAM molecules in homophilic adhesions. Mol Cell Biol 21(7):2570–2580. doi:10.1128/MCB.21.7.2570-2580.2001

Batchelder EL, Hollopeter G, Campillo C, Mezanges X, Jorgensen EM, Nassoy P, Sens P, Plastino J (2011) Membrane tension regulates motility by controlling lamellipodium organization. Proc Natl Acad Sci USA 108(28):11429–11434. doi:1010481108 [pii] 10.1073/pnas.1010481108

Bembenek JN, Richie CT, Squirrell JM, Campbell JM, Eliceiri KW, Poteryaev D, Spang A, Golden A, White JG (2007) Cortical granule exocytosis in *C. elegans* is regulated by cell cycle components including separase. Development 134(21):3837–3848. doi:dev.011361 [pii] 10.1242/dev.011361

Bodding M (2007) TRP proteins and cancer. Cell Signal 19(3):617–624. doi:S0898-6568(06)00206-3 [pii] 10.1016/j.cellsig.2006.08.012

Bowerman B, Kurz T (2006) Degrade to create: developmental requirements for ubiquitin-mediated proteolysis during early *C. elegans* embryogenesis. Development 133(5):773–784. doi:133/5/773 [pii] 10.1242/dev.02276

Browning H, Strome S (1996) A sperm-supplied factor required for embryogenesis in *C. elegans*. Development 122(1):391–404

Bullock TL, Parthasarathy G, King KL, Kent HM, Roberts TM, Stewart M (1996a) New crystal forms of the motile major sperm protein (MSP) of *Ascaris suum*. J Struct Biol 116(3):432–437. doi:S1047-8477(96)90061-1 [pii] 10.1006/jsbi.1996.0061

Bullock TL, Roberts TM, Stewart M (1996b) 2.5 A resolution crystal structure of the motile major sperm protein (MSP) of *Ascaris suum*. J Mol Biol 263(2):284–296. doi:S0022-2836(96)90575-6 [pii] 10.1006/jmbi.1996.0575

Buttery SM, Ekman GC, Seavy M, Stewart M, Roberts TM (2003) Dissection of the *Ascaris* sperm motility machinery identifies key proteins involved in major sperm protein-based amoeboid locomotion. Mol Biol Cell 14(12):5082–5088. doi:10.1091/mbc.E03-04-0246 E03-04-0246 [pii]

Chakrabarti R, Cheng L, Puri P, Soler D, Vijayaraghavan S (2007) Protein phosphatase PP1 gamma 2 in sperm morphogenesis and epididymal initiation of sperm motility. Asian J Androl 9(4):445–452. doi:10.1111/j.1745-7262.2007.00307.x

Chatterjee I, Richmond A, Putiri E, Shakes DC, Singson A (2005) The *Caenorhabditis elegans spe-38* gene encodes a novel four-pass integral membrane protein required for sperm function at fertilization. Development 132(12):2795–2808. doi:132/12/2795 [pii] 10.1242/dev.01868

Cheng KC, Klancer R, Singson A, Seydoux G (2009) Regulation of MBK-2/DYRK by CDK-1 and the pseudophosphatases EGG-4 and EGG-5 during the oocyte-to-embryo transition. Cell 139(3):560–572. doi:S0092-8674(09)01187-8 [pii] 10.1016/j.cell.2009.08.047

Chu DS, Shakes DC (2012) Spermatogenesis. Advances in Experimental Medicine and Biology 757:171–203. (Chap. 7, this volume) Springer, New York

Chung MK, Jung SJ, Oh SB (2011) Role of TRP channels in pain sensation. Advances in Experimental Medicine and Biology 704:615–636. doi:10.1007/978-94-007-0265-3_33

Cowan CR, Hyman AA (2004) Centrosomes direct cell polarity independently of microtubule assembly in *C. elegans* embryos. Nature 431(7004):92–96. doi:10.1038/nature02825 nature02825 [pii]

Dammermann A, Maddox PS, Desai A, Oegema K (2008) SAS-4 is recruited to a dynamic structure in newly forming centrioles that is stabilized by the gamma-tubulin-mediated addition of centriolar microtubules. J Cell Biol 180(4):771–785. doi:jcb.200709102 [pii] 10.1083/jcb.200709102

Davis ES, Wille L, Chestnut BA, Sadler PL, Shakes DC, Golden A (2002) Multiple subunits of the *Caenorhabditis elegans* anaphase-promoting complex are required for chromosome segregation during meiosis I. Genetics 160(2):805–813

Detwiler MR, Reuben M, Li X, Rogers E, Lin R (2001) Two zinc finger proteins, OMA-1 and OMA-2, are redundantly required for oocyte maturation in *C. elegans*. Dev Cell 1(2):187–199. doi:S1534-5807(01)00026-0 [pii]

Dietrich A, Chubanov V, Gudermann T (2010) Renal TRPathies. J Am Soc Nephrol 21(5):736–744. doi:ASN.2009090948 [pii] 10.1681/ASN.2009090948

Eder P, Molkentin JD (2011) TRPC channels as effectors of cardiac hypertrophy. Circ Res 108(2):265–272. doi:108/2/265 [pii] 10.1161/CIRCRESAHA.110.225888

Edmonds JW, Prasain JK, Dorand D, Yang Y, Hoang HD, Vibbert J, Kubagawa HM, Miller MA (2010) Insulin/FOXO signaling regulates ovarian prostaglandins critical for reproduction. Dev Cell 19(6):858–871. doi:S1534-5807(10)00532-0 [pii] 10.1016/j.devcel.2010.11.005

Eppig JJ (1996) Coordination of nuclear and cytoplasmic oocyte maturation in eutherian mammals. Reprod Fertil Dev 8(4):485–489

Fraire-Zamora JJ, Broitman-Maduro G, Maduro M, Cardullo RA (2011) Evidence for phosphorylation in the MSP cytoskeletal filaments of amoeboid spermatozoa. Int J Biochem Mol Biol 2(3):263–273

Furuta T, Tuck S, Kirchner J, Koch B, Auty R, Kitagawa R, Rose AM, Greenstein D (2000) EMB-30: an APC4 homologue required for metaphase-to-anaphase transitions during meiosis and mitosis in *Caenorhabditis elegans*. Mol Biol Cell 11(4):1401–1419

Geldziler B, Kadandale P, Singson A (2004) Molecular genetic approaches to studying fertilization in model systems. Reproduction 127(4):409–416. doi:10.1530/rep. 1.00009 127/4/409 [pii]

Geldziler B, Marcello MR, Shakes DC, Singson A (2011) The genetics and cell biology fertilization. In: Rothman JL, Singson A (eds) *Caenorhabditis elegans*: molecular genetics and development, methods in cell biology, vol 106. Elsevier, Oxford

Golden A, Sadler PL, Wallenfang MR, Schumacher JM, Hamill DR, Bates G, Bowerman B, Seydoux G, Shakes DC (2000) Metaphase to anaphase (mat) transition-defective mutants in *Caenorhabditis elegans*. J Cell Biol 151(7):1469–1482

Grant B, Hirsh D (1999) Receptor-mediated endocytosis in the *Caenorhabditis elegans* oocyte. Mol Biol Cell 10(12):4311–4326

Grant RP, Buttery SM, Ekman GC, Roberts TM, Stewart M (2005) Structure of MFP2 and its function in enhancing MSP polymerization in *Ascaris* sperm amoeboid motility. J Mol Biol 347(3):583–595. doi:S0022-2836(05)00099-9 [pii] 10.1016/j.jmb.2005.01.054

Guven-Ozkan T, Nishi Y, Robertson SM, Lin R (2008) Global transcriptional repression in *C. elegans* germline precursors by regulated sequestration of TAF-4. Cell 135(1):149–160. doi:S0092-8674(08)00960-4 [pii] 10.1016/j.cell.2008.07.040

Haaf A, Butler PJ, Kent HM, Fearnley IM, Roberts TM, Neuhaus D, Stewart M (1996) The motile major sperm protein (MSP) from *Ascaris suum* is a symmetric dimer in solution. J Mol Biol 260(2):251–260. doi:S0022283696903964 [pii]

Hamill DR, Severson AF, Carter JC, Bowerman B (2002) Centrosome maturation and mitotic spindle assembly in *C. elegans* require SPD-5, a protein with multiple coiled-coil domains. Dev Cell 3(5):673–684. doi:S1534580702003271 [pii]

Han SM, Cottee PA, Miller MA (2010) Sperm and oocyte communication mechanisms controlling *C. elegans* fertility. Dev Dyn 239(5):1265–1281. doi:10.1002/dvdy.22202

Harris MT, Lai K, Arnold K, Martinez HF, Specht CA, Fuhrman JA (2000) Chitin synthase in the filarial parasite, Brugia malayi. Mol Biochem Parasitol 111(2):351–362. doi:S0166685100003285 [pii]

Hill KL, L'Hernault SW (2001) Analyses of reproductive interactions that occur after heterospecific matings within the genus *Caenorhabditis*. Dev Biol 232(1):105–114. doi:10.1006/dbio.2000.0136 S0012-1606(00)90136-2 [pii]

Hill DP, Shakes DC, Ward S, Strome S (1989) A sperm-supplied product essential for initiation of normal embryogenesis in *Caenorhabditis elegans* is encoded by the paternal-effect embryonic-lethal gene, *spe-11*. Dev Biol 136(1):154–166. doi:0012-1606(89)90138-3 [pii]

Hirsh D, Oppenheim D, Klass M (1976) Development of the reproductive system of *Caenorhabditis elegans*. Dev Biol 49(1):200–219. doi:0012-1606(76)90267-0 [pii]

Horner VL, Wolfner MF (2008) Transitioning from egg to embryo: triggers and mechanisms of egg activation. Dev Dyn 237(3):527–544. doi:10.1002/dvdy.21454

Huang Z, Vijayaraghavan S (2004) Increased phosphorylation of a distinct subcellular pool of protein phosphatase, PP1gamma2, during epididymal sperm maturation. Biol Reprod 70(2):439–447. doi:10.1095/biolreprod.103.020024 biolreprod.103.020024 [pii]

Hwang HY, Horvitz HR (2002) The SQV-1 UDP-glucuronic acid decarboxylase and the SQV-7 nucleotide-sugar transporter may act in the Golgi apparatus to affect *Caenorhabditis elegans* vulval morphogenesis and embryonic development. Proc Natl Acad Sci USA 99(22):14218–14223. doi:10.1073/pnas.172522199

Italiano JE Jr, Roberts TM, Stewart M, Fontana CA (1996) Reconstitution *in vitro* of the motile apparatus from the amoeboid sperm of *Ascaris* shows that filament assembly and bundling move membranes. Cell 84(1):105–114. doi:S0092-8674(00)80997-6 [pii]

Italiano JE Jr, Stewart M, Roberts TM (2001) How the assembly dynamics of the nematode major sperm protein generate amoeboid cell motility. Int Rev Cytol 202:1–34

Jenkins N, Saam JR, Mango SE (2006) CYK-4/GAP provides a localized cue to initiate anteroposterior polarity upon fertilization. Science 313(5791):1298–1301. doi:1130291 [pii] 10.1126/science.1130291

Johnston WL, Krizus A, Dennis JW (2006) The eggshell is required for meiotic fidelity, polar-body extrusion and polarization of the *C. elegans* embryo. BMC Biol 4:35. doi:1741-7007-4-35 [pii] 10.1186/1741-7007-4-35

Johnston WL, Krizus A, Dennis JW (2010) Eggshell chitin and chitin-interacting proteins prevent polyspermy in *C. elegans*. Curr Biol 20(21):1932–1937. doi:S0960-9822(10)01216-9 [pii] 10.1016/j.cub.2010.09.059

Kadandale P, Stewart-Michaelis A, Gordon S, Rubin J, Klancer R, Schweinsberg P, Grant BD, Singson A (2005) The egg surface LDL receptor repeat-containing proteins EGG-1 and EGG-2 are required for fertilization in *Caenorhabditis elegans*. Curr Biol 15(24):2222–2229. doi:S0960-9822(05)01296-0 [pii] 10.1016/j.cub.2005.10.043

Kaitna S, Mendoza M, Jantsch-Plunger V, Glotzer M (2000) Incenp and an aurora-like kinase form a complex essential for chromosome segregation and efficient completion of cytokinesis. Curr Biol 10(19):1172–1181. doi:S0960-9822(00)00721-1 [pii]

Kim S, Spike CA, Greenstein D (2012) Control of oocyte growth and meiotic maturation in *C. elegans*. Advances in Experimental Medicine and Biology 757:277–320. (Chap. 10, this volume) Springer, New York

Kimble J, Hirsh D (1979) The postembryonic cell lineages of the hermaphrodite and male gonads in *Caenorhabditis elegans*. Dev Biol 70(2):396–417. doi:0012-1606(79)90035-6 [pii]

King KL, Essig J, Roberts TM, Moerland TS (1994) Regulation of the *Ascaris* major sperm protein (MSP) cytoskeleton by intracellular pH. Cell Motil Cytoskeleton 27(3):193–205. doi:10.1002/cm.970270302

Kiontke K, Fitch DH (2005) The phylogenetic relationships of *Caenorhabditis* and other rhabditids. WormBook:1–11. doi:10.1895/wormbook.1.11.1

Kiselyov K, Soyombo A, Muallem S (2007) TRPpathies. J Physiol 578(Pt 3):641–653. doi:jphysiol.2006.119024 [pii] 10.1113/jphysiol.2006.119024

Klass M, Wolf N, Hirsh D (1976) Development of the male reproductive system and sexual transformation in the nematode *Caenorhabditis elegans*. Dev Biol 52(1):1–18. doi:0012-1606(76)90002-6 [pii]

Kosinski M, McDonald K, Schwartz J, Yamamoto I, Greenstein D (2005) *C. elegans* sperm bud vesicles to deliver a meiotic maturation signal to distant oocytes. Development 132(15):3357–3369. doi:dev.01916 [pii] 10.1242/dev.01916

Kroft TL, Gleason EJ, L'Hernault SW (2005) The *spe-42* gene is required for sperm-egg interactions during *C. elegans* fertilization and encodes a sperm-specific transmembrane protein. Dev Biol 286(1):169–181. doi:S0012-1606(05)00476-8 [pii] 10.1016/j.ydbio.2005.07.020

Kubagawa HM, Watts JL, Corrigan C, Edmonds JW, Sztul E, Browse J, Miller MA (2006) Oocyte signals derived from polyunsaturated fatty acids control sperm recruitment *in vivo*. Nat Cell Biol 8(10):1143–1148. doi:ncb1476 [pii] 10.1038/ncb1476

LaMunyon CW, Ward S (1998) Larger sperm outcompete smaller sperm in the nematode *Caenorhabditis elegans*. Proc Biol Sci 265(1409):1997–2002. doi:10.1098/rspb.1998.0531

LeClaire LL 3rd, Stewart M, Roberts TM (2003) A 48 kDa integral membrane phosphoprotein orchestrates the cytoskeletal dynamics that generate amoeboid cell motility in *Ascaris* sperm. J Cell Sci 116(Pt 13):2655–2663. doi:10.1242/jcs.00469 jcs.00469 [pii]

Lee MH, Schedl T (2001) Identification of in vivo mRNA targets of GLD-1, a maxi-KH motif containing protein required for *C. elegans* germ cell development. Genes Dev 15(18):2408–2420. doi:10.1101/gad.915901

L'Hernault SW (2006) Spermatogenesis. WormBook:1–14. doi:10.1895/wormbook.1.85.1

L'Hernault SW, Roberts TM (1995) Cell biology of nematode sperm. In: *Caenorhabditis elegans*: modern biological analysis of an organism, vol 48. Academic Press, San Diego, pp 273–301

L'Hernault SW, Shakes DC, Ward S (1988) Developmental genetics of chromosome I spermatogenesis-defective mutants in the nematode *Caenorhabditis elegans*. Genetics 120(2):435–452

Liu J, Maller JL (2005) Calcium elevation at fertilization coordinates phosphorylation of XErp1/Emi2 by Plx1 and CaMK II to release metaphase arrest by cytostatic factor. Curr Biol 15(16):1458–1468. doi:S0960-9822(05)00774-8 [pii] 10.1016/j.cub.2005.07.030

Lu C, Mains PE (2007) The *C. elegans* anaphase promoting complex and MBK-2/DYRK kinase act redundantly with CUL-3/MEL-26 ubiquitin ligase to degrade MEI-1 microtubule-severing activity after meiosis. Dev Biol 302(2):438–447. doi:S0012-1606(06)01269-3 [pii] 10.1016/j.ydbio.2006.09.053

Madl JE, Herman RK (1979) Polyploids and sex determination in *Caenorhabditis elegans*. Genetics 93(2):393–402

Maeda I, Kohara Y, Yamamoto M, Sugimoto A (2001) Large-scale analysis of gene function in *Caenorhabditis elegans* by high-throughput RNAi. Curr Biol 11(3):171–176. doi:S0960-9822(01)00052-5 [pii]

Marcello MR, Singson A (2010) Fertilization and the oocyte-to-embryo transition in *C. elegans*. BMB Rep 43(6):389–399

Maruyama R, Velarde NV, Klancer R, Gordon S, Kadandale P, Parry JM, Hang JS, Rubin J, Stewart-Michaelis A, Schweinsberg P, Grant BD, Piano F, Sugimoto A, Singson A (2007) EGG-3 regulates cell-surface and cortex rearrangements during egg activation in *Caenorhabditis elegans*. Curr Biol 17(18):1555–1560. doi:S0960-9822(07)01787-3 [pii]10.1016/j.cub.2007.08.011

Masui Y, Clarke HJ (1979) Oocyte maturation. Int Rev Cytol 57:185–282

McCarter J, Bartlett B, Dang T, Schedl T (1999) On the control of oocyte meiotic maturation and ovulation in *Caenorhabditis elegans*. Dev Biol 205(1):111–128. doi:S0012-1606(98)99109-6 [pii] 10.1006/dbio.1998.9109

McNally KL, McNally FJ (2005) Fertilization initiates the transition from anaphase I to metaphase II during female meiosis in *C. elegans*. Dev Biol 282(1):218–230. doi:S0012-1606(05)00177-6 [pii] 10.1016/j.ydbio.2005.03.009

Mensah KA, Ritchlin CT, Schwarz EM (2010) RANKL induces heterogeneous DC-STAMP(lo) and DC-STAMP(hi) osteoclast precursors of which the DC-STAMP(lo) precursors are the master fusogens. J Cell Physiol 223(1):76–83. doi:10.1002/jcp. 22012

Miao L, Vanderlinde O, Stewart M, Roberts TM (2003) Retraction in amoeboid cell motility powered by cytoskeletal dynamics. Science 302(5649):1405–1407. doi:10.1126/science.1089129 302/5649/1405 [pii]

Miller MA, Nguyen VQ, Lee MH, Kosinski M, Schedl T, Caprioli RM, Greenstein D (2001) A sperm cytoskeletal protein that signals oocyte meiotic maturation and ovulation. Science 291(5511):2144–2147. doi:10.1126/science.1057586 291/5511/2144 [pii]

Miyamoto T (2006) The dendritic cell-specific transmembrane protein DC-STAMP is essential for osteoclast fusion and osteoclast bone-resorbing activity. Mod Rheumatol 16(6):341–342. doi:10.1007/s10165-006-0524-0

Mizushima N (2007) Autophagy: process and function. Genes Dev 21(22):2861–2873. doi:21/22/2861 [pii] 10.1101/gad.1599207

Morsci NS, Haas LA, Barr MM (2011) Sperm status regulates sexual attraction in *Caenorhabditis elegans*. Genetics. doi:genetics.111.133603 [pii] 10.1534/genetics.111.133603

Nakatogawa H, Suzuki K, Kamada Y, Ohsumi Y (2009) Dynamics and diversity in autophagy mechanisms: lessons from yeast. Nat Rev Mol Cell Biol 10(7):458–467. doi:nrm2708 [pii] 10.1038/nrm2708

Nigon V, Dougherty EC (1949) Reproductive patterns and attempts at reciprocal crossing of *Rhabditis elegans* Maupas, 1900, and *Rhabditis briggsae* Dougherty and Nigon, 1949 (Nematoda: Rhabditidae). J Exp Zool 112(3):485–503

Nishi Y, Rogers E, Robertson SM, Lin R (2008) Polo kinases regulate *C. elegans* embryonic polarity via binding to DYRK2-primed MEX-5 and MEX-6. Development 135(4):687–697. doi:dev.013425 [pii] 10.1242/dev.013425

Nishimura H, L'Hernault SW (2010) Spermatogenesis-defective (spe) mutants of the nematode *Caenorhabditis elegans* provide clues to solve the puzzle of male germline functions during reproduction. Dev Dyn 239(5):1502–1514. doi:10.1002/dvdy.22271

Nykjaer A, Willnow TE (2002) The low-density lipoprotein receptor gene family: a cellular Swiss army knife? Trends Cell Biol 12(6):273–280. doi:S0962892402022821 [pii]

O'Connell KF, Maxwell KN, White JG (2000) The spd-2 gene is required for polarization of the anteroposterior axis and formation of the sperm asters in the *Caenorhabditis elegans* zygote. Dev Biol 222(1):55–70. doi:S0012-1606(00)99714-8 [pii] 10.1006/dbio.2000.9714

Olson SK, Bishop JR, Yates JR, Oegema K, Esko JD (2006) Identification of novel chondroitin proteoglycans in *Caenorhabditis elegans*: embryonic cell division depends on CPG-1 and CPG-2. J Cell Biol 173(6):985–994. doi:jcb.200603003 [pii] 10.1083/jcb.200603003

Oppedisano L, Haines G, Hrabchak C, Fimia G, Elliott R, Sassone-Corsi P, Varmuza S (2002) The rate of aneuploidy is altered in spermatids from infertile mice. Hum Reprod 17(3):710–717

Pang KM, Ishidate T, Nakamura K, Shirayama M, Trzepacz C, Schubert CM, Priess JR, Mello CC (2004) The minibrain kinase homolog, *mbk-2*, is required for spindle positioning and asymmetric cell division in early *C. elegans* embryos. Dev Biol 265(1):127–139. doi:S0012160603005761 [pii]

Parry JM, Singson A (2011) EGG molecules couple the oocyte-to-embryo transition with cell cycle progression. Results Probl Cell Differ 53:135–151. doi:10.1007/978-3-642-19065-0_7

Parry JM, Velarde NV, Lefkovith AJ, Zegarek MH, Hang JS, Ohm J, Klancer R, Maruyama R, Druzhinina MK, Grant BD, Piano F, Singson A (2009) EGG-4 and EGG-5 link events of the oocyte-to-embryo transition with meiotic progression in *C. elegans*. Curr Biol 19(20):1752–1757. doi:S0960-9822(09)01692-3 [pii] 10.1016/j.cub.2009.09.015

Pelletier L, O'Toole E, Schwager A, Hyman AA, Muller-Reichert T (2006) Centriole assembly in *Caenorhabditis elegans*. Nature 444(7119):619–623. doi:nature05318 [pii] 10.1038/nature05318

Pellettieri J, Reinke V, Kim SK, Seydoux G (2003) Coordinate activation of maternal protein degradation during the egg-to-embryo transition in *C. elegans*. Dev Cell 5(3):451–462. doi:S1534580703002314 [pii]

Portereiko MF, Mango SE (2001) Early morphogenesis of the *Caenorhabditis elegans* pharynx. Dev Biol 233(2):482–494. doi:10.1006/dbio.2001.0235 S0012-1606(01)90235-0 [pii]

Portereiko MF, Saam J, Mango SE (2004) ZEN-4/MKLP1 is required to polarize the foregut epithelium. Curr Biol 14(11):932–941. doi:10.1016/j.cub.2004.05.052 S0960982204003872 [pii]

Putiri E, Zannoni S, Kadandale P, Singson A (2004) Functional domains and temperature-sensitive mutations in SPE-9, an EGF repeat-containing protein required for fertility in *Caenorhabditis elegans*. Dev Biol 272(2):448–459. doi:10.1016/j.ydbio.2004.05.014 S0012160604003586 [pii]

Quintin S, Mains PE, Zinke A, Hyman AA (2003) The *mbk-2* kinase is required for inactivation of MEI-1/katanin in the one-cell *Caenorhabditis elegans* embryo. EMBO Rep 4(12):1175–1181. doi:10.1038/sj.embor.7400029 7400029 [pii]

Rappleye CA, Paredez AR, Smith CW, McDonald KL, Aroian RV (1999) The coronin-like protein POD-1 is required for anterior-posterior axis formation and cellular architecture in the nematode *Caenorhabditis elegans*. Genes Dev 13(21):2838–2851

Rauh NR, Schmidt A, Bormann J, Nigg EA, Mayer TU (2005) Calcium triggers exit from meiosis II by targeting the APC/C inhibitor XErp1 for degradation. Nature 437(7061):1048–1052. doi:nature04093 [pii] 10.1038/nature04093

Roberts TM, Stewart M (2000) Acting like actin. The dynamics of the nematode major sperm protein (msp) cytoskeleton indicate a push-pull mechanism for amoeboid cell motility. J Cell Biol 149(1):7–12

Roberts TM, Streitmatter G (1984) Membrane-substrate contact under the spermatozoon of *Caenorhabditis elegans*, a crawling cell that lacks filamentous actin. J Cell Sci 69:117–126

Roberts TM, Ward S (1982) Centripetal flow of pseudopodial surface components could propel the amoeboid movement of *Caenorhabditis elegans* spermatozoa. J Cell Biol 92(1):132–138

Robertson S, Lin R (2012) The oocyte-to-embryo transition. Advances in Experimental Medicine and Biology, 757:351–372. (Chap. 12, this volume) Springer, New York

Samuel AD, Murthy VN, Hengartner MO (2001) Calcium dynamics during fertilization in *C. elegans*. BMC Dev Biol 1:8

Sato M, Sato K (2011) Degradation of paternal mitochondria by fertilization-triggered autophagy in *C. elegans* embryos. Science. doi:science.1210333 [pii] 10.1126/science.1210333

Sato M, Grant BD, Harada A, Sato K (2008) Rab11 is required for synchronous secretion of chondroitin proteoglycans after fertilization in *Caenorhabditis elegans*. J Cell Sci 121(Pt 19):3177–3186. doi:jcs.034678 [pii]10.1242/jcs.034678

Schubert CM, Lin R, de Vries CJ, Plasterk RH, Priess JR (2000) MEX-5 and MEX-6 function to establish soma/germline asymmetry in early *C. elegans* embryos. Mol Cell 5(4):671–682. doi:S1097-2765(00)80246-4 [pii]

Seidel HS, Rockman MV, Kruglyak L (2008) Widespread genetic incompatibility in *C. elegans* maintained by balancing selection. Science 319(5863):589–594. doi:1151107 [pii] 10.1126/science.1151107

Seidel HS, Ailion M, Li J, van Oudenaarden A, Rockman MV, Kruglyak L (2011) A novel sperm-delivered toxin causes late-stage embryo lethality and transmission ratio distortion in *C. elegans*. PLoS Biol 9(7):e1001115. doi:10.1371/journal.pbio.1001115 PBIOLOGY-D-10-00587 [pii]

Shakes DC, Ward S (1989) Initiation of spermiogenesis in *C. elegans*: a pharmacological and genetic analysis. Dev Biol 134(1):189–200. doi:0012-1606(89)90088-2 [pii]

Shakes DC, Allen AK, Albert KM, Golden A (2011) *emb-1* encodes the APC16 subunit of the *Caenorhabditis elegans* anaphase-promoting complex. Genetics 189(2):549–560. doi:genetics.111.131714 [pii] 10.1534/genetics.111.131714

Shimabukuro K, Noda N, Stewart M, Roberts TM (2011) Reconstitution of amoeboid motility *in vitro* identifies a motor-independent mechanism for cell body retraction. Curr Biol. doi:S0960-9822(11)00953-5 [pii] 10.1016/j.cub.2011.08.047

Singaravelu G, Singson A (2011) New insights into the mechanism of fertilization in nematodes. Int Rev Cell Mol Biol 289:211–238. doi:B9780123860392000067 [pii] 10.1016/B978-0-12-386039-2.00006-7

Singaravelu G, Chatterjee I, Rahimi S, Druzhinina MK, Kang L, Xu XZ, Singson A (2012) The sperm surface localization of the TRP-3/SPE-41 Ca(2+)-permeable channel depends on SPE-38 function in *Caenorhabditis elegans*. Dev Biol 365(2):376–383

Singson A (2001) Every sperm is sacred: fertilization in *Caenorhabditis elegans*. Dev Biol 230(2):101–109. doi:10.1006/dbio.2000.0118 S0012-1606(00)90118-0 [pii]

Singson A, Mercer KB, L'Hernault SW (1998) The *C. elegans spe-9* gene encodes a sperm transmembrane protein that contains EGF-like repeats and is required for fertilization. Cell 93(1):71–79. doi:S0092-8674(00)81147-2 [pii]

Singson A, Hang JS, Parry JM (2008) Genes required for the common miracle of fertilization in *Caenorhabditis elegans*. Int J Dev Biol 52(5–6):647–656. doi:072512as [pii] 10.1387/ijdb.072512as

Siomos MF, Badrinath A, Pasierbek P, Livingstone D, White J, Glotzer M, Nasmyth K (2001) Separase is required for chromosome segregation during meiosis I in *Caenorhabditis elegans*. Curr Biol 11(23):1825–1835. doi:S0960-9822(01)00588-7 [pii]

Smith H (2006) Sperm motility and MSP. WormBook:1–8. doi:10.1895/wormbook.1.68.1

Soler DC, Kadunganattil S, Ramdas S, Myers K, Roca J, Slaughter T, Pilder SH, Vijayaraghavan S (2009) Expression of transgenic PPP1CC2 in the testis of Ppp 1 cc-null mice rescues spermatid viability and spermiation but does not restore normal sperm tail ultrastructure, sperm motility, or fertility. Biol Reprod 81(2):343–352. doi:biolreprod.109.076398 [pii] 10.1095/biolreprod.109.076398

Stanfield GM, Villeneuve AM (2006) Regulation of sperm activation by SWM-1 is required for reproductive success of *C. elegans* males. Curr Biol 16(3):252–263. doi:S0960-9822(06)01024-4 [pii] 10.1016/j.cub.2005.12.041

Stitzel ML, Seydoux G (2007) Regulation of the oocyte-to-zygote transition. Science 316(5823):407–408. doi:316/5823/407 [pii] 10.1126/science.1138236

Stitzel ML, Pellettieri J, Seydoux G (2006) The *C. elegans* DYRK kinase MBK-2 marks oocyte proteins for degradation in response to meiotic maturation. Curr Biol 16(1):56–62. doi:S0960-9822(05)01473-9 [pii] 10.1016/j.cub.2005.11.063

Stitzel ML, Cheng KC, Seydoux G (2007) Regulation of MBK-2/Dyrk kinase by dynamic cortical anchoring during the oocyte-to-zygote transition. Curr Biol 17(18):1545–1554. doi:S0960-9822(07)01905-7 [pii] 10.1016/j.cub.2007.08.049

Sudhaus W, Kiontke K (2007) Comparison of the cryptic nematode species *Caenorhabditis brenneri sp. n.* and *C. remanei* (Nematoda: Rhabditidae) with stem species pattern of the *Caenorhabditis Elegans* group. Zootaxa 1456:45–62

Sudhaus S, Fricke B, Schneider S, Stachon A, Klein H, von During M, Hasenbring M (2007) [The cortisol awakening response in patients with acute and chronic low back pain. Relations with psychological risk factors of pain chronicity]. Schmerz 21(3):202–204, 206–211. doi:10.1007/s00482-006-0521-4

Tunquist BJ, Schwab MS, Chen LG, Maller JL (2002) The spindle checkpoint kinase bub1 and cyclin e/cdk2 both contribute to the establishment of meiotic metaphase arrest by cytostatic factor. Curr Biol 12(12):1027–1033. doi:S0960982202008941 [pii]

Tsukamoto S, Kuma A, Murakami M, Kishi C, Yamamoto A, Mizushima N (2008) Autophagy is essential for preimplantation development of mouse embryos. Science 321(5885):117–120. doi:321/5885/117 [pii] 10.1126/science.1154822

Varkey JP, Jansma PL, Minniti AN, Ward S (1993) The *Caenorhabditis elegans* spe-6 gene is required for major sperm protein assembly and shows second site non-complementation with an unlinked deficiency. Genetics 133(1):79–86

Varmuza S, Jurisicova A, Okano K, Hudson J, Boekelheide K, Shipp EB (1999) Spermiogenesis is impaired in mice bearing a targeted mutation in the protein phosphatase 1cgamma gene. Dev Biol 205(1):98–110. doi:S0012-1606(98)99100-X [pii] 10.1006/dbio.1998.9100

Veronico P, Gray LJ, Jones JT, Bazzicalupo P, Arbucci S, Cortese MR, Di Vito M, De Giorgi C (2001) Nematode chitin synthases: gene structure, expression and function in *Caenorhabditis elegans* and the plant parasitic nematode *Meloidogyne artiellia*. Mol Genet Genomics 266(1):28–34

Wallenfang MR, Seydoux G (2000) Polarization of the anterior-posterior axis of *C. elegans* is a microtubule-directed process. Nature 408(6808):89–92. doi:10.1038/35040562

Wang T, Jiao Y, Montell C (2005) Dissecting independent channel and scaffolding roles of the *Drosophila* transient receptor potential channel. J Cell Biol 171(4):685–694. doi:jcb.200508030 [pii] 10.1083/jcb.200508030

Ward S, Carrel JS (1979) Fertilization and sperm competition in the nematode *Caenorhabditis elegans*. Dev Biol 73(2):304–321

Ward S, Hogan E, Nelson GA (1983) The initiation of spermiogenesis in the nematode *Caenorhabditis elegans*. Dev Biol 98(1):70–79. doi:0012-1606(83)90336-6 [pii]

Watts JL, Browse J (2002) Genetic dissection of polyunsaturated fatty acid synthesis in *Caenorhabditis elegans*. Proc Natl Acad Sci USA 99(9):5854–5859. doi:10.1073/pnas.092064799 092064799 [pii]

Wessel GM, Brooks JM, Green E, Haley S, Voronina E, Wong J, Zaydfudim V, Conner S (2001) The biology of cortical granules. Int Rev Cytol 209:117–206

Wilson LD, Sackett JM, Mieczkowski BD, Richie AL, Thoemke K, Rumbley JN, Kroft TL (2011) Fertilization in *C. elegans* requires an intact C-terminal RING finger in sperm protein SPE-42. BMC Dev Biol 11:10. doi:1471-213X-11-10 [pii] 10.1186/1471-213X-11-10

Wu JC, Go AC, Samson M, Cintra T, Mirsoian S, Wu TF, Jow MM, Routman EJ, Chu DS (2011) PP1 phosphatases regulate multiple stages of sperm development and motility in *Caenorhabditis elegans*. Genetics. doi:genetics.111.135376 [pii] 10.1534/genetics.111.135376

Xie Z, Klionsky DJ (2007) Autophagosome formation: core machinery and adaptations. Nat Cell Biol 9(10):1102–1109. doi:ncb1007-1102 [pii] 10.1038/ncb1007-1102

Xu XZ, Sternberg PW (2003) A *C. elegans* sperm TRP protein required for sperm-egg interactions during fertilization. Cell 114(3):285–297. doi:S0092867403005658 [pii]

Yamamoto I, Kosinski ME, Greenstein D (2006) Start me up: cell signaling and the journey from oocyte to embryo in *C. elegans*. Dev Dyn 235(3):571–585. doi:10.1002/dvdy.20662

Yi K, Buttery SM, Stewart M, Roberts TM (2007) A Ser/Thr kinase required for membrane-associated assembly of the major sperm protein motility apparatus in the amoeboid sperm of *Ascaris*. Mol Biol Cell 18(5):1816–1825. doi:E06-08-0741 [pii] 10.1091/mbc.E06-08-0741

Yi K, Wang X, Emmett MR, Marshall AG, Stewart M, Roberts TM (2009) Dephosphorylation of major sperm protein (MSP) fiber protein 3 by protein phosphatase 2A during cell body retraction in the MSP-based amoeboid motility of *Ascaris* sperm. Mol Biol Cell 20(14):3200–3208. doi:E09-03-0240 [pii] 10.1091/mbc.E09-03-0240

Zanetti S, Puoti A (2012) Sex determination in the *C. elegans* germline. Advances in Experimental Medicine and Biology 757:41–69. (Chap. 3, this volume) Springer, New York

Zannoni S, L'Hernault SW, Singson AW (2003) Dynamic localization of SPE-9 in sperm: a protein required for sperm-oocyte interactions in *Caenorhabditis elegans*. BMC Dev Biol 3:10. doi:10.1186/1471-213X-3-10 1471-213X-3-10 [pii]

Zhang Y, Foster JM, Nelson LS, Ma D, Carlow CK (2005) The chitin synthase genes *chs-1* and *chs-2* are essential for *C. elegans* development and responsible for chitin deposition in the eggshell and pharynx, respectively. Dev Biol 285(2):330–339. doi:S0012-1606(05)00443-4 [pii]10.1016/j.ydbio.2005.06.037

Chapter 12
The Oocyte-to-Embryo Transition

Scott Robertson and Rueyling Lin

Abstract The oocyte-to-embryo transition refers to the process whereby a fully grown, relatively quiescent oocyte undergoes maturation, fertilization, and is converted into a developmentally active, mitotically dividing embryo, arguably one of the most dramatic transitions in biology. This transition occurs very rapidly in *Caenorhabditis elegans*, with fertilization of a new oocyte occurring every 23 min and the first mitotic division occurring 45 min later. Molecular events regulating this transition must be very precisely timed. This chapter reviews our current understanding of the coordinated temporal regulation of different events during this transition. We divide the oocyte-to-embryo transition into a number of component processes, which are coordinated primarily through the MBK-2 kinase, whose activation is intimately tied to completion of meiosis, and the OMA-1/OMA-2 proteins, whose expression and functions span multiple processes during this transition. The oocyte-to-embryo transition occurs in the absence of de novo transcription, and all the factors required for the process, whether mRNA or protein, are already present within the oocyte. Therefore, all regulation of this transition is posttranscriptional. The combination of asymmetric partitioning of maternal factors, protein modification-mediated functional switching, protein degradation, and highly regulated translational repression ensure a smooth oocyte-to-embryo transition. We will highlight protein degradation and translational repression, two posttranscriptional processes which play particularly critical roles in this transition.

Keywords Oocyte maturation • Oocyte-to-embryo transition • OMA-1 • MBK-2 • Asymmetric partitioning • Protein degradation • Translational repression • ZIF-1 • 3′UTR • RNA binding

---

S. Robertson (✉) • R. Lin
Department of Molecular Biology, University of Texas Southwestern Medical Center, 5323 Harry Hines Blvd, Dallas, TX 75390-9148, USA
e-mail: scott.robertson@utsouthwestern.edu

T. Schedl (ed.), *Germ Cell Development in C. elegans*, Advances in Experimental Medicine and Biology 757, DOI 10.1007/978-1-4614-4015-4_12,

12.1 Introduction

The oocyte-to-embryo transition refers to the process whereby a fully grown, relatively quiescent oocyte undergoes maturation, fertilization, and is converted into a developmentally active, mitotically dividing embryo. These events occur in rapid succession and without any apparent delay in *Caenorhabditis elegans*, suggesting that the molecular events controlling the oocyte-to-embryo transition must be very precisely regulated. The details of oocyte maturation, ovulation, and fertilization are described elsewhere in this issue (Kim et al. 2012, Chap. 10; Marcello et al. 2012, Chap. 11). Our aim in this chapter is not to repeat describing each event occurring during this transition, but instead to focus more on the coordinated temporal regulation of these events. We will discuss selected events associated with oocyte maturation, fertilization, and early embryonic development in order to highlight our current understanding of the complex regulation of this rapid transition and the coordination between processes. We will also take a somewhat "extended" view into embryonic development, up to approximately the 4-cell embryo (Fig. 12.1), in order to incorporate a brief discussion of the transition from maternal-to-zygotic control of development, which we consider the final phase of the oocyte-to-embryo transition.

In *C. elegans*, an oocyte matures, and then is ovulated and fertilized approximately every 23 min in young hermaphrodites (McCarter et al. 1999). Soon after fertilization, the oocyte-derived nucleus completes two rounds of meiotic division and then replicates its haploid genome (Fig. 12.2) (Begasse and Hyman 2011). The oocyte-derived pronucleus fuses with the sperm-derived pronucleus, which has also just replicated its haploid genome, and the resulting nucleus immediately enters metaphase of the first mitotic cycle. Because *C. elegans* oocytes do not undergo an arrest in meiosis II after maturation, a stage equivalent to the vertebrate "egg" does

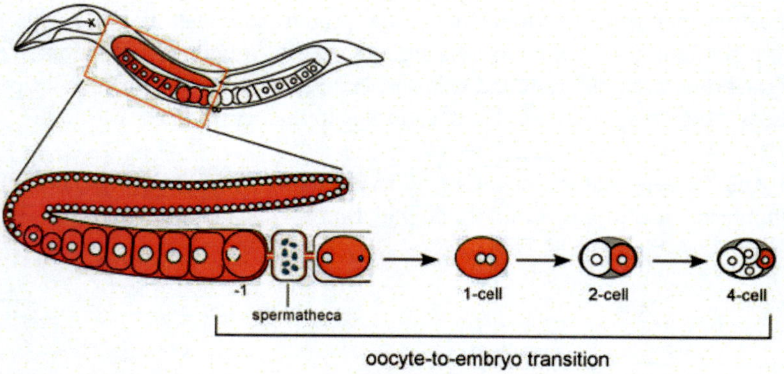

Fig. 12.1 Schematic of the oocyte-to-embryo transition in *C. elegans*. One arm of the bilobed adult gonad is expanded below a cartoon of an adult hermaphrodite. In this chapter, the oocyte-to-embryo transition refers to the conversion of a −1 oocyte to a 4-cell stage embryo. Germline and germline blastomeres are shaded in *red*

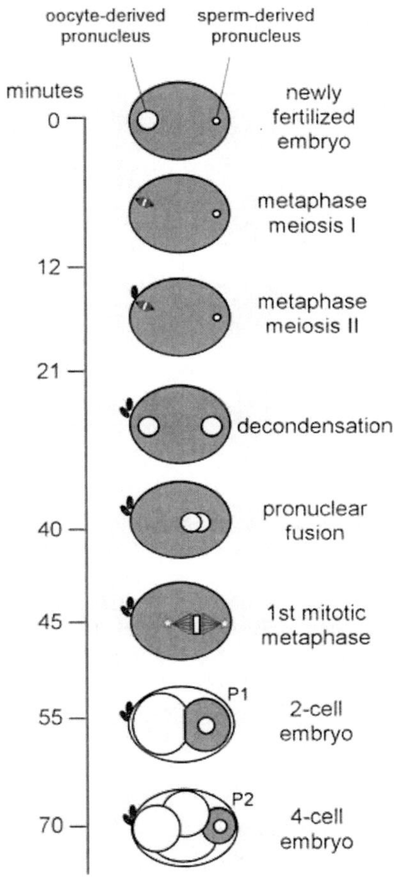

Fig. 12.2 Schematic of various events in newly fertilized *C. elegans* embryos. Morphologically distinct stages between the newly fertilized embryo and the 4-cell embryo are displayed beside a timeline indicating minutes post-fertilization. Germline blastomeres are shaded in *red*. *Small black ovals* = polar bodies; *stars* = centrosomes. Astral microtubules in the first mitotic metaphase embryo are not shown. The time it takes for each step depends on the temperature. Times shown here are at 20–22°C and are derived from Albertson (1984) and McCarter et al. (1999). All embryos are orientated with the anterior to the left in all figures

not exist. Therefore, it is more appropriate to use the term "oocyte-to-embryo," rather than "egg-to-embryo," transition to describe events in this chapter. Within less than 30 min of fertilization, the 1-cell embryo switches from meiotic divisions to mitotic divisions. Certain meiotic spindle-specific proteins are "toxic" for mitotic spindle formation. Therefore, the proper transition from meiosis to mitosis requires precisely timed turnover of meiosis-specific regulators and synthesis of mitosis-specific regulators. This requires the reproducible execution of a number of interconnected processes in precisely the right sequence within a short period of time, necessitating very tight regulation and coordination.

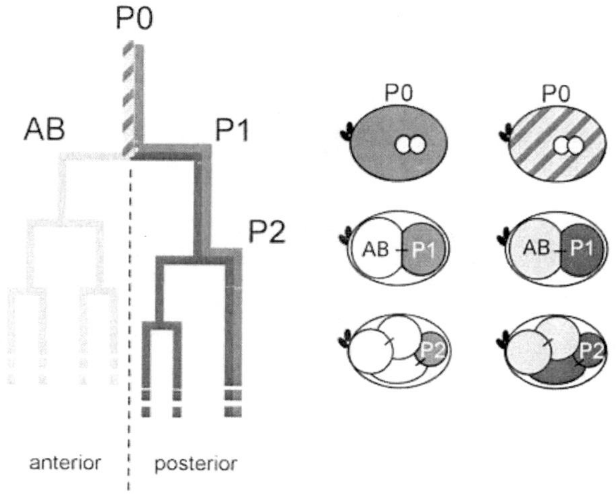

Fig. 12.3 Partitioning of developmental fate during the first two mitotic cycles. *Left*: A lineage diagram of the first few embryonic divisions. Lineage branches expressing anterior proteins are shown in *yellow*, posterior proteins in *blue*, and germline proteins in *red*. Two sets of schematic drawing of early blastomeres are shown on the *right*. The first set highlights the separation of germline blastomeres (*red*) from somatic blastomeres (*white*). The second set highlights the separation of AB-derived, anterior blastomere fates (*yellow*) from the P1-derived, posterior blastomere fates (*blue*). Sister blastomeres are connected by a *short black line*. Germline blastomeres P0, P1, and P2, as well as the anterior blastomere AB are labeled

While zygotic transcription can be detected as early as the 4-cell stage, embryos depleted of the large subunit of RNA polymerase II, AMA-1, exhibit no observable defects in cell divisions until the 28-cell stage (Powell-Coffman et al. 1996). Therefore, maternally provided proteins and RNAs control the characteristic asymmetric early cleavages, orientation of cleavage planes, and lineage-specific timing of early divisions, as well as all events during the oocyte-to-embryo transition. The first mitotic division occurs at about 45 min post fertilization and is asymmetric: it always aligns along the long embryonic axis (the anterior-posterior axis) and gives rise to two daughters of different size, molecular make up, and cell fate (Gönczy and Rose 2005) (Fig. 12.3). The site of sperm entry determines the posterior end (Goldstein and Hird 1996). The sperm provides a cue(s) for the asymmetric localization of cortical polarity proteins (PAR), which asymmetrically localize in complexes at the cortex (Wang and Seydoux 2012, Chap 2). Asymmetric distribution of PAR protein complexes determines the position of the mitotic spindle along the A-P axis, as well as the differential localization of many maternally provided proteins. Among these are key regulators for the specification or differentiation of individual tissues as well as regulators guiding cell division patterns. Following the first asymmetric cell division, the posterior daughter, P1, also divides asymmetrically and gives rise to all body-wall muscles but one, the entire intestine, pharyngeal tissues, and germ cells. The anterior daughter, AB, on the other hand, divides symmetrically and goes on to produce mostly skin and neuronal cells. Mislocalization of these key maternal regulators usually results in abnormal cell specification and embryonic lethality (Draper et al. 1996; Guedes and Priess 1997; Kemphues et al. 1988; Mello et al. 1992; Schubert et al. 2000; Tabara et al. 1999).

The first mitosis is also the first segregation of strictly somatic (AB) versus germline/somatic fate (P1) (Fig. 12.3). The single germline precursor in the *C. elegans* embryo, P4, is specified very early (reviewed by Wang and Seydoux 2012, Chap. 2; Strome 2005; Strome and Lehmann 2007). After four rounds of asymmetric divisions that begin with the 1-cell embryo, with each division generating a germline blastomere and a somatic sister, P4, along with intestinal precursors, moves into the center of the embryo during gastrulation. Following gastrulation, P4 divides one more time, symmetrically, to produce Z2 and Z3, at the ~100-cell stage. Z2 and Z3 do not divide further until halfway through the first larval stage, and will eventually give rise to the ~2,000 germ cells in the adult. As in all animals, primordial germ cells in *C. elegans* are subject to transcriptional repression. This repression begins with the first germline blastomere, the 1-cell embryo (Seydoux et al. 1996). Only after the first mitotic division is a blastomere generated (AB) with strictly somatic developmental fate. Therefore, the 1-cell embryo has to retain germline fate (totipotency), which requires that it be transcriptionally silenced, while simultaneously preparing its somatic daughter for activation of lineage-specific zygotic transcription.

Before the first mitotic division, potent regulators for anterior blastomere fates, posterior blastomere fates, and germline blastomere fates coexist within a common cytoplasm. In fact, these maternally provided regulators, with a few important exceptions, are proteins translated in oocytes from maternally provided mRNAs and deposited into the newly fertilized embryo. The 1-cell embryo is therefore faced with the unique problem of keeping the activity of these potent regulators in check before they are segregated to their appropriate blastomere(s) or lineage. The solution to this problem seems to shape much of how *C. elegans* regulates its oocyte-to-embryo transition.

Molecular events regulating the oocyte-to-embryo transition must be very precisely timed. While we do not have a complete understanding for how these events are coordinated, what has emerged over the last several years is that a relatively small number of key players regulate this process, as well as a clear understanding of the importance of both translational control and protein degradation in regulating this transition. In addition, protein phosphorylation by several maternally supplied kinases plays a pivotal role at multiple points in this transition. These phosphorylation events not only mark several proteins for immediate proteasomal degradation, but also coordinate events by regulating the timing of degradation relative to other parallel processes, such as the cell cycle, during the oocyte-to-embryo transition. Finally, a protein with multiple distinct functions throughout the oocyte-to-embryo transition is switched from one state to another by a specific phosphorylation.

In this chapter, we divide the oocyte-to-embryo transition into three key components: (1) oocyte maturation, ovulation, and fertilization; (2) the transition from meiosis to mitosis: degradation of MEI-1; and (3) transition from a single-cell embryo to a multicell embryo, and we review our current understanding of these processes.

We will emphasize two key factors whose activities are crucial not just for individual processes, but also for coordinating multiple steps within the oocyte-to-embryo transition. In addition, we summarize our current understanding of the roles that protein degradation and translational regulation play during this transition.

12.2 Three Key Components of the Oocyte-to-Embryo Transition

12.2.1 Oocyte Maturation, Ovulation, and Fertilization

These topics are described in considerable detail elsewhere in this volume (Kim et al. 2012, Chap. 10; Marcello et al. 2012, Chap. 11), and we direct readers there for a more complete discussion on these topics. Following induction to undergo maturation by the sperm MSP (major sperm protein) signal, the oocyte immediately adjacent to the spermatheca enters meiotic metaphase I, is ovulated through the spermatheca, and is fertilized. After fertilization, the oocyte-derived nucleus completes both meiotic divisions. Two molecular events occur during this stage that are crucial for a proper oocyte-to-embryo transition. First, MBK-2 kinase, a key coordinator for the oocyte-to-embryo transition (see Sect. 12.3.2), is activated (Cheng et al. 2009; Pellettieri et al. 2003; Stitzel et al. 2006). The activation of MBK-2 is dependent upon the completion of meiosis I, and not sperm entry. Upon activation, MBK-2 phosphorylation of several substrates during meiosis II is critical in coordinating the oocyte-to-embryo transition. Sperm entry does trigger many other events, including the block to polyspermy, cortical vesicle release, calcium fluxes, and the second meiotic division (reviewed in Chap. 11, Marcello et al. 2012). Sperm entry also provides the first polarity cue in the embryo (Goldstein and Hird 1996). *C. elegans* oocytes do not have inherent polarity. It was shown that sperm entry, by destabilizing the actomyosin network in the surrounding cortex, initiates a flow of cortically localized non-muscle myosin and actin. This cortical flow carries other cortical proteins, including some PAR proteins, to the opposite cortex (Munro et al. 2004). Establishment and maintenance of opposing PAR complexes on the cortex (see Wang and Seydoux 2012, Chap. 2) is crucial for a proper oocyte-to-embryo transition.

12.2.2 Transitioning from Meiosis to Mitosis: Degradation of MEI-1 and MEI-2

Meiotic divisions initiate upon oocyte maturation. For each meiotic division, a short acentrosomal meiotic spindle sets up very close to the cortex, enabling the extrusion of the polar bodies with minimal cytoplasm (Fig. 12.2) (Albertson and Thomson 1993).

MEI-1 and MEI-2 form a heterodimeric meiosis-specific katanin and function in oocyte meiotic spindle formation, association of the meiosis I and II spindles with the oocyte cortex, and microtubule severing as part of the process of chromosome segregation during meiotic anaphase (Clark-Maguire and Mains 1994a, b; Mains et al. 1990; Srayko et al. 2000). Subsequently, the much larger mitotic spindle requires the generation of long arrays of astral microtubules from the sperm-supplied centrosomes, which are also critical for asymmetric spindle positioning. Continued presence of MEI-1/2 in the embryo interferes with the formation of the mitotic spindle (Clandinin and Mains 1993; Clark-Maguire and Mains 1994a; Mains et al. 1990). As the first mitotic spindle begins to form within approximately 20 min of the completion of meiosis (Fig. 12.2) (McCarter et al. 1999), MEI-1/2 must be degraded rapidly at the end of meiosis II.

Three levels of regulation ensure the rapid and timely degradation of MEI-1/2 at completion of meiosis. First, the E3 ligase that degrades MEI-1 is not activated until completion of meiosis II. MEI-1 is degraded by a CUL-3-containing E3 ubiquitin ligase that contains the MATH and BTB/POZ domain-containing protein MEL-26 as the substrate-binding component (Dow and Mains 1998; Pintard et al. 2003; Xu et al. 2003). MEL-26 levels are low during meiosis I and II, but increase rapidly following meiosis and remain high throughout the early mitotic divisions (Johnson et al. 2009). Second, a phosphorylation event that marks MEI-1 for degradation is also developmentally regulated. Degradation of MEI-1 requires phosphorylation at serine 92 by the MBK-2 kinase, which itself is activated at meiosis II (more on MBK-2 activation below) (Cheng et al. 2009; Pang et al. 2004; Pellettieri et al. 2003; Quintin et al. 2003; Stitzel et al. 2006). Third, translation of maternal *mei-1* mRNA, which is still present in the early embryo, is actively repressed in order to prevent more MEI-1 from being made. Reduction of *spn-2* function results in mitotic spindle defects due to ectopic MEI-1 expression during embryonic divisions (Li et al. 2009). MEL-26 is present at normal levels in *spn-2* mutant embryos, suggesting that the MEI-1/2 degradation pathway is functional. *spn-2* encodes an eIF4E-binding protein that localizes to the cytoplasm and to P granules. SPN-2 binds to the RNA-binding protein OMA-1, which in turn binds to the *mei-1* 3′UTR (Li et al. 2009). This suggests that SPN-2 and OMA-1 function to negatively regulate translation of *mei-1*.

12.2.3 Transition from a Single-Cell Embryo to a Multi-cell Embryo

Due to its complete dependence upon maternally supplied factors, with many developmental regulators already present as proteins in oocytes (Fig. 12.4), the *C. elegans* 1-cell embryo is faced with something of a developmental conundrum. It is a precursor for both somatic cells and germline blastomeres (Fig. 12.3). In addition, it is a precursor for both anterior blastomeres and posterior blastomeres. The one cell embryo contains key maternal regulators for the specification of somatic lineages,

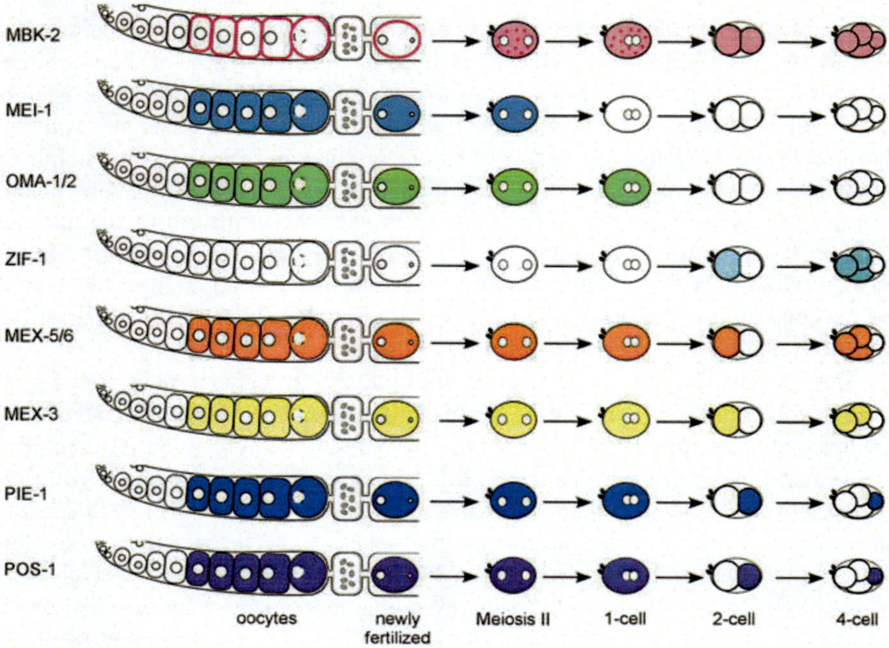

Fig. 12.4 Maternal protein dynamics during the oocyte-to-embryo transition in *C. elegans*. Schematic of germline and early embryonic development to show spatiotemporal dynamics of selected maternally supplied proteins (each shown as a *different shade of gray*). MBK-2 activation begins at completion of meiosis II (cytoplasmic localization plus speckles) and remains activated. MEI-1 is degraded prior to initiation of the first mitosis. OMA proteins are present throughout the 1-cell stage, but are degraded at the first mitosis. ZIF-1 protein is absent from oocytes and the 1-cell embryo, but is present at the 2-cell and 4-cell stage. All other proteins shown are present in oocytes, the 1-cell, 2-cell, and 4-cell embryos. Activated MBK-2 is present in all blastomeres, while ZIF-1 and MEX-5/6 are present in all somatic blastomeres (i.e., absent in the germline blastomere). PIE-1 and POS-1 proteins are restricted to the germline blastomeres, while MEX-3 is restricted to AB-derived blastomeres

germline blastomeres, anterior blastomeres, and posterior blastomeres, all within a common cytoplasm. Only after the first mitotic division are many key regulators for AB-derived cells versus P1 derived cells, and somatic cells versus germline blastomeres asymmetrically segregated. One unique dilemma the *C. elegans* 1-cell embryo faces is how to keep these potent maternal regulators in check.

While many details are still missing, recent studies have shed light on certain strategies that *C. elegans* employs to restrict or exclude certain activities to, or from, the 1-cell embryo, respectively. The most striking is the identification of two proteins, OMA-1 and OMA-2, which are present in the embryo only during the 1-cell stage and are degraded at the first mitosis (Fig. 12.4) (OMA proteins are discussed further in Sect. 12.3.1). This restricted embryonic expression to only the 1-cell embryo is unique, not being observed for any other *C. elegans* protein to date. OMA-1 and OMA-2 encode closely related proteins with tandem CCCH zinc fingers, a motif shown to bind RNA (Detwiler et al. 2001; Lai et al. 1999; Pagano

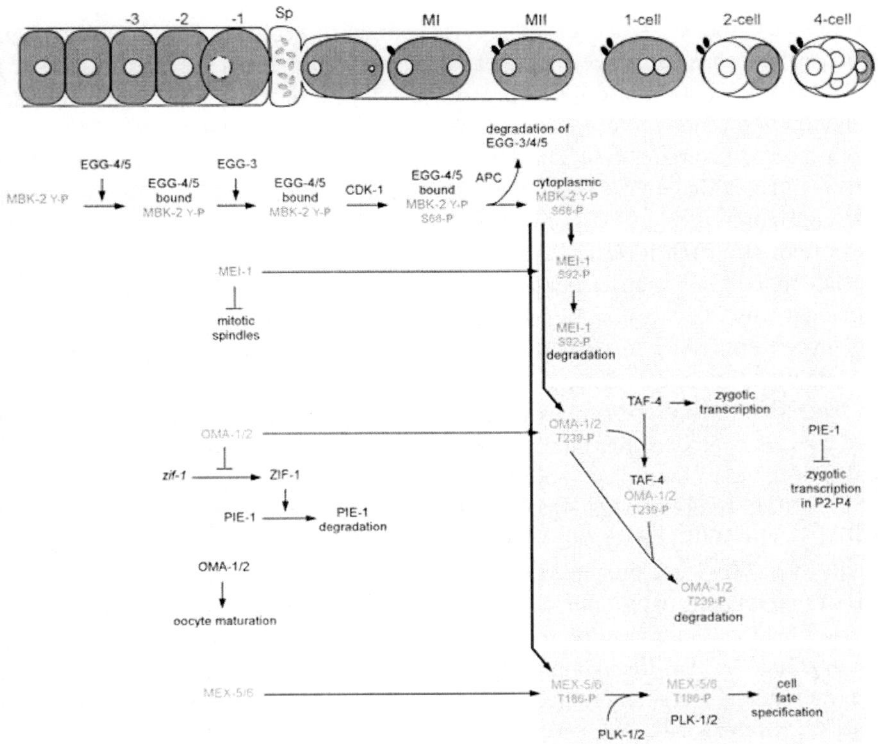

Fig. 12.5 Summary of the oocyte-to-embryo transition in *C. elegans*. Four parallel panels showing the functions of, or regulation of, MBK-2, MEI-1, OMA-1/OMA-2, and MEX-5/MEX-6 are outlined below the schematic of the oocyte-to-embryo transition to highlight their relative timing as well as their regulatory interdependency. A series of phosphorylation events coupled with pseudophosphatase (EGG-3/4/5) interactions and degradation specifies the activation of MBK-2 at the end of meiosis II. MBK-2 is known to phosphorylate three sets of substrates. (1) MEI-1. Phosphorylation at S92 results in its rapid degradation, clearing the way for mitotic-specific spindle components. (2) OMA-1. Phosphorylation at T239 has three consequences. (a) It interferes with the translational repression of *zif-1* by OMA-1. (b) It enhances OMA-1 binding to TAF-4, repressing transcription. (c) It earmarks OMA-1 for degradation after the first mitotic division. In 4-cell embryos, *zif-1* is translated in somatic blastomeres, but not the germline blastomere, resulting in PIE-1 being stable in the germline blastomere and repression of transcription. (3) MEX-5. Phosphorylation at T186 primes MEX-5 for phosphorylation by PLK-1 and -2 and subsequent function in embryos. OMA-2 and MEX-6 have redundant functions with OMA-1 and MEX-5, respectively, and are believed to be regulated in a similar fashion as their corresponding redundant counterpart. Two different shades of *gray* are used to emphasize events prior to (*light gray*) and after (*dark gray*) MBK-2 activation. See text for more details

et al. 2007; Shimada et al. 2002). OMA proteins have indeed been shown to bind to sequences in the 3′UTR of target mRNAs and repress their translation (Guven-Ozkan et al. 2010; Jadhav et al. 2008). OMA proteins have a second function that is completely independent of RNA binding: they bind to an essential component of the transcription initiation complex, TAF-4, sequestering TAF-4 in the cytoplasm, and globally repressing transcription (Fig. 12.5) (Guven-Ozkan et al. 2008). Continued repression of transcription in the subsequent germline blastomeres (P2-P4) is dependent upon another tandem CCCH zinc finger containing protein, PIE-1 (Seydoux et al. 1996). PIE-1 repression of transcription is also independent of RNA binding and does not function like OMA proteins, instead inhibiting other molecular events required for both the initiation and elongation of RNA polymerase II (see Wang and Seydoux 2012, Chap. 2; Batchelder et al. 1999; Ghosh and Seydoux 2008; Zhang et al. 2003). One intriguing facet of PIE-1 function is that PIE-1 protein is made in oocytes and is present at a high level in P0 and P1 (Fig. 12.4) (Mello et al. 1996), but is not sufficient to repress transcription in P0 or P1 (Guven-Ozkan et al. 2008). While it is not clear why PIE-1 is not utilized in P0, or why OMA proteins and a second mechanism are employed to repress transcription in 1-cell embryos, this observation nonetheless highlights the uniqueness of the 1-cell embryo.

Transcriptional repression by OMA proteins or PIE-1 is readily reversible. OMA proteins are degraded after the 1-cell stage, whereas PIE-1 is segregated asymmetrically to the germline blastomere at each P lineage blastomere division (Detwiler et al. 2001; Lin 2003; Mello et al. 1996). However, the asymmetric segregation of PIE-1 is not 100% effective. At each germline blastomere division, whereas the majority of PIE-1 is segregated to the new germline blastomere, there is a portion of PIE-1 that remains in the somatic sister (Reese et al. 2000). Therefore, PIE-1, which could function as a transcriptional repressor in these somatic blastomeres, needs to be degraded so that lineage-appropriate transcription and translation can initiate. Degradation of PIE-1 in somatic cells is carried out by a CUL-2-containing E3 ligase (DeRenzo et al. 2003). The substrate-binding subunit of this E3 ligase, ZIF-1, binds to the first CCCH zinc finger of PIE-1 (DeRenzo et al. 2003). Although *zif-1* RNA is provided maternally to the oocytes and is present throughout the early embryo, ZIF-1 protein is not made until the 4-cell stage, and then only in somatic cells (Fig. 12.4) (Guven-Ozkan et al. 2010). This then ensures stability of PIE-1 in oocytes and germline blastomeres, and PIE-1 degradation in the somatic sisters of the germline blastomeres (discussed further in Sect. 12.3.1).

For maternal regulators deposited into the oocyte as mRNAs, one way to keep them in check in 1-cell embryos is to delay their translation until after the first mitotic division. We have shown recently that translational repression of *zif-1* in 1-cell embryos versus later germline blastomeres is achieved by two separate mechanisms (Oldenbroek et al. 2012). In later germline blastomeres P2-P4, translational repression of *zif-1* is achieved by a germline-blastomere-specific RNA-binding protein. However, in the 1-cell embryo, translational repression of *zif-1* requires the combined action of two other RNA-binding proteins, with neither protein alone being sufficient to repress translation. These two RNA-binding proteins are differentially segregated after the first mitotic division and therefore no longer co-localize within the same cell, effectively restricting their *zif-1* translational repression to only the 1-cell embryo (see also Sect. 12.4.2). The general strategy of delaying translation of maternal mRNAs has also been employed to restrict the expression of a potent transcription factor, SKN-1, primarily to P1 descendants and the Notch receptor, GLP-1, to only the AB descendants (Bowerman et al. 1993; Evans et al. 1994).

12.3 Two Key Coordinators for the Oocyte-to-Embryo Transition

12.3.1 OMA-1 and OMA-2

12.3.1.1 Expression

OMA-1 and OMA-2 are functionally redundant proteins that contain tandem CCCH Tis-11-like zinc fingers, C-x_8-C-x_5-C-x_3-H, a motif shown to function in RNA binding (Detwiler et al. 2001; Lai et al. 1999; Pagano et al. 2007; Shimada et al. 2002). The OMA protein expression pattern is unique, having not been observed for any other *C. elegans* protein (Detwiler et al. 2001; Shimada et al. 2002). OMA mRNA is synthesized maternally and the transcript is found throughout the germline. OMA proteins, which are exclusively cytoplasmic, are expressed in oocytes and reach their peak level in the most proximal −1 oocyte (Fig. 12.4). OMA proteins are also present at a high level in the 1-cell embryo and are then degraded immediately after the first mitotic cycle. OMA proteins are expressed only in developing oocytes and the 1-cell embryo, with no other expression observed in larval or adult animals. OMA proteins are not expressed in the male at any stage of development.

12.3.1.2 Functions

The *oma-1* and *-2* genes were identified as the first genetic mutants defective specifically in the process of oocyte maturation (Detwiler et al. 2001). The two genes encode closely related proteins that are functionally redundant. Whereas mutations in either *oma-1* or *oma-2* alone generate no discernable phenotype, mutants with presumed null alleles of both *oma-1* and *oma-2* are sterile with abnormally large oocytes (Detwiler et al. 2001). These oocytes are arrested at diakinesis in prophase of meiosis I and are not ovulated or fertilized. No sperm defect was detected in *oma-1;oma-2* double mutants. Oocytes in double mutant animals show initial signs of oocyte maturation—that is, partial nuclear envelope breakdown and cortical rearrangement occur—but the maturation process is not completed. Two sperm-dependent molecular events that normally occur in the −1 oocyte, maintenance of activation of the MAP kinase MPK-1, and chromosomal association of the aurora-like kinase AIR-2, do not occur in Oma oocytes (Detwiler et al. 2001). These results suggest that OMA-1 and OMA-2 are required for the response to the sperm signal for oocyte maturation. The prophase arrest of Oma oocytes can be partially suppressed by RNAi of the *C. elegans* MYT-1 homologue, *wee-1.3* (Detwiler et al. 2001). MYT-1 has been shown in other systems to be a negative regulator of maturation promoting factor (MPF) and therefore is a negative regulator of meiotic progression (Burrows et al. 2006; Nakajo et al. 2000; Nebreda and Ferby 2000). This result suggests that OMA-1 and OMA-2 function downstream of MSP to promote prophase progression. The molecular mechanism for the meiotic arrest and oocyte morphology defect in Oma animals, as well as the functional relationship between OMA proteins and other known regulators of oocyte maturation is not known.

In addition to repressing translation of *mei-1* in the 1-cell embryo, OMA-1/2 also function as specific translational repressors in oocytes. Recent studies have demonstrated that OMA-1/2 repress translation of *zif-1* in oocytes via direct binding to the *zif-1* 3′UTR (Fig. 12.5) (Guven-Ozkan et al. 2010). It was shown that repression of *zif-1* by OMA also requires the eIF4E-binding and OMA-binding protein SPN-2, generating a bridge between the *zif-1* 3′UTR and 5′CAP and a circularized, translationally repressed transcript. However, defective translational repression of *zif-1* does not account for the oocyte maturation defective phenotype associated with *oma-1;oma-2* mutant animals. Depletion of *zif-1* in *oma-1;oma-2* animals does not rescue the Oma phenotype. The demonstration that OMA proteins function as translational repressors in oocytes suggests that aberrant translation of a yet to be identified protein or a combination of proteins may underlie the Oma phenotype.

Two OMA protein functions, one in developing and maturing oocytes, where they repress translation of *zif-1* mRNA, and the other in the 1-cell embryo, where they bind and sequester TAF-4 protein in the cytoplasm, are clearly separable genetically and spatiotemporally (Guven-Ozkan et al. 2010). Yet both OMA functions, in end effect, work toward the same goal: maintaining transcriptional repression in germline blastomeres. TAF-4 binding in the 1-cell embryo directly represses global RNA polymerase II transcription, whereas *zif-1* translational repression in oocytes protects PIE-1 (and other maternally supplied ZIF-1 targets) from premature degradation, thereby indirectly promoting transcriptional repression in the later germline lineage blastomeres.

12.3.1.3 Regulation of OMA Proteins

OMA-1 is phosphorylated on T239 by MBK-2 in vitro (Nishi and Lin 2005; Stitzel et al. 2006). Using an antibody to OMA-1 phospho-T239, it was shown that OMA-1 is phosphorylated at meiosis II, precisely the time when MBK-2 is activated (Fig. 12.5, more below) (Nishi and Lin 2005; Pellettieri et al. 2003). Phosphorylation by MBK-2, therefore, could create two populations of OMA proteins that are separated temporally during the oocyte-to-embryo transition: one population not phosphorylated at T239 restricted to oocytes and another population phosphorylated at T239 restricted to the 1-cell embryo. Whereas OMA-1 phosphorylated by the MBK-2 kinase at T239 binds better to TAF-4, translational repression of *zif-1* appears to require OMA-1 unphosphorylated at T239 (Guven-Ozkan et al. 2008, 2010). This means that phosphorylation by MBK-2 serves as a molecular switch that converts OMA function from specific translational repressor to global transcriptional repressor. It was shown that SPN-2 binding to the *zif-1* 3′UTR is not compatible with MBK-2 phosphorylation of OMAs (and possibly additional proteins), which provides a mechanistic explanation for MBK-2-dependent inhibition of *zif-1* translation (Guven-Ozkan et al. 2010). It is interesting that translational repression of *mei-1* by OMA-1 and SPN-2 appears to occur only in the 1-cell embryo and not in oocytes (Li et al. 2009). It is therefore not clear whether or how MBK-2 phosphorylation affects repression of *mei-1* translation by OMA-1 and SPN-2.

Phosphorylation at T239 by MBK-2 also marks OMA-1 for degradation immediately after the first mitotic division (Fig. 12.5) (Nishi and Lin 2005). Mutations that reduce OMA-1 phosphorylation at T239 result in OMA-1 persisting past the 1-cell stage and embryonic lethality (Lin 2003; Nishi and Lin 2005; Shirayama et al. 2006). One such mutation, *oma-1(zu405)*, results in a change of the residue immediately adjacent to T239, P240, to L (Lin 2003). In this mutant, OMA-1 unphosphorylated at T239 persists past the 1-cell stage (Nishi and Lin 2005). This is a complex mutation, owing to the fact that OMA proteins have multiple functions which depend on their phosphorylation state. Therefore, the interpretation of the *oma-1(zu405)* mutation varies depending on which function is being referred to. *oma-1(zu405)* is a reduction-of-function allele with respect to TAF-4-binding but a gain-of-function allele with respect to translational repression of *zif-1,* as persisting unphosphorylated OMA-1 P240L results in ectopic repression of *zif-1* (Guven-Ozkan et al. 2008, 2010). The precise cause for the embryonic lethality associated with *oma-1(zu405)* is not known, but is likely to result from a combination of defects.

12.3.1.4 How OMA Proteins Coordinate the Oocyte-to-Embryo Transition

The importance of OMA proteins in the oocyte-to-embryo transition is underscored by the fact that they are required for all three key components of this transition described above. They are required for oocyte maturation, degradation of the meiotic spindle, and proper transition from a 1-cell to a multicell embryo. The switch from one OMA-1 function to another by MBK-2 phosphorylation at OMA-1T239 is particularly interesting. MBK-2 phosphorylation not only promotes OMA-1 binding to TAF-4 (Guven-Ozkan et al. 2008), but also simultaneously inactivates the *zif-1* translational repression exhibited by OMA-1 in oocytes (Guven-Ozkan et al. 2010). The dual OMA protein functions regulate completely different biochemical processes, and these two functions are switched by a reversible posttranslational modification. The mutual exclusivity of the two OMA functions suggests that both functions must not overlap within the organism, but why should this be the case? There might be a developmental requirement that the first function be completed before the second function initiates, or that the second function must initiate immediately upon termination of the first function. Such a robust functional switch is readily achieved via a single dual-function protein whose modification, which is also stringently timed, terminates the first function while simultaneously activating the second function. Furthermore, phosphorylation of the OMA proteins by MBK-2 not only results in their switch in function, but also marks the proteins for proteasomal degradation (Nishi and Lin 2005). This very effectively delimits the second OMA protein function, sequestration of TAF-4, to the 1-cell embryo only, without invoking, for example, an additional phosphatase to squelch this function. OMA proteins, in collaboration with MBK-2 (below), play key roles in orchestrating the oocyte-to-embryo transition in *C. elegans*.

12.3.2 MBK-2

12.3.2.1 Expression and Activation

MBK-2 kinase, which is activated at meiosis II, is required for degradation of a subclass of maternal proteins following fertilization, posterior localization of germline lineage factors, and activation of polarity factors acting downstream of the initial PAR protein polarity cues (see further below) (Nishi and Lin 2005; Nishi et al. 2008; Pang et al. 2004; Pellettieri et al. 2003; Quintin et al. 2003).

MBK-2 kinase is a dual-specificity kinase belonging to the DYRK family. DYRK kinases undergo co-translational self-phosphorylation, when an intermediate form, present only during translation, phosphorylates itself on tyrosine within the YTY motif present in the kinase domain (Lochhead et al. 2005). This self-phosphorylation is required for enzymatic activity. Mature DYRKs, however, are no longer capable of tyrosine phosphorylation and only phosphorylate serine and threonine residues within their targets (Lochhead et al. 2005). MBK-2 protein is maternally supplied and is present at a high level in oocytes and embryos (Pellettieri et al. 2003). However, MBK-2 kinase activation is very tightly regulated during the oocyte-to-embryo transition, with phosphorylated substrates first detectable in 1-cell embryos in anaphase of MI and peaking following the completion of meiosis (extrusion of the second polar body) (Nishi and Lin 2005; Stitzel et al. 2006).

MBK-2 is initially localized at the cortex prior to the sperm signal that triggers oocyte maturation and the completion of meiosis I, whereupon MBK-2 initiates relocalization to puncta within the cytoplasm (Fig. 12.4) (Pellettieri et al. 2003; Stitzel et al. 2006). This cytoplasmic relocalization continues through meiosis II, which is triggered by fertilization. Recent studies have elegantly shown how MBK-2 activity is regulated during this transition, and how this regulation is intimately associated with cell cycle control. This process depends upon meiotic cell cycle regulators CDK-1 and APC/C, along with a small set of pseudo-tyrosinephosphatases (Fig. 12.5) (Cheng et al. 2009; Maruyama et al. 2007; Parry and Singson 2011; Parry et al. 2009). Pseudophosphatases are proteins that resemble phosphatases and bind specific phosphorylated motifs found within their targets, but lack enzymatic activity. It is believed that one of the functions of pseudophosphatases is to shield phosphorylated proteins from the action of bone fide phosphatases.

MBK-2, as a result of the autophosphorylation on tyrosine within the kinase catalytic domain, is "activated" immediately following translation. Furthermore, MBK-2 has been produced prior to when it is required to be functional, and therefore it must be "held in check" or restrained from phosphorylating its substrates, which are also present in the oocytes. At the same time, phosphorylated MBK-2 must be protected from any endogenous tyrosine phosphatase activity present in the oocyte. This is achieved via the binding of two pseudo tyrosine phosphatases, EGG-4 and -5, which bind to MBK-2 by interacting with the tyrosine-phosphorylated YTY motif (Cheng et al. 2009; Parry et al. 2009). Complexes of MBK-2 bound by EGG-4/5 are bound by a third pseudo tyrosinephosphatase, EGG-3, which also functions to tether the complex to the oocyte cortex (Cheng et al. 2009; Maruyama et al. 2007; Parry et al. 2009). It has been proposed that another protein also interacts with this complex, repressing MBK-2 activity by an unknown mechanism (Cheng et al. 2009). MBK-2 is phosphorylated on serine 68 by CDK-1, the kinase that drives oocyte maturation and the meiotic divisions, and this phosphorylation is thought to result in dissociation of the unknown repressor (Cheng et al. 2009). However, MBK-2 remains bound and inactivated by EGG-3/4/5 and localized to the cortex away from its cytoplasmic targets. APC, the anaphase-promoting complex required for the transition from metaphase to anaphase in MI, stimulates release of the MBK-2/EGG-3/4/5 complex and subsequent proteasomal degradation of EGG-3 and EGG-4/5 (Cheng et al. 2009; Maruyama et al. 2007; Parry et al. 2009). All three tyrosine pseudophosphatases contain putative RxxL destruction boxes. Active MBK-2, phosphorylated at the YTY motif in the catalytic domain as well as at S68, is released into the cytoplasm where it can phosphorylate its targets.

12.3.2.2 Substrates

Two in vivo substrates of MBK-2 have been identified: OMA-1 and MEI-1 (Fig. 12.5) (Nishi and Lin 2005; Stitzel et al. 2006). Using phospho-specific antibodies, MBK-2 phosphorylation of both proteins was first detected immediately following the proposed activation/cytoplasmic translocation of MBK-2. Phosphorylation by MBK-2 is required for the developmentally regulated degradation of MEI-1 and OMA-1 (Nishi and Lin 2005; Shirayama et al. 2006; Stitzel et al. 2006). As mentioned earlier, failure to degrade either MEI-1 or OMA-1 results in embryonic lethality (Clandinin and Mains 1993; Lin 2003; Mains et al. 1990). In addition, MBK-2 phosphorylation of OMA-1 at T239 serves as a molecular switch for OMA function from *zif-1* translational repression to TAF-4 binding and transcriptional repression (Guven-Ozkan et al. 2010). MBK-2 activation is complex in large part because it needs to be very precisely timed. Precocious activation of MBK-2 could lead to precocious degradation of MEI-1, OMA-1/2, or mistimed conversion of OMA functions. Whereas MBK-2 activity needs to be kept in check in oocytes and during the completion of meiosis, it needs to be activated following the completion of meiosis and before the first mitosis. There is only approximately 20 min between the completion of meiosis II and first mitosis.

Two additional proteins, MEX-5 and MEX-6, are also likely to be MBK-2 substrates (Fig. 12.5), although there is no direct in vivo evidence as yet (Nishi et al. 2008). MEX-5 and MEX-6 are closely related CCCH zinc finger proteins that, in response to PAR asymmetry, localize toward the anterior cytoplasm of the 1-cell embryo, and preferentially localize to the somatic daughters following germline blastomere (P lineage) divisions (Schubert et al. 2000). MEX-5/6 function ensures that germline proteins, such as PIE-1, POS-1 and MEX-1, are localized to the germline blastomeres. Degradation of ZIF-1 substrates in somatic blastomeres is dependent on MEX-5/6 (DeRenzo et al. 2003).

The polo kinases PLK-1 and -2 promote MEX-5/6 activity. Polo kinases require a polo-docking site on their substrates that has been primed by phosphorylation by another kinase [reviewed in (Archambault and Glover 2009)]. MBK-2 phosphorylates MEX-5 at T186 within a polo-docking site, and this modification has been shown to prime MEX-5 for subsequent PLK-1/2 phosphorylation and activation (Nishi et al. 2008).

12.4 Posttranscriptional Regulation of the Oocyte-to-Embryo Transition

The oocyte-to-embryo transition occurs in the absence of de novo transcription, and all factors required for the process, whether mRNA or protein, are already present within the oocyte. Therefore, all regulation of the transition is posttranscriptional. Two posttranscriptional processes, in particular, play a significant role in the oocyte-to-embryo transition: protein degradation and translational repression (Bowerman and Kurz 2006; Evans and Hunter 2005; Seydoux 1996).

12.4.1 Protein Degradation

The role that protein degradation plays in the regulation of OMA1/2 and MEI-1 function has already been discussed. In addition, protein degradation is critical for the following processes during the oocyte-to-embryo transition.

As in mitotic cell cycles, protein degradation plays a crucial role during meiotic divisions [reviewed in (Bowerman and Kurz 2006; DeRenzo and Seydoux 2004; Pesin and Orr-Weaver 2008; Peters 2002; Stitzel et al. 2007; Verlhac et al. 2010)]. Cyclin-dependent kinase (CDK) activities, determined by association with regulatory cyclin subunits and modified by positive and negative regulation, drive the cell cycle. Entry and progression through meiosis require high CDK activity associated with interaction with A- and B-type cyclins. CDK activity peaks at metaphase with all chromosomes attached to the spindle and aligned at the metaphase plate. The transition from metaphase to anaphase requires the E3 ubiquitin ligase anaphase-promoting complex. APC initiates (1) chromosome segregation through proteolytic degradation of securin, which leads to activation of separase resulting in cleavage of cohesin, and (2) destruction of B-type cyclins, which leads to downregulation of CDK activity and eventual exit from meiosis [reviewed in (Pesin and Orr-Weaver 2008; Peters 2002)]. In addition, APC regulates degradation of EGG-3, EGG-4, and EGG-5, which is key to the timing of MBK-2 activation (Cheng et al. 2009; Stitzel et al. 2007).

The asymmetric segregation of PIE-1, MEX-1, and POS-1, three germline blastomere-specific CCCH finger proteins, is, in part, regulated by the degradation of these proteins in non-germ cell precursors (Reese et al. 2000). Degradation of all three proteins is dependent upon ZIF-1 and MEX-5/6 (DeRenzo et al. 2003).

12.4.2 Translational Repression

Recent studies have shown that expression of the majority of maternally supplied proteins in *C. elegans* is regulated by the corresponding 3′UTR sequence (Merritt et al. 2008). In most cases, the 3′UTR is sufficient to confer the correct temporal and spatial expression pattern to a reporter protein. In addition, many of the key regulators, identified either molecularly or genetically, that regulate this transition contain an RNA-binding motif (Detwiler et al. 2001; Draper et al. 1996; Guedes and Priess 1997; Li et al. 2009; Mello et al. 1996; Ogura et al. 2003; Schubert et al. 2000; Tabara et al. 1999). Translational repression appears to be the most widely used and important among all posttranscriptional regulatory mechanisms in the *C. elegans* germline. Regulation of translational repression during the mitosis-to-meiosis transition in the adult germline and during the various stages of meiotic progression will not be repeated here (see Nousch and Eckmann 2012, Chap. 8). We highlight below some key instances of translational repression during the oocyte-to-embryo transition.

In the distal arm of the gonad, *oma-1* translation is repressed by GLD-1, a STAR domain RNA-binding protein (see Hansen and Schedl 2012, Chap. 4; Lee and Schedl 2001). GLD-1 levels drop sharply as oocytes progress from pachytene to diplotene, and are undetectable in diakinesis oocytes in the germline proximal region (Jones et al. 1996). Since *gld-1* mRNA remains abundant, the drastic disappearance of GLD-1 is likely a result of translational repression and rapid protein degradation. *oma-1* and *oma-2* transcripts, along with several other (but not all) GLD-1 targets, are relieved of GLD-1 translational repression and their protein levels increase (Lee and Schedl 2001).

Translation of OMA-1 and -2 in proximal oocytes, along with a number of other RNA-binding proteins, results in the repression of certain maternal transcripts that should not be translated until after fertilization. One example, already discussed, is the repression of *zif-1* translation (Guven-Ozkan et al. 2010). Similarly, maternally supplied *nos-2* mRNA is only translated in the P4 blastomere. Translational repression of *nos-2* in oocytes also depends on OMA-1 and OMA-2 (Jadhav et al. 2008).

As noted earlier, *zif-1* translation needs to be continuously repressed in 1-cell embryos in order to maintain a high level of PIE-1 (Guven-Ozkan et al. 2010). After the first mitotic division, *zif-1* translational repression can be relieved in AB, the first blastomere with strictly somatic developmental fate, but must remain repressed in P1, the germline blastomere. In the 1-cell embryo, OMA proteins have been phosphorylated by MBK-2 and presumably no longer repress *zif-1* translation (Guven-Ozkan et al. 2010; Nishi and Lin 2005). However, *zif-1* remains translationally repressed. Our recent studies show that embryos employ a simple but clever way to maintain repression of *zif-1* translation in 1-cell embryos (Oldenbroek et al. 2012). Two RNA-binding proteins, MEX-3 and SPN-4, function together to repress translation of *zif-1*. What makes this regulation elegant is that neither MEX-3 nor SPN-4 alone is able to repress translation of *zif-1*, and both proteins are only co-expressed in the 1-cell and (briefly) 2-cell embryo (Fig. 12.4), thereby limiting repression of *zif-1* by MEX-3 and SPN-4 only to these very early stages. After the 2-cell stage, an additional set of RNA-binding proteins restrict the translation of *zif-1* only to somatic blastomeres, while maintaining translational repression in the germline lineage (Oldenbroek et al. 2012). The separate modes of translational repression for *zif-1* in 1-cell embryos and in later germline blastomeres further highlights the uniqueness of the 1-cell embryo as described earlier (Sect. 12.2.3).

12.5 Concluding Remarks

The oocyte-to-embryo transition in *C. elegans* is a prime example of how the power of genetics, cell biology, and biochemistry can be brought to bear upon a highly complex developmental process in this model organism. While genetic screens led to the isolation of mutations defective in individual processes, cell biological and biochemical analyses allow us to look at how these separate events are coordinated. Much of this progress in *C. elegans* relies upon the fact that all of these processes occur in a highly ordered linear sequence within a short time span, and can be observed live in the transparent adult hermaphrodite.

The degree to which these component processes are coordinately regulated and timed during this transition is remarkable—for example, how MBK-2 activation, so tightly coordinated with the completion of meiosis, sets in motion various pathways critical to the oocyte-to-embryo transition, or how the OMA proteins are switched so dramatically in function as a result of MBK-2 phosphorylation, which also sets them ultimately on the path to destruction. The complicated interplay between maternal regulatory factors over both space and time is quite amazing. In one example, an activated kinase is held in check until precisely the right moment in the oocyte-to-embryo transition, and in another example, one protein functions in the oocyte to ensure that another maternally supplied protein is protected from degradation, and then both proteins function later in the embryo to sequentially maintain transcriptional quiescence in germline precursors. As each individual facet of the oocyte-to-embryo transition is currently the subject of intense investigation, there is little doubt that this critical developmental transition will be dissected in ever-greater detail in the near future.

How the oocyte-to-embryo transition is coordinated in mammals is less well understood. Molecular events that take place in individual processes, such as oocyte maturation, fertilization, and meiotic divisions, are very similar in outline with minor differences between *C. elegans* and mammals. However, there are three significant differences between worms and mammals regarding the oocyte-to-embryo transition. First, the time span it takes for the transition to occur differs greatly. In mammals, oocytes remain arrested in prophase of meiosis I for extended periods, eventually being triggered to resume meiosis and be ovulated by a hormonal signal. The ovulated oocytes arrest again at metaphase of meiosis II and complete meiosis if fertilized. Because of the arrest before fertilization for mammalian eggs, precisely timed tight coordination prior to fertilization is not as critical. Second, the dependence on proteins prepackaged into the oocytes differs.

In mammals, zygotic transcription can start as early as the 1-cell embryo, and cell fate specification begins significantly later than in the *C. elegans* embryo. Therefore, early embryonic development in mammals is not as dependent on maternal factors in the oocyte. Third, the early cleavage patterns and the mechanism by which primordial germ cells are specified differ. The first embryonic divisions in mammals are symmetrical and blastomeres remain totipotent up to the blastocyst stage. In the mouse, primordial germ cells are not specified until much later, around embryonic day 6.5, when pluripotent epiblast cells respond to signals from neighboring extra-embryonic tissue. In addition, the mechanism by which a pool of primordial germ cells in the mouse are specified is quite different from the very early specification of the single germline precursor in worms. These differences suggest that coordination of the oocyte-to-embryo transition is likely to be regulated quite differently between mammals and *C. elegans*, even if component processes share considerable similarity.

Acknowledgments Work conducted in the authors' laboratory was supported by NIH grants HD37933 and GM84198.

References

Albertson DG (1984) Formation of the first cleavage spindle in nematode embryos. Dev Biol 101(1):61–72

Albertson DG, Thomson JN (1993) Segregation of holocentric chromosomes at meiosis in the nematode, *Caenorhabditis elegans*. Chromosome Res 1(1):15–26

Archambault V, Glover DM (2009) Polo-like kinases: conservation and divergence in their functions and regulation. Nat Rev Mol Cell Biol 10(4):265–275

Batchelder C, Dunn MA, Choy B, Suh Y, Cassie C, Shim EY, Shin TH, Mello C, Seydoux G, Blackwell TK (1999) Transcriptional repression by the *Caenorhabditis elegans* germ-line protein PIE-1. Genes Dev 13(2):202–212

Begasse ML, Hyman AA (2011) The first cell cycle of the *Caenorhabditis elegans* embryo: spatial and temporal control of an asymmetric cell division. Results Probl Cell Differ 53:109–133

Bowerman B, Kurz T (2006) Degrade to create: developmental requirements for ubiquitin-mediated proteolysis during early *C. elegans* embryogenesis. Development 133(5):773–784

Bowerman B, Draper BW, Mello CC, Priess JR (1993) The maternal gene *skn-1* encodes a protein that is distributed unequally in early *C. elegans* embryos. Cell 74(3):443–452

Burrows AE, Sceurman BK, Kosinski ME, Richie CT, Sadler PL, Schumacher JM, Golden A (2006) The *C. elegans* Myt1 ortholog is required for the proper timing of oocyte maturation. Development 133(4):697–709

Cheng KC, Klancer R, Singson A, Seydoux G (2009) Regulation of MBK-2/DYRK by CDK-1 and the pseudophosphatases EGG-4 and EGG-5 during the oocyte-to-embryo transition. Cell 139(3):560–572

Clandinin TR, Mains PE (1993) Genetic studies of *mei-1* gene activity during the transition from meiosis to mitosis in *Caenorhabditis elegans*. Genetics 134(1):199–210

Clark-Maguire S, Mains PE (1994a) Localization of the *mei-1* gene product of *Caenorhaditis elegans*, a meiotic-specific spindle component. J Cell Biol 126(1):199–209

Clark-Maguire S, Mains PE (1994b) *mei-1*, a gene required for meiotic spindle formation in *Caenorhabditis elegans*, is a member of a family of ATPases. Genetics 136(2):533–546

DeRenzo C, Seydoux G (2004) A clean start: degradation of maternal proteins at the oocyte-to-embryo transition. Trends Cell Biol 14(8):420–426

DeRenzo C, Reese KJ, Seydoux G (2003) Exclusion of germ plasm proteins from somatic lineages by cullin-dependent degradation. Nature 424(6949):685–689

Detwiler MR, Reuben M, Li X, Rogers E, Lin R (2001) Two zinc finger proteins, OMA-1 and OMA-2, are redundantly required for oocyte maturation in *C. elegans*. Dev Cell 1(2):187–199

Dow MR, Mains PE (1998) Genetic and molecular characterization of the *Caenorhabditis elegans* gene, *mel-26*, a postmeiotic negative regulator of *mei-1*, a meiotic-specific spindle component. Genetics 150(1):119–128

Draper BW, Mello CC, Bowerman B, Hardin J, Priess JR (1996) MEX-3 is a KH domain protein that regulates blastomere identity in early *C. elegans* embryos. Cell 87(2):205–216

Evans TC, Hunter CP (2005) Translational control of maternal RNAs. WormBook, ed. The *C. elegans* Research Community, WormBook. doi:/10.1895/wormbook.1.34.1, http://www.wormbook.org.

Evans TC, Crittenden SL, Kodoyianni V, Kimble J (1994) Translational control of maternal *glp-1* mRNA establishes an asymmetry in the *C. elegans* embryo. Cell 77(2):183–194

Ghosh D, Seydoux G (2008) Inhibition of transcription by the *Caenorhabditis elegans* germline protein PIE-1: genetic evidence for distinct mechanisms targeting initiation and elongation. Genetics 178(1):235–243

Goldstein B, Hird SN (1996) Specification of the anteroposterior axis in *Caenorhabditis elegans*. Development 122(5):1467–1474

Gönczy P, Rose LS (2005) Asymmetric cell division and axis formation in the embryo. WormBook, ed. The *C. elegans* Research Community, WormBook. doi:/10.1895/wormbook.1.30.1, http://www.wormbook.org

Guedes S, Priess JR (1997) The *C. elegans* MEX-1 protein is present in germline blastomeres and is a P granule component. Development 124(3):731–739

Guven-Ozkan T, Nishi Y, Robertson SM, Lin R (2008) Global transcriptional repression in *C. elegans* germline precursors by regulated sequestration of TAF-4. Cell 135:149–160

Guven-Ozkan T, Robertson SM, Nishi Y, Lin R (2010) *zif-1* translational repression defines a second, mutually exclusive OMA function in germline transcriptional repression. Development 137(20):3373–3382

Hansen D, Schedl T (2012) Stem cell proliferation versus meiotic fate decision in *C. elegans*. Advances in Experimental Medicine and Biology 757:71–99. (Chap. 4, this volume) Springer, New York

Jadhav S, Rana M, Subramaniam K (2008) Multiple maternal proteins coordinate to restrict the translation of *C. elegans* nanos-2 to primordial germ cells. Development 135(10):1803–1812

Johnson JL, Lu C, Raharjo E, McNally K, McNally FJ, Mains PE (2009) Levels of the ubiquitin ligase substrate adaptor MEL-26 are inversely correlated with MEI-1/katanin microtubule-severing activity during both meiosis and mitosis. Dev Biol 330(2):349–357

Jones AR, Francis R, Schedl T (1996) GLD-1, a cytoplasmic protein essential for oocyte differentiation, shows stage- and sex-specific expression during *Caenorhabditis elegans* germline development. Dev Biol 180(1):165–183

Kemphues KJ, Priess JR, Morton DG, Cheng NS (1988) Identification of genes required for cytoplasmic localization in early *C. elegans* embryos. Cell 52(3):311–320

Kim S, Spike CA, Greenstein D (2012) Control of oocyte growth and meiotic maturation in *C. elegans*. Advances in Experimental Medicine and Biology 757:277–320. (Chap. 10, this volume) Springer, New York

Lai WS, Carballo E, Strum JR, Kennington EA, Phillips RS, Blackshear PJ (1999) Evidence that tristetraprolin binds to AU-rich elements and promotes the deadenylation and destabilization of tumor necrosis factor alpha mRNA. Mol Cell Biol 19(6):4311–4323

Lee MH, Schedl T (2001) Identification of in vivo mRNA targets of GLD-1, a maxi-KH motif containing protein required for *C. elegans* germ cell development. Genes Dev 15(18):2408–2420

Li W, DeBella LR, Guven-Ozkan T, Lin R, Rose LS (2009) An eIF4E-binding protein regulates katanin protein levels in *C. elegans* embryos. J Cell Biol 187(1):33–42

Lin R (2003) A gain-of-function mutation in *oma-1*, a *C. elegans* gene required for oocyte maturation, results in delayed degradation of maternal proteins and embryonic lethality. Dev Biol 258(1):226–239

Lochhead PA, Sibbet G, Morrice N, Cleghon V (2005) Activation-loop autophosphorylation is mediated by a novel transitional intermediate form of DYRKs. Cell 121(6):925–936

Mains PE, Kemphues KJ, Sprunger SA, Sulston IA, Wood WB (1990) Mutations affecting the meiotic and mitotic divisions of the early *Caenorhabditis elegans* embryo. Genetics 126(3):593–605

Marcello MR, Singaravelu G, Singson A (2012) Fertilization. Advances in Experimental Medicine and Biology 757:321–350. (Chap. 11, this volume) Springer, New York

Maruyama R, Velarde NV, Klancer R, Gordon S, Kadandale P, Parry JM, Hang JS, Rubin J, Stewart-Michaelis A, Schweinsberg P, Grant BD, Piano F, Sugimoto A, Singson A (2007) EGG-3 regulates cell-surface and cortex rearrangements during egg activation in *Caenorhabditis elegans*. Curr Biol 17(18):1555–1560

McCarter J, Bartlett B, Dang T, Schedl T (1999) On the control of oocyte meiotic maturation and ovulation in *Caenorhabditis elegans*. Dev Biol 205(1):111–128

Mello CC, Draper BW, Krause M, Weintraub H, Priess JR (1992) The *pie-1* and *mex-1* genes and maternal control of blastomere identity in early *C. elegans* embryos. Cell 70(1):163–176

Mello CC, Schubert C, Draper B, Zhang W, Lobel R, Priess JR (1996) The PIE-1 protein and germline specification in *C. elegans* embryos. Nature 382(6593):710–712

Merritt C, Rasoloson D, Ko D, Seydoux G (2008) 3' UTRs are the primary regulators of gene expression in the *C. elegans* germline. Curr Biol 18(19):1476–1482

Munro E, Nance J, Priess JR (2004) Cortical flows powered by asymmetrical contraction transport PAR proteins to establish and maintain anterior-posterior polarity in the early *C. elegans* embryo. Dev Cell 7(3):413–424

Nakajo N, Yoshitome S, Iwashita J, Iida M, Uto K, Ueno S, Okamoto K, Sagata N (2000) Absence of Wee1 ensures the meiotic cell cycle in *Xenopus* oocytes. Genes Dev 14(3):328–338

Nebreda AR, Ferby I (2000) Regulation of the meiotic cell cycle in oocytes. Curr Opin Cell Biol 12(6):666–675

Nishi Y, Lin R (2005) DYRK2 and GSK-3 phosphorylate and promote the timely degradation of OMA-1, a key regulator of the oocyte-to-embryo transition in *C. elegans*. Dev Biol 288(1):139–149

Nishi Y, Rogers E, Robertson SM, Lin R (2008) Polo kinases regulate *C. elegans* embryonic polarity via binding to DYRK2-primed MEX-5 and MEX-6. Development 135(4):687–697

Nousch M, Eckmann CR (2012) Translational control in the *C. elegans* germ line. Advances in Experimental Medicine and Biology 757:205–247. (Chap. 8, this volume) Springer, New York

Ogura K, Kishimoto N, Mitani S, Gengyo-Ando K, Kohara Y (2003) Translational control of maternal *glp-1* mRNA by POS-1 and its interacting protein SPN-4 in *Caenorhabditis elegans*. Development 130(11):2495–2503

Oldenbroek M, Robertson SM, Guven-Ozkan T, Gore S, Nishi Y, Lin R (2012) Multiple RNA-binding proteins function combinatorially to control the soma-restricted expression pattern of the E3 ligase subunit ZIF-1. Dev Biol 363(2):388–398

Pagano JM, Farley BM, McCoig LM, Ryder SP (2007) Molecular basis of RNA recognition by the embryonic polarity determinant MEX-5. J Biol Chem 282(12):8883–8894

Pang KM, Ishidate T, Nakamura K, Shirayama M, Trzepacz C, Schubert CM, Priess JR, Mello CC (2004) The minibrain kinase homolog, *mbk-2*, is required for spindle positioning and asymmetric cell division in early *C. elegans* embryos. Dev Biol 265(1):127–139

Parry JM, Singson A (2011) EGG molecules couple the oocyte-to-embryo transition with cell cycle progression. Results Probl Cell Differ 53:135–151

Parry JM, Velarde NV, Lefkovith AJ, Zegarek MH, Hang JS, Ohm J, Klancer R, Maruyama R, Druzhinina MK, Grant BD, Piano F, Singson A (2009) EGG-4 and EGG-5 link events of the oocyte-to-embryo transition with meiotic progression in *C. elegans*. Curr Biol 19(20):1752–1757

Pellettieri J, Reinke V, Kim SK, Seydoux G (2003) Coordinate activation of maternal protein degradation during the egg-to-embryo transition in *C. elegans*. Dev Cell 5(3):451–462

Pesin JA, Orr-Weaver TL (2008) Regulation of APC/C activators in mitosis and meiosis. Annu Rev Cell Dev Biol 24:475–499

Peters JM (2002) The anaphase-promoting complex: proteolysis in mitosis and beyond. Mol Cell 9(5):931–943

Pintard L, Willis JH, Willems A, Johnson JL, Srayko M, Kurz T, Glaser S, Mains PE, Tyers M, Bowerman B, Peter M (2003) The BTB protein MEL-26 is a substrate-specific adaptor of the CUL-3 ubiquitin-ligase. Nature 425(6955):311–316

Powell-Coffman JA, Knight J, Wood WB (1996) Onset of *C. elegans* gastrulation is blocked by inhibition of embryonic transcription with an RNA polymerase antisense RNA. Dev Biol 178(2):472–483

Quintin S, Mains PE, Zinke A, Hyman AA (2003) The *mbk-2* kinase is required for inactivation of MEI-1/katanin in the one-cell *Caenorhabditis elegans* embryo. EMBO Rep 4(12):1175–1181

Reese KJ, Dunn MA, Waddle JA, Seydoux G (2000) Asymmetric segregation of PIE-1 in *C. elegans* is mediated by two complementary mechanisms that act through separate PIE-1 protein domains. Mol Cell 6(2):445–455

Schubert CM, Lin R, de Vries CJ, Plasterk RH, Priess JR (2000) MEX-5 and MEX-6 function to establish soma/germline asymmetry in early *C. elegans* embryos. Mol Cell 5(4):671–682

Seydoux G (1996) Mechanisms of translational control in early development. Curr Opin Genet Dev 6(5):555–561

Seydoux G, Mello CC, Pettitt J, Wood WB, Priess JR, Fire A (1996) Repression of gene expression in the embryonic germ lineage of *C. elegans*. Nature 382(6593):713–716

Shimada M, Kawahara H, Doi H (2002) Novel family of CCCH-type zinc-finger proteins, MOE-1, -2 and –3, participates in *C. elegans* oocyte maturation. Genes Cells 7(9):933–947

Shirayama M, Soto MC, Ishidate T, Kim S, Nakamura K, Bei Y, van den Heuvel S, Mello CC (2006) The conserved kinases CDK-1, GSK-3, KIN-19, and MBK-2 promote OMA-1 destruction to regulate the oocyte-to-embryo transition in *C. elegans*. Curr Biol 16(1):47–55

Srayko M, Buster DW, Bazirgan OA, McNally FJ, Mains PE (2000) MEI-1/MEI-2 katanin-like microtubule severing activity is required for *Caenorhabditis elegans* meiosis. Genes Dev 14(9):1072–1084

Stitzel ML, Pellettieri J, Seydoux G (2006) The *C. elegans* DYRK kinase MBK-2 marks oocyte proteins for degradation in response to meiotic maturation. Curr Biol 16(1):56–62

Stitzel ML, Cheng KC, Seydoux G (2007) Regulation of MBK-2/Dyrk kinase by dynamic cortical anchoring during the oocyte-to-zygote transition. Curr Biol 17(18):1545–1554

Strome S (2005) Specification of the germ line. WormBook, ed. The *C. elegans* Research Community, WormBook. doi:/10.1895/wormbook.1.9.1, http://www.wormbook.org

Strome S, Lehmann R (2007) Germ versus soma decisions: lessons from flies and worms. Science 316(5823):392–393

Tabara H, Hill RJ, Mello CC, Priess JR, Kohara Y (1999) *pos-1* encodes a cytoplasmic zinc-finger protein essential for germline specification in *C. elegans*. Development 126(1):1–11

Verlhac MH, Terret ME, Pintard L (2010) Control of the oocyte-to-embryo transition by the ubiquitin-proteolytic system in mouse and *C. elegans*. Curr Opin Cell Biol 22(6):758–763

Wang JT, Seydoux S (2012) Germ cell specification. Advances in Experimental Medicine and Biology 757:17–39. (Chap. 2, this volume) Springer, New York

Xu L, Wei Y, Reboul J, Vaglio P, Shin TH, Vidal M, Elledge SJ, Harper JW (2003) BTB proteins are substrate-specific adaptors in an SCF-like modular ubiquitin ligase containing CUL-3. Nature 425(6955):316–321

Zhang F, Barboric M, Blackwell TK, Peterlin BM (2003) A model of repression: CTD analogs and PIE-1 inhibit transcriptional elongation by P-TEFb. Genes Dev 17(6):748–758

Chapter 13
Epigenetic Control of Germline Development

Priscilla M. Van Wynsberghe and Eleanor M. Maine

Abstract Dynamic regulation of histone modifications and small noncoding RNAs is observed throughout the development of the *C. elegans* germ line. Histone modifications are differentially regulated in the mitotic vs meiotic germ line, on X chromosomes vs autosomes and on paired chromosomes vs unpaired chromosomes. Small RNAs function in transposon silencing and developmental gene regulation. Histone modifications and small RNAs produced in the germ line can be inherited and impact embryonic development. Disruption of histone-modifying enzymes or small RNA machinery in the germ line can result in sterility due to degeneration of the germ line and/or an inability to produce functional gametes.

Keywords Epigenetic • *C. elegans* • Germ line • Chromatin • MES complex • Histone • Small RNA • siRNA • piRNA • Meiotic silencing

13.1 Introduction

The term epigenetics is commonly used to describe mechanisms that regulate gene expression in a heritable manner without altering the DNA sequence. That said, different writers interpret this rather vague definition more or less broadly when

P.M. Van Wynsberghe
Department of Biology, Syracuse University, 107 College Place, Syracuse, NY 13244, USA

Department of Biology, Colgate University, 13 Oak Drive, Hamilton, NY 11346, USA
e-mail: pvanwynsberghe@colgate.edu

E.M. Maine (✉)
Department of Biology, Syracuse University, 107 College Place, Syracuse, NY 13244, USA
e-mail: emmaine@syr.edu

T. Schedl (ed.), *Germ Cell Development in C. elegans*, Advances in Experimental Medicine and Biology 757, DOI 10.1007/978-1-4614-4015-4_13,

deciding what mechanisms to classify as epigenetic (see Bird 2007). One point of debate is that some researchers consider heritable to mean a change that persists through cell generations (i.e., passed through mitotic divisions), while other researchers more narrowly define it to mean a change that persists through organismal generations (i.e., passed through meiosis). Two generally accepted mechanisms of epigenetic inheritance are histone modification and DNA modification, both of which may persist during mitotic cell division and during gametogenesis. In addition, gene regulation via noncoding RNA is sometimes described as epigenetic because many noncoding RNAs are heritable. In writing this chapter, we chose to discuss both chromatin modification and small RNA function.

Studies in many organisms demonstrate the importance of epigenetic regulation in development. Examples of epigenetic phenomena include imprinting, X chromosome dosage compensation, and gene silencing. X chromosome dosage compensation mechanisms, for example, utilize chromatin modifications and noncoding RNAs to heritably inactivate one X chromosome in female placental mammals and up-regulate the male X chromosome in *Drosophila* (Arthold et al. 2011; Ilik and Akhtar 2009). Phenomena such as position effect variegation and paramutation exemplify how epigenetic mechanisms can be inappropriately triggered to silence gene expression (Erhard and Hollick 2011; Eissenberg and Reuter 2009). Position effect variegation, for example, arises when a chromosomal rearrangement places what should be an active gene into or near a region of transcriptionally inactive chromatin (heterochromatin).

A dynamic chromatin structure accompanies germline development (Sasaki and Matsui 2008; Feng et al. 2010; Schaner and Kelly 2006). In the very early embryo, the chromatin state of the newly formed germ cell precursors (primordial germ cells, PGCs) is thought to be important for maintaining totipotency and preventing these cells from taking on a somatic fate. Epigenetic changes observed as PGCs proliferate to form the germ line may be important for rapid proliferation and/or for subsequent gametogenesis. During gametogenesis, distinct patterns of chromatin modifications are observed in male vs female germ cells. In this chapter, we describe what is known about the epigenetic regulation of the developing *Caenorhabditis elegans* germ line and compare this process to what occurs in two other model organisms, mouse and *Drosophila*.

13.2 Epigenetic Regulation by Histone Modifications

Histone modifications influence many biological processes by altering chromatin structure or the recruitment of nonhistone proteins. These changes ultimately determine the transcription state of the gene and thus a particular biological outcome. Multiple classes of histone modifications have been identified including: acetylation, methylation, phosphorylation, deimination, ubiquitylation, sumoylation, ADP ribosylation, and the non-covalent structural modification proline isomerization (Kouzarides 2007). Considerable effort has been made to describe the distributions of histone marks across the genome, and differential distribution of histone marks

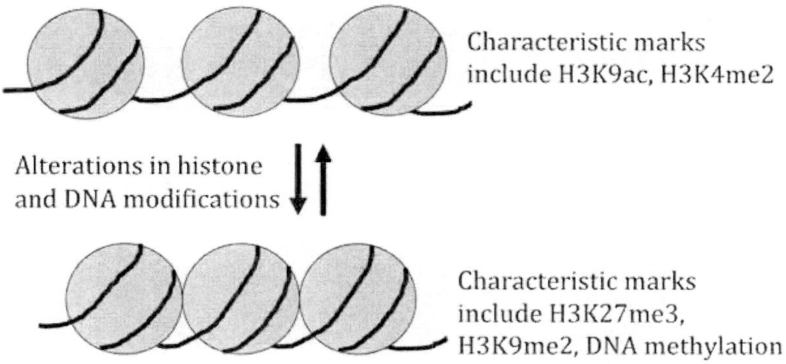

Fig. 13.1 Histone and DNA modifications regulate chromatin compaction. Highly simplified drawing depicts nucleosomes, each of which contains DNA (*line*) wrapped around a histone octamer core (*circles*). The presence of certain histone modifications and/or DNA methylation promote tighter compaction (*lower drawing*) and transcriptional repression, whereas the presence of certain histone modifications and the absence of DNA methylation promote a more open configuration allowing transcriptional activation

has been observed at both the transcription start sites and internal introns and/or exons of expressed vs non-expressed genes (Barski et al. 2007; Gerstein et al. 2010; Li et al. 2007, Andersson et al. 2009; Kolasinska-Zwierz et al. 2009). Here, we describe the types and sites of histone modifications in the *C. elegans* germline as well as their modes of regulation.

13.2.1 Descriptions and Sites of Modifications

Histone methylation is a common type of modification that can cause distinct outcomes on gene expression depending on the extent or location of methylation (see Table 13.1). Methylation of lysine 4 on histone 3 (H3K4me) is associated with transcriptional activation, while methylation of lysine 9 or lysine 27 on histone 3 (H3K9me or H3K27me) is most often associated with transcriptional repression (Kouzarides 2007). In contrast, histone acetylation, another common modification, is associated with transcriptionally active genes (Fig. 13.1) (Kouzarides 2007).

With the advent of chromatin immunoprecipitation coupled deep sequencing (ChIP-seq) and microarray analysis (ChIP-chip) technology, multiple studies have analyzed the sites of specific histone modifications throughout the *C. elegans* genome (Gu and Fire 2010; Liu et al. 2011; Gerstein et al. 2010). As part of the *C. elegans* modENCODE (Model Organism Encyclopedia Of DNA Elements) project, Liu et al. (2011) and Gerstein et al. (2010) examined chromatin isolated from the embryo, which contains primarily somatic cells, and L3 larvae, where the germline is substantially smaller than the soma. In contrast, Gu and Fire (2010) examined chromatin isolated from young adults, where germ cells are more abundant but nevertheless comprise fewer than half the cells in the body. In order to analyze germ cells in a focused way, researchers have relied primarily on antibody labeling

Table 13.1 Histone modifications discussed in this chapter

Modification	Transcriptional state of associated chromatin[a]
H3K4me2/3	Active
H3K9me1/2/3	Inactive
H3K9ac	Active
H3S10phos	Active
H3K27me1	Active
H3K27me2/3	Inactive
H3K27ac	Active
H3K36me2/3	Active
H3K79me2/3	Active
H4K20me1	Active

[a]Table lists the typical transcriptional state as described by genome-wide studies. See text for further details. *Me* methyl; *ac* acetyl; *phos* phosphate

experiments (as described below). Overall, global studies found the common activation marks H3K27ac and H3K4me2/3 to be enriched and displaying nearly identical profiles on the promoter regions of highly expressed genes (Liu et al. 2011; Gu and Fire 2010). The common repressive marks H3K27me3 and H3K9me1/2/3 were enriched in transcriptionally silent regions (Liu et al. 2011; Gu and Fire 2010). On a global scale, the autosomal arms and left arm of the X, regions enriched for repetitive sequences and transposable elements, were enriched for H3K9me marks, while chromosome centers and the right arm of the X, regions enriched for expressed genes, tended to be enriched for H3K4 methylation and H3K27 acetylation marks (Liu et al. 2011; Gu and Fire 2010; Gerstein et al. 2010). Enrichment for H3K9me was even higher in the vicinity of the meiotic pairing center on each chromosome (Liu et al. 2011; Gu and Fire 2010). However, it should be noted that these broad domains do not have sharp boundaries, and the transition from an H3K9 methylation-poor to methylation-rich state, for example, happens gradually over many hundred kilobases (Liu et al. 2011; Gu and Fire 2010). At the level of the individual gene, H3K79me2/3 and H3K36me3 were, respectively, enriched in regions near the transcription start site or throughout the body of highly expressed genes, respectively (Liu et al. 2011).

To balance X-chromosome gene expression in male and hermaphrodite somatic tissues, *C. elegans* uses a process called dosage compensation to reduce gene expression from both hermaphrodite X chromosomes (Meyer 2010). However, dosage compensation is not active in the germ line, and instead other mechanisms regulate X chromosome expression. Overall, germline-expressed genes are underrepresented on the X in both males and hermaphrodites (see below) (Reinke et al. 2000, 2004). Consistent with a low level of X-linked gene expression in the germ line, global analysis of histone modifications found that active marks were more enriched on autosomal genes while repressive marks were more enriched on X chromosome genes (Fig. 13.2) (Liu et al. 2011; Gerstein et al. 2010). In somatic cells, H3K27me1 and H4K20me1, two marks associated with transcriptional activation (Barski et al. 2007),

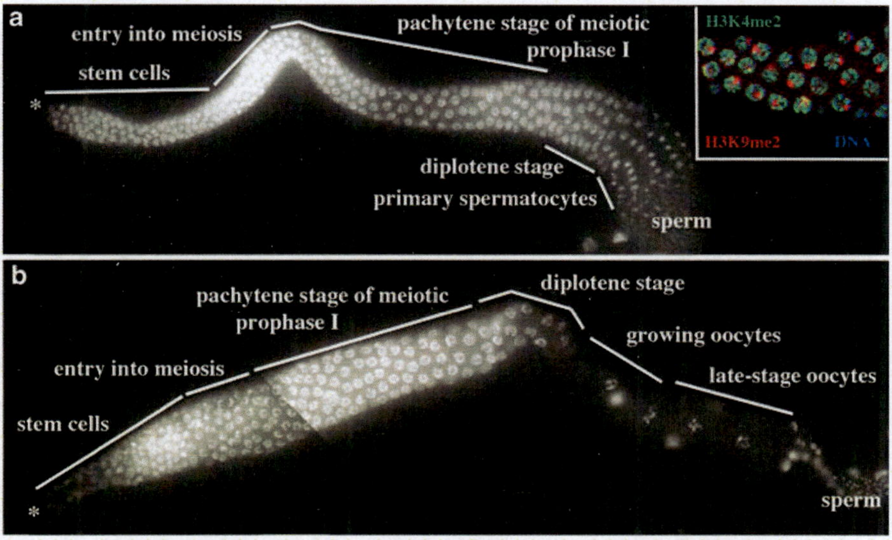

c

Histone modifications showing differential distribution within germline nuclei

Males	stem cells (mitotic)	entry into meiosis	pachytene	diplotene	primary spermatocytes
Active marks, e.g., H3K4me2, H3K9ac	A: high X: low	A: high X: low	A: high X: low	A: high X: low	A: high X: low
H3K27me3	A: moderate X: high	A: moderate X: high	A: moderate X: high	A: transient X: high	A: moderate X: high
H3K9me2	A: low X: low	A: low X: moderate	A: low X: high	A: low X: high	A: low X: moderate

Hermaphrodites	stem cells (mitotic)	entry into meiosis	early-mid pachytene	late pachytene/ diplotene	developing oocytes (diplotene)
Active marks, e.g., H3K4me2, H3K9ac	A: high X: low	A: high X: low	A: high X: low	A: high X: moderate	A: high X: low
H3K27me3	A: moderate X: high	A: moderate X: high	A: moderate X: high	A: transient X: high	A: moderate X: high
H3K9me2	A: low X: low	A: low X: low	A: low X: low	A: low X: transient	A: low X: low

Fig. 13.2 Distribution of histone modifications in the adult *C. elegans* germ line. (**a**, **b**) Dissected male and hermaphrodite germ lines labeled with DAPI to visualize DNA. Proliferating germ cells are present at the distal end of the gonad arm (*); as they exit mitosis and move proximally, germ cells progress through meiosis/gametogenesis in assembly line manner. *Inset in panel A* contains a set of pachytene nuclei from a male germline labeled via indirect immunofluorescence to visualize H3K9me2 (*red*) and H3K4me2 (*green*). H3K9me2 is enriched on the X, and H3K4me2 is enriched on the autosomes. DNA is labeled in blue. (**c**) Table summarizes the differential distribution of certain histone modifications across the developing germ line. Column headings correspond to labeled regions in panels (**a**) and (**b**)

tend to accumulate at sites where the dosage compensation proteins accumulate (Liu et al. 2011; Gerstein et al. 2010). Moreover, these two marks tend to be more highly enriched on transcribed regions of highly expressed X-linked compared to autosomal genes. The two marks show different patterns of enrichment over developmental time, perhaps reflecting different functions. H4K20me1 marks were particularly enriched in (L3) larvae where they were also present on silent genes, while H3K27me1 marks were strongly enriched on highly expressed X-linked genes in embryos. It was suggested that embryonic H3K27me1 marks may be remnants of germline X chromosome regulation that persist into embryogenesis, while H4K20 methylation may be linked to the somatic dosage compensation process that is fully established by L3 stage (Liu et al. 2011).

In general, genes specific to the soma or germline display intermediate levels of active and repressive histone marks compared to ubiquitously expressed genes and silent genes, which respectively show high levels of active or repressive marks (Liu et al. 2011). One exception to this pattern is the distribution of H3K27me1, which was more enriched on soma- and germline-specific genes than on ubiquitously expressed genes (Liu et al. 2011). This distribution suggests a role for H3K27 methylation in tissue-specific gene expression (Liu et al. 2011).

Examination of histone marks specifically in the germline via indirect immunofluorescence reveals the presence of many marks of active chromatin, e.g., H3K4 methylation and H3K9 acetylation, on the autosomes and the near absence of those marks on the X in all or most germ cells (Kelly et al. 2002; Fong et al. 2002). Conversely, marks associated with gene silencing are observed on both autosomes and X chromosomes, although certain marks are enriched on hermaphrodite and male X chromosomes or specifically on the male X (Kelly et al. 2002; Bender et al. 2004). This general pattern correlates well with microarray data indicating expression of few X-linked genes in the germ line (Reinke et al. 2000, 2004).

13.2.2 Mechanisms of Regulation

Multiple proteins and protein complexes modify histones. Addition or elimination of a histone mark can dramatically alter the ability of effector proteins to interact with a particular histone residue causing a change in the transcriptional state of a gene. Below we discuss some of the main protein complexes and histone methyltransferases (HMTases) responsible for histone modification in the *C. elegans* germ line.

13.2.2.1 MES Proteins and H3K27 Methylation

During germ cell mitosis and early meiosis both X chromosomes in hermaphrodites and the single X chromosome in males are silenced by histone modification in the germline (Kelly et al. 2002). The MES complex, composed of the Polycomb group chromatin repressors MES-2 and MES-6 as well as the MES-3 protein, is responsible for this silencing (Bender et al. 2004). Together these MES proteins cause silencing via H3K27me2/me3 in the adult germline and early embryos (Bender et al. 2004). H3K27me3 marks are concentrated on the X chromosome, and loss of MES protein function causes maternal-effect sterility due to germ cell underproliferation and death (Bender et al. 2004; Capowski et al. 1991; Garvin et al. 1998). The SET domain of MES-2 is crucial for the HMTase activity of the MES complex (Bender et al. 2004). Together, the MES-2/-3/-6 complex functions as the *C. elegans* PRC2 (Polycomb repressive complex 2). Parenthetically, we note that *C. elegans* apparently lacks a PRC1.

13.2.2.2 MES Proteins and H3K36 Methylation

Another SET domain protein, MES-4, also has HMTase activity that it uses to di- and tri-methylate lysine 36 on histone 3 (H3K36me2, H3K36me3) in mitotic and early meiotic germline nuclei and early embryos (Bender et al. 2006; Furuhashi et al. 2010). In the embryonic soma, H3K36 methylation also depends on activity of the SET domain protein, MET-1 (ortholog of yeast Set2) (Furuhashi et al. 2010). Genome-wide analysis of histone modifications in a variety of species has determined that H3K36 methyl marks tend to be enriched in the body of expressed genes (Li et al. 2007; Shilatifard 2008). In the *C. elegans* germ line, H3K36 methylation is enriched on autosomes and at the left end of the X chromosome, and very low elsewhere on the X chromosome, consistent with autosomal linkage of most germline-expressed genes (Bender et al. 2006). Germline MES-4 protein is concentrated at the sites of H3K36me accumulation, as expected for a protein with a direct role in depositing the mark (Bender et al. 2006). This pattern of MES-4 activity is in striking contrast the MES-2/3/6 complex, which is active across the X chromosome (Bender et al. 2006). Intriguingly, exclusion of MES-4 activity from (most of) the X chromosome depends on the MES-2/3/6 complex since MES-2/3/6 mutants contain both MES-4 and H3K36me2/3 marks across the X chromosome (Bender et al. 2006). Moreover, despite the main presence of MES-4 on autosomes, MES-4 activity is important for silencing X-linked genes in the germline, and *mes-4* mutants exhibit a maternal-effect sterile phenotype similar to that of *mes-2*, *mes-3*, and *mes-6* mutants (Bender et al. 2006; Capowski et al. 1991). Interestingly, the memory of genes last expressed in the parental germline, and marked by H3K36me, is transferred from parent to offspring by MES-4 activity (Rechtsteiner et al. 2010; Furuhashi et al. 2010). This epigenetic inheritance is crucial for germline viability (Rechtsteiner et al. 2010).

13.2.2.3 The MLL Complex and H3K4 Methylation

As described above, H3K4 methyl marks are commonly associated with actively transcribed genes of multiple species (Kouzarides 2007). The MLL complex is responsible for H3K4 methylation in *C. elegans* and many other species. The canonical MLL complex contains four components: MLL, a SET domain protein with HMTase activity; WDR-5; ASH-2/Ash2L; and RBBP-5. In *C. elegans*, MLL complexes containing different sets of components are responsible for H3K4 di- and tri-methylation in the embryo and in much of the germ line. In the early embryo, the MLL complex components WDR-5.1, RBBP-5, and ASH-2 are essential for H3K4 methylation (Li and Kelly 2011; Xiao et al. 2011). Though *wdr-5.1* and *rbbp-5* mutants do not exhibit strong phenotypes, when grown at 25°C successive generations have progressively smaller brood sizes and exhibit a variety of germline developmental defects (Li and Kelly 2011; Xiao et al. 2011). The HMTase responsible for H3K4 trimethylation in the embryo is SET-2, while the HMTase responsible for H3K4 dimethylation is unknown (Li and Kelly 2011; Xiao et al. 2011). Though little RNA polymerase II transcription occurs in early dividing blastomeres and germline precursors of *C. elegans* embryos, high H3K4me2 levels, but not H3K4me3 levels, are maintained throughout multiple cell divisions by the MLL complex (Li and Kelly 2011).

In adult germ cells, H3K4me2 and H3K4me3 marks are present in high abundance across all autosomes, but not on the X chromosome (Kelly et al. 2002; Reuben and Lin 2002). The presence of H3K4me3 marks in the germline stem cells (GSCs) depends on SET-2, WDR-5.1, and RBBP-5 (Li and Kelly 2011; Xiao et al. 2011). H3K4me2 marks in the mitotic germ line clearly depend on WDR-5.1 and RBBP-5 activity; however, there is debate about the importance of SET-2 activity in deposition of these marks. In the absence of SET-2 activity, Li and Kelly (2011) observed moderately reduced H3K4me2 levels while Xiao et al. (2011) observed virtually no H3K4me2 signal. Interestingly, maintenance of H3K4 methylation in the GSCs is independent of active transcription (Li and Kelly 2011). In contrast to the GSCs, maintenance of H3K4 methylation in meiotic germ cells is partially independent of SET-2, WDR-5.1, RBBP-5, and ASH-2 activity (Li and Kelly 2011; Xiao et al. 2011). Taken together, these results suggest that MLL complexes containing WDR-5.1 and RBBP-5 are required for H3K4 di- and tri-methylation in early embryos and GSCs, while other proteins are required for H3K4 methylation during meiosis (Li and Kelly 2011; Xiao et al. 2011).

13.2.2.4 SET Domain Proteins Implicated in H3K9 Methylation

As previously described, H3K9 methylation commonly correlates with heterochromatin and silenced genes (Kouzarides 2007). However, despite only a single methyl difference between H3K9me2 and H3K9me3, in *C. elegans* these two marks exhibit distinct localization patterns, functions, and require different HMTases (Bessler et al. 2010). H3K9me2 is present in a gradient pattern throughout the adult meiotic germline with low levels found in early pachytene and progressively higher levels found throughout late pachytene and diplotene stages (Kelly et al. 2002; Bessler et al. 2010). The H3K9me2 mark is highly enriched on the unpaired X chromosome in XO males and in *him-8* mutant hermaphrodites, as well as other unpaired regions such as free chromosomal duplications and extrachromosomal arrays (Fig. 13.3) (Kelly et al. 2002; Bean et al. 2004). The HMTase MET-2 is essential for germline H3K9 dimethylation in both males and hermaphrodites (Bessler et al. 2010; Andersen and Horvitz 2007). MET-2 is homologous to the SETDB1 family of histone methyltransferases that have been shown to methylate lysine 9 of histone 3 (Schultz et al. 2002).

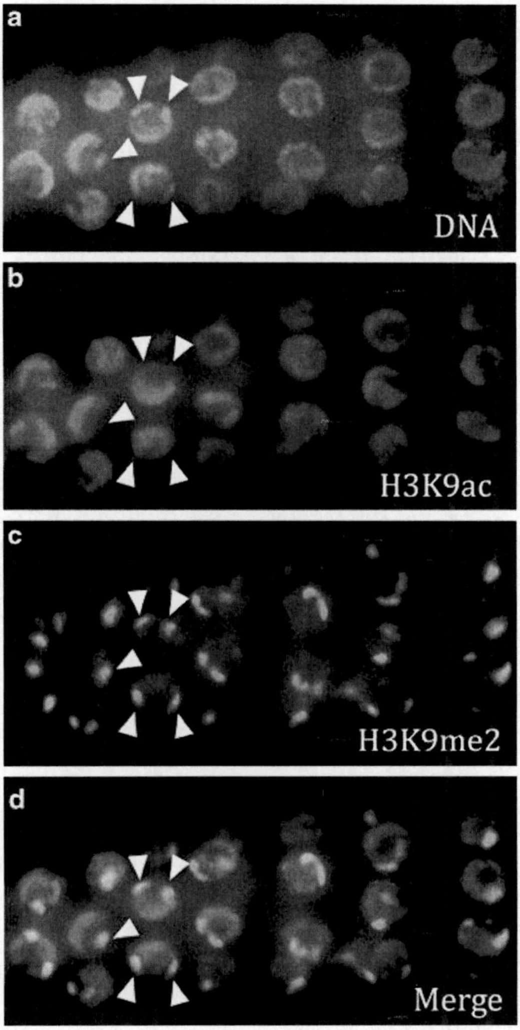

Fig. 13.3 Accumulation of H3K9me2 on unpaired homologs in the hermaphrodite germ line. Pachytene nuclei from a dissected *him-8* hermaphrodite germ line with histone modifications labeled as indicated. HIM-8 protein (not shown) associates with the X chromosome pairing center and promotes pairing/synapsis of the homologous chromosomes. In the *him-8* mutant, X chromosomes typically fail to pair or synapse. (**a**) DNA stained with DAPI. (**b**) H3K9ac marks are enriched on autosomes and not detected on the X chromosomes (*arrowheads*), as also observed in wild-type hermaphrodites. Two X homologs are visually distinct from each other in many nuclei, while in other nuclei only one X is visible in this focal plane. (**c**) The two unpaired X chromosomes are enriched for H3K9me2 marks. (**d**) Merged H3K9ac and H3K9me2 images

Although H3K9 methylation marks generally correlate with heterochromatin and reduced transcription, this correlation is not as strong as for H3K27 methylation (e.g., see Lienert et al. 2011). Nonetheless, evidence suggests the accumulation of H3K9me2 on the single X may have a transcriptional repressive effect. During spermatogenesis, essentially no X-linked genes are transcribed while, during oogenesis in XX animals, a number of X-linked genes are transcribed in late pachytene/diplotene stage (as monitored by in situ hybridization). Interestingly, in XO hermaphrodites or females (produced as a result of *her-1* or *fem-3* mutations, respectively), the single X accumulates H3K9me2 and the burst of oogenesis-specific X-linked transcription is not observed (Bean et al. 2004; Jaramillo-Lambert and Engebrecht 2010). Hence, enrichment for H3K9me2 on the single X correlates with transcriptional repression.

In contrast to H3K9me2 marks, H3K9me3 marks are present on all chromosomes in all germ cells throughout the *C. elegans* gonad (Bessler et al. 2010). The H3K9me3 mark is also enriched on high-copy transgene arrays present in germ cell nuclei (Bessler et al. 2010). The HMTase MES-2, but not MET-2, is required for germline accumulation of H3K9me3 in conjunction with at least one additional, unidentified enzyme (Bessler et al. 2010). These results suggest that the same MES complex required for H3K27 methylation, described above, may also be needed for acquiring or maintaining the H3K9me3 mark (Bessler et al. 2010).

13.2.2.5 The Small RNA Pathway and H3K9 Methylation

In addition to MET-2, activity of the RNA-dependent RNA polymerase (RdRP) EGO-1 is required for H3K9me2 enrichment on unpaired chromosomes (Maine et al. 2005). Male worms mutant for the Argonaute protein CSR-1, the Tudor domain protein EKL-1, or the DEAH/D-box helicase DRH-3 also exhibit reduced H3K9me2 enrichment on unsynapsed chromosomes and increased H3K9me2 accumulation on synapsed autosomes (She et al. 2009). Mutation of *ego-1*, *csr-1*, *ekl-1*, or *drh-3* causes various germline defects, and double mutants display even stronger phenotypes in both hermaphrodites and male worms (She et al. 2009). Because of the roles that EGO-1, CSR-1, DRH-3, and EKL-1 play in small RNA production and function (Gu et al. 2009; Claycomb et al. 2009), these results suggest that H3K9me2 enrichment on unpaired X chromosomes may be driven by small RNAs (She et al. 2009). It is unclear at present whether the small RNA pathway is involved in the initiation and/or maintenance of H3K9me2 marks (see Future Directions, below).

13.2.3 Erasure of Histone Modifications

The removal of specific histone modifications is just as important as their addition for eliciting the appropriate gene expression response. However, until recently many histone modifications, including methylation, were presumed irreversible. RBR-2, the *C. elegans* homolog of the Jumonji C domain-containing JARID1 protein, specifically demethylates H3K4me3 and H3K4me2 in vitro (Christensen et al. 2007). In addition, *rbr-2* mutants displayed increased H3K4me3 levels at all developmental stages and vulval defects, suggesting that RBR-2 regulates vulval development through demethylation of H3K4me3 (Christensen et al. 2007).

H3K4 dimethylation has been proposed to act as an epigenetic memory mark of transcriptional activity. This would allow stable transmission of gene expression patterns in developing somatic cells, but could cause inappropriate expression in the germline. The *C. elegans* homolog of the H3K4me2 demethylase, LSD1/KDM1, is SPR-5. Mutants for *spr-5* exhibit phenotypes, an egg-laying defect and reduced brood size, which become progressively worse over multiple generations (Katz et al. 2009). H3K4me2 levels also increase after many generations, and most spermatogenesis-expressed genes are misregulated in *spr-5* mutants (Katz et al. 2009). These results suggest that SPR-5 normally demethylates H3K4me2 to prevent transmission of this mark to successive generations and thus inappropriate overexpression of spermatogenesis-expressed genes and sterility (Katz et al. 2009) (see below).

Histone modifications appear to be removed in germ cell precursors and reestablished as the germline develops. The *C. elegans* germline precursor cell, P4, forms at the fourth embryonic cleavage division, and later divides once to form two primordial germ cells, Z2 and Z3 (see Wang and Seydoux 2012, Chap. 2). The levels of many histone modifications associated with active chromatin, e.g., H3K4 methylation and H4K8 acetylation, are severely reduced (or masked) in Z2 and Z3 such that they cannot be detected by indirect immunofluorescence (Schaner et al. 2003).

13.2.4 Transgenerational Maintenance of Germline Viability

Maintenance of germline size and viability over succeeding generations is critical for continuation of the species. In *C. elegans*, progressive germline loss (termed a mortal [Mrt] germ line defect) occurs over succeeding generations in animals with mutations in many DNA damage response and repair proteins (Ahmed and Hodgkin 2000; Meier et al. 2009). In some cases, e.g., *mrt-1* and *mrt-2*, telomere shortening occurs suggesting that the Mrt phenotype is triggered by the loss of genome integrity (Ahmed and Hodgkin 2000; Meier et al. 2009). Interestingly, an Mrt germline is also associated with mutations in certain histone-modifying enzymes, e.g., the demethylase SPR-5/LSD1, the HMTase MET-2, and components of the MLL complex described above (Bessler et al. 2010; Katz et al. 2009; Li and Kelly 2011; Xiao et al. 2011). In these mutants, there is no evidence of telomere shortening or other chromosomal abnormalities, e.g., chromosome segregation defects. Therefore, inappropriate gene expression levels, resulting from incorrect patterns of H3K4me2 and H3K9me2 marks may be responsible for the Mrt phenotype. The Mrt phenotype has not been described in other study organisms, e.g., *Drosophila* and mouse, perhaps because of their longer generation times. Nonetheless, it has provided an opportunity to identify mechanisms essential for long-term fertility and highlights the critical importance of histone regulation in the establishment of the germline.

13.3 Epigenetic Regulation by Small RNA-Mediated Silencing

Different classes of small interfering RNAs (siRNAs) are commonly used by RNAi-like pathways in *C. elegans* to guide sequence-specific regulation of gene silencing and chromatin structure (see Table 13.2). The siRNA molecule provides specificity by interacting with an Argonaute (AGO) protein and targeting it to specific RNAs based on sequence complementarity (Ketting 2011). Classes of siRNAs differ in their source, structural features like size and 5' or 3' modifications, mechanism of biogenesis, and means of function. They also differ in their associated AGO. Exogenous siRNAs (exo-siRNAs) are ~22 nt siRNAs made from processed exogenous dsRNAs, while endogenous siRNAs (endo-siRNAs) are mainly formed from short RdRP transcripts. RdRP products are produced as part of the response to exogenous dsRNA (exogenous RNAi) and as part of endogenous gene-regulatory mechanisms. Endo-siRNAs in *C. elegans* are mainly classified as 22G-RNAs or 26G-RNAs (Han et al. 2009). 22G-RNAs are ~22 nt in length and have a triphosphorylated 5'G, while 26G-RNAs are ~26 nt in length and have a monophosphorylated 5'G (Han et al. 2009). These two classes of siRNAs are produced by related, but distinct mechanisms and function together with different Argonuate proteins. *C. elegans* produce another class of small RNA, 21U-RNA, which function together with associated Argonaute proteins to silence certain transposons. 21U-RNAs are 21 nucleotides in length and contain a 5' U (Ruby et al. 2006). 21U-RNAs are analogous to piRNAs described in other organisms (Castaneda et al. 2011) and, like piRNAs, their mechanism of biogenesis appears to be very distinct from that of other siRNAs. In this section we discuss the role of endo-siRNAs, 21U-RNAs, and their associated Argonaute proteins in transposon silencing and germline development.

Table 13.2 Endogenous small RNAs implicated in germline epigenetic control in *C. elegans*

Type	Origin	Associated argonaute	Function
22G-RNAs	EGO-1/RRF-1 RdRP activity	WAGOs	Transposon, pseudogene, aberrant RNA silencing
22G-RNAs	EGO-1 RdRP activity	CSR-1	MSUC; embryonic chromosome segregation
26G-RNAs, class I	RRF-3 RdRP and DCR-1 activity	T22B3.2, ZK757.3	Translational repression (sperm)
26G-RNAs, class II	RRF-3 RdRP and DCR-1 activity	ERGO-1, other?	Translational repression (oocytes, embryos)
21U-RNAs (piRNAs)	Genome-encoded; biogenesis unknown	PRG-1, -2	Transposon silencing

See text for further details

13.3.1 Role of 22G-RNAs in Transposon Silencing

22G-RNAs can associate with two distinct AGOs: WAGO-1 or CSR-1 (Gu et al. 2009; Claycomb et al. 2009; Maniar and Fire 2011). In addition to WAGO-1, *C. elegans* contains an additional 11 WAGO proteins (members of the worm-specific Argonaute clade) that also can associate with 22G-RNAs. The WAGO-dependent 22G-RNAs are essential for silencing their target genes, which include transposable elements, pseudogenes, and aberrant transcripts (Gu et al. 2009). Most 22G-RNAs are expressed in the germline and many are maternally inherited. WAGO-1 is also expressed in the germline where it localizes to perinuclear foci called P granules (Gu et al. 2009). P granules are ribonucleoprotein particles located on the cytoplasmic side of nuclear pores and enriched in polyadenylated mRNAs (Updike and Strome 2010). WAGO-1 mutants contain reduced germline 22G-RNAs levels, and worms mutant for all WAGO proteins express no detectable germline 22G-RNAs, suggesting that WAGO-1 is essential for 22G-RNA production and function (Gu et al. 2009). Nearly identical populations of germline 22G-RNAs were also depleted in *mut-16*, *mut-7* or *rde-3*, but not *rde-4* mutants, suggesting that MUT-16, MUT-7 and RDE-3 also function in the 22G-RNA silencing pathway (Gu et al. 2009; Zhang et al. 2011). All transposon classes were depleted of 22G-RNAs in *rde-3*, *mut-7* or WAGO mutants, and these 22G-RNAs apparently associate with WAGO-1 as they are recovered in anti-WAGO-1 co-immunoprecipitation experiments (Gu et al. 2009). Consequently, protein complexes composed of WAGO, RDE-3, and MUT-7 use 22G-RNAs to guide transposon silencing in the germline (Gu et al. 2009).

Levels of WAGO-associated 22G-RNAs are also reduced in *rrf-1 ego-1* double mutant worms, but not single *rrf-1* or *ego-1* mutants (Gu et al. 2009). Thus, these RdRPs likely function redundantly in the germline to produce the class of 22G-RNAs that associates with WAGOs (Gu et al. 2009). Both RRF-1 and EGO-1 can physically interact with the Dicer-related helicase (DRH-3), and germline 22G-RNAs are absent in *drh-3* mutant worms (Gu et al. 2009). Consistent with a germline function, *drh-3* mutants exhibit a variety of phenotypes including sterility, embryonic lethality, and high incidence of males (Nakamura et al. 2007; She et al. 2009; Claycomb et al. 2009). DRH-3 also physically interacts with the Tudor-domain protein EKL-1 (Gu et al. 2009), and *ekl-1* mutants also exhibit phenotypes similar to *drh-3* mutants (She et al. 2009; Claycomb et al. 2009). Thus, DRH-3, EGO-1, and EKL-1 likely interact to form a core RdRP complex essential for 22G-RNA biogenesis, while WAGO, RDE-3 and MUT-7 participate with a subset of these 22G-RNAs in the germline to mediate transposon silencing (Gu et al. 2009).

13.3.2 Role of 22G-RNAs in Germline Development

As discussed above, 22G-RNAs can associate with two distinct types of AGO protein to mediate different functions regulating development. 22G-RNAs associated with the AGO CSR-1 are expressed in the germline and target germline-expressed protein-coding genes, not transposons or pseudogenes (Claycomb et al. 2009, Maniar and Fire 2011). Like WAGO-1, CSR-1, DRH-3, EGO-1, and EKL-1 all colocalize with P granules, which are normally located on the cytoplasmic side of nuclear pores and are enriched in polyadenylated mRNAs (Gu et al. 2009; Claycomb et al. 2009, E. Maine and X. Xu, unpublished data). However, though WAGO-1 has no effect on the intracellular position of P granules, CSR-1, EGO-1, DRH-3, and EKL-1 are important for perinuclear localization of P granules (Vought et al. 2005; Claycomb et al. 2009; Updike and Strome 2009). Further evidence that EGO-1, DRH-3, EKL-1, and CSR-1 function in a common pathway is provided by genetic analysis of siRNA level. *ego-1*, *drh-3*, and *ekl-1* mutant worms have reduced levels of CSR-1-associated 22G-RNAs (Claycomb et al. 2009; Maniar and Fire 2011). Accordingly, these mutants have phenotypes similar to *csr-1* mutants (Claycomb et al. 2009; She et al. 2009; Rocheleau et al. 2008).

There is debate in the literature about the extent to which interaction of CSR-1-associated 22G-RNAs with their antisense target sequences causes target mRNA degradation (Claycomb et al. 2009, Maniar and Fire 2011). Claycomb et al. (2009) observe little change in target mRNA levels and instead posit that CSR-1-associated 22G-RNAs interact with targets to mediate proper organization of the holocentric chromosomes of *C. elegans* during metaphase (Claycomb et al. 2009). In the absence of *csr-1*, worms exhibit defects in chromosome segregation that cause various abnormalities including aberrant chromosome numbers, a high incidence of males (*him*) phenotype, and ultimately reduced fertility (Claycomb et al. 2009). In contrast, Maniar and Fire (2011) observe an increase in the level of most mRNAs targeted by EGO-1-dependent 22G-RNAs and CSR-1, and they hypothesize that EGO-1 activity is critical for negative regulation of developmentally important genes in the germ line.

To limit the accumulation of CSR-1-associated 22G-RNAs and thus inappropriate gene silencing and chromosome segregation defects, the nucleotidyl transferase CDE-1 localizes to mitotic chromosomes in an EGO-1- and CSR-1-dependent manner where it uridylates these RNAs at the 3' end triggering their degradation (Claycomb et al. 2009; van Wolfswinkel et al. 2009). CDE-1 physically associates with EGO-1, and its activity may be critical for targeting specific EGO-1 products to CSR-1 (Van Wolfswinkel et al. 2009). It is unclear what mechanism targets CDE-1 activity to a particular subset of EGO-1 products. Nonetheless, multiple proteins function with CSR-1 and CSR-1-associated 22G-RNAs to mediate proper chromosome segregation (Claycomb et al. 2009).

13.3.3 Role of 26G-RNAs in Germline Development

26G-RNAs are enriched in the germline of *C. elegans* where they regulate gene expression of mature mRNAs during spermatogenesis (Han et al. 2009). Moreover, maternally inherited 26G-RNAs regulate gene expression in the zygote (Han et al. 2009).

There are two subclasses of 26G-RNAs whose pattern of expression and associated Argonautes differ (Han et al. 2009). Class I 26G-RNA expression in sperm coincides with spermatogenesis in the L4 and young adult stages. Class II 26G-RNAs are expressed in oocytes and embryos. Both types of 26G-RNAs are perfectly complementary to their target genes.

26G-RNAs are produced via a different mechanism than 22G-RNAs. Their expression depends on the RdRP RRF-3, the exonuclease ERI-1, DCR-1/Dicer, and the dsRNA-binding protein RDE-4 (Han et al. 2009; Vasale et al. 2010). Together, this biogenesis pathway is referred to as the ERI (enhanced RNAi) pathway because exogenous RNAi is enhanced when the ERI pathway is disabled (e.g., Simmer et al. 2002). Consistent with a role for some 26G-RNAs in spermatogenesis, both *rrf-3* and *eri-1* single mutant worms are temperature-sensitive sterile due to spermatogenesis defects (Gent et al. 2009; Simmer et al. 2002; Kennedy et al. 2004). Class I 26G-RNAs associate with AGOs T22B3.2 or ZK757.3 that are enriched during spermatogenesis, while class II 26G-RNAs use the AGO ERGO-1 (Han et al. 2009; Conine et al. 2010; Vasale et al. 2010).

The relationship between 26G- and 22G- RNAs is complex. Many more 22G-RNA species have been described than 26G-RNA species, and distinct RdRPs are linked to production of 22G- vs 26G- RNAs. However, 26G-RNAs appear to function in a two-step mechanism that also involves 22G-RNAs. In the soma, certain mRNAs are targeted by both 26G- and 22G-RNAs, and activity of the 26G-RNA machinery promotes accumulation of this particular subset of 22G-RNAs (Gent et al. 2010). Hence, production of these particular 26G- and 22G- RNAs is coordinated. A similar relationship is observed in the germ line between certain 26G- and 22G-RNAs (Vasale et al. 2010). These results are consistent with a coordinated mechanism in the germ line whereby certain mRNAs are targeted first by 26G-RNA/ERGO-1 activity and later by 22G-RNA/WAGO activity (Vasale et al. 2010).

13.3.4 Role of 21U RNAs in Transposon Silencing and Development

Members of the Piwi subfamily of Argonaute proteins function in germline development and transposon silencing in diverse animals. In most systems, Piwi proteins interact with 24–30 nucleotide piRNAs (Castaneda et al. 2011). In *C. elegans,* a different class of small RNA, initially termed the 21U-RNA, physically interacts with the Piwi protein, PRG-1 (Wang and Reinke 2008; Batista et al. 2008). The vast majority of 21U-RNA/piRNA sequences are present in clusters located in intergenic or intronic regions on chromosome IV (Ruby et al. 2006; Batista et al. 2008; Kato et al. 2009). Although biogenesis of 21U-RNAs is not clear, it appears to require substantially different machinery than does biogenesis of miRNAs and siRNAs (Batista et al. 2008; Das et al. 2008). 21U-RNAs are largely absent in *prg-1* mutants, suggesting that they are stabilized by PRG-1 activity (Wang and Reinke 2008; Batista et al. 2008). PRG-1 and PRG-2, an Argonaute sharing ~90% amino acid sequence identity with PRG-1, are implicated in transposon silencing: excision rates of Tc3 elements are elevated approximately 100-fold in *prg-1 prg-2* double mutants (Das et al. 2008). Evidence suggests that the 21U-RNA pathway acts upstream of MUT-7 (Das et al. 2008), hence this pathway may feed into the WAGO pathway described above.

PRG-1 activity is required for fertility at elevated temperatures, although the loss of *prg-1* function partially impairs germline development at a range of culture temperatures (Batista et al. 2008; Wang and Reinke 2008). Mutations in *prg-1* are associated with many germline defects, particularly in mitotic proliferation (Batista et al. 2008) and spermatogenesis (Wang and Reinke 2008). PRG-1 associates with P granules, although there is debate about whether this occurs strictly in the spermatogenic germline (Wang and Reinke 2008) or also in the oogenic germline (Batista et al. 2008). In *prg*-1 mutants, the levels of many spermatogenesis-enriched mRNAs are reduced while the levels of other germline-enriched mRNAs are not substantially changed (Wang and Reinke 2008; Batista et al. 2008).

13.4 Epigenetic Regulation of Germline Development in Other Animals

Epigenetic regulation during germline development in other animals shares some broad similarities with *C. elegans*, including the importance of histone modifications and small RNAs and sex-specific reorganization of chromatin structure during gametogenesis. A major mechanistic difference in many organisms is the use of DNA methylation, in addition to histone modification, to limit transcription. Here, we discuss the general features of epigenetic regulation in the developing germ lines of two common study organisms, mouse and *Drosophila*.

13.4.1 Mechanisms of Epigenetic Regulation in the Murine Germ Line

Chromatin regulation in the mouse germ line involves extensive DNA methylation in addition to histone modifications (Feng et al. 2010; Zamudio et al. 2008, 2011; Sasaki and Matsui 2008). Methylation occurs at cytosine residues via one of the three different mechanisms that are active in different DNA sequence contexts (Feng et al. 2010). In mammals, methylation at CG sites is maintained by a DNA methyltransferase called DNMT1 in conjunction with a co-factor, UHRF1. At some sites, additional DNA methyltransferases, DNMT3A and Dnmt3b, are required to maintain CG methylation. Methylation can also occur at CHG and CHH sites (where H represents A, T, or G), although in animals CG is by far the most common site of methylation. Certain genes are methylated ("imprinted") during either oogenesis or spermatogenesis to ensure expression of only the paternal or maternal allele in the embryo (Feng et al. 2010; Hudson et al. 2010). This regulation is developmentally important: in at least some cases, defects in imprinting adversely impact development and health of the offspring (Surani et al. 1986; Feng et al. 2010).

The extensive chromatin reorganization in the developing mouse germline involves the removal/reestablishment of DNA methylation and many histone modifications (Hajkova et al. 2008; Farthing et al. 2008; Sasaki and Matsui 2008; Feng et al. 2010). When the PGCs form at embryonic day 7.25 (E7.25), their pattern of repressive chromatin marks (DNA methylation, H3K27me3, and H3K9me2) resembles that in adjacent somatic cells. The levels of these repressive marks begin to change very soon thereafter, with DNA methylation and H3K9me2 decreasing and H3K27me3 increasing. There is a brief period of time (E7.5-E8.25) when the levels of repressive marks are relatively low and transcription might be expected to initiate; however, RNA polymerase II is inactive during this time due to another (unknown) mechanism. During this period, X chromosome dosage compensation is reversed in the PGCs, as described below.

Once the migrating PGCs reach the developing somatic gonad, they proliferate and, in females, eventually initiate gametogenesis. In males, gametogenesis does not initiate until after birth. Changes in chromatin modifications continue through this period, including removal of parental imprints (Sasaki and Matsui 2008; Feng et al. 2010). This reprogramming is important for regulating PGC-specific gene expression and for the eventual establishment of sex-specific chromatin modifications. During gametogenesis, sex-specific patterns of de novo DNA methylation and histone modification are observed, presumably reflecting the very extensive differences in sperm vs oocyte formation (Sasaki and Matsui 2008; Feng et al. 2010). In males, DNA methylation imprints are established in mitotic germ cells prior to entry into meiosis. Upon entry into meiosis, widespread changes in histone modification are observed, as well as incorporation of numerous histone variants (Godman et al. 2009; Kageyama et al. 2007; Sasaki and Matsui 2008). In female germ cells, DNA methylation imprinting occurs during diplotene stage of meiotic prophase I when oocytes are in the growth phase. Global analysis of histone modifications revealed a general increase in the number of histone modifications in female germ cells as oogenesis proceeds, as well as incorporation of a histone H1 variant, although overall the observed chromatin reorganization during meiosis is much less dramatic than that observed in males (Gu et al. 2010; Sasaki and Matsui 2008). The most obvious pattern of altered histone modifications is an increase in acetylated H3 and H4 during prophase of meiosis I, which is then reversed later as oocytes proceed through the meiotic divisions at fertilization.

Genetic analysis has underscored the functional importance of chromatin reorganization during murine gametogenesis: mutations in components of the chromatin regulatory machinery are associated with sterility. For example, mutations in Prdm9 H3K4 tri-methyltransferase and Ehmt2 H3K9 mono- and di- methyltransferase, two HMTases normally active in the male and female germ line, cause extensive defects including meiotic arrest and an incorrect pattern of gene expression (Sasaki and Matsui 2008).

13.4.1.1 Meiotic Sex Chromosome Inactivation

In mouse, the male sex chromosomes are transcriptionally silenced for a portion of first meiotic prophase (Turner 2007). This process, called meiotic sex chromosome inactivation (MSCI), is an example of a larger phenomenon called meiotic silencing of unpaired chromosomes (MSUC) that also targets asynapsed autosomes and large chromosomal translocations (Schimenti 2005; Turner 2007). MSUC is thought to be analogous to meiotic silencing of unpaired chromatin in *C. elegans.* In male germ cells, the X and Y chromosomes form a distinct structure called the XY-body. During first meiotic prophase, the XY-body accumulates a specific set of histone variants, e.g., H3.1 and H3.2 are replaced with H3.3, and an altered pattern of histone modifications (Turner 2007; Sasaki and Matsui 2008). Changes in histone modification include elevated H3K9me2, H2A ubiquitination, and H2AX phosphorylation, and reduced H3K9ac (Turner 2007; Payer et al. 2011). Similar histone modifications and replacement also occur on asynapsed autosomes (van der Heijden et al. 2007). Interestingly, the reverse situation is observed in *C. elegans* where synapsed chromosomes accumulate H3.3 and the single X does not (Ooi et al. 2006). Global analysis of gene expression in the mouse identified a phase during early meiosis where X-linked gene expression is down-regulated relative to autosomal gene expression (Wang et al. 2005). Disruption of MSCI results in up-regulation of X-linked genes and arrest of male meiosis suggesting that the differential regulation of XY chromatin silences gene expression in a manner necessary for meiosis (Turner 2007; Zamudio et al. 2008; Royo et al. 2010).

The mechanism of meiotic silencing in mouse differs (at least to some extent) from that in nematodes, although asynapsis appears to trigger the process in each species (Turner 2007; Maine 2010). MSCI in mouse requires components of the DNA damage response machinery and accumulation of H2A variants that are associated with the DNA damage response (Turner 2007; Sasaki and Matsui 2008; Payer et al. 2011). Histone variant H2AX, which localizes to meiotic double strand breaks (DSBs), also localizes to the XY-body. Initial steps in the meiotic silencing process include the association of BRCA1 protein with asynapsed chromosomes and subsequent recruitment of the checkpoint kinase, ATR, which then phosphorylates H2AX at late zygotene/early pachytene stage (Payer et al. 2011). H2AX located at DSBs is also phosphorylated, although this occurs earlier, and these marks are no longer detected at pachytene when synapsis is complete. In *C. elegans*, components of the DNA damage response machinery apparently do not have a role in meiotic silencing (Maine 2010). Numerous other histone regulatory proteins either associate with or are excluded from the XY-body, and mutations in many of these factors lead to defects in XY-body formation (Sasaki and Matsui 2008). A role for the small RNA machinery in meiotic silencing in mouse has not been ruled out, and it will be very interesting to see if this aspect of the process is conserved.

Relevant to X chromosome regulation in the germ line is the process of X chromosome dosage compensation. Ultimately, one X chromosome is randomly inactivated in cells of the early female embryo. In order for X chromosome inactivation to be random, the inherited paternal X that was silenced in the male germline by MSCI must be activated. There has been substantial debate about the details of paternal X regulation in the female embryo, and the bulk of the evidence in mouse now suggests a complex series of events, as follows. MSCI is reversed after fertilization and then the paternal X is quickly re-silenced such that only the maternal X is expressed in extraembryonic tissues. The paternal X is reactivated in the inner cell mass of the blastocyst, and subsequently one X chromosome is randomly inactivated in each cell of the early embryo (epiblast) by the dosage compensation machinery (Payer et al. 2011). The inactive X is eventually reactivated in the early PGCs (Sasaki and Matsui 2008).

13.4.1.2 Transposon Silencing and Other siRNA-Mediated Mechanisms in the Germ Line

In mouse, the repression of transposon activity is especially important in males where actively dividing germline stem cells are maintained. Moreover, the global reduction in silencing marks in early PGCs might provide an opportunity for elevated transposon activity. In mouse, as in *C. elegans*, transposon activity is limited in the germ line via a small RNA-mediated mechanism involving Argonaute proteins of the Piwi clade and associated piRNAs (analogous to *C. elegans* 21U RNAs) (Castaneda et al. 2011; Sasaki and Matsui 2008). PiRNA pathway activity leads to degradation of transposon-encoded mRNA and methylation of transposable element DNA. Accordingly, inactivation of the piRNA pathway is associated with very high expression of transposable elements during meiosis, in turn leading to myriad meiosis defects and eventual sterility (Casteneda et al. 2011). In addition to regulating transposons, there is some evidence that the small RNA machinery may directly regulate developmental gene expression in the germ line. For example, endogenous siRNAs are produced from dsRNA in developing oocytes and limit the accumulation of cognate mRNAs (Watanabe et al. 2008).

13.4.2 Mechanisms of Epigenetic Regulation in the Drosophila Germ Line

In *Drosophila*, as in *C. elegans* and mouse, a repressive chromatin structure is responsible for maintaining transcriptional quiescence in the PGCs (often called pole cells) (Nakamura and Seydoux 2008). In *Drosophila*, as in nematodes, DNA methylation is absent and chromatin modifications strictly involve histones. Global analysis of histone modifications in *Drosophila* indicates that active histone marks are absent from (or present at very low levels in) newly formed pole cells. These studies were performed by indirect immunofluorescence analysis of histone modifications in pole cells of female embryos (Schaner et al. 2003; Rudolph et al. 2007) and ChIP-chip and ChIP-seq analysis of undifferentiated germ cells derived from *bam* (bag of marbles) mutant males (Gan et al. 2010a, b). Activity of the H3K4 demethylase, SU(VAR)3-3, restricts accumulation of H3K4 methyl (activation) marks in pole cells and is required for accumulation of H3K9me2 silencing marks and formation of heterochromatin (Rudolph et al. 2007). In comparison, marks of active chromatin are present in the *C. elegans* germ line founder cell, P4, but are globally removed as P4 divides during embryogenesis to form the initial PGCs (Z2 and Z3) (Schaner et al. 2003). No further PGC divisions occur until larval development, at which time active modifications are detected on all chromosomes except the X (Schaner et al. 2003; Nakamura and Seydoux 2008). This regulation resembles the situation in mouse described above where the global pattern of chromatin marks in the initial PGCs is similar to that of the surrounding somatic cells, but quickly becomes distinctive as PGCs begin to divide and migrate.

In *Drosophila*, chromatin regulation is critical for maintenance of the GSCs in the adult gonad. Scrawny (Scny) is a ubiquitin-specific protease that deubiquitylates H2B and is essential for maintenance of several types of stem cells, including GSCs (Buszczak et al. 2009). In *scny* mutants, GSCs have elevated levels of ubiquitinylated H2B and H3K4me3 and, as a consequence, transcription. It is hypothesized that Scny activity maintains the stem cell fate by preventing expression of differentiation genes. Sex-specific histone modifiers function in the GSCs, as well. In the female germline, H3K9 methyltransferase activity is critical for fertility. The activities of three distinct H3K9 methyltransferases, dSETDB1/Eggless, SU(VAR)3-9, and dG9a, produce H3K9me3 modifications during oogenesis (Yoon et al. 2008; Lee et al. 2010). However, dSETDB1/Eggless activity in GSCs and early in oogenesis is required for female fertility, while SU(VAR)3-9 activity later in oogenesis is nonessential for fertility (Clough et al. 2007; Yoon et al. 2008). Expression of dG9a is required in germline support cells (nurse cells) during oogenesis (Lee et al. 2010). Genetic studies suggest partial functional redundancy among these three H3K9 MTases (Lee et al. 2010 and references therein). H3K9me3 levels also depend on activity of the heterochromatin-associated protein, Stonewall (Stwl) (Yi et al. 2009). Stwl activity maintains the female GSC fate and prevents premature germ cell differentiation (Maines et al. 2007; Yi et al. 2009). In *stwl* mutants, levels of H3K9me3 and H3K27me3 are reduced, and these changes presumably contribute to the inappropriate expression of differentiation genes and loss of the GSC fate in *stwl* mutants (Yi et al. 2009).

Male-specific regulators of GSC chromatin also have been identified. Nclb (No child left behind) is a chromatin-binding protein whose function is essential for maintaining the GSC fate in males but not in females (Casper et al. 2011). Levels of H3S10 phosphorylation, a histone modification associated with transcriptional elongation, are very reduced in *nclb* mutants. This and other evidence suggest that transcription is reduced in *nclb* mutants, and therefore Nclb activity is likely to ensure transcription of genes necessary for maintenance of the GSC fate in males.

As in mouse and *C. elegans*, chromatin regulation is critical during *Drosophila* gametogenesis and changes in male germ cell chromatin are more substantial than those in female germ cell chromatin. As in other species, the incorporation of histone variants into germ cell chromatin during meiosis is important for chromosome condensation. In the absence of H3.3 expression, visible defects in chromosome morphology are observed beginning in meiosis; chromosomes fail to condense properly and later fail to segregate correctly during the meiotic divisions (Ooi et al. 2006; Sakai et al. 2009). Chromatin compaction in the mature sperm head late in spermatogenesis (during spermiogenesis) requires accumulation of histone H4 acetylation marks, which then promote the replacement of histones by small basic proteins called protamines (Awe and Renkawitz-Pohl 2010).

13.4.2.1 Meiotic Sex Chromosome Inactivation

There is contradictory evidence as to whether meiotic sex chromosome inactivation occurs in *Drosophila*. One difficulty in answering this question may have to do with technical problems in examining nuclei of the correct meiotic stage in the *Drosophila* testis, which includes a heterogeneous population of somatic and germ cells. In addition, autosomes do not synapse during meiosis in *Drosophila* males, so such a distinction between autosomes and sex chromosomes is not present. Nonetheless, several studies have examined global analysis of gene expression in the testis. Gene expression analysis of spermatogenic arrest mutants and developing (wild-type) testes failed to detect evidence of MSCI (Sturgill et al. 2007; Mikhaylova and Nurminsky 2011), whereas analysis of dissected regions of the testis found a very mildly reduced level of X-linked relative to autosomal gene expression in cells enriched for meiotic as opposed to mitotic or post-meiotic cells (Vibranovski et al. 2009). The best evidence for silencing of the *Drosophila* male X chromosome was provided by transgene studies showing that expression of autosomal spermatogenesis genes becomes down-regulated when these genes are incorporated into the X chromosome as transgenes (Hense et al. 2007; Meiklejohn et al. 2011). This phenomenon was initially interpreted as a sign that X-linked genes are silenced during male meiosis (Hense et al. 2007). However, recent evidence indicates that transcription from the male X is reduced relative to autosomes even prior to meiotic entry (i.e., in mitotic germ cells), and therefore the observed transgene silencing is not strictly meiotic (Meiklejohn et al. 2011). Indeed, genes with a male-biased expression pattern are severely underrepresented on the X, and therefore X-linked gene expression should be relatively low in the male germ line (Parisi et al. 2003; Sturgill et al. 2007). So far, there are no reports suggesting that unsynapsed regions other than the male X and Y, e.g., autosomes or translocations, are silenced, as would be expected for general meiotic silencing of unpaired chromatin.

13.4.2.2 Transposon Silencing and Other siRNA-Mediated Processes in the Germ Line

As in other organisms, repression of transposon activity in the *Drosophila* germ line is important for maintenance of genome integrity. A piRNA pathway analogous to that present in mouse functions to repress transposon expression (Khurana and Theurkauf 2010). This pathway is active in both the male and female germline, and Piwi/piRNA complexes are transmitted maternally in order to prevent transposable element activity in the progeny. Up-regulation of transposon production in the female germline can cause sterility, possibly by triggering a checkpoint that would normally eliminate germ cells with a high potential for carrying mutations (Chen et al. 2007). Interestingly, activity of dSETDB1/Eggless is required for transcription of piRNA clusters (Rangan et al. 2011). Since dSETDB1 activity promotes heterochromatin assembly, and piRNA clusters are located within heterochromatic regions, it is hypothesized that expression of piRNAs is triggered in some way by the presence of heterochromatin (Rangan et al. 2011). In addition, piRNAs (originally called repeat-associated RNAs, rasiRNAs) appear to be important for germline development as mutations that disrupt their accumulation cause female sterility (Pane et al. 2007).

13.5 Implications for the Embryo

Evidence from many organisms suggests that the epigenetic state of gamete chromatin directly influences gene expression in the embryo. Imprinted DNA methylation marks regulate the expression of maternal vs paternal genes in the early embryo prior to global erasure of such imprinted marks. Moreover, although chromatin modifications are removed from many sites in the early embryo, other marks are reported to escape removal. Therefore, epigenetic regulation established in the germ line can be inherited by and influence gene expression in the offspring. This phenomenon, termed transgenerational epigenetic inheritance, has been described in *C. elegans*, mouse, and *Drosophila* as well as in many other animals and in plants (Daxinger and Whitelaw 2010). Mechanisms of epigenetic transgenerational inheritance involve histone modifications and small RNAs. In *C. elegans*, gene activity in the parental germ line contributes to epigenetic regulation in both of the embryonic germ cell precursors (Rechtsteiner et al. 2010; Furuhashi et al. 2010) and in the soma (Arico et al. 2011). Moreover, effects can be observed well beyond embryogenesis: the loss of H3K4 methylation complex activity in the parent (caused by mutations in *set-2*, *ash-2*, or *wdr-5*) will extend lifespan in descendants for up to three generations (Greer et al. 2011). This effect is suppressed by the loss of RBR-2 demethylase activity. Therefore, in this case, the longevity phenotype is presumably caused by insufficient H3K4 methylation.

Heritability of RNAi via either maternally or paternally transmitted factors was demonstrated more than 10 years ago (Grishok et al. 2000), and RNAi was recently shown to trigger the heritable expression of siRNAs in the progeny of animals treated with dsRNA (Burton et al. 2011). Other recent studies suggest that inherited 26G RNAs regulate gene expression during embryogenesis (Gent et al. 2009; Han et al. 2009). Gent et al. (2009) demonstrated that RRF-3 activity during spermatogenesis is required for normal embryonic development, suggesting that paternally inherited 26G RNAs may function in embryogenesis. In a complementary study, Han et al. (2009) demonstrated maternal inheritance of 26G RNAs whose presence correlates with reduced expression of target genes. Using genetic assays, Alcazar et al. (2008) demonstrated the ability of RNAi to persist over multiple generations in the absence of the original targeted allele. This effect could be passed through both oocytes and sperm and was independent of the original targeted allele. These data are consistent with the inherited factor being an siRNA and with the ability of inherited 26G RNAs to repress gene expression. Similar to siRNA, antiviral RNAs (viRNAs) produced in response to viral infection can be inherited (Rechavi et al. 2011). ViRNAs function in silencing viral gene expression via an RNAi-like mechanism (Rechavi et al. 2011). Individuals who lack the machinery to generate viRNAs can mount an antiviral response utilizing inherited viRNAs. These observations further substantiate the hypothesis that inherited small RNAs are critical regulators of gene expression during development.

In their analysis of heritable gene silencing, Burton et al. (2011) demonstrated that RNAi triggers the heritable expression not only of siRNAs but also of H3K9me3 marks. Accumulation of H3K9me3 marks was observed at the target locus, a somatically expressed gene called *dpy-11*, in dsRNA-treated (P0) animals and in their F1 progeny. Evidence suggests that *dpy-11* siRNAs are inherited, while H3K9me3 marks at the *dpy-11* locus are not inherited and instead are reestablished in the F1 progeny. Activity of the nuclear RNAi (NRDE) pathway in the F1 progeny is necessary for expression of *dpy-11* siRNAs and reestablishment of H3K9me3. The nuclear RNAi mechanism involves the Argonaute NRDE-3 and siRNAs generated by RdRP activity in the exogenous RNAi process (termed secondary [2° siRNAs]). The NRDE-3/2° siRNA complex enters the nucleus and recruits two other factors, NRDE-1 and NRDE-3, to chromatin-associated transcripts having homology to the 2° siRNAs (Guang et al. 2008, 2010). A fourth component, NRDE-4, associates with NRDE-1 in the nucleus, and together the NRDE proteins repress transcription of the RNAi-targeted locus by inhibiting RNA polymerase II elongation and directing the deposition of H3K9me3 marks (Guang et al. 2010; Burkhart et al. 2011). In their analysis of heritable RNAi, Burton et al. (2011) observed NRDE-dependent RNAi inheritance phenotypes for only a single generation, whereas other groups have described examples of inherited RNAi of germline-expressed genes extending over many generations (e.g., Grishok et al. 2000; Alcazar et al. 2008; Vastenhouw et al. 2006). It is not yet known if the NRDE pathway functions to maintain the inheritance of germline RNAi and if only germline-expressed genes can be heritably silenced over many generations. Moreover, the function of heritable RNAi is not known, although Burton and colleagues suggest it may be a way for the individual to transmit an environmental, gene-regulatory signal from one generation to the next. In addition, these findings demonstrate that siRNAs can participate in the establishment/maintenance of a heritable pattern of histone modifications at a specific locus.

Recently, Johnson and Spence (2011) described a new phenomenon, termed epigenetic licensing, whereby the presence of a maternally inherited transcript is essential for expression of the cognate gene in the embryonic germ line. This phenomenon was described with respect to *fem-1*, a gene expressed in XO animals and in the larval XX germ line to allow development of the male fate. Johnson and Spence (2011) observed that the inheritance of either a complete or partial *fem-1* mRNA was required for transcription of embryonic *fem-1* in the PGCs. Even inheritance of a non-protein coding, partial transcript was sufficient to allow transcription of embryonic *fem-1*; hence, RNA appears to be the critical inherited factor. The authors propose that *fem-1* may be subject to a form of epigenetic silencing in the PGCs (but not the soma), and the presence of inherited transcripts may override this silencing. The mechanism of this regulation is unknown; however, epigenetic licensing appears to be a new form of regulation distinct from previously described mechanisms.

13.6 Future Directions

Many questions remain as to the mechanisms and developmental importance of epigenetic control in the *C. elegans* germ line. Despite global chromatin analysis that has already been done, in most cases we do not yet know the specific sites of histone modification, and more importantly, the mechanisms responsible for differential distribution of chromatin marks. For example, what mechanisms ensure preferential H3K27me3 accumulation on the X chromosome and H3K9me2 accumulation on unpaired chromosomes? Do all genes on a chromosome receive these marks or only a subset of genes? Another task is to identify the protein complexes responsible for recognizing specific histone modifications and responding to them, e.g., to control gene expression. A related question is the extent to which certain modifications function to regulate chromatin on a chromosomal level, e.g., in order to establish structure important for chromosome segregation. Finally, what is the developmental importance to the embryo of inherited chromatin marks and small RNAs from the sperm and/or oocyte?

Different mechanisms are likely to target MET-2 activity to unpaired chromosomes vs MES2/3/6 activity to X chromosomes. Because MET-2 activity is highest on unpaired chromosomes, it may be recruited by a factor associated with unpaired/unsynapsed chromosomes or excluded by a factor associated with synapsed chromosomes. MES-2/3/6 activity, in contrast, associates with the X chromosome independent of its pairing status and is presumably regulated via a different mechanism that requires MES-4 activity. In other organisms, PRC2 targets regions containing Polycomb response elements (PRE), several DNA-binding proteins have been shown to promote PRC2 binding, and long noncoding RNAs have been implicated as regulators of PRC2 function (Margueron and Reinberg 2011). *C. elegans* MES-2/3/6 activity may likewise be governed by a complex interplay of *cis*-regulatory sites, *trans*-acting proteins, and noncoding RNAs. Recent work also indicates that H3K27 methylation is inhibited by the presence of modifications associated with active chromatin (Schmitges et al. 2011). Perhaps the low density of active marks on the X, reflecting the paucity of germline-essential genes, creates an appropriate environment for H3K27 methylation.

The developmental importance of histone modifications and histone variants will become clearer as researchers identify the factors that "read" and respond to specific patterns of modifications. The extensive chromatin biology literature has established two distinct roles for histone modifications in (1) directly regulating structural conformation of the chromatin, e.g., physically blocking chromatin compaction, and (2) functioning as binding sites for proteins such as histone-modifying enzymes, transcription factors, and chromatin remodeling proteins (Oliver and Denu 2011). Many modifications appear to function in a context-dependent manner, and a major goal of the field now is to understand the dynamic interactions occurring among chromatin-binding proteins and histones in different chromatin domains. H3K27me3 accumulation at promoter regions is widely observed to correlate with transcriptional repression (Justin et al. 2010). Therefore the elevated level of H3K27me3 marks on *C. elegans* germline X chromosomes presumably reflects the lower numbers of actively expressed genes on the X as compared with autosomes. In contrast, H3K9me2 is broadly distributed over genomic DNA corresponding to facultative heterochromatin regions and, while absent from active regions in some cell types (e.g., mouse ES cells and neuronal stem cells), is detected within the bodies of some active genes in differentiated cell types (Wen et al. 2009; Lienert et al. 2011). This pattern is interesting given the situation in the *C. elegans* germ line where the presence of elevated H3K9me2 marks on unpaired autosomes, e.g., on unpaired chromosome V in *zim-2* mutants, does not disrupt development and, therefore, presumably does not indicate a dramatic reduction in transcription. This result is consistent with accumulation of H3K9me2 marks that do not disrupt expression of active genes. Instead, these marks might serve another function, for example relevant to segregation of non-synapsed homologs. An additional level of complexity is added by the recent report that MET-2 activity, and hence H3K9me2 accumulation, is important for repressing the pachytene checkpoint in XO animals, but not in *him-8* hermaphrodites (Checchi and Engebrecht 2011). Hence, the chromatin state of heterogametic sex chromosomes may serve a purpose distinct from that of unsynapsed homologs.

Although the embryo clearly inherits histone modifications, the developmental importance of these marks is not completely understood. As already discussed, MES-4 function (H3K36 methylation) is important for setting up the correct pattern of transcription in the embryo. The importance of other inherited histone modifications to embryonic development, including differentially distributed H3K27me3 and H3K9me2 marks, remains to be determined.

Acknowledgments We thank Xingyu She for the images included in Fig. 13.3, and Bill Kelly, Michael Cosgrove, Yiqing Guo, and Tim Schedl for comments on the manuscript. This work was supported by NIH funding (1R01GM089818) to EMM.

References

Ahmed S, Hodgkin J (2000) MRT-2 checkpoint protein is required for germline immortality and telomere replication in *C. elegans*. Nature 403:159–164

Alcazar RM, Lin R, Fire AZ (2008) Transmission dynamics of heritable silencing induced by double-stranded RNA in *Caenorhabditis elegans*. Genetics 180:1275–1288

Andersen E, Horvitz H (2007) Two *C. elegans* histone methyltransferases repress *lin-3* EGF transcription to inhibit vulval development. Development 134:2991–2999

Andersson R, Enroth S, Rada-Iglesias A, Wadelius C et al (2009) Nucleosomes are well positioned in exons and carry characteristic histone modifications. Genome Res 19:1732–1741

Arico JK, Katz DJ, van der Vlag J, Kelly WG (2011) Epigenetic patterns maintained in early *Caenorhabditis elegans* embryos can be established by gene activity in the parental germ cells. PLoS Genet 7(6):e1001391

Arthold S, Kurowski A, Wutz A (2011) Mechanistic insights into chromosome-wide silencing in X inactivation. Hum Genet 130:295–305

Awe S, Renkawitz-Pohl R (2010) Histone H4 acetylation is essential to proceed from a histone- to a protamine-based chromatin structure in spermatid nuclei of *Drosophila melanogaster*. Syst Biol Reprod Med 56:44–61

Barski A, Cuddapah S, Cui K, Roh T-Y et al (2007) High-resolution profiling of histone methylations in the human genome. Cell 129:823–837

Batista PJ, Ruby JG, Claycomb JM, Chiang R, Fahlgren N et al (2008) PRG-1 and 21U-RNAs interact for form the piRNA complex required for fertility in *C. elegans*. Mol Cell 31:67–78

Bean CJ, Schaner CE, Kelly WG (2004) Meiotic pairing and imprinted X chromatin assembly in *Caenorhabditis elegans*. Nat Genet 36:100–105

Bender LB, Cao R, Zhang Y, Strome S (2004) The MES-2/MES-3/MES-6 complex and regulation of histone H3 methylation in *C. elegans*. Curr Biol 14:1639–1643

Bender LB, Suh J, Carroll CR, Fong Y et al (2006) MES-4: an autosome-associated histone methyltransferase that participates in silencing the X chromosomes in the *C. elegans* germ line. Development 133:3907–3917

Bessler JB, Andersen EC, Villeneuve AM (2010) Differential localization and independent acquisition of the H3K9me2 and H3K9me3 chromatin modifications in the *Caenorhabditis elegans* adult germ line. PLoS Genet 6:1–16

Bird A (2007) Perceptions of epigenetics. Nature 447:396–398

Burkhart KB, Guang S, Buckley BA, Wong L et al (2011) A pre-mRNA-associating factor links endogenous siRNAs to chromatin regulation. PLoS Genet 7:e1002249

Burton NO, Burkhart KB, Kennedy S (2011) Nuclear RNAi maintains heritable gene silencing in *Caenorhabditis elegans*. Proc Natl Acad Sci USA 108:19683–19688

Buszczak M, Paterno S, Spradling AC (2009) Drosophila stem cells share a common requirement for the histone HB ubiquitin protease scrawny. Science 323:248–251

Capowski EE, Martin P, Garvin C, Strome S (1991) Identification of grandchildless loci whose products are required for normal germ-line development in the nematode *Caenorhabditis elegans*. Genetics 129:1061–1072

Casper AL, Baxter K, Van Doren M (2011) No child left behind encodes a novel chromatin factor required for germline stem cell maintenance in males but not females. Development 138:3357–3366

Casteneda J, Genzor P, Bortvin A (2011) piRNAs, transposon silencing, and germline genome integrity. Mutation Res 714:95–104

Checchi P, Engebrecht J (2011) *Caenorhabditis elegans* histone methyltransferase MET-2 shields the male X chromosome from checkpoint machinery and mediates meiotic sex chromosome inactivation. PLoS 9:e1002267

Chen Y, Pane A, Schupbach T (2007) Cutoff and aubergine mutations result in retrotransposon upregulation and checkpoint activation in Drosophila. Curr Biol 17:637–642

Christensen J, Agger K, Cloos PA, Pasini D et al (2007) RBP2 belongs to a family of demethylases, specific for tri- and dimethylated lysine 4 on histone 3. Cell 128:1063–1076
Claycomb JM, Batista PJ, Pang KM, Gu W et al (2009) The Argonaute CSR-1 and its 22 G-RNA cofactors are required for holocentric chromosome segregation. Cell 139:123–134
Clough E, Moon W, Wang S, Smith K, Hazelrigg T (2007) Histone methylation is required for oogenesis in *Drosophila*. Development 134:157–165
Conine CC, Batista PJ, Gu W, Claycomb JM, Chaves DA et al (2010) Argonautes ALG-3 and ALG-4 are required for spermatogenesis-specific 26 G-RNAs and thermotolerant sperm in *Caenorhabditis elegans*. Proc Natl Acad Sci USA 107:3588–3593
Das PP, Bagijn MP, Goldstein LD, Woolford JR, Lehrbach NJ et al (2008) Piwi and piRNAs act upstream of an endogenous siRNA pathway to suppress Tc3 transposon mobility in the *Caenorhabditis elegans* germline. Mol Cell 31:79–90
Daxinger L, Whitelaw E (2010) Transgenerational epigenetic inheritance: more questions than answers. Genome Res 20:1623–1628
Eissenberg JC, Reuter G (2009) Cellular mechanism for targeting heterochromatin formation in Drosophila. Int Rev Cell Mol Biol 273:1–47
Erhard KF, Hollick JB (2011) Paramutation: a process for acquiring trans-generational regulatory states. Curr Opin Plant Biol 14:210–216
Farthing CR, Ficz G, Ng RK, Chan CF et al (2008) Global mapping of DNA methylation in mouse promoters reveals epigenetic reprogramming of pluripotency genes. PLoS Genet 4:e1000116
Feng S, Cokus SJ, Zhang X, Chen PY et al (2010) Conservation and divergence of methylation patterning in plants and animals. Proc Natl Acad Sci USA 107:8689–8694
Fong Y, Bender L, Wang W, Strome S (2002) Regulation of the different chromatin states of autosomes and X chromosomes in the germ line of *C. elegans*. Science 296:2235–2238
Furuhashi H, Takasaki T, Rechtsteiner A, Le T, Kimura H, Checchi PM, Strome S, Kelly WG (2010) Transgenerational epigenetic regulation of *C. elegans* primordial germ cells. Epigenetics Chromatin 3:15
Gan Q, Schones DE, Ho Eun S, Wei G, Cui K, Zhao K, Chen X (2010a) Monovalent and unpoised status of most genes in undifferentiated cell-enriched *Drosophila* testis. Genome Biol 11:R42
Gan Q, Chepelev I, Wei G, Tarayrah L, Cui K, Zhao K, Chen X (2010b) Dynamic regulation of alternative splicing and chromatin structure in *Drosophila* gonads revealed by RNA-seq. Cell Res 20:763–783
Garvin C, Holdeman R, Strome S (1998) The phenotype of *mes-2*, *mes-3*, *mes-4*, and *mes-6*, maternal-effect genes required for survival of the germline in *Caenorhabditis elegans*, is sensitive to chromosome dosage. Genetics 148:167–185
Gent JI, Schvarzstein M, Villeneuve AM, Gu SG, Jantsch V, Fire AZ, Baudrimont A (2009) A *Caenorhabditis elegans* RNA-directed RNA polymerase in sperm development and endogenous RNA interference. Genetics 183:1297–1314
Gent JI, Lamm AT, Pavelec DM, Maniar JM, Parameswaran P, Tao L, Kennedy S, Fire AZ (2010) Distinct phases of siRNA synthesis in an endogenous RNAi pathway in *C. elegans* soma. Mol Cell 37:679–689
Gerstein MB, Lu ZJ, Van Nostrand EL, Cheng C et al (2010) Integrative analysis of the *Caenorhabditis elegans* genome by the modENCODE project. Science 330:1775–1787
Godman M, Lambrot R, Kimmins S (2009) The dynamic epigenetic program in male germ cells: its role in spermatogenesis, testis cancer, and its response to the environment. Microsc Res Tech 72:603–619
Greer EL, Maures TJ, Ucar D, Hauswirth AG et al (2011) Transgenerational epigenetic inheritance of longevity *in Caenorhabditis elegans*. Nature 479:365–371
Grishok A, Tabara H, Mello CC (2000) Genetic requirements for inheritance of RNAi in *C. elegans*. Science 287:2494–2497
Gu S, Fire A (2010) Partitioning the *C. elegans* genome by nucleosome modification, occupancy, and positioning. Chromosoma 119:73–87

Gu W, Shirayama M, Conte D Jr, Vasale J et al (2009) Distinct argonaute-mediated 22 G-RNA pathways direct genome surveillance in the *C. elegans* germline. Mol Cell 36:231–244

Gu L, Want Q, Sun QY (2010) Histone modifications during mammalian oocyte maturation. Dynamics, regulation and functions. Cell Cycle 9:194201950

Guang S, Bochner AF, Pavelec DM, Burkhart KB et al (2008) An Argonaute transports siRNAs from the cytoplasm to the nucleus. Sicence 321:537–541

Guang S, Bochner AF, Burkhart KB, Burton N et al (2010) Small regulator RNAs inhibit RNA polymerase II during the elongation phase of transcription. Nature 465:1097–1837

Hajkova P, Ancelin K, Waldmann T et al (2008) Chromatin dynamics during epigenetic reprogramming in the mouse germ line. Nature 452:877–881

Han T, Manoharan AP, Harkins TT, Bouffard P et al (2009) 26G endo-siRNAs regulate spermatogenic and zygotic gene expression in *Caenorhabditis elegans*. Proc Natl Acad Sci USA 106:18674–18679

Hense W, Baines JF, Parsch J (2007) X chromosome inactivation during *Drosophila* spermatogenesis. PLoS Biol 5:e273

Hudson QJ, Kulinski TM, Huetter SP, Barlow DP (2010) Genomic imprinting mechanisms in embryonic and extraembryonic mouse tissues. Heredity 105:45–56

Ilik I, Akhtar A (2009) roX RNAs: non-coding regulators of the male X chromosome in flies. RNA Biol 6:113–121

Jaramillo-Lambert A, Engebrecht J (2010) A single unpaired and transcriptionally silenced X chromosome locally precludes checkpoint signaling in the *Caenorhabditis elegans* germ line. Genetics 184:613–628

Johnson CL, Spence AM (2011) Epigenetic licensing of germline gene expression by maternal RNA in *C. elegans*. Science 333:1311–1314

Justin N, De Marco V, Aasland R, Gamblin SJ (2010) Reading, writing, and editing methylated lysines on histone tails: new insights from recent structural studies. Curr Opin Struct Biol 20:730–738

Kageyama S, Liu H, Kaneko N, Ooga M, Nagata M, Aoki F (2007) Alternations in epigenetic modification during oocyte growth in mice. Reproduction 133:85–94

Kato M, de Lencastre A, Pincus Z, Slack FJ (2009) Dynamic expression of small non-coding RNAs, including novel microRNAs and piRNAs/21U-RNAs, during *Caenorhabditis elegans* development. Genome Biol 10:R54. doi:10.1.1186/gb-2009-10-5-r54

Katz DJ, Edwards TM, Reinke V, Kelly WG (2009) A *C. elegans* LSD1 demethylase contributes to germline immortality by reprogramming epigenetic memory. Cell 137:308–320

Kelly WG, Schaner CE, Dernburg AF, Lee MH et al (2002) X-chromosome silencing in the germline of *C. elegans*. Development 129:479–492

Kennedy S, Wang D, Ruvkun G (2004) A conserved siRNA-degrading RNase negatively regulates RNA interference in *C. elegans*. Nature 427:645–649

Ketting RF (2011) The many faces of RNAi. Dev Cell 20:148–161

Khurana JS, Theurkauf W (2010) piRNAs, transposon silencing, and *Drosophila* germline development. J Cell Biol 191:905–913

Kolasinska-Zwierz P, Down T, Latorre I, Liu T, Liu XS, Ahringer J (2009) Differential chromatin marking of introns and expressed exons by H3K36me3. Nat Genet 41:376–381

Kouzarides T (2007) Chromatin modifications and their function. Cell 128:693–705

Lee KS, Yoon J, Park JS, Kang YK (2010) *Drosophila* G9a is implicated in germ cell development. Insect Mol Biol 19:131–139

Li T, Kelly WG (2011) A role for Set1/MLL-related components in epigenetic regulation of the *Caenorhabditis elegans* germ line. PLoS Genet 7:1–20

Li B, Carey M, Workman JL (2007) The role of chromatin during transcription. Cell 128:707–719

Lienert F, Mohn F, Tiwari VK, Baubec T, Roloff TC et al (2011) Genomic prevalence of heterochromatic H3K9me2 and transcription do not discriminate pluripotent from terminally differentiated cells. PLoS Genet 7:e1002090

Liu T, Rechtsteiner A, Egelhofer TA, Vielle A et al (2011) Broad chromosomal domains of histone modification patterns in *C. elegans*. Genome Res 21:227–236

Maine EM (2010) Meiotic silencing in *Caenorhabditis elegans*. Intl Rev Cell Mol Biol 282: 91–134

Maine EM, Hauth J, Radcliff T, Vought VE, She X, Kelly WG (2005) EGO-1, a putative RNA-dependent RNA polymerase, is required for heterochromatin assembly on unpaired DNA during *C. elegans* meiosis. Curr Biol 15:1972–1978

Maines JZ, Park JK, Williams M, McKearin DM (2007) Stonewalling *Drosophila* stem cell differentiation by epigenetic controls. Development 134:1471–1479

Maniar JM, Fire AZ (2011) EGO-1, a *C. elegans* RdRP, modulates gene expression via production of mRNA-templated short antisense RNAs. Curr Biol 21:449–459

Margueron R, Reinberg D (2011) The Polycomb complex PRC2 and its mark in life. Nature 469:343–349

Meier B, Barber LJ, Shtessel L, Boulton SJ et al (2009) The MRT-1 nuclease is required for DNA crosslink repair and telomerase activity in vivo in *Caenorhabditis elegans*. EMBO J 28: 3549–3563

Meiklejohn CD, Landeen EL, Cook JM, Kingan SB, Presgraves DC (2011) Sex chromosome-specific regulation in the *Drosophila* male germ line but little evidence for chromosomal dosage compensation of meiotic inactivation. PLoS Biol 9:e1001126

Meyer BJ (2010) X-chromosome dosage compensation. Curr Opin Genet Dev 20:179–189

Mikhaylova LM, Nurminsky DI (2011) Lack of global meiotic sex chromosome inactivation, and paucity of tissue-specific gene expression on the *Drosophila* X chromosome. BMC Biol 9:29

Nakamura A, Seydoux G (2008) Less is more: specification of the germline by transcriptional repression. Development 135:3817–3827

Nakamura M, Ando R, Nakazawa T, Yudazono T, Tsutsumi N, Hatanaka N, Ohgake T, Hanaoka F, Eki T (2007) Dicer-related *drh-3* gene functions in germ-line development by maintenance of chromosomal integrity in *Caenorhabditis elegans*. Genes Cells 12:997–1010

Oliver SS, Denu JM (2011) Dynamic interplay between histone H3 modifications and protein interpreters: emerging evidence for a "histone language". Chembiochem 12:299–307

Ooi SL, Priess JR, Henikoff S (2006) Histone H3.3 variant dynamics in the germline of *Caenorhabditis elegans*. PLoS Genet 2:e97

Pane A, Wehr K, Schupbach T (2007) Zucchini and squash encode two putative nucleases required for rasiRNA production in the *Drosophila* germline. Dev Cell 12:851–862

Parisi M, Nuttall R, Naiman D, Bouffard G, Malley J et al (2003) Paucity of genes on the *Drosophila* X chromosome showing male-biased expression. Science 299:697–700

Payer B, Lee JT, Namekawa SH (2011) X-inactivation and X-reactivation: epigenetic hallmarks of mammalian reproduction and plurpotent stem cells. Hum Genet 130:265–280

Rangan P, Malone CD, Navarro C, Newbold SP, Hayes PS, Sachidanandam R, Hannon GJ, Lehmann R (2011) piRNA productions requires heterochromatin formation in *Drosophila*. Curr Biol 21:1373–1379

Rechavi O, Minevich G, Hobert O (2011) Transgenerational inheritance of an acquired small RNA-based antiviral response in *C. elegans*. Cell 147:1248–1256

Rechtsteiner A, Ercan S, Takasaki T, Phippen TM et al (2010) The histone H3K36 methyltransferase MES-4 acts epigenetically to transmit the memory of germline gene expression to progeny. PLoS Genet 6:1–15

Reinke V, Smith HE, Nance J, Wang J, Van Doren C et al (2000) A global profile of germline gene expression in *C. elegans*. Mol Cell 6:605–616

Reinke V, Gil IS, Ward S, Kazmer K (2004) Genome-wide germline-enriched and sex-biased expression profiles in *Caenorhabditis elegans*. Development 131:311–323

Reuben M, Lin R (2002) Germline X chromosomes exhibit contrasting patterns of histone H3 methylation in *Caenorhabditis elegans*. Dev Biol 245:71–82

Rocheleau CE, Cullison K, Huang K, Bernstein Y, Spilker AC, Sundaram MV (2008) The *Caenorhabditis elegansi ekl* (enhance of *ksr-1* lethality) genes include putative components of a germline small RNA pathway. Genetics 178:1431–1443

Royo H, Polikiewicz G, Mahadevaiah SK et al (2010) Evidence that meiotic sex chromosome inactivation is essential for male fertility. Curr Biol 20:2117–2123

Ruby JG, Jan C, Player C, Axtell MJ, Lee W, Nusbaum C, Ge H, Bartel DP (2006) Large-scale sequencing reveals 21U-RNAs and additional microRNAs and endogenous siRNAs in *C. elegans*. Cell 127:1193–1207

Rudolph T, Yonezawa M, Lein S et al (2007) Heterochromatin formation in *Drosophila* is initiated through active removal of H3K4 methylation by the LSD1 homolog of SU(VAR)3-3. Mol Cell 26:103–115

Sakai A, Schwartz BE, Goldstein S, Ahmad K (2009) Transcriptional and developmental functions of the H3.3 histone variant in *Drosophila*. Curr Biol 19:1816–1820

Sasaki H, Matsui Y (2008) Epigenetic events in mammalian germ-cell development: reprogramming and beyond. Nat Rev Genet 9:129–140

Schaner CE, Deshpande G, Schedl PD, Kelly WG (2003) A conserved chromatin architecture marks and maintains the restricted germ cell lineage in worms and flies. Dev Cell 5:747–757

Schaner CE, Kelly WG (2006) Germline chromatin (January 24, 2006), *WormBook*, ed. The *C. elegans* Research Community, WormBook, doi/10.1895/wormbook.1.73.1, http://www.wormbook.org

Schimenti J (2005) Synapsis or silence. Nat Genet 37:11–13

Schmitges FW, Prusty AB, Faty M et al (2011) Histone methylation by PRC2 is inhibited by active chromatin marks. Mol Cell 42:330–341

Schultz DC, Ayyanathan K, Negorev D, Maul GG et al (2002) SETDB1: a novel KAP-1-associated histone H3, lysine 9-specific methylatransferase that contributes to HP1-mediated silencing of euchromatic genes by KRAB zinc-finger proteins. Genes Dev 16:919–932

She X, Xu W, Fedotov A, Kelly WG, Maine EM (2009) Regulation of heterochromatin assembly on unpaired chromosomes during *Caenorhabditis elegans* meiosis by components of a small RNA-mediated pathway. PLoS Genet 5:1–15

Shilatifard A (2008) Molecular implementation and physiological roles for histone H3 lysine 4 (H3K4) methylation. Curr Op Cell Biol 20:341–348

Simmer F, Tijsterman M, Parrish S, Koushika SP, Nonet ML, Fire A, Ahringer J, Plasterk RH (2002) Loss of the putative RNA-directed RNa polymerase RRF-3 makes *C. elegans* hypersensitive to RNAi. Curr Biol 12:1317–1319

Sturgill D, Zhang Y, Parisi M, Oliver B (2007) Demasculinization of X chromosomes in the *Drosophila* genus. Nature 450:238–241

Surani MA, Reik W, Norris ML, Barton SC (1986) Influence of germline modifications of homologous chromosomes on mouse development. J Embyol Exp Morphol 90:267–285

Turner JMA (2007) Meiotic sex chromosome inactivation. Development 134:1823–1831

Updike DL, Strome S (2009) A genomewide RNAi screen for genes that affect the stability, distribution, and function of P granules in *Caenorhabditis elegans*. Genetics 183:1397–1419

Updike DL, Strome S (2010) P granule assembly and function in *Caenorhabditis elegans* germ cells. J Androl 31:53–60

Van der Heijden GW, Derijck AA, Posfai E, Giele M et al (2007) Chromosome-wide nucleosome replacement and H3.3 incorporation during mammalian meiotic sex chromosome inactivation. Nat Genet 39:251–258

Van Wolfswinkel JC, Claycomb JM, Batista PJ, Mello CC, Berezikov E, Ketting RF (2009) CDE-1 affects chromosome segregation through uridylation of CSR-1-bound siRNAs. Cell 139:135–148

Vasale JJ, Gu W, Thivierge C, Batista PJ, Claycomb JM, Youngman EM, Duchaine TF et al (2010) Sequential rounds of RNA-dependent RNA transcription drive endogenous small-RNA biogenesis in the ERGO-1/Argonaute pathway. Proc Natl Acad Sci USA 107:3582–3587

Vastenhouw NL, Brunschwig K, Okihara KL, Muller F et al (2006) Gene expression: long-term gene silencing by RNAi. Nature 442:882

Vibranovski MD, Lopes HF, Karr TL, Long M (2009) Stage-specific expression profiling of *Drosophila* spermatogenesis suggests that meiotic sex chromosome inactivation drives genomic relocation of testis-expressed genes. PLoS Genet 5:e1000731

Vought VE, Ohmachi M, Lee MH, Maine EM (2005) EGO-1, a putative RNA-directed RNA polymerase, promotes germline proliferation in parallel with GLP-1/Notch signaling and regulates the spatial organization of nuclear pore complexes and germline P granules in *C. elegans*. Genetics 170:1121–1132

Wang G, Reinke V (2008) A *C. elegans* Piwi, PRG-1, regulates 21U-RNAs during spermatogenesis. Curr Biol 18:861–867

Wang JT, Seydoux S (2012) Germ cell specification. Advances in Experimental Medicine and Biology 757:17–39. (Chap. 2, this volume) Springer, New York

Wang PJ, Page DC, McCarrey JR (2005) Differential expression of sex-linked and autosomal germ-cell-specific genes during spermatogenesis in the mouse. Hum Mol Genet 14:2911–2918

Watanabe T, Totoki Y, Toyoda A, Kaneda M et al (2008) Endogenous siRNAs from naturally formed dsRNAs regulate transcripts in mouse oocytes. Nature 453:539–543

Wen B, Wu H, Shinkai Y, Irizarry RA, Feinberg AP (2009) Large histone H3 lysine 9 demethylated chromatin blocks distinguish differentiated from embryonic stem cells. Nature Genet 41:246–250

Xiao Y, Bedet C, Robert VJP, Simonet T, Dunkelbarger S et al (2011) SET-2 and ASH-2 are differentially required for histone H3 Lys 4 methylation in embryos and adult germ cells. Proc Natl Acad Sci USA 108:8305–8310

Yi X, de Vries HI, Siudeja K, Rana A et al (2009) Stwl modifies chromatin compaction and is required to maintain DNA integrity in the presence of perturbed DNA replication. Mol Biol Cell 20:983–994

Yoon J, Lee KS, Park JS, Yu K, Paik SG, Kang YK (2008) dSETDB1 and SU(VAR)3-9 sequentially function during germline-stem cell differentiation in *Drosophila melanogaster*. PLoS One 3:e2234

Zamudio NM, Chong S, O'Bryan MK (2008) Epigenetic regulation in male germ cells. Reproduction 136:131–146

Zamudio NM, Scott HS, Wolski K, Lo CY et al (2011) DNMT3L is a regulator of X chromosome compaction and post-meiotic gene transcription. PLoS One 6:e18276

Zhang C, Montgomery TA, Gabel HW, Fischer SE et al (2011) *mut-16* and other mutator class genes modulate 22G and 26G siRNA pathways in *Caenorhabditis elegans*. Proc Natl Acad Sci USA 108:1201–1208

Chapter 14
Using *Caenorhabditis* to Explore the Evolution of the Germ Line

Eric S. Haag and Qinwen Liu

Abstract Germ cells share core attributes and homologous molecular components across animal phyla. Nevertheless, abrupt shifts in reproductive mode often occur that are mediated by the rapid evolution of germ cell properties. Studies of *Caenorhabditis* nematodes show how the otherwise conserved RNA-binding proteins (RBPs) that regulate germline development and differentiation can undergo surprisingly rapid functional evolution. This occurs even as the narrow biochemical tasks performed by the RBPs remain constant. The biological roles of germline RBPs are thus highly context-dependent, and the inference of archetypal roles from isolated models in different phyla may therefore be premature.

Keywords RNA-binding protein • Translation • PUF proteins • GLD-1

14.1 Comparative Biology of Germline Development

14.1.1 Evolutionary Overview

In this chapter, we seek to put attributes of the *C. elegans* germ line covered by other authors in this volume in an evolutionary context. Two major themes run through it. First, *C. elegans* germ cells share many features with those of other animals and are thus a useful model system for inferring general principles. Second, we discuss how comparisons with other closely related nematodes offer an important window onto the role germ cells play in the evolution of important new reproductive adaptations.

E.S. Haag (✉) • Q. Liu
Department of Biology, University of Maryland, Building 144,
College Park, MD 20742, USA
e-mail: ehaag@umd.edu

T. Schedl (ed.), *Germ Cell Development in C. elegans*, Advances in Experimental Medicine and Biology 757, DOI 10.1007/978-1-4614-4015-4_14,

Before these main issues are addressed, however, it is appropriate to take the widest possible view.

As the cells that undergo meiosis and produce the haploid gametes that fuse to produce sexual diploids, germ cells are the essential mediators of sexual reproduction in animals. But sex itself is found not only in the multicellular animal, fungal, and plant lineages, but also in the unicellular protists that collectively gave rise to them (Leonard 2010; Ramesh et al. 2005). Thus, the meiotic aspects of germ cells have very ancient roots. However, in protists the haploid phase can dominate the life cycle, and highly differentiated gametes are often not produced. For example, in the unicellular alga *Chlamydomonas*, cryptically different haploid cells of opposite mating types fuse to produce diploid zygotes. These zygotes immediately enter meiosis, and mitotic diploid cell cycles do not normally occur (Goodenough et al. 2007). The elegant molecular machinery that regulates mating type in *Chlamydomonas* may have evolved, at least in part, to allow reliable detection of diploidy in the zygote (Haag 2007).

Extensive haploidy and an absence of gametes also characterize the life cycles of fungi, where multicellular haploid mycelia (or, in the case of yeasts, individualized cells) fuse with others to produce diploid mycelia (Lee et al. 2010). Plants in the broad sense (including multicellular algae, ferns, etc.) are variable, but most make gametes. Fern gametes are produced by a detached haploid gametophyte, while Charophytic algae and flowering plants produce gametes in haploid organs that are retained in the diploid plant's body (McCourt et al. 2004; Tanurdzic and Banks 2004). However, despite producing specialized gametes, no evidence for (and considerable evidence against) a dedicated germ line exists in plants. Thus, animals are unique among multicellular organisms in producing sexually dimorphic gametes from a dedicated germ cell population. That the different groups would show such variation seems at first puzzling, but makes more sense when the independent origin of multicellularity in animals, plants, and fungi from distinct protistan ancestors is taken into account (Steenkamp et al. 2006).

The conservation of germ cells across all metazoa and the ample knowledge about *C. elegans* germline biology serve as a solid foundation to identify both core and taxon-specific attributes of germ cells, as well as how these features have evolved. We start by summarizing what is known about *C. elegans,* and then contrast it with other well-characterized developmental systems.

14.1.2 Specification of Germ Cell Lineage

In *C. elegans*, germline fate is separated from the somatic fate by the 16- to 24-cell stage of embryogenesis (Wood et al. 1982). Through asymmetric localization and cell divisions, the germline blastomeres inherit electron-dense P granules, and give rise to the primordial germ cell P4 (Hird et al. 1996). During this early germline specification, the zinc finger protein PIE-1 is specifically enriched in the germline blastomeres, and functions as a transcriptional repressor to prevent somatic differentiation (Mello et al. 1992, 1996; Seydoux et al. 1996). For roughly two decades it was believed that P granules themselves were stable, classical cytoplasmic determinants that directly specified germline fate (Strome and Wood 1983). More recent studies using real-time imaging, however, indicate that P granules are actually fluid structures that are constantly disassembled and reformed during early development (Brangwynne et al. 2009). In addition, partitioning of P granules to the anterior is not sufficient for germline fate (Gallo et al. 2010). This suggests that, rather than acting as static determinants, P granules are responding to the global anterior–posterior axis that independently specifies germ cell fate (also see Wang and Seydoux 2012, Chap. 2).

In the fly *Drosophila melanogaster*, germ cells form by cellularization of nuclei and surrounding germ plasm at the posterior pole of the embryo (Huettner 1923), and these cells have the unique potential to form the germ line (Beer et al. 1987; Illmensee and Mahowald 1974; Williamson and Lehmann 1996). Similar to PIE-1 in *C. elegans,* the protein Pgc acts as a germ cell transcriptional repressor during early germline development (Hanyu-Nakamura et al. 2008; Timinszky et al. 2008).

The African clawed toad, *Xenopus laevis*, is another widely studied model organism. In *Xenopus* oocytes, the vegetal cortex contains electron-dense granules known as mitochondria cloud nuage. After fertilization, these granules are segregated asymmetrically into a few cells that become the primordial germ cells (Heasman et al. 1984; Whitington and Dixon 1975; Zhou and King 1996). Remarkably, transcriptional activity in the PGCs is repressed through the same general mechanisms described above for flies and nematodes, preventing the nascent germ line from adopting a somatic fate (Venkatarama et al. 2010).

Germline specification in the teleost zebrafish *Danio rerio* resembles *Xenopus*, in that germ plasm is localized to the vegetal hemisphere of the oocyte (Komiya et al. 1994; Yoon et al. 1997). In zebrafish, the germ plasm moves from the vegetal hemisphere to the cleavage furrows in the animal hemisphere during early development (Raz 2003). Germ plasm carrying a *vasa*-like gene product and other germline determinants is segregated into four cells that eventually development to PGCs (Hashimoto et al. 2004; Olsen et al. 1997; Yoon et al. 1997).

The phylogenetic breadth of the above cases might suggest that maternally deposited germ plasm (whether a strict determinant or not) is a universal metazoan attribute. However, though germ cells in different phyla do indeed share many characteristics (Extavour 2007), this "preformation" mechanism is far from universal (Extavour and Akam 2003). Indeed, in mammals and other (i.e., non-*Xenopus*) amphibians, germ cells are induced later in development, as a result of signaling events from surrounding tissues (Tam and Zhou 1996). During gastrulation, a small cluster of cells in the epiblast are transformed to PGCs by BMP signals from nearby tissue (Lawson et al. 1999; Ying et al. 2001; Ying and Zhao 2001), and then migrate to somatic gonads later on. In the axolotl *Ambystoma mexicanum*, neither a mitochondrial cloud nor obviously localized molecular determinants are present in oocytes (Johnson et al. 2001, 2003). PGCs arise in the lateral plate mesoderm as a result of inductive signals from the ventral endoderm (Boterenbrood and Nieuwkoop 1973; Humphrey 1925, 1927; Ikenishi and Nieuwkoop 1978). Interestingly, the identification of somatic genes downregulated in isolated mice PGCs indicates that transcriptional repression is a common mode of germ cell specification across vertebrates and invertebrates (Kurimoto et al. 2008).

14.1.3 Germline Maintenance: Germline Stem Cells and Their Niche

Germline stem cells (GSCs) are long-lived and capable of self-renewal, and provide a sustained source of sperm or eggs. In *C. elegans*, GSCs reside at the distal end of the gonad, where they are surrounded by processes of a distal tip cell (DTC) (Kimble and White 1981). This physical interaction of germ cells with the DTC processes likely helps maintain stem cell identity (Cinquin et al. 2010; Crittenden et al. 2006). Activation of Notch signaling in germ cells is necessary and sufficient for GSC identity (McGovern et al. 2009).

The GSC niches have also been identified in *Drosophila melanogaster*. In the ovary, the cap and escort stem cells (Kirilly and Xie 2007; Lin 2002) together form the germline stem cell niche (Song and Xie 2002; Xie and Spradling 2000). In the testis, GSCs contact the hub cells, adjacent somatic cells, and cyst progenitors to form the male niche (Gonczy and DiNardo 1996; Xie and Li 2007). Unlike *C. elegans*, in *D. melanogaster* each sex uses distinct signaling pathways for regulation of GSCs. Although both sexes need BMP signal to control GSC self-renewal (Cox et al. 1998; Kawase et al. 2004; King and Lin 1999; Song et al. 2004), ovarian GSCs require the Yb/Piwi-regulated signal (Cox et al. 1998; King and Lin 1999), while *Drosophila* testicular GSCs need the JAK-STAT signal (Brawley and Matunis 2004).

The best-characterized vertebrate model for GSC–soma interactions is the mouse testis. Here, spermatogonial stem cells (SSCs) are located at the basement membrane of seminiferous tubules, predominantly adjacent to interstitial spaces between tubules. In this niche, SSCs receive glial cell line-derived neurotrophic factor (GDNF) (Meng et al. 2000) and FGF2 (Goriely et al. 2005). These signaling factors are produced by surrounding somatic Sertoli cells to promote self-renewal (de Rooij 2009). Why GSCs are not uniformly distributed around the tubule circumference remains unclear.

14.1.4 The Planarian Flatworm: An Emerging Model for Germ/Stem Cell Biology

Planarians are free-living platyhelminths that can regenerate any part of the body. Their regenerative abilities depend on a proliferating population of somatic stem cells called neoblasts (Salo and Baguna 1984; Shibata et al. 2010). During natural physiological tissue turnover, neoblasts generate progeny cells to compensate for cell loss. When body transection occurs, the neoblasts are induced to migrate and proliferate, giving rise to a regeneration blastema that heals the wound (Newmark and Sanchez Alvarado 2002; Reddien and Sanchez Alvarado 2004; Sanchez Alvarado 2006). Beyond their remarkable capacity for somatic tissue regeneration, planarians are also able to regenerate germ cells (Morgan 1901). The neoblasts are thought to be the source cell underlying a somatic to germ cell transition (Handberg-Thorsager and Salo 2007; Sato et al. 2006; Wang et al. 2007). This notion is further supported by the discovery in neoblasts of electron-dense ribonucleoprotein particles found in the germ line in other organisms (Coward 1974; Hori and Kishida 2003; Sato et al. 2001). Neoblasts also express genes that are important for germline development (Guo et al. 2006; Salvetti et al. 2005; Shibata et al. 1999). Understanding inductive germ cell specification in planarians is thus an exciting emerging field of research.

14.2 *C. elegans* as a Starting Point for Studying Germline Biology and Its Evolution

The *C. elegans* hermaphrodite gonad is a flattened U-shaped tube with two symmetric arms, while the male only has one J-shaped arm (Kimble and White 1981). In both sexes, the adult germ line shows distal–proximal polarity, with the distal end of the gonad hosting a mitotic GSC population, while meiotic cells are gradually displaced proximally (Schedl 1997). The polarity is actively maintained throughout the life span to sustain reproductive capacity. Adult *C. elegans* have about 1,000 somatic cells (Sulston and Horvitz 1977), yet the single PGC in the early embryo (P4) will give rise to more than twice this number of germ cells (Crittenden et al. 2006). Its large cell number and continuous activity when food is plentiful render the germ line the most metabolically active tissue in adult animals. The abundant genetic tools, rapid assembly-line development, and transparent cuticle all facilitate detailed studies (many of which are highlighted elsewhere in this volume). Thus, *C. elegans* provides an excellent system to address germline biology, including control of the balance between proliferation and meiosis and sex determination. Brief summaries of these processes are provided below.

In *C. elegans*, the GLP-1/Notch signaling pathway controls germline cell proliferation during larval development and stem cell maintenance in adults (Kimble and Simpson 1997). The Notch receptor GLP-1 is expressed in the germ line, and receives a juxtacrine signal from the distal tip cell (DTC) to promote mitotic divisions at the expense of entry into meiosis (Austin and Kimble 1989; Crittenden et al. 1994; Yochem and Greenwald 1989). The DTC signal is mediated by a ligand, LAG-2, whose expression strength affects germline proliferation (Henderson et al. 1994; Tax et al. 1994). GLP-1/Notch signaling is both necessary and sufficient for germline mitotic proliferation (Berry et al. 1997). AC/VU precursor cells and the sheath/spermathecal (SS) precursor cells or their descendants are also important to promote germline proliferation (McCarter et al. 1997; Pepper et al. 2003a, b). As germ cells move out of the stem cell niche into the transition zone, they exit the mitotic cell cycle and enter meiosis (Crittenden et al. 2006). Various RNA-binding proteins are required to regulate the transition between cell proliferation and differentiation. In the mitotic zone adjacent to the DTC, the mitotic cell cycle is maintained in part by repression of meiosis-promoting mRNAs by the PUF family FBF proteins (Crittenden et al. 2002; Lamont et al. 2004). Just prior to when germ cells move into transition zone, the influence of GLP-1/Notch signaling decreases, which in turn leads to the expression and/or altered activity of the GLD proteins and meiosis (Hansen et al. 2004; Schmid et al. 2009; Suh et al. 2006).

Germline sex determination in *C. elegans* is controlled by both the global sex determination pathway and by germline-specific regulators (Ellis 2008; Zarkower 2006; Zanetti and Puoti 2012, Chap. 3). Germline sex determination and the mitosis/meiosis decision are both ongoing processes linked in space, as indicated by sex-specific gene expression at distal mitotic region (Jones et al. 1996; Segal et al. 2001; Thompson et al. 2005) and the fact that germ cell sexual fate can be reversed in adult animals by certain conditional sex determination mutations (Barton et al. 1987; Chen and Ellis 2000; Otori et al. 2006) and chemical compounds (Morgan et al. 2010).

14.3 *Caenorhabditis* as a System for Studying Recent Events in Germline Evolution

Caenorhabditis nematode species use one of the two modes of reproduction (Fig. 14.1). Most retain the ancestral gonochoristic mode, with XO males and XX females. Self-fertile species like *C. elegans* produce an XX hermaphrodite instead and have evolved independently from gonochoristic species multiple times (Cho

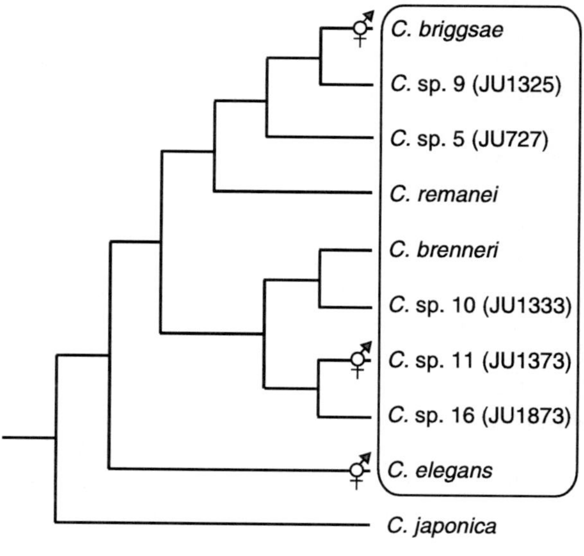

Fig. 14.1 Reproductive mode variation in *Caenorhabditis* nematodes. Inferred shifts from gonochorism to self-fertility (as indicated by the hermaphrodite designation) in the *Elegans* group of *Caenorhabditis* species (*boxed*) are mapped onto a recent phylogenetic hypothesis (modified from Kiontke et al. 2011)

et al. 2004; Kiontke et al. 2004, 2011). The novel trait of male gamete production in female body is the major anatomical difference between hermaphrodites and females, and is rooted in germline development. Below we discuss three aspects of *Caenorhabditis* that makes it an attractive system to study germline evolution.

14.3.1 Tools for Work with Non-Elegans Species Now Extensive and Powerful

Twenty-three *Caenorhabditis* species are currently in culture (Kiontke and Sudhaus 2006; Kiontke et al. 2011). Moreover, their phylogenetic relationships are now well understood (Cho et al. 2004; Kiontke et al. 2004, 2011), which establishes a framework for evolutionary comparisons. To facilitate and expand genomic resources for more of these species, the Caenorhabditis Genome Analysis Consortium (http://wormgenomes.caltech.edu/consortium.html) has sequenced, or is currently sequencing, the genomes from seven gonochoristic (*C. remanei, C. brenneri, C. japonica, C.* sp. 5 JU800, *C.* sp. 7 JU1286, *C.* sp. 9 JU1422, and *C. angaria* PS1010), and three hermaphroditic (*C. elegans, C. briggsae*, and the newly discovered *C.* sp. 11 JU1373) species.

Like *C. elegans, C. briggsae* is androdioecious (with XX hermaphrodite and XO male sexes). A number of genetic and genomic tools have been developed for it, sparked by the production of a high quality draft shotgun genome assembly (Stein et al. 2003). Later, the physical map was organized into chromosomes by recombination mapping of single nucleotide polymorphisms (Hillier et al. 2007). More recently, a larger number of genotyped SNPs were combined with an advanced-intercross design to further refine the recombination map and genome assembly (Ross et al. 2011). In addition, as of 2011, more than 200 mutant strains have been generated to allow characterization of the affected loci and to serve as markers in linkage mapping (http://www.briggsae.org).

Beyond being a resource for comparative biology, *C. briggsae* offers its own unique attributes that make it worthy of study. Relative to *C. elegans, C. briggsae* has increased levels of single nucleotide and insertion–deletion polymorphisms and greater population genetic structure (Cutter et al. 2006; Dolgin et al. 2008; Graustein et al. 2002). It is thus an excellent choice for genome-wide association studies and forward genetic mapping. More than 30,000 polymorphisms identified between two *C. briggsae* strains, AF16 and HK104 (Koboldt et al. 2010) have supported bulk segregant and single animal-based PCR genotyping assays (Zhao et al. 2010). Such tools greatly accelerate genetic mapping of induced mutants, including those affecting somatic and germ cell sex determination (Beadell et al. 2011; Guo et al. 2009; Kelleher et al. 2008). In addition, both RNA interference (by injection) and the isolation of deletion mutations have been successfully employed to examine *C. briggsae* gene functions (Beadell et al. 2011; Hill et al. 2006; Kuwabara 1996). Moreover, microparticle bombardment-mediated gene transformation technique (Praitis et al. 2001) has been successfully applied to the species to study cell lineage and gene expression and enables production of germline-expressing transgenes (Zhao et al. 2010).

14.3.2 *How Did Hermaphrodites Evolve? Are Convergent Hermaphrodites Made the Same Way?*

Caenorhabditis species are morphologically very similar, and germline anatomy and development are no exception. However, the most obvious difference is that hermaphroditic *Caenorhabditis* have each acquired limited XX spermatogenesis. This allows self-fertility, a trait with enormous implications for ecology and population genetics. The tools described above allow comparisons between independently evolved hermaphrodite species and between hermaphrodites and females of closely related gonochoristic species. These contrasts, in turn, make *Caenorhabditis* an excellent system for assessing the reproducibility of the evolution of an adaptive trait at the level of developmental genetics.

14.3.2.1 *C. elegans* Hermaphrodites Evolved by Fine-Tuning a Conserved Sex Determination Pathway

Hermaphrodite spermatocytes have the XX karyotype characteristic of female cells, suggesting that hermaphrodites evolved by modifying the regulation of sex determination downstream of X dosage. The core sex determination pathway in *C. elegans* is conserved in other *Caenorhabditis* species. In *C. elegans*, *C. briggsae*, and *C. remanei*, *tra-1* (de Bono and Hodgkin 1996; Kelleher et al. 2008), *tra-2* (Haag and Kimble 2000; Kelleher et al. 2008; Kuwabara 1996), *tra-3* (Kelleher et al. 2008), *fem-2* (Hansen and Pilgrim 1998; Hill et al. 2006; Stothard et al. 2002; Stothard and Pilgrim 2006), *fem-3* (Haag et al. 2002; Hill et al. 2006), and *fog-3* (Chen et al. 2001) play conserved roles in somatic sex determination. Moreover, the physical interaction between TRA-2 and FEM-3 (Haag et al. 2002) and FEM-2 and FEM-3 (Stothard and Pilgrim 2006) is conserved among *C. elegans*, *C. briggsae*, and *C. remanei*, and the TRA-2-TRA-1 interaction has been documented in both *C. elegans* and *C. briggsae* (Wang and Kimble 2001).

The above studies might suggest that sex determination in generally congruent in all *Caenorhabditis*, and for the soma this is a fair statement. However, in the germ line more dynamic evolution is seen, and this correlates with the rapid phenotypic evolution of this tissue in hermaphroditic lineages. *C. elegans* is the best characterized hermaphrodite species, and has been the subject of genetic and molecular studies for many years (Kimble and Crittenden 2007). We thus turn our attention to the control of germline sex in *C. elegans* hermaphrodites and how it may have evolved.

In *C. elegans*, initiation of XX spermatogenesis requires repression of *tra-2* translation (Goodwin et al. 1993). This is accomplished by the RNA-binding protein GLD-1 (Jan et al. 1999) and its co-factor FOG-2, which form a heterodimer that binds to a cluster of GLD-1-binding elements (GBEs) residing in a short direct repeat element (DRE) in the 3′UTR of the *tra-2* mRNA (Clifford et al. 2000; Francis et al. 1995; Jones et al. 1996; Schedl and Kimble 1988). GLD-1 repression of *tra-2* mRNA may be a conserved regulatory linkage (Haag and Kimble 2000; Jan et al.

1997). However, the multimerized GBEs in *C. elegans tra-2* are all necessary for XX spermatogenesis (Goodwin et al. 1993), are not found in other *Caenorhabditis* species, and correlate with unusually strong interaction between *tra-2* mRNA and GLD-1 in *C. elegans* (Beadell et al. 2011). Thus, it is likely that recent changes in the *cis*-regulatory RNA sequences of *tra*-2 were important for the evolution of self-fertility in *C. elegans.*

While *gld-1* is highly conserved (Nayak et al. 2005), *fog-2* is the product of recent tandem duplications (Clifford et al. 2000). Further, *fog-2* is the only gene in the large *C. elegans* the F-box family that carries the C-terminal sequences necessary to mediate an interaction with GLD-1 (Nayak et al. 2005). Therefore, *fog-2* is likely a new gene with a new function, and thus represents another key step in the evolution of XX spermatogenesis in the *C. elegans* lineage.

A second important translational regulatory mechanism implicated in *C. elegans* hermaphroditism affects *fem-3*. As with *fem-1* and *fem-2*, *fem-3* promotes male development throughout the body, including XX spermatogenesis (Hodgkin 1986). However, *fem-3* activity must also be kept in check to allow the switch to oogenesis (Ahringer and Kimble 1991). This is mediated through post-transcriptional control on its 3′UTR by the PUF family FBF proteins and their co-factors (Kraemer et al. 1999; Zhang et al. 1997), and the *mog* genes (Gallegos et al. 1998). Therefore, post-transcriptional regulation of *fem-3* regulation is another candidate mechanism for how *C. elegans* evolved limited XX spermatogenesis. Surprisingly, in *C. remanei,* while *Cr-fem-3(RNAi)* feminizes the XO male soma, it does not feminize the germ line, nor does it suppress the germline masculinization of XX *Cr-tra-2(RNAi)* (Haag et al. 2002). This suggests (but does not prove) that the role of *fem-3* in germline sex is not constant, something we return to again below.

The above results lead to a tentative model for the evolution of *C. elegans* hermaphrodite germline patterning, in which changes in multiple genes altered translational regulation of key sex-determining genes. Recently, it was demonstrated that hermaphrodites dependent upon the seminal fluid of males for partial self-fertility could be produced via downregulation of a single sex determination gene (Baldi et al. 2009). In the gonochoristic *C. remanei* female, partial RNAi knockdown of *Cre-tra-2* produces a bisexual germ cell population, but such animals are not self-fertile (Haag and Kimble 2000). By simultaneously reducing the function of a repressor of sperm activation, *swm-1*, such intersexual XX animals can produce active sperm autonomously, and as a result become self-fertile to a limited extent (Baldi et al. 2009). Baldi et al. proposed that the evolution of hermaphroditism in *Caenorhabditis* probably required two steps: a mutation in the sex-determination pathway that initiated spermatogenesis and a mutation that allowed these spermatids to self-activate. This hypothesis is very plausible, but we note that these lab manipulations may or may not affect the same loci or have the same overall genetic architecture as in actual historical transitions.

Relevant to the above issue of lab manipulation versus historical reality is a recent study employing interspecies hybrids between *C. briggsae* and its close gonochoristic relative, *C.* sp. 9 (Woodruff et al. 2010). Woodruff et al. found that XX spermatogenesis is recessive in F1 hybrids, and the inability of backcrosses of hybrids to

C. briggsae to produce hermaphrodites is consistent with a polygenic basis of self-fertility. However, hybrid incompatibilities are also pervasive and *C. briggsae* alleles are preferentially lost in some parts of the hybrid genome. This leaves open the possibility that a key selfing locus cannot be made homozygous for the *C. briggsae* allele. The exact genetic architecture of the trait therefore remains unclear.

14.3.2.2 Hermaphrodites Have Evolved Using Distinct Genetic Paths

In *C. elegans* and *C. briggsae*, hermaphrodite development is similar in extent and timing, yet has evolved convergently (Cho et al. 2004; Kiontke et al. 2004). As described above, genetic comparisons between *C. elegans* and *C. briggsae* reveal that the global sex determination pathway is conserved. However, species-specific germline sex determination genes and gene regulation have been described. Both *C. elegans* and *C. briggsae* utilize F-box genes (*fog-2* and *she-1*, respectively) to promote XX spermatogenesis. However, each of these genes is a recent, species-specific gene duplicate (Guo et al. 2009; Nayak et al. 2005). Another example is the role of the *fem* genes. Although they promote male somatic fate in both species, their germline sex determination function differs. In *C. elegans*, *fem* mutations transform germ cells to female mode in both males and hermaphrodites (Hodgkin 1986), while XX *C. briggsae Cbr-fem-2* and *Cbr-fem-3* mutants (single or double) are normal hermaphrodites, while XO counterparts are transformed to hermaphrodites, not to females as in *C. elegans* (Hill et al. 2006). In addition, double mutants with *Cbr-tra-2* and any of the *Cbr-fem* genes produce normal, self-fertile hermaphrodites, as opposed to the true females seen in the equivalent mutants of *C. elegans*. Thus, while the FEM complex has a conserved role in the soma, its role is different in the germ line of the two convergent hermaphrodites. More precisely, these results suggest that *C. briggsae* regulates XX spermatogenesis via factors downstream of the FEM complex. These controls remain unknown, but an important component is likely to be factors other than Cbr-TRA-1 that are regulated by FEM-3 (Hill and Haag 2009).

Two recent studies in the authors' lab provide more evidence of flexibility in the convergent evolution of germline sex determination. The first study (Beadell et al. 2011) examined why loss of *gld-1* function has opposite effects on germline sex determination in *C. elegans* and *C. briggsae* (Nayak et al. 2005). While *C. elegans* XX *gld-1* loss-of-function mutants are feminized (and tumorous), the equivalent mutants in *C. briggsae* are strongly masculinized. Cross-species rescue experiments indicate that these different roles are not a consequence of evolution in *gld-1* itself, but instead result from distinct contexts for GLD-1 action that evolved in each hermaphrodite lineage. Further, while Beadell et al. (2011) found the expected strong in vivo association between *tra-2* mRNA and GLD-1, it is not observed in *C. briggsae*. This is most likely because the latter lacks the DREs and the associated multimerized GBEs (Wright et al. 2010).

If *Cbr-tra*-2 is not a target of Cbr-GLD-1, then this can explain why *Cbr-gld-1* mutants are not feminized—presumably no major increase in TRA-2 expression occurs. However, this is not sufficient to explain why are they masculinized. A genome-wide search for Cbr-GLD-1 target mRNAs identified the Puf family gene

Cbr-puf-8 as a direct sperm-promoting target gene, and reduction of *Cbr-puf-8* function can suppress the masculinization of germline (Mog) phenotype of *Cbr-gld-1* mutants (Beadell et al. 2011). This led us to propose a model in which *tra-2* hyperactivity largely explains the germline feminization of *C. elegans gld-1* mutants, while *Cbr-puf-8* hyperactivity causes masculinization of *C. briggsae gld-1* mutants.

In a second study (Liu et al. 2012), we characterized homologs of the FBF proteins that limit XX spermatogenesis in *C. elegans* hermaphrodites. A comprehensive *Caenorhabditis* PUF family phylogeny defined nine distinct sub-families whose origins predate the common *Caenorhabditis* ancestor. However, the lineages leading to *C. elegans* and *C. briggsae* experienced complementary losses in the FBF and PUF-2 subfamilies, such that *C. elegans* has FBFs but not PUF-2 orthologs, while *C. briggsae* lacks FBF orthologs but retains three duplicated PUF-2 orthologs (*Cbr-puf-1.1, Cbr-puf-1.2,* and *Cbr-puf-2*). This is intriguing because, similar to the *fbf* genes, *Cbr-puf-1.2* and *Cbr-puf-2* have redundant roles in hermaphrodite germline sex determination. However, as with *gld-1* above, these roles are opposite: While loss of *C. elegans fbf-1/2* function creates a Mog animal, simultaneous knockdown of *Cbr-puf-1.2/2* produces a strong feminization of germline (Fog) phenotype. This feminization is not seen males of *C. briggsae* or when *fbf* and PUF-2 sub-family members are knocked down in related gonochoristic species (though other abnormalities are observed). This suggests that the *fbf-1/2* and *Cbr-puf-1.2/2* genes were independently co-opted into germline sex determination in *C. briggsae* and *C. elegans*.

One feature the FBF and PUF-2 share, however, is the repression of *gld-1* mRNA through a conserved binding site (Crittenden et al. 2002; Liu et al. 2012). Liu et al. (2012) provided molecular and genetic evidence consistent with *Cbr-gld-1* being the major oocyte-promoting target of repression by Cbr-PUF-2/1.2. Thus, the opposite roles of both *gld-1* and *fbf* homologs in *C. briggsae* and *C. elegans* are likely not coincidental. We suggest that the PUF family genes are a case of "cooption by association," due to a conserved negative regulatory linkage connecting them to *gld-1*.

In summary, the generally conserved *Caenorhabditis* sex determination pathway has repeatedly provided the raw material for adaptive evolution of germ cells. Its germline regulation is modified post-transcriptionally in the two characterized hermaphrodites by both species-specific gene co-option and altered target gene linkages of otherwise conserved mRNA-binding proteins. Thus, though homologous players often act to pattern the hermaphrodite germ line in both *C. elegans* and *C. briggsae*, their exact roles are idiosyncratic.

14.3.3 Translational Control: An Emerging Regulatory Domain in Evolutionary Developmental Biology

The above studies suggest that translation regulation is an important emerging area of evolutionary developmental biology, especially for tissues that depend heavily upon it. Germline gene regulation is dominated by control at the 3′UTR through RNA-binding proteins (RBPs; Merritt et al. 2008). It is therefore not surprising, in retrospect, that RBPs dominate germline-specific regulators of sexual fate and cell cycle control. However, germline regulatory networks are complex, as RBPs (including those regulating sex) are often pleiotropic and regulate many downstream targets (Kershner and Kimble 2010; Lee and Schedl 2001, 2010; Merritt and Seydoux 2010; Wright et al. 2010), which complicates genetic analysis. For example, in addition to sexual fate, the entry into meiosis and oocyte maturation is also regulated by GLD-1, and the loss-of-function phenotypes include a wide variety of germline defects (Beadell et al. 2011; Francis et al. 1995; Jones et al. 1996). Germline regulators also often belong to gene families, in which members may have both redundant and specific functions (Crittenden et al. 2002; Kraemer et al. 1999; Lamont et al. 2004; Zhang et al. 1997). For example, while FBF-1 and FBF-2 act redundantly to repress spermatogenesis in *C. elegans* hermaphrodites (Kraemer et al. 1999; Zhang et al. 1997), loss of FBF-1 or FBF-2 individually produces weak but opposite effects in sex determination and the meiosis/mitosis decision (Lamont et al. 2004). Despite these challenges, however, general classes of evolutionary phenomena affecting translational controls can still be discerned. We summarize two of these below:

14.3.3.1 RBP Co-Option

The loss-of-function phenotypes of *gld-1* (Beadell et al. 2011) and *fbf* (Liu et al. 2012) homologs in gonochoristic *Caenorhabditis* suggest that the XX female ancestors of *C. briggsae* and *C. elegans* relied on them for meiotic commitment and oocyte differentiation, but not for regulation of gamete sex. This repeated repurposing may reflect two developmental constraints on adaptive evolution of the germ line. First, the PUF and GLD-1 proteins are pleiotropic germline mRNA-binding proteins (Ariz et al. 2009; Lublin and Evans 2007; Subramaniam and Seydoux 2003; Wickens et al. 2002), and are thus *a priori* on a short list of candidates for mediating germline gene regulation, including sex determination. Second, the spatial and temporal overlap between the events regulating germline cell cycle control and sexual fate positions genes that regulate conserved aspects of XX germline development to be repeatedly co-opted into hermaphrodite patterning. The alteration of the affinity for mRNA targets may result in target gain or loss, or quantitative strengthening or weakening of preexisting regulation.

14.3.3.2 Novel Genes

In addition to recycling of conserved RBPs, novel genes also have important roles in the evolution of hermaphroditism. In *C. elegans, fog-2* evolved through a series of lineage-specific duplications in the F-box family (Nayak et al. 2005). The GLD-1-binding domain in FOG-2 was probably created by a unique frameshift mutation (Nayak et al. 2005). Therefore, FOG-2 acquired its germline sex determination function in *C. elegans* recently. *C. briggsae* also utilizes a species-specific F-box genes, *she-1*, to promote spermatogenesis (Guo et al. 2009). SHE-1 does not bind Cbr-GLD-1, nor has it been implied in translational control, so its molecular function remains unknown. As F-box genes encode components of E3 ubiquitin ligases typically involved in degradation of specific target proteins, a role for SHE-1 in proteolysis is likely.

14.4 Conclusions and Prospects for Future Research

The research summarized here paints a picture of the germ line as an evolutionary mosaic. Its reliance upon post-transcriptional, mRNA-level regulation through conserved RBPs is probably universal, as are some features of the gametes it eventually produces. Indeed, the most widely used markers for germ cell fate across the metazoa are RBPs like Vasa and Nanos (e.g., Wu et al. 2011). However, layered upon this conservatism is a remarkable capacity for rapid change. As the studies of both the STAR family protein GLD-1 and the PUF family proteins FBF-1/2 and Cbr-PUF-1.2/2 indicate, well-conserved proteins with evolutionarily stable binding sites can nevertheless come to play highly variable roles in different taxa. In being pleiotropic regulators of many targets, and in having the ability to gain and lose individual target genes through changes in *cis*-regulatory elements, the functional evolution of germline RBPs is reminiscent of transcription factors (Carroll 2008). Also similar to transcription factors (Heffer et al. 2010; Mann et al. 2009), lineage-restricted cofactors (like FOG-2) play important roles in modulating the activities of RBPs. However, at this point we know very little about how such cofactors accomplish this.

In this review, we have illustrated how *Caenorhabditis* serves a model genus to study germline biology and evolution. The evolution of mating system in this genus is a fundamental organismal phenomenon, but is also rooted in the dynamics of germline gene regulation. Future research will likely focus on the ways that RBP-target gene linkages are modified by *cis*-regulatory modifications to mRNAs, and on how novel proteins interact with deeply conserved RBPs to alter their function. Understanding these processes will help link organismal phenomena that evolve over large spatial scales and millions of years with the details of molecular biology that underpin them, details that act on the molecular scale to alter gene expression in minutes to hours. Building this link is a grand challenge, but one well worth undertaking.

Acknowledgments We thank the community of *Caenorhabditis* researchers for fostering a collegial and collaborative research culture, and the National Institutes of Health and National Science Foundation for research funding.

References

Ahringer J, Kimble J (1991) Control of the sperm-oocyte switch in *Caenorhabditis elegans* hermaphrodites by the *fem-3* 3' untranslated region. Nature 349(6307):346–348

Ariz M, Mainpal R, Subramaniam K (2009) *C. elegans* RNA-binding proteins PUF-8 and MEX-3 function redundantly to promote germline stem cell mitosis. Dev Biol 326(2):295–304

Austin J, Kimble J (1989) Transcript analysis of *glp-1* and *lin-12*, homologous genes required for cell interactions during development of *C. elegans*. Cell 58(3):565–571. doi:0092-8674(89)90437-6 [pii]

Baldi C, Cho S, Ellis RE (2009) Mutations in two independent pathways are sufficient to create hermaphroditic nematodes. Science 326(5955):1002–1005. doi:326/5955/1002 [pii] 10.1126/science.1176013

Barton MK, Schedl TB, Kimble J (1987) Gain-of-function mutations of *fem-3*, a sex-determination gene in *Caenorhabditis elegans*. Genetics 115(1):107–119

Beadell AV, Liu Q, Johnson DM, Haag ES (2011) Independent recruitments of a translational regulator in the evolution of self-fertile nematodes. Proc Natl Acad Sci USA 108:19672–19677. doi:10.1073/pnas.1108068108

Beer J, Technau GM, Campos-Ortega JA (1987) Lineage analysis of transplanted individual cells in embryos of *Drosophila melanogaster*. Dev Genes Evol 196(4):222–230. doi:10.1007/bf00376346

Berry LW, Westlund B, Schedl T (1997) Germ-line tumor formation caused by activation of *glp-1*, a *Caenorhabditis elegans* member of the Notch family of receptors. Development 124(4):925–936

Boterenbrood EC, Nieuwkoop PD (1973) The formation of the mesoderm in urodelean amphibians. Dev Genes Evol 173(4):319–332. doi:10.1007/bf00575837

Brangwynne CP, Eckmann CR, Courson DS, Rybarska A, Hoege C, Gharakhani J, Julicher F, Hyman AA (2009) Germline P granules are liquid droplets that localize by controlled dissolution/condensation. Science 324(5935):1729–1732

Brawley C, Matunis E (2004) Regeneration of male germline stem cells by spermatogonial dedifferentiation in vivo. Science 304(5675):1331–1334. doi:10.1126/science.1097676

Carroll SB (2008) Evo-devo and an expanding evolutionary synthesis: a genetic theory of morphological evolution. Cell 134(1):25–36

Chen P, Ellis RE (2000) TRA-1A regulates transcription of *fog-3*, which controls germ cell fate in *C. elegans*. Development 127(14):3119–3129

Chen PJ, Cho S et al (2001) Specification of germ cell fates by FOG-3 has been conserved during nematode evolution. Genetics 158(4):1513–1525

Cho S, Jin SW, Cohen A, Ellis RE (2004) A phylogeny of *Caenorhabditis* reveals frequent loss of introns during nematode evolution. Genome Res 14(7):1207–1220

Cinquin O, Crittenden SL, Morgan DE, Kimble J (2010) Progression from a stem cell-like state to early differentiation in the *C. elegans* germ line. Proc Natl Acad Sci USA 107(5):2048–2053. doi:0912704107 [pii] 10.1073/pnas.0912704107

Clifford R, Lee MH, Nayak S, Ohmachi M, Giorgini F, Schedl T (2000) FOG-2, a novel F-box containing protein, associates with the GLD-1 RNA binding protein and directs male sex determination in the *C. elegans* hermaphrodite germline. Development 127(24):5265–5276

Coward SJ (1974) Chromatoid bodies in somatic cells of the planarian: observations on their behavior during mitosis. Anat Rec 180(3):533–545. doi:10.1002/ar.1091800312

Cox DN, Chao A, Baker J, Chang L, Qiao D, Lin H (1998) A novel class of evolutionarily conserved genes defined by *piwi* are essential for stem cell self-renewal. Genes Dev 12(23):3715–3727

Crittenden SL, Troemel ER, Evans TC, Kimble J (1994) GLP-1 is localized to the mitotic region of the *C. elegans* germ line. Development 120(10):2901–2911

Crittenden SL, Bernstein DS, Bachorik JL, Thompson BE, Gallegos M, Petcherski AG, Moulder G, Barstead R, Wickens M, Kimble J (2002) A conserved RNA-binding protein controls germline stem cells in *Caenorhabditis elegans*. Nature 417(6889):660–663. doi:10.1038/nature754

Crittenden SL, Leonhard KA, Byrd DT, Kimble J (2006) Cellular analyses of the mitotic region in the *Caenorhabditis elegans* adult germ line. Mol Biol Cell 17(7):3051–3061. doi:E06-03-0170 [pii] 10.1091/mbc.E06-03-0170

Cutter AD, Felix MA, Barriere A, Charlesworth D (2006) Patterns of nucleotide polymorphism distinguish temperate and tropical wild isolates of *Caenorhabditis briggsae*. Genetics 173(4):2021–2031 de Bono M, Hodgkin J (1996) Evolution of sex determination in Caenorhabditis: unusually high divergence of *tra-1* and its functional consequences. Genetics 144:587–595

de Rooij DG (2009) The spermatogonial stem cell niche. Microsc Res Tech 72(8):580–585. doi:10.1002/jemt.20699

Dolgin ES, Felix MA, Cutter AD (2008) Hakuna Nematoda: genetic and phenotypic diversity in African isolates of *Caenorhabditis elegans* and *C. briggsae*. Heredity 100(3):304–315

Ellis RE (2008) Sex determination in the *Caenorhabditis elegans* germ line. Curr Top Dev Biol 83:41–64

Extavour CG (2007) Evolution of the bilaterian germ line: lineage origin and modulation of specification mechanisms. Integr Comp Biol 47(5):770–785

Extavour CG, Akam M (2003) Mechanisms of germ cell specification across the metazoans: epigenesis and preformation. Development 130(24):5869–5884

Francis R, Maine E, Schedl T (1995) Analysis of the multiple roles of *gld-1* in germline development: interactions with the sex determination cascade and the *glp-1* signaling pathway. Genetics 139(2):607–630

Gallegos M, Ahringer J, Crittenden S, Kimble J (1998) Repression by the 3' UTR of *fem-3*, a sex-determining gene, relies on a ubiquitous *mog*-dependent control in *Caenorhabditis elegans*. EMBO J 17(21):6337–6347

Gallo CM, Wang JT, Motegi F, Seydoux G (2010) Cytoplasmic partitioning of P granule components is not required to specify the germline in *C. elegans*. Science 330(6011):1685–1689

Gonczy P, DiNardo S (1996) The germ line regulates somatic cyst cell proliferation and fate during *Drosophila* spermatogenesis. Development 122(8):2437–2447

Goodenough U, Lin H, Lee JH (2007) Sex determination in *Chlamydomonas*. Semin Cell Dev Biol 18(3):350–361

Goodwin EB, Okkema PG, Evans TC, Kimble J (1993) Translational regulation of *tra-2* by its 3' untranslated region controls sexual identity in *C. elegans*. Cell 75(2):329–339

Goriely A, McVean GA et al (2005) Gain-of-function amino acid substitutions drive positive selection of FGFR2 mutations in human spermatogonia. Proc Natl Acad Sci USA 102(17):6051–6056

Graustein A, Gaspar JM, Walters JR, Palopoli MF (2002) Levels of DNA polymorphism vary with mating system in the nematode genus *Caenorhabditis*. Genetics 161(1):99–107

Guo T, Peters AH, Newmark PA (2006) A Bruno-like gene is required for stem cell maintenance in planarians. Dev Cell 11(2):159–169. doi:S1534-5807(06)00260-7 [pii] 10.1016/j.devcel.2006.06.004

Guo Y, Lang S, Ellis RE (2009) Independent recruitment of F box genes to regulate hermaphrodite development during nematode evolution. Curr Biol 19(21):1853–1860

Haag ES (2007) Why two sexes? Sex determination in multicellular organisms and protistan mating types. Semin Cell Dev Biol 18(3):348–349

Haag ES, Kimble J (2000) Regulatory elements required for development of *Caenorhabditis elegans* hermaphrodites are conserved in the *tra-2* homologue of *C. remanei*, a male/female sister species. Genetics 155(1):105–116

Haag ES, Wang S, Kimble J (2002) Rapid coevolution of the nematode sex-determining genes *fem-3* and *tra-2*. Curr Biol 12(23):2035–2041

Handberg-Thorsager M, Salo E (2007) The planarian nanos-like gene *Smednos* is expressed in germline and eye precursor cells during development and regeneration. Dev Genes Evol 217(5):403–411. doi:10.1007/s00427-007-0146-3

Hansen D, Pilgrim D (1998) Molecular evolution of a sex determination protein. FEM-2 (pp 2c) in Caenorhabditis. Genetics 149:1353–1362

Hansen D, Wilson-Berry L, Dang T, Schedl T (2004) Control of the proliferation versus meiotic development decision in the *C. elegans* germline through regulation of GLD-1 protein accumulation. Development 131:93–104

Hanyu-Nakamura K, Sonobe-Nojima H, Tanigawa A, Lasko P, Nakamura A (2008) *Drosophila* Pgc protein inhibits P-TEFb recruitment to chromatin in primordial germ cells. Nature 451(7179):730–733. doi:nature06498 [pii] 10.1038/nature06498

Hashimoto Y, Maegawa S, Nagai T, Yamaha E, Suzuki H, Yasuda K, Inoue K (2004) Localized maternal factors are required for zebrafish germ cell formation. Dev Biol 268(1):152–161. doi:10.1016/j.ydbio.2003.12.013 S0012160603007942 [pii]

Heasman J, Quarmby J, Wylie CC (1984) The mitochondrial cloud of *Xenopus* oocytes: the source of germinal granule material. Dev Biol 105(2):458–469. doi:0012-1606(84)90303-8 [pii]

Heffer A, Shultz JW, Pick L (2010) Surprising flexibility in a conserved Hox transcription factor over 550 million years of evolution. Proc Natl Acad Sci USA 107(42):18040–18045

Henderson ST, Gao D, Lambie EJ, Kimble J (1994) *lag-2* may encode a signaling ligand for the GLP-1 and LIN-12 receptors of *C. elegans*. Development 120(10):2913–2924

Hill RC, Haag ES (2009) A sensitized genetic background reveals evolution near the terminus of the *Caenorhabditis* germline sex determination pathway. Evol Dev 4:333–341

Hill RC, de Carvalho CE, Salogiannis J, Schlager B, Pilgrim D, Haag ES (2006) Genetic flexibility in the convergent evolution of hermaphroditism in *Caenorhabditis* nematodes. Dev Cell 10(4):531–538

Hillier LW, Miller RD, Baird SE, Chinwalla A, Fulton LA, Koboldt DC, Waterston RH (2007) Comparison of *C. elegans* and *C. briggsae* genome sequences reveals extensive conservation of chromosome organization and synteny. PLoS Biol 5(7):e167

Hird SN, Paulsen JE, Strome S (1996) Segregation of germ granules in living *Caenorhabditis elegans* embryos: cell-type-specific mechanisms for cytoplasmic localisation. Development 122(4):1303–1312

Hodgkin J (1986) Sex determination in the nematode *C. elegans*: analysis of *tra-3* suppressors and characterization of *fem* genes. Genetics 114(1):15–52

Hori I, Kishida Y (2003) Quantitative changes in nuclear pores and chromatoid bodies induced by neuropeptides during cell differentiation in the planarian *Dugesia japonica*. J Submicrosc Cytol Pathol 35(4):439–444

Huettner AF (1923) The origin of the germ cells in *Drosophila melanogaster*. J Morphol 37(2): 385–423. doi:10.1002/jmor.1050370204

Humphrey RR (1925) The primordial germ cells of *Hemidactylium* and other amphibia. J Morphol 41(1):1–43. doi:10.1002/jmor.1050410103

Humphrey RR (1927) Extirpation of the primordial germ cells of *Amblystoma*: its effect upon the development of the gonad. J Exp Zool 49(2):363–399. doi:10.1002/jez.1400490207

Ikenishi K, Nieuwkoop PD (1978) Location and ultrastructure of primordial germ cells (PGCs) in *Ambystoma mexicanum*. Dev Growth Differ 20(1):1–9. doi:10.1111/j.1440-169X.1978.00001.x

Illmensee K, Mahowald AP (1974) Transplantation of posterior polar plasm in *Drosophila*. Induction of germ cells at the anterior pole of the egg. Proc Natl Acad Sci USA 71(4):1016–1020

Jan E, Yoon JW, Walterhouse D, Iannaccone P, Goodwin EB (1997) Conservation of the *C. elegans tra-2* 3'UTR translational control. EMBO J 16(20):6301–6313

Jan E, Motzny CK, Graves LE, Goodwin EB (1999) The STAR protein, GLD-1, is a translational regulator of sexual identity in *Caenorhabditis elegans*. EMBO J 18(1):258–269. doi:10.1093/emboj/18.1.258

Johnson AD, Bachvarova RF, Drum M, Masi T (2001) Expression of axolotl DAZL RNA, a marker of germ plasm: widespread maternal RNA and onset of expression in germ cells approaching the gonad. Dev Biol 234(2):402–415. doi:10.1006/dbio.2001.0264 S0012-1606(01)90264-7 [pii]

Johnson AD, Crother B, White ME, Patient R, Bachvarova RF, Drum M, Masi T (2003) Regulative germ cell specification in axolotl embryos: a primitive trait conserved in the mammalian lineage. Philos Trans R Soc Lond B Biol Sci 358(1436):1371–1379. doi:10.1098/rstb.2003.1331

Jones AR, Francis R, Schedl T (1996) GLD-1, a cytoplasmic protein essential for oocyte differentiation, shows stage- and sex-specific expression during *Caenorhabditis elegans* germline development. Dev Biol 180(1):165–183. doi:S0012-1606(96)90293-6 [pii] 10.1006/dbio.1996.0293

Kawase E, Wong MD, Ding BC, Xie T (2004) Gbb/Bmp signaling is essential for maintaining germline stem cells and for repressing bam transcription in the *Drosophila* testis. Development 131(6):1365–1375. doi:10.1242/dev.01025 dev.01025 [pii]

Kelleher DF, de Carvalho CE, Doty AV, Layton M, Cheng AT, Mathies LD, Pilgrim D, Haag ES (2008) Comparative genetics of sex determination: masculinizing mutations in *Caenorhabditis briggsae*. Genetics 178(3):1415–1429

Kershner AM, Kimble J (2010) Genome-wide analysis of mRNA targets for *Caenorhabditis elegans* FBF, a conserved stem cell regulator. Proc Natl Acad Sci USA 107(8):3936–3941. doi:1000495107 [pii] 10.1073/pnas.1000495107

Kimble J, Crittenden SL (2007) Controls of germline stem cells, entry into meiosis, and the sperm/oocyte decision in *Caenorhabditis elegans*. Annu Rev Cell Dev Biol 23:405–433. doi:10.1146/annurev.cellbio.23.090506.123326

Kimble J, Simpson P (1997) The LIN-12/Notch signaling pathway and its regulation. Annu Rev Cell Dev Biol 13:333–361. doi:10.1146/annurev.cellbio.13.1.333

Kimble JE, White JG (1981) On the control of germ cell development in *Caenorhabditis elegans*. Dev Biol 81(2):208–219. doi:0012-1606(81)90284-0 [pii]

King FJ, Lin H (1999) Somatic signaling mediated by fs(1)Yb is essential for germline stem cell maintenance during *Drosophila* oogenesis. Development 126(9):1833–1844

Kiontke K, Sudhaus W (2006) Ecology of *Caenorhabditis* species. WormBook:1–14. doi:10.1895/wormbook.1.37.1

Kiontke K, Gavin NP, Raynes Y, Roehrig C, Piano F, Fitch DHA (2004) *Caenorhabditis* phylogeny predicts convergence of hermaphroditism and extensive intron loss. Proc Natl Acad Sci USA 101(24):9003–9008

Kiontke K, Félix M-A, Ailion M, Rockman M, Braendle C, Penigault J-B, Fitch DHA (2011) A phylogeny and molecular barcodes for *Caenorhabditis*, with numerous new species from rotting fruits. BMC Evol Biol 11:339. doi:10.1186/1471-2148-11-339

Kirilly D, Xie T (2007) The *Drosophila* ovary: an active stem cell community. Cell Res 17(1): 15–25. doi:7310123 [pii] 10.1038/sj.cr.7310123

Koboldt DC, Staisch J, Thillainathan B, Haines K, Baird SE, Chamberlin HM, Haag ES, Miller RD, Gupta BP (2010) A toolkit for rapid gene mapping in the nematode *Caenorhabditis briggsae*. BMC Genomics 11:236. doi:1471-2164-11-236 [pii] 10.1186/1471-2164-11-236

Komiya T, Itoh K, Ikenishi K, Furusawa M (1994) Isolation and characterization of a novel gene of the DEAD box protein family which is specifically expressed in germ cells of *Xenopus laevis*. Dev Biol 162(2):354–363. doi:10.1006/dbio.1994.1093 S0012-1606(84)71093-1 [pii]

Kraemer B, Crittenden S, Gallegos M, Moulder G, Barstead R, Kimble J, Wickens M (1999) NANOS-3 and FBF proteins physically interact to control the sperm-oocyte switch in *Caenorhabditis elegans*. Curr Biol 9(18):1009–1018

Kurimoto K, Yabuta Y, Ohinata Y, Shigeta M, Yamanaka K, Saitou M (2008) Complex genome-wide transcription dynamics orchestrated by Blimp1 for the specification of the germ cell lineage in mice. Genes Dev 22(12):1617–1635. doi:22/12/1617 [pii] 10.1101/gad.1649908

Kuwabara PE (1996) Interspecies comparison reveals evolution of control regions in the nematode sex-determining gene *tra-2*. Genetics 144(2):597–607

Lamont LB, Crittenden SL, Bernstein D, Wickens M, Kimble J (2004) FBF-1 and FBF-2 regulate the size of the mitotic region in the *C. elegans* germline. Dev Cell 7(5):697–707. doi:S1534580704003338 [pii] 10.1016/j.devcel.2004.09.013

Lawson KA, Dunn NR, Roelen BA, Zeinstra LM, Davis AM, Wright CV, Korving JP, Hogan BL (1999) Bmp4 is required for the generation of primordial germ cells in the mouse embryo. Genes Dev 13(4):424–436

Lee MH, Schedl T (2001) Identification of in vivo mRNA targets of GLD-1, a maxi-KH motif containing protein required for *C. elegans* germ cell development. Genes Dev 15(18): 2408–2420. doi:10.1101/gad.915901

Lee MH, Schedl T (2010) *C. elegans* star proteins, GLD-1 and ASD-2, regulate specific RNA targets to control development. Advances in Experimental Medicine and Biology 693:106–122

Lee SC, Ni M, Li W, Shertz C, Heitman J (2010) The evolution of sex: a perspective from the fungal kingdom. Microbiol Mol Biol Rev 74(2):298–340. doi:74/2/298 [pii] 10.1128/MMBR.00005-10

Leonard J (2010) The evolution of sexes, anisogamy, and sexual systems: natural versus sexual selection. In: Leonard J, Cordoba-Aguilar A (eds) The evolution of primary sexual characters in animals. Oxford University Press, Oxford

Lin H (2002) The stem-cell niche theory: lessons from flies. Nat Rev Genet 3(12):931–940. doi:10.1038/nrg952 nrg952 [pii]

Liu Q, Stumpf C, Wickens M, Haag ES (2012) Context-dependent function of a conserved translational regulatory module. Development 139:ppTBA

Lublin AL, Evans TC (2007) The RNA-binding proteins PUF-5, PUF-6, and PUF-7 reveal multiple systems for maternal mRNA regulation during *C. elegans* oogenesis. Dev Biol 303(2):635–649

Mann RS, Lelli KM, Joshi R (2009) Hox specificity unique roles for cofactors and collaborators. Curr Top Dev Biol 88:63–101

McCarter J, Bartlett B, Dang T, Schedl T (1997) Soma-germ cell interactions in *Caenorhabditis elegans*: multiple events of hermaphrodite germline development require the somatic sheath and spermathecal lineages. Dev Biol 181(2):121–143. doi:10.1006/dbio.1996.8429 S0012-1606(96)98429-8 [pii]

McCourt RM, Delwiche CF, Karol KG (2004) Charophyte algae and land plant origins. Trends Ecol Evol 19(12):661–666

McGovern M, Voutev R, Maciejowski J, Corsi AK, Hubbard EJ (2009) A "latent niche" mechanism for tumor initiation. Proc Natl Acad Sci USA 106(28):11617–11622. doi:0903768106 [pii] 10.1073/pnas.0903768106

Mello CC, Draper BW, Krause M, Weintraub H, Priess JR (1992) The pie-1 and mex-1 genes and maternal control of blastomere identity in early *C. elegans* embryos. Cell 70(1):163–176. doi:0092-8674(92)90542-K [pii]

Mello CC, Schubert C, Draper B, Zhang W, Lobel R, Priess JR (1996) The PIE-1 protein and germline specification in *C. elegans* embryos. Nature 382(6593):710–712. doi:10.1038/382710a0

Meng X, Lindahl M, Hyvonen ME, Parvinen M, de Rooij DG, Hess MW, Raatikainen-Ahokas A, Sainio K, Rauvala H, Lakso M, Pichel JG, Westphal H, Saarma M, Sariola H (2000) Regulation of cell fate decision of undifferentiated spermatogonia by GDNF. Science 287(5457): 1489–1493. doi:8297 [pii]

Merritt C, Seydoux G (2010) The Puf RNA-binding proteins FBF-1 and FBF-2 inhibit the expression of synaptonemal complex proteins in germline stem cells. Development 137(11): 1787–1798. doi:dev.050799 [pii] 10.1242/dev.050799

Merritt C, Rasoloson D, Ko D, Seydoux G (2008) 3' UTRs are the primary regulators of gene expression in the *C. elegans* germline. Curr Biol 18(19):1476–1482

Morgan T (1901) Growth and regeneration in *Planaria lugubris*. Dev Genes Evol 13(1):179–212. doi:10.1007/bf02161982

Morgan CT, Lee M-H et al (2010) Chemical reprogramming of *Caenorhabditis* elegans germ cell fate. Nat Chem Biol 6(2):102–104

Nayak S, Goree J, Schedl T (2005) *fog-2* and the evolution of self-fertile hermaphroditism in *Caenorhabditis*. PLoS Biol 3(1):e6

Newmark PA, Sanchez Alvarado A (2002) Not your father's planarian: a classic model enters the era of functional genomics. Nat Rev Genet 3(3):210–219. doi:10.1038/nrg759 nrg759 [pii]

Olsen LC, Aasland R, Fjose A (1997) A *vasa*-like gene in zebrafish identifies putative primordial germ cells. Mech Dev 66(1–2):95–105

Otori M, Karashima T, Yamamoto M (2006) The *Caenorhabditis elegans* homologue of deleted in azoospermia is involved in the sperm/oocyte switch. Mol Biol Cell 17(7):3147–3155. doi:E05-11-1067 [pii] 10.1091/mbc.E05-11-1067

Pepper AS, Killian DJ, Hubbard EJ (2003a) Genetic analysis of *Caenorhabditis elegans* glp-1 mutants suggests receptor interaction or competition. Genetics 163(1):115–132

Pepper AS, Lo TW, Killian DJ, Hall DH, Hubbard EJ (2003b) The establishment of *Caenorhabditis elegans* germline pattern is controlled by overlapping proximal and distal somatic gonad signals. Dev Biol 259(2):336–350. doi:S0012160603002033 [pii]

Praitis V, Casey E, Collar D, Austin J (2001) Creation of low-copy integrated transgenic lines in *Caenorhabditis elegans*. Genetics 157(3):1217–1226

Ramesh MA, Malik SB, Logsdon JM Jr (2005) A phylogenomic inventory of meiotic genes; evidence for sex in *Giardia* and an early eukaryotic origin of meiosis. Curr Biol 15(2):185–191

Raz E (2003) Primordial germ-cell development: the zebrafish perspective. Nat Rev Genet 4(9):690–700. doi:10.1038/nrg1154

Reddien PW, Sanchez Alvarado A (2004) Fundamentals of planarian regeneration. Annu Rev Cell Dev Biol 20:725–757. doi:10.1146/annurev.cellbio.20.010403.095114

Ross JA, Koboldt DC, Staisch JE, Chamberlin HM, Gupta BP, Miller RD, Baird SE, Haag ES (2011) *Caenorhabditis briggsae* recombinant inbred line genotypes reveal inter-strain incompatibility and the evolution of recombination. PLoS Genet 7(7):e1002174

Salo E, Baguna J (1984) Regeneration and pattern formation in planarians. I. The pattern of mitosis in anterior and posterior regeneration in *Dugesia* (G) *tigrina*, and a new proposal for blastema formation. J Embryol Exp Morphol 83:63–80

Salvetti A, Rossi L, Lena A, Batistoni R, Deri P, Rainaldi G, Locci MT, Evangelista M, Gremigni V (2005) DjPum, a homologue of *Drosophila* Pumilio, is essential to planarian stem cell maintenance. Development 132(8):1863–1874. doi:dev.01785 [pii] 10.1242/dev.01785

Sanchez Alvarado A (2006) Planarian regeneration: its end is its beginning. Cell 124(2):241–245. doi:S0092-8674(06)00060-2 [pii] 10.1016/j.cell.2006.01.012

Sato K, Sugita T, Kobayashi K, Fujita K, Fujii T, Matsumoto Y, Mikami T, Nishizuka N, Nishizuka S, Shojima K, Suda M, Takahashi G, Himeno H, Muto A, Ishida S (2001) Localization of mitochondrial ribosomal RNA on the chromatoid bodies of marine planarian polyclad embryos. Dev Growth Differ 43(2):107–114. doi:dgd558 [pii]

Sato K, Shibata N, Orii H, Amikura R, Sakurai T, Agata K, Kobayashi S, Watanabe K (2006) Identification and origin of the germline stem cells as revealed by the expression of nanos-related gene in planarians. Dev Growth Differ 48(9):615–628. doi:DGD897 [pii] 10.1111/j.1440-169X.2006.00897.x

Schedl T (1997) Developmental genetics of the germ line. doi:NBK20133 [bookaccession]

Schedl T, Kimble J (1988) *fog-2*, a germ-line-specific sex determination gene required for hermaphrodite spermatogenesis in *Caenorhabditis elegans*. Genetics 119(1):43–61

Schmid M, Kuchler B, Eckmann CR (2009) Two conserved regulatory cytoplasmic poly(A) polymerases, GLD-4 and GLD-2, regulate meiotic progression in *C. elegans*. Genes Dev 23(7): 824–836. doi:23/7/824 [pii] 10.1101/gad.494009

Segal SP, Graves LE, Verheyden J, Goodwin EB (2001) RNA-Regulated TRA-1 nuclear export controls sexual fate. Dev Cell 1(4):539–551. doi:S1534-5807(01)00068-5 [pii]

Seydoux G, Mello CC, Pettitt J, Wood WB, Priess JR, Fire A (1996) Repression of gene expression in the embryonic germ lineage of *C. elegans*. Nature 382(6593):713–716. doi:10.1038/382713a0

Shibata N, Umesono Y, Orii H, Sakurai T, Watanabe K, Agata K (1999) Expression of vasa(vas)-related genes in germline cells and totipotent somatic stem cells of planarians. Dev Biol 206(1):73–87. doi:S0012-1606(98)99130-8 [pii] 10.1006/dbio.1998.9130

Shibata N, Rouhana L, Agata K (2010) Cellular and molecular dissection of pluripotent adult somatic stem cells in planarians. Dev Growth Differ 52(1):27–41. doi:DGD1155 [pii] 10.1111/j.1440-169X.2009.01155.x

Song X, Xie T (2002) DE-cadherin-mediated cell adhesion is essential for maintaining somatic stem cells in the *Drosophila* ovary. Proc Natl Acad Sci USA 99(23):14813–14818. doi:10.1073/pnas.232389399 232389399 [pii]

Song X, Wong MD, Kawase E, Xi R, Ding BC, McCarthy JJ, Xie T (2004) Bmp signals from niche cells directly repress transcription of a differentiation-promoting gene, *bag of marbles*, in germline stem cells in the *Drosophila* ovary. Development 131(6):1353–1364. doi:10.1242/dev.01026 dev.01026 [pii]

Steenkamp ET, Wright J, Baldauf SL (2006) The protistan origins of animals and fungi. Mol Biol Evol 23(1):93–106. doi:msj011 [pii] 10.1093/molbev/msj011

Stein LD, Bao Z, Blasiar D, Blumenthal T, Brent MR, Chen N, Chinwalla A, Clarke L, Clee C, Coghlan A (2003) The genome sequence of *Caenorhabditis briggsae*: a platform for comparative genomics. PLoS Biol 1(2):E45

Stothard P, Pilgrim D (2006) Conspecific and interspecific interactions between the FEM-2 and the FEM-3 sex-determining proteins despite rapid sequence divergence. J Mol Evol 62(3):281–291
Stothard P, Hansen D, Pilgrim D (2002) Evolution of the PP2C family in *Caenorhabditis*: rapid divergence of the sex-determining protein FEM-2. J Mol Evol 54(2):267–282
Strome S, Wood WB (1983) Generation of asymmetry and segregation of germ-line granules in early *C. elegans* embryos. Cell 35(1):15–25
Subramaniam K, Seydoux G (2003) Dedifferentiation of primary spermatocytes into germ cell tumors in *C. elegans* lacking the pumilio-like protein PUF-8. Curr Biol 13(2):134–139
Suh N, Jedamzik B, Eckmann CR, Wickens M, Kimble J (2006) The GLD-2 poly(A) polymerase activates *gld-1* mRNA in the *Caenorhabditis elegans* germ line. Proc Natl Acad Sci USA 103(41):15108–15112. doi:0607050103 [pii] 10.1073/pnas.0607050103
Sulston JE, Horvitz HR (1977) Post-embryonic cell lineages of the nematode *Caenorhabditis elegans*. Dev Biol 56(1):110–156. doi:0012-1606(77)90158-0 [pii]
Tam PP, Zhou SX (1996) The allocation of epiblast cells to ectodermal and germ-line lineages is influenced by the position of the cells in the gastrulating mouse embryo. Dev Biol 178(1): 124–132. doi:S0012-1606(96)90203-1 [pii] 10.1006/dbio.1996.0203
Tanurdzic M, Banks JA (2004) Sex-determining mechanisms in land plants. Plant Cell 16(Suppl):S61–S71. doi:10.1105/tpc.016667
Tax FE, Yeargers JJ, Thomas JH (1994) Sequence of *C. elegans lag-2* reveals a cell-signalling domain shared with Delta and Serrate of *Drosophila*. Nature 368(6467):150–154. doi:10.1038/368150a0
Thompson BE, Bernstein DS, Bachorik JL, Petcherski AG, Wickens M, Kimble J (2005) Dose-dependent control of proliferation and sperm specification by FOG-1/CPEB. Development 132(15):3471–3481
Timinszky G, Bortfeld M, Ladurner AG (2008) Repression of RNA polymerase II transcription by a *Drosophila* oligopeptide. PLoS One 3(6):e2506. doi:10.1371/journal.pone.0002506
Venkatarama T, Lai F, Luo X, Zhou Y, Newman K, King ML (2010) Repression of zygotic gene expression in the *Xenopus* germline. Development 137(4):651–660. doi:137/4/651 [pii] 10.1242/dev.038554
Wang S, Kimble J (2001) The TRA-1 transcription factor binds TRA-2 to regulate sexual fates in *Caenorhabditis elegans*. EMBO J 20(6):1363–1372. doi:10.1093/emboj/20.6.1363
Wang JT, Seydoux S (2012) Germ cell specification. Advances in Experimental Medicine and Biology 757:17–39. (Chap. 2, this volume) Springer, New York
Wang Y, Zayas RM, Guo T, Newmark PA (2007) *nanos* function is essential for development and regeneration of planarian germ cells. Proc Natl Acad Sci USA 104(14):5901–5906. doi:0609708104 [pii] 10.1073/pnas.0609708104
Whitington PM, Dixon KE (1975) Quantitative studies of germ plasm and germ cells during early embryogenesis of *Xenopus* laevis. J Embryol Exp Morphol 33(1):57–74
Wickens M, Bernstein DS, Kimble J, Parker R (2002) A PUF family portrait: 3'UTR regulation as a way of life. Trends Genet 18(3):150–157
Williamson A, Lehmann R (1996) Germ cell development in *Drosophila*. Annu Rev Cell Dev Biol 12:365–391. doi:10.1146/annurev.cellbio.12.1.365
Wood WB, Laufer JS, Strome S (1982) Developmental determinants in embryos of *Caenorhabditis elegans*. J Nematol 14(2):267–273
Woodruff GC, Eke O, Baird SE, Felix MA, Haag ES (2010) Insights into species divergence and the evolution of hermaphroditism from fertile interspecies hybrids of Caenorhabditis nematodes. Genetics 186(3):997–1012. doi:genetics.110.120550 [pii] 10.1534/genetics.110.120550
Wright JE, Gaidatzis D, Senften M, Farley BM, Westhof E, Ryder SP, Ciosk R (2010) A quantitative RNA code for mRNA target selection by the germline fate determinant GLD-1. EMBO J 30(3):533–545. doi:emboj2010334 [pii] 10.1038/emboj.2010.334
Wu H-R, Chen Y-T, Su Y-J, Holland L, Yu J-K (2011) Asymmetric localization of germline markers Vasa and Nanos during early development in the amphioxus *Branchiostoma floridae*. Dev Biol 353(147–59)
Xie T, Li L (2007) Stem cells and their niche: an inseparable relationship. Development 134(11):2001–2006. doi:134/11/2001 [pii] 10.1242/dev.002022

Xie T, Spradling AC (2000) A niche maintaining germ line stem cells in the *Drosophila* ovary. Science 290(5490):328–330. doi:8892 [pii]

Ying Y, Zhao GQ (2001) Cooperation of endoderm-derived BMP2 and extraembryonic ectoderm-derived BMP4 in primordial germ cell generation in the mouse. Dev Biol 232(2):484–492. doi:10.1006/dbio.2001.0173 S0012-1606(01)90173-3 [pii]

Ying Y, Qi X, Zhao GQ (2001) Induction of primordial germ cells from murine epiblasts by synergistic action of BMP4 and BMP8B signaling pathways. Proc Natl Acad Sci USA 98(14): 7858–7862. doi:10.1073/pnas.151242798

Yochem J, Greenwald I (1989) *glp-1* and *lin-12*, genes implicated in distinct cell-cell interactions in *C. elegans*, encode similar transmembrane proteins. Cell 58(3):553–563. doi:0092-8674(89)90436-4 [pii]

Yoon C, Kawakami K, Hopkins N (1997) Zebrafish *vasa* homologue RNA is localized to the cleavage planes of 2- and 4-cell-stage embryos and is expressed in the primordial germ cells. Development 124(16):3157–3165

Zanetti S, Puoti A (2012) Sex determination in the *C. elegans* germline. Advances in Experimental Medicine and Biology 757:41–69. (Chap. 3, this volume) Springer, New York

Zarkower D (2006) Somatic sex determination. WormBook:1–12 (http://www.wormbook.org)

Zhang B, Gallegos M, Puoti A, Durkin E, Fields S, Kimble J, Wickens MP (1997) A conserved RNA-binding protein that regulates sexual fates in the *C. elegans* hermaphrodite germ line. Nature 390(6659):477–484. doi:10.1038/37297

Zhao Z, Flibotte S, Murray JI, Blick D, Boyle TJ, Gupta B, Moerman DG, Waterston RH (2010) New tools for investigating the comparative biology of *Caenorhabditis briggsae* and *C. elegans*. Genetics 184(3):853–863. doi:genetics.109.110270 [pii]

Zhou Y, King ML (1996) Localization of Xcat-2 RNA, a putative germ plasm component, to the mitochondrial cloud in *Xenopus* stage I oocytes. Development 122:2947–2953

Index

A
AMPK, 110–112
Apoptosis, 7, 29, 77, 149, 175, 213, 250, 284
Asymmetric partitioning, 24, 27

C
Caenorhabditis elegans, 1–13, 28, 41–62, 71–94, 133–158, 172–177, 205–238, 277–308, 322–324, 352
Cell fate, 4, 10, 23, 26–29, 31, 33, 34, 45–47, 72, 74, 75, 77, 78, 87, 89, 93, 94, 111, 206, 207, 217, 224–226, 228, 230, 232–238, 354, 369, 392, 407, 417
Chromatin, 7, 21, 22, 31–34, 91, 136, 140, 141, 143, 145, 150, 151, 153–155, 157, 158, 172, 174–187, 190, 194, 195, 284, 285, 289, 291, 300, 301, 303, 374–376, 378, 380, 383, 384, 388–394, 396, 397
Chromosome segregation, 133–147, 153, 156–158, 183, 185, 186, 188, 195, 196, 212, 225, 280, 285, 286, 290, 302, 307, 308, 338, 357, 383, 384, 386, 396
Cohesion, 134–141, 148, 153, 156, 185, 258, 285, 286, 307

D
Deadenylation, 25, 54, 215–217, 237
Diapauses, 9, 12
Differentiation, 2, 8, 11, 12, 26, 72, 73, 76, 80, 81, 87, 90–94, 105, 113, 115, 116, 124, 148, 158, 172, 177–184, 186, 196, 206, 212, 213, 216–218, 221, 232, 250, 253, 260, 284, 286, 299, 324, 329, 354, 392, 407, 410, 416
DNA damage response, 217, 249–270, 296, 383, 390
DNA repair, 255–257, 259, 261, 262

E
Egg activation, 194, 278, 300, 322, 324, 333–340
Eggshell, 8, 23, 104, 194, 305, 323, 331, 334–338, 340–342
Epigenetic, 12–13, 32, 33, 61, 172, 178, 194, 195, 373–397

F
Fertilization, 8, 42, 91, 116, 133, 172, 206, 250, 278, 322–324, 352, 389, 407

G
Gametogenesis, 5, 26, 77, 102, 105, 109, 116, 119, 123, 206, 212, 213, 217, 220, 231, 237, 238, 295, 296, 299, 374, 377, 388, 389, 392
Germ cell, 2, 18, 43, 72, 104, 135, 172, 206, 250, 287, 354, 374, 405
Germ granules, 19, 23, 26
Germline
  blastomeres, 22, 25–28, 30, 31, 33, 352–354, 358, 360, 368, 406–407
  stem cell, 111, 115, 295, 408
Germ plasm, 18–21, 23–30, 33, 34, 407
GLD-1, 45, 49, 52, 54, 56–58, 61, 76, 80–83, 85–91, 113, 120, 223, 224, 231, 234, 235, 237, 261, 262, 292, 413–417
GLD-1 pathway, 81–86, 88, 89
GLD-2 pathway, 81–89

T. Schedl (ed.), *Germ Cell Development in C. elegans*, Advances in Experimental Medicine and Biology 757, DOI 10.1007/978-1-4614-4015-4,

GLP-1 notch, 57, 80–86, 88–90, 113, 115, 231, 282, 295, 299
G-protein signaling, 288

H
Histone, 7, 13, 32, 91, 154, 155, 178, 195, 301, 374–384, 388–397

I
Insulin, 105–107, 110, 112, 114, 124, 307

L
Lifespan, 102, 103, 106, 108, 109, 116, 117, 119–122, 394

M
Major sperm protein (MSP), 11, 174, 177, 278, 282–286, 288–291, 293, 295–299, 322–323, 325–329
Maternal effect sterile (MES) complex, 378, 379, 382
Maternal RNA, 33, 61
MBK-2, 25, 30, 292, 300, 331, 334, 335, 337, 356–359, 362–368
Meiosis, 11, 42, 73, 105, 133, 173, 206, 251, 278–279, 323, 352–353, 374, 406
Meiosis checkpoints, 266, 270, 306
Meiotic chromosome segregation, 157–158, 188, 195, 225, 285, 308
Meiotic fate, 10, 71–94, 225
Meiotic maturation signaling, 306
Meiotic recombination, 10, 11, 89, 138, 149–155, 175, 250, 251, 254, 256, 257, 262, 266–270, 278, 306
Meiotic resumption, 280, 283, 290, 322, 334–336
Meiotic silencing, 390, 393
Meiotic spindle, 157, 214, 278, 282, 285, 290. 300–304, 337, 353, 356, 357, 363
miRNA, 183, 217–218

N
Notch, 10, 57, 80–86, 88–90, 104, 111, 113–115, 231, 282, 295, 298, 299, 408
Nutrient, 11, 108, 111, 115

O
OMA-1, 25, 29–31, 212, 214, 220, 223, 228, 291, 292, 331, 337, 357–359, 361–363, 365, 367
Oocyte maturation, 10, 185, 206, 214, 231, 280, 282, 329, 352, 355, 356, 361–365, 368
Oocyte-to-embryo transition, 11, 30, 290, 300–305, 323, 331, 335, 342, 351–369
Oogenesis, 4, 9–13, 29, 30, 45, 46, 50, 51, 53–55, 58–61, 77, 80, 119, 123, 174–175, 183–185, 194, 206, 213, 218, 220, 236, 254, 283, 292, 295–296, 300, 302, 306–308, 322, 324, 329, 382, 389, 392, 413
Organelle morphogenesis, 186

P
Pairing, 10, 74, 137, 141–144, 146–148, 151, 154, 157, 158, 174, 183, 184, 258, 266–270, 376, 381, 396
Paternal epigenetics, 172, 178
piRNA, 181, 182, 195, 218, 384, 387, 391, 393, 394
P lineage, 2, 18–23, 25, 29, 33
Polarity, 10, 18, 20, 23, 24, 28, 33, 54, 73, 80, 194, 300, 304, 305, 337–340, 354, 356, 364, 409
Polyadenylation, 56, 58, 215, 217, 225, 237, 281
Polyspermy, 300, 305, 322, 331, 341–342, 356
Post transcriptional RNA regulation, 52–54, 206, 208, 413, 417
Primordial germ cells, 8, 18, 27, 33, 221, 355, 369, 374, 383, 406, 407
Proliferative fate, 5, 10, 73, 78–91, 93, 104, 113
Protein degradation, 227, 334, 336, 355, 356, 366, 367
PUFA-derived signals, 328
PUF proteins, 219–221

R
Recombination, 6, 10, 11, 89, 135, 137, 138, 141–143, 146–155, 158, 174, 175, 184, 250, 251, 254, 256–259, 262, 263, 266–270, 278, 306, 340, 411
Reproduction, 2, 9, 12, 13, 42, 73, 102, 103, 106, 107, 109, 116–117, 119–122, 124, 149, 278, 283, 306, 308, 322, 328, 342, 406, 410
RNA-binding, 29, 30, 53, 54, 83, 91, 212, 217, 219–222, 225, 226, 235, 367
RNA-binding protein, 13, 23–26, 28, 29, 33, 47, 52–59, 85, 207, 208, 211, 214, 215, 218–226, 228, 234, 299, 357, 360, 367, 368, 387, 409, 412, 415

RNA decay, 208
RNA processing, 215
RNA regulatory network, 232–234, 237–238

S
Sex determination, 11, 13, 41–62, 83, 116, 173, 234, 236, 268, 409–416
siRNA, 181, 183–184, 195, 217, 218, 384, 386, 387, 391, 393–395
Small RNA, 177, 181–184, 195, 281, 374, 382, 384–388, 390, 391, 394–396
Somatic gonad, 4, 5, 8–10, 12, 21, 29, 44, 46, 49, 73, 78, 80, 104, 121, 122, 266, 287, 289, 290, 307, 407
Sperm activation, 172, 179, 182, 187, 189, 191–193, 195, 196, 322–325, 327–328, 330, 332, 413
Spermiogenesis, 10, 174, 176–177, 187, 189–193, 195, 322, 324, 325, 393
Sperm maturation, 176, 185, 189–190
Sperm migration, 329
Stem cell, 10, 12, 46, 50, 57, 71–94, 111, 115, 116, 124, 177, 196, 221, 250, 255, 260, 292, 295, 306–308, 391, 392, 397, 408–409
Synapsis, 10, 74, 135, 137, 140, 143–151, 153, 158, 174, 184, 250, 254, 266, 267, 269, 270, 381, 390

T
TGF-β, 79
TOR, 88, 105, 112, 114–116, 124, 180, 214, 291, 361
Transcriptional regulation, 179, 262
Transcriptional repression, 29, 50, 292, 355, 363, 375, 382, 397, 408
Translation, 25, 28–30, 33, 34, 53, 54, 82–86, 89, 186, 187, 207–213, 215, 222, 231, 236, 238, 280, 281, 291–292, 357, 359, 360, 362, 364, 367, 368
Translational control, 13, 20, 52, 83, 205–238, 281, 292, 355, 415–417
Translational repression, 54, 87, 181, 208, 212, 214, 217, 234, 236, 237, 292, 359, 360, 362–363, 366–368, 384

U
3'Untranslated regions (3'UTR), 13, 25, 29, 52–54, 56, 82, 85, 86, 214, 217, 222–226, 232, 234–237, 261, 291, 357, 359, 362, 367, 412, 413, 415

Z
ZIF-1, 25, 28, 214, 291, 292, 358–360, 362, 363, 365–368